THE FIRST SYMPOSIUM ON THE INFRARED CIRRUS AND DIFFUSE INTERSTELLAR CLOUDS

Steward Observatory Publication No. 1188
NRAO Workshop Proceedings No. 24

A SERIES OF BOOKS ON RECENT DEVELOPMENTS IN ASTRONOMY AND ASTROPHYSICS

Printed by BookCrafters, Inc.

First published 1994

Library of Congress Catalog Card Number: 94-71216
ISBN 0-937707-77-5

D. Harold McNamara, Managing Editor of Conference Series
408 ESC Brigham Young University
Provo, UT 84602
801-378-2298

A SERIES OF BOOKS ON RECENT DEVELOPMENTS IN ASTRONOMY AND ASTROPHYSICS

Vol. 1-Progress and Opportunities in Southern Hemisphere Optical Astronomy: The CTIO 25th Anniversary Symposium
ed. V. M. Blanco and M. M. Phillips ISBN 0-937707-18-X

Vol. 2-Proceedings of a Workshop on Optical Surveys for Quasars
ed. P. S. Osmer, A. C. Porter, R. F. Green, and C. B. Foltz ISBN 0-937707-19-8

Vol. 3-Fiber Optics in Astronomy
ed. S. C. Barden ISBN 0-937707-20-1

Vol. 4-The Extragalactic Distance Scale: Proceedings of the ASP 100th Anniversary Symposium
ed. S. van den Bergh and C. J. Pritchet ISBN 0-937707-21-X

Vol. 5-The Minnesota Lectures on Clusters of Galaxies and Large-Scale Structure
ed. J. M. Dickey ISBN 0-937707-22-8

Vol. 6-Synthesis Imaging in Radio Astronomy: A Collection of Lectures from the Third NRAO Synthesis Imaging Summer School
ed. R. A. Perley, F. R. Schwab, and A. H. Bridle ISBN 0-937707-23-6

Vol. 7-Properties of Hot Luminous Stars: Boulder-Munich Workshop
ed. C. D. Garmany ISBN 0-937707-24-4

Vol. 8-CCDs in Astronomy
ed. G. H. Jacoby ISBN 0-937707-25-2

Vol. 9-Cool Stars, Stellar Systems, and the Sun. Sixth Cambridge Workshop
ed. G. Wallerstein ISBN 0-937707-27-9

Vol. 10-The Evolution of the Universe of Galaxies. The Edwin Hubble Centennial Symposium
ed. R. G. Kron ISBN 0-937707-28-7

Vol. 11-Confrontation Between Stellar Pulsation and Evolution
ed. C. Cacciari and G. Clementini ISBN 0-937707-30-9

Vol. 12-The Evolution of the Interstellar Medium
ed. L. Blitz ISBN 0-937707-31-7

Vol. 13-The Formation and Evolution of Star Clusters
ed. K. Janes ISBN 0-937707-32-5

Vol. 14-Astrophysics with Infrared Arrays
ed. R. Elston ISBN 0-937707-33-3

Vol. 15-Large-Scale Structures and Peculiar Motions in the Universe
ed. D. W. Latham and L. A. N. da Costa ISBN 0-937707-34-1

Vol. 16-Atoms, Ions and Molecules: New Results in Spectral Line Astrophysics
ed. A. D. Haschick and P. T. P. Ho ISBN 0-937707-35-X

Vol. 17-Light Pollution, Radio Interference, and Space Debris
ed. D. L. Crawford ISBN 0-937707-36-8

Vol. 18-The Interpretation of Modern Synthesis Observations of Spiral Galaxies
ed. N. Duric and P. C. Crane ISBN 0-937707-37-6

Vol. 19-Radio Interferometry: Theory, Techniques, and Applications, IAU Colloquium 131
ed. T. J. Cornwell and R. A. Perley ISBN 0-937707-38-4

Vol. 20-Frontiers of Stellar Evolution, celebrating the 50th Anniversary of McDonald Observatory
ed. D. L. Lambert ISBN 0-937707-39-2

Vol. 21-The Space Distribution of Quasars
ed. D. Crampton ISBN 0-937707-40-6

Vol. 22-Nonisotropic and Variable Outflows from Stars
ed. L. Drissen, C. Leitherer, and A. Nota ISBN 0-937707-41-4

Vol. 23-Astronomical CCD Observing and Reduction Techniques
ed. S. B. Howell ISBN 0-937707-42-4

Vol. 24-Cosmology and Large-Scale Structure in the Universe
ed. R. R. de Carvalho ISBN 0-937707-43-0

Vol. 25-Astronomical Data Analysis Software and Systems I
ed. D. M. Worrall, C. Biemesderfer, and J. Barnes ISBN 0-937707-44-9

Vol. 26-Cool Stars, Stellar Systems, and the Sun, Seventh Cambridge Workshop
ed. M. S. Giampapa and J. A. Bookbinder ISBN 0-937707-45-7

Vol. 27-The Solar Cycle
ed. K. L. Harvey ISBN 0-937707-46-5

Vol. 28-Automated Telescopes for Photometry and Imaging
ed. S. J. Adelman, R. J. Dukes, Jr., and C. J. Adelman ISBN 0-937707-47-3

Vol. 29-Workshop on Cataclysmic Variable Stars
ed. N. Vogt ISBN 0-937707-48-1

Vol. 30-Variable Stars and Galaxies, in honor of M. S. Feast on his retirement
ed. B. Warner ISBN 0-937707-49-X

Vol. 31-Relationships Between Active Galactic Nuclei and Starburst Galaxies
ed. A. V. Filippenko ISBN 0-937707-50-3

Vol. 32-Complementary Approaches to Double and Multiple Star Research, IAU Colloquium 135
ed. H. A. McAlister and W. I. Hartkopf ISBN 0-937707-51-1

Vol. 33-Research Amateur Astronomy
ed. S. J. Edberg ISBN 0-937707-52-X

Vol. 34-Robotic Telescopes in the 1990s
ed. A. V. Filippenko ISBN 0-937707-53-8

Vol. 35-Massive Stars: Their Lives in the Interstellar Medium
ed. J. P. Cassinelli and E. B. Churchwell ISBN 0-937707-54-6

Vol. 36-Planets and Pulsars
ed. J. A. Phillips, S. E. Thorsett, and S. R. Kulkarni ISBN 0-937707-55-4

Vol. 37-Fiber Optics in Astronomy II
ed. P. M. Gray ISBN 0-937707-56-2

Vol. 38-New Frontiers in Binary Star Research
ed. K. C. Leung and I. S. Nha ISBN 0-937707-57-0

Vol. 39-The Minnesota Lectures on the Structure and Dynamics of the Milky Way
ed. Roberta M. Humphreys ISBN 0-937707-58-9

Vol. 40-Inside the Stars, IAU Colloquium 137
ed. Werner W. Weiss and Annie Baglin ISBN 0-937707-59-7

Vol. 41-Astronomical Infrared Spectroscopy: Future Observational Directions
ed. Sun Kwok ISBN 0-937707-60-0

Vol. 42-GONG 1992: Seismic Investigation of the Sun and Stars
ed. Timothy M. Brown ISBN 0-937707-61-9

Vol. 43-Sky Surveys: Protostars to Protogalaxies
ed. B. T. Soifer ISBN 0-937707-62-7

Vol. 44-Peculiar Versus Normal Phenomena in A-Type and Related Stars
ed. M. M. Dworetsky, F. Castelli, and R. Faraggiana ISBN 0-937707-63-5

Vol. 45-Luminous High-Latitude Stars
ed. D. D. Sasselov ISBN 0-937707-64-3

Vol. 46-The Magnetic and Velocity Fields of Solar Active Regions, IAU Colloquium 141
ed. H. Zirin, G. Ai, and H. Wang ISBN 0-937707-65-1

Vol. 47-Third Decinnial US-USSR Conference on SETI
ed. G. Seth Shostak ISBN 0-937707-66-X

Vol. 48-The Globular Cluster-Galaxy Connection
ed. Graeme H. Smith and Jean P. Brodie ISBN 0-937707-67-8

Vol. 49-Galaxy Evolution: The Milky Way Perspective
ed. Steven R. Majewski ISBN 0-937707-68-6

Vol. 50-Structure and Dynamics of Globular Clusters
ed. S. G. Djorgovski and G. Meylan ISBN 0-937707-69-4

Vol. 51-Observational Cosmology
ed. G. Chincarini, A. Iovino, T. Maccacaro, and D. Maccagni ISBN 0-937707-70-8

Vol. 52-Astronomical Data Analysis Software and Systems II
ed. R. J. Hanisch, J. V. Brissenden, and Jeannette Barnes ISBN 0-937707-71-6

Vol. 53-Blue Stragglers
ed. Rex A. Saffer ISBN 0-937707-72-4

Vol. 54-The First Stromlo Symposium: The Physics of Active Galaxies
ed. Geoffrey V. Bicknell, Michael A. Dopita, and Peter J. Quinn ISBN 0-937707-73-2

Vol. 55-Optical Astronomy from the Earth and Moon,
ed. Diane M. Pyper and Ronald J. Angione ISBN 0-937707-74-0

Vol. 56-Interacting Binary Stars
ed. Allen W. Shafter ISBN 0-937707-75-9

Vol. 57-Stellar and Circumstellar Astrophysics
ed. George Wallerstein and Alberto Noriega-Crespo ISBN 0-937707-76-7

Vol. 58-The First Symposium on the Infrared Cirrus and Diffuse Interstellar Clouds
ed. Roc M. Cutri and William B. Latter ISBN 0-937707-77-5

Inquiries concerning these volumes should be directed to the:
Astronomical Society of the Pacific
CONFERENCE SERIES
390 Ashton Avenue
San Francisco, CA 94112-1722
415-337-1100

ASTRONOMICAL SOCIETY OF THE PACIFIC
CONFERENCE SERIES

Volume 58

THE FIRST SYMPOSIUM ON THE INFRARED CIRRUS AND DIFFUSE INTERSTELLAR CLOUDS

Conference held 7-9 April 1993
Tucson, Arizona

Edited by
Roc M. Cutri and William B. Latter

TABLE OF CONTENTS

IV. CIRRUS IN EXTERNAL GALAXIES

V. DUST AND ORGANIC MATTER IN DIFFUSE INTERSTELLAR CLOUDS

VI. CLOUD CHEMISTRY

VII. CLOUD ENERGETICS AND DYNAMICS

PREFACE

Astronomical investigations of the interstellar medium have been ongoing for nearly a century. The discovery of diffuse gas in interstellar space is usually credited to J.F. Hartmann who in 1904, while observing the spectrum of the binary star δ Ori over its orbital phase, noticed very narrow CaII H and K lines that did not share the periodically varying doppler shifts of the stellar lines. The first convincing demonstration of attenuation due to dust in interstellar space came from Trumpler's 1930 study of the sizes, brightnesses and distances of open clusters. In 1926, Eddington made one of the earliest significant attempts at a theoretical analysis of the physical conditions in the interstellar medium. Despite this lengthy heritage, a true renaissance in the study of the diffuse interstellar medium was triggered by the all-sky radio and infrared surveys of the late 1970's and early 1980's. In particular, Heiles' 21-cm survey and the 1983 mid- and far-infrared space survey by the Infrared Astronomical Satellite (IRAS) revealed vast complexes of neutral hydrogen and warm dust that covered virtually the entire sky. These intricate filamentary clouds, commonly referred to as infrared cirrus and diffuse interstellar clouds, are now known to be condensations of atomic and molecular gas and dust that are illuminated primarily by the ambient interstellar radiation field.

In the years since the 21-cm and IRAS surveys, a wealth of new observations of the infrared cirrus and diffuse clouds have been carried out at wavelengths ranging from the radio to the X-ray. Radio continuum and molecular line observations have provided detailed kinematic information about diffuse clouds, insight into the chemistry of the gas therein, and indirect probing of the interstellar radiation field. Analyses of the mid- and far-infrared data from IRAS have revealed the temperature and size distribution of dust grains and macro-molecules in the clouds. Measurements of optical continuum emission have been used to trace the scattered light from the dust and fluorescence of small grains and/or molecules. Optical and ultraviolet absorption line studies have probed the extinction and elemental composition within the diffuse structures. Even X-ray measurements have been used to probe the column density of neutral gas in the clouds, and to trace hot gas associated with shocks that have been postulated to shape the filamentary structures. Although there has been an impressive variety of new research in the last two decades, there was no good compendium of the results in this growing body of work. To rectify this, the *First Symposium on the Infrared Cirrus and Diffuse Interstellar Clouds* was held on April 7-9, 1993 in Tucson, Arizona.

The goal of the *Symposium* was to promote discussion in three broad subject areas: the general phenomenology of the cirrus and diffuse clouds, including descriptions of the general morphology, taxonomy and spectral energy distributions of the cirrus and diffuse clouds as determined by measurements from radio to X-ray wavelengths; the composition and chemistry of the interstellar clouds, including the interpretation and modelling of the observed cloud properties such as gas and dust abundances and the interaction of the gas and dust with the interstellar radiation field; and the general energetics, dynamics and evolution of the cirrus and diffuse clouds, including the origin and fate of the diffuse clouds and how the clouds are related to the denser components of the interstellar medium. These *Proceedings* contain the invited and contributed papers presented at the *Symposium*. As is evidenced by the diversity and quality of these works, our goals were well met. But what is also obvious from this volume is that there remain many questions about these components of the diffuse interstellar medium.

The *First Symposium on the Infrared Cirrus and Diffuse Interstellar Clouds* was co-sponsored by the Steward Observatory of the University of Arizona and the National Radio Astronomy Observatory, Tucson[1]. We particularly wish to acknowledge the support and encouragement of Drs. Peter Strittmatter, Darrel Emerson and Robert Brown. The *Symposium* was also supported by a generous grant from the National Aeronautics and Space Administration (NAGW-3455 and NAGW-18294), and for that we especially thank Dr. Michael Bicay.

The success of the *Symposium* was due in a large part to the sound advice of the Scientific Organizing Committee: Drs. Lou Allamandola, Chas Beichman, John Black, François Boulanger, Gillian Knapp, Frank Low, Peter Martin, Kristen Sellgren, Barry Turner and Ewine van Dishoeck. It is also our pleasure to acknowledge the invaluable assistance of Jennifer Neighbours and Esther Torres who smoothly handled the many registration and clerical duties that accompany such an endeavor. We also thank Sharon White, Gopal Narayanan, Nadine Dinshaw and Aditya Dayal for their assistance during the *Symposium*, and Tom Balonek who is responsible (and fully-so) for the candid photographs in this book. Finally, and most importantly, we wish to thank Elaine Cutri and Diane Engler, without whose help, patience and constant support this *Symposium* would not have been possible.

Roc M. Cutri and William B. Latter
Tucson, Arizona 1993

[1] The National Radio Astronomy Observatory is operated by Associated Universities, Inc., under cooperative agreement with the National Science Foundation.

CONFERENCE PARTICIPANTS

LOU ALLAMANDOLA
NASA/Ames Research Ctr.

JAAP BAARS
SMTO/Steward Obs.

EMMA BAKES
Princeton University

THOMAS BALONEK
Colgate Univ./NRAO

CHAS BEICHMAN
IPAC/Caltech

JEAN-PHILIPPE BERNARD
Nagoya University

JOHN BIEGING
Steward Observatory

JOHN BLACK
Steward Observatory

JAMES J. BOCK
Univ. of California Berkeley

FRANÇOIS BOULANGER
Inst. d'Astrophysique Sp.

HAROLD M. BUTNER
NASA/Ames Research Ctr.

JASON CARDELLI
University of Wisconsin

EMMANUEL CAUX
CESR

FRANK CLARK
Phillips Laboratory

GEOFFREY CLAYTON
CASA/Univ. of Colorado

ROC M. CUTRI
Steward Observatory

ADITYA DAYAL
Steward Observatory

NICK DEVEREUX
New Mexico State Univ.

LOUIS d'HENDECOURT
Inst. d'Astrophysique Sp.

NADINE DINSHAW
Steward Observatory

STEVEN DOTY
Rensselaer Polytechnic Inst.

BRUCE T. DRAINE
Princeton University Obs.

BRUCE G. ELMEGREEN
IBM/T.J. Watson Res. Ctr.

EDITH FALGARONE
École Normale Supérieure

STEVEN FEDERMAN
University of Toledo

MIKE FICH
University of Waterloo

DAVID R. FLOWER
University of Durham

JAMES R. FOWLER
Apache Point Observatory

GARY FULLER
NRAO/Tucson

NICK GAUTIER
Jet Propulsion Laboratory

STEVEN J. GIBSON
University of Wisconsin

ALYSSA A. GOODMAN
Harvard University

PURAGRA GUHATHAKURTA
Princeton University

MARGARET M. HANSON
CASA/Univ. of Colorado

ANDREAS HEITHAUSEN
Universitaet zu Köln

GEORGE HELOU
IPAC/Caltech

UWE HERBSTMEIER
Universität Bonn

DOUGLAS HUDGINS
NASA/Ames Research Ctr.

PETER JENNISKENS
NASA/Ames Research Ctr.

CHRISTINE JOBLIN
NASA/Ames Research Ctr.

GILLES JONCAS
Université Laval

SANG-HEE KIM
Univ. of Toronto/Caltech

GILLIAN KNAPP
Princeton University

JACEK KREŁOWSKI
U. Texas/Copernicus Inst.

TED LA ROSA
NASA/Marshall Sp. Flt. Ctr.

WILLIAM B. LATTER
NRAO/Tucson

RENE LAUREIJS
ESA Space Science

WEI LEE
U. of Alabama Birmingham

CHUN MING LEUNG
Rensselaer Polytechnic Inst.

FRANK LOW
Steward Observatory

BARRY L. LUTZ
Northern Arizona Univ.

LORIS MAGNANI
University of Georgia

SANGEETA MALHOTRA
Princeton University

JEFF MANGUM
SMTO/Steward Obs.

PETER MARTIN
Caltech/CITA

ULRICH MEBOLD
Universität Bonn

DAVID MEYER
Northwestern University

JAYANT MURTHY
Johns Hopkins University

GOPAL NARAYANAN
Steward Observatory

YVONNE PENDLETON
NASA/Ames Research Ctr.

BRYAN PENPRASE
IPAC/Caltech

BILL PETERS
Steward Observatory

JENS K. PETERSOHN
University of Toledo

ROB PRICE
University of New Mexico

WILLIAM T. REACH
NASA/Goddard Sp. Flt. Ctr.

CHRIS ROGERS
DRAO

KATHERINE C. ROTH
Space Telescope Sci. Inst.

FRANÇOIS ROULEAU
University of Toronto

FARID SALAMA
NASA/Ames Research Ctr.

KRIS SELLGREN
Ohio State University

DAVID SHUPE
Caltech

THEODORE SNOW
CASA/Univ. of Colorado

MARCO SPAANS
Sterrewacht Leiden

RONALD STARK
Sterrewacht Leiden

BARRY TURNER
NRAO/Charlottesville

DAVE VAN BUREN
IPAC/Caltech

EWINE VAN DISHOECK
Sterrewacht Leiden

GERRIT L. VERSCHUUR
Rhodes College

BART P. WAKKER
University of Illinois

CONNIE WALKER
University of Texas

WILLIAM H. WALLER
NASA/Goddard Sp. Flt. Ctr.

ZHONG WANG
IPAC/Caltech

THOMAS J. WDOWIAK
U. of Alabama Birmingham

DANIEL WELTY
University of Chicago

MICHAEL WERNER
Jet Propulsion Laboratory

DOUG WHITTET
Rensselaer Polytechnic Inst.

ADOLF N. WITT
University of Toledo

MICHAEL J. WOLFF
University of Wisconsin

CHAPTER I

ABSORPTION AND SCATTERING IN DIFFUSE INTERSTELLAR CLOUDS

The First Symposium on the Infrared Cirrus and Diffuse Interstellar Clouds
ASP Conference Series, Vol. 58, 1994
Roc M. Cutri and William B. Latter (eds.)

HIGH RESOLUTION HST AND CCD OBSERVATIONS OF INTERSTELLAR ABSORPTION LINES

DAVID M. MEYER
Department of Physics and Astronomy, Northwestern University, Evanston, IL 60208

ABSTRACT The study of interstellar absorption lines is currently undergoing a renaissance at ultraviolet wavelengths with the Goddard High Resolution Spectrograph (GHRS) on the *Hubble Space Telescope* and at optical wavelengths with efficient high resolution spectrographs and a new generation of CCD detectors. These instruments have made it possible to explore new regimes in measuring the velocity structure of interstellar lines, determining the abundances of trace atomic and molecular species in diffuse clouds, and probing the translucent and halo ISM toward faint stars and extragalactic objects. Of particular interest are recent GHRS results on depletion effects within interstellar line profiles, the detection of heavy elements such as germanium and tin in the ISM, and accurate measures of the interstellar carbon abundance. At optical wavelengths, velocity resolutions of 0.6 km s^{-1} have recently been achieved in measuring the complicated component structure and hyperfine splitting of the interstellar Na I D lines. When combined with the high signal-to-noise ratios ($S/N > 2000$) attainable with CCD detectors, this degree of resolution makes it feasible to attack such difficult problems as measuring the interstellar ^{7}Li/^{6}Li isotope ratio. Other applications include searches for new molecular species in diffuse clouds, studies of small-scale ISM structure through observations of the interstellar absorption lines toward both components of resolvable binary star systems, and investigations of the high-velocity cloud phenomenon in the Galactic halo.

INTRODUCTION

Diffuse interstellar clouds contain appreciable amounts of gas and dust through which stars are visible. In the cold, tenuous gas of diffuse clouds, essentially all of the atoms, ions, and molecules will be in their ground electronic states. By studying the absorption lines produced in stellar spectra by the resonance transitions arising from these states, the composition and physical conditions of the intervening cloud(s) can be deduced. In conducting these studies, the following criteria constitute a "wish list" for any intrepid observer of interstellar absorption lines:

1. Since the lines are typically quite narrow, high spectral resolution is essential with optimal values greater than $R = \lambda/\Delta\lambda = 10^5$ or, in terms of velocity, less than 3 km s^{-1}. This degree of resolution is necessary to determine the velocity structure of interstellar line profiles and possibly resolve individual com-

ponents. High spectral resolution is also essential for specific applications such as the measurement of interstellar isotope ratios and the rotational excitation of interstellar molecules.

2. Since many important interstellar lines are weak ($W_\lambda \leq 10$ mÅ), the acquisition of high signal-to-noise ($S/N \geq 500$) spectra at high resolution is necessary to make convincing detections. The motivation in accurately measuring such weak lines is that they are usually optically thin and therefore easily interpreted in terms of the column density of the absorbing atom, ion, or molecule. Since rare species and the low oscillator strength transitions of common species both produce weak lines, high S/N observations are vital to determining accurate abundances in diffuse clouds.

3. The ability to obtain high resolution, high S/N spectra of stars and extragalactic objects down to increasingly faint limiting magnitudes ($V \geq 10$) is important in that it opens up a larger and more diverse range of sightlines to studies of interstellar absorption. Such sightlines include those through dustier regions such as translucent clouds and long pathlengths through the Galactic halo and across adjacent spiral arms in the Galactic disk.

In this review, I will discuss some recent observations of interstellar absorption lines that have opened new research avenues by pushing up to and beyond these resolution, S/N, and brightness goals at optical and ultraviolet wavelengths. Although this short discussion cannot do justice to all of the recent work done by the many investigators in this field, I hope that it illustrates the robust character and exciting potential of current interstellar absorption line studies.

CURRENT OBSERVATIONAL CAPABILITIES

Ultraviolet

Most of the ground-state transitions of the atoms, ions, and molecules prevalent in diffuse clouds occur in the ultraviolet spectral region (912–3100 Å). Much of what we know about diffuse clouds has come from observations of these UV lines with the *Copernicus* and *International Ultraviolet Explorer* (*IUE*) satellites (Cowie and Songaila 1986). With the launch of the *Hubble Space Telescope* (*HST*) in 1990, the Goddard High Resolution Spectrograph (GHRS) has brought new capabilities to these studies (Gilliland et al. 1992). In particular, the spectral resolution attainable with the echelle grating on GHRS ($R \approx 10^5$) is appreciably higher than the best achievable with *Copernicus* ($R \approx 3 \times 10^4$) and *IUE* ($R \approx 10^4$). Also, the GHRS Digicon detectors are more sensitive and capable of producing net spectra with S/N ratios in excess of 500 (Federman et al. 1993). In terms of speed, the GHRS is about 10^2 times faster than *IUE* when compared at similar resolutions of $R \approx 10^4$. The major drawbacks of GHRS are the limited spectral coverage provided by the linear (512 pixel) Digicon arrays and the very small amount of *HST* observing time available due to heavy oversubscription. Also, since GHRS lost its far-UV echelle capability early in the mission, only a few high resolution observations of interstellar absorption lines have been made shortward of 1700 Å. However, the *HST* servicing mission scheduled for late 1993 should recover this capability as well as improve the throughput for small aperture GHRS echelle observations by removing the spherical aberration of the *HST* focal image with corrective optics.

Optical

Except for the Ca II, Na I, and K I doublets, the interstellar lines observable at optical wavelengths with ground-based telescopes are typically very weak. However, optical instrumentation consisting of coudé and echelle spectrographs and CCD detectors is now very well-equipped to detect interstellar features with equivalent widths less than 1 mÅ. In particular, very high resolution spectra have recently been obtained with the 0.9 m coudé feed telescope ($R = 3 \times 10^5$) at Kitt Peak National Observatory (KPNO) (Meyer, Hawkins, and Wright 1993), the coudé spectrograph ($R = 6 \times 10^5$) on the McDonald Observatory 2.7 m telescope (Lambert, Sheffer, and Crane 1990), and the Ultra-High Resolution Facility ($R = 9 \times 10^5$) on the Anglo-Australian Observatory 3.9 m telescope (Diego et al. 1993). In terms of detectors, the latest generation of CCDs has high quantum efficiency ($QE \geq 60\%$) from 3100 Å in the near-ultraviolet to 9000 Å in the near-infrared. With careful flat-fielding and summation techniques, CCDs can produce net spectra with photon-noise-limited S/N ratios of 2000 or more (Meyer 1990a). Also, the low readout noise ($rms < 10$ electrons/pixel) of CCD detectors makes fainter objects accessible and smaller telescopes viable for high resolution, high S/N interstellar absorption line work.

ULTRAVIOLET STUDIES OF DIFFUSE CLOUDS WITH HST

Only a few *HST* studies of interstellar absorption lines have been published to date. This initial work is important in that it illustrates the capabilities and potential of the GHRS and it represents essentially all of the far-ultraviolet echelle data on interstellar absorption that will appear before the *HST* servicing mission.

Multicomponent Absorption Line Structure

Savage, Cardelli, and Sofia (1992) have recently analyzed GHRS echelle observations of the interstellar absorption produced by 14 dominant ions such as C II, O I, Mg II, Mn II, Fe II, and Ni II in the diffuse clouds toward ζ Oph. The 3.5 km s^{-1} velocity resolution of the GHRS echelle data allows them to evaluate the gas phase column densities in the principal absorbing components observed at heliocentric velocities of -27 and -15 km s^{-1}. Although higher resolution can be obtained at optical wavelengths, the ultraviolet spectral region contains the numerous transitions needed to conduct a rigorous abundance study of many ions. Savage et al. find a striking contrast in abundances between the two ζ Oph clouds in that highly refractory elements such as Fe and Ni are more heavily depleted than elements such as O and Mg in the -15 km s^{-1} cloud as compared to the -27 km s^{-1} cloud. They suggest that this is the result of sampling two distinct types of interstellar regions where each element has one value of depletion in the low-density medium represented by the -27 km s^{-1} component and a different, but higher, value in the cold cloud represented by the -15 km s^{-1} component. This work is a good example of why high spectral resolution is needed to avoid line-of-sight blending effects in interpreting interstellar absorption line data.

Detection of Rare Atomic Species

In addition to providing higher quality data on the strong UV interstellar lines originally studied with *Copernicus* and *IUE*, the greater sensitivity of the GHRS makes it possible to detect the trace transitions of much rarer atomic species. Cardelli, Savage, and Ebbets (1991) have made the first stroke in this effort by identifying weak ($W_\lambda \leq 10$ mÅ) lines of Ga II, Ge II, and Kr I in their GHRS echelle spectra of ζ Oph. With atomic numbers of 31, 32, and 36, gallium, germanium, and krypton represented the heaviest elements detected in the ISM at the time of this discovery. The detection of Kr I is particularly significant since, as a noble gas, it is not expected to form chemical or mechanical bonds and should reside primarily in the gas phase. Since the abundance of Kr reflects a mixture of $s-$ and $r-$process nucleosynthesis, further observations of this atom will provide important information about interstellar enrichment processes. More recently, Hobbs et al. (1993) have identified the 1400.45 Å line of Sn II in moderate resolution ($R = 2 \times 10^4$) GHRS spectra of three stars. With $Z = 50$, tin is the first element from the fifth row of the periodic table to be detected in the interstellar medium. Hobbs et al. find that the depletions of Cu, Ga, Ge, Kr, and Sn generally decrease monotonically with increasing atomic number toward the stars in their sample, and Sn is essentially undepleted within the observational errors. Their data also imply that the depletions of 26 elements from the interstellar gas in an average dense interstellar cloud appear to correlate with the elemental condensation temperatures more closely than with the first ionization potentials. As these depletion uncertainties are better understood and the oscillator strengths of certain transitions are better determined, it should become possible to carefully examine the homogeneity of different products of nucleosynthesis in the ISM with interstellar absorption lines.

Abundance of Interstellar Carbon

After hydrogen and helium, carbon and oxygen are the most abundant constituents of the interstellar gas. Carbon is especially important to the thermal balance of interstellar clouds because of its contribution to the cooling rate either in its atomic form or as part of molecules such as CO. Carbon also constitutes a significant fraction of the mass residing in interstellar dust grains and polycyclic aromatic hydrocarbons (PAHs). Unfortunately, the abundance of carbon in C II, the dominant C ion in diffuse clouds, has been reliably measured in only a few sightlines due to the heavy saturation of the strong C II line at 1335 Å ($\log f\lambda = 2.232$). Recently, however, Cardelli et al. (1993) have detected the weak intersystem transition of C II at 2325 Å ($\log f\lambda = -3.853$) with GHRS echelle spectra of ζ Oph. They measure an equivalent width of 0.52 ± 0.12 mÅ for this feature from a net spectrum characterized by a S/N ratio of 450. This line strength corresponds to a C/H abundance ratio toward ζ Oph that is about a factor of 2 smaller than that measured toward ξ Per by Cardelli et al. (1991). Although some of this difference may be attributable to inhomogeneities in the cosmic abundance of carbon, Cardelli et al. (1993) suggest that at least part of it could be due to the conversion of carbon grains and/or PAHs to gas-phase

C II in the ξ Per sightline. As the carbon abundance in other sightlines is nailed down with high S/N GHRS observations of the C II 2325 Å line, it will be possible to firm up this conclusion and rigorously test the possibility of C/H inhomogeneities in the interstellar medium.

OPTICAL STUDIES OF DIFFUSE CLOUDS WITH CCDS

The rapid advances in optical instrumentation over the past few years have opened new vistas in and increased the utility of interstellar absorption line studies. As discussed below, some of these advances have been applied to basic questions concerning the nature of diffuse clouds. Others have utilized diffuse cloud absorption as a laboratory to probe related topics. As highlighted below, studies of the interstellar ^{7}Li/^{6}Li isotope ratio, rare molecular species in the diffuse ISM, and high-velocity clouds in the Galactic halo have particularly pushed the S/N ratios, near-UV sensitivity, and brightness limits of the best instrumentation currently available.

Very High Resolution Observations of Interstellar Na I

By observing the component structure of an interstellar absorption line at a spectral resolution near the limit set by the thermal widths of the individual components, one can not only isolate the number of kinematically distinct gas clouds in the sightline but also place an upper limit on the temperature and turbulent motion of each cloud. The interstellar Na I D doublet near 5890 Å is especially well-suited to such observations since these lines exhibit a hyperfine splitting (1.05 km s^{-1}) that is about three times greater than the thermal width expected for a typical $T \approx 80$ K diffuse cloud. At an instrumental resolution of $R = 5 \times 10^5$ or $\Delta v = 0.6$ km s^{-1}, it is possible to partially resolve a thermally broadened pair of Na I hyperfine components in a cloud with $T \leq 500$ K and thus explicitly identify the presence of cold gas. Using the double-pass echelle configuration of the coudé spectrograph on the McDonald Observatory 2.7 m telescope, Hobbs and Welty (1991) have recently obtained such data for six sightlines. They detect at least 42 kinematically distinct clouds toward the five most distant stars in addition to a single cloud seen toward the star δ Cyg. Thirteen of these clouds, including the one toward δ Cyg, are sufficiently narrow and unblended to clearly reveal the Na I hyperfine structure. The most stringent upper limits on temperature and turbulent velocity from any one of these clouds are $T < 160$ K and $v_t < 0.24$ km s^{-1}. From this data set, it is certainly clear that interstellar line profiles are typically very complicated when observed at very high spectral resolution. With at least 12 Na I components over a velocity interval of 25 km s^{-1}, the case of the ϵ Ori sightline in particular raises an important question concerning the nature of these components. Namely, do they represent multiple clouds or clumps within a cloud? The key to this issue is gaining the spatial insight to better understand the velocity structure.

Small-Scale ISM Structure

The range in size of diffuse interstellar clouds and the scales at which matter is clumped in these clouds is not well known. The question of structure is especially important in that almost all theoretical interpretations of diffuse clouds are based on slab models with either constant or slowly varying physical properties. Although most of the existing work on diffuse cloud structure has involved radio observations, not much has been done at small scales (< 0.2 pc) due to aperture limitations and the difficulty of interferometric studies. Observations of interstellar absorption lines toward the components of resolvable binary and multiple star systems provide another approach to the problem of diffuse ISM structure at these scales. Since the absorption-line beam toward a star corresponds to a tiny fraction of an arc second, this method can probe scales down to a few arc seconds or less depending on the smallest-separation binary that the seeing allows. Among the initial results of such a program (Meyer 1990b) has been the finding that the components of 5 of 6 binaries with projected linear separations of 700–5700 AU display similar interstellar absorption line strengths. The one case (κ CrA) that exhibits evidence for small-scale structure (< 2800 AU) may be due to an intervening cloud associated with the prominent Loop I shell in the local ISM. More recently, O'Dell et al. (1993) have obtained high (3.5 km s^{-1}) resolution observations of the interstellar Ca II absorption toward the 4 Trapezium stars (θ^1 Ori) with the 0.9 m coudé feed telescope at KPNO. They find significant Ca II line strength and profile differences among these stars that are indicative of ISM structure on scales less than 6000 AU. Such detail would be difficult to observe at millimeter wavelengths with the NRAO 12 m antenna at Kitt Peak, for example, because the millimeter beam would be larger than the separations of the Trapezium stars. Since the optical ISM structure work can now be performed on well-equipped small telescopes such as the KPNO coudé feed, it will be possible to build up a database suitable for statistical analysis. The preliminary conclusions from the Meyer (1993) survey of a dozen sightlines are that small-scale ISM structure appears to be more common when probed with binaries whose separations are greater than 2500 AU and more pronounced in sightlines through shocked and interphase gas such as that in the Trapezium and the Loop I cloud toward κ CrA.

The Interstellar ^{7}Li/^{6}Li Isotope Ratio

The interstellar ^{7}Li/^{6}Li isotope ratio is an important parameter governed by the chemical evolution of the Galaxy and primordial nucleosynthesis during the Big Bang (Audouze et al. 1983). The ^{7}Li I lines at 6707.761 and 6707.912 Å and the corresponding ^{6}Li I doublet redshifted by 0.160 Å are the only Li resonance lines accessible to ground-based observations. Among the few stars toward which interstellar ^{7}Li I has been detected (White 1986), the lines are very weak with total equivalent widths of about 1 mÅ. Consequently, any realistic attempt at measuring the interstellar ^{7}Li/^{6}Li ratio requires a spectrograph capable of attaining very high spectral resolution and a detector capable of achieving very high S/N spectra. Meyer, Hawkins, and Wright (1993) have recently obtained such observations ($R = 3 \times 10^5$; $S/N \approx 2000$) of the interstellar Li I absorption toward the stars ζ Oph and ζ Per using the KPNO coudé feed telescope. Each of these spectra involved a net exposure of 40 hours and represent, in terms of resolution and S/N ratio, the most sensitive spectra ever recorded of a star other than the Sun. The Li I line profiles indicate the presence of both ^{7}Li and ^{6}Li in these two sightlines with a resultant best-fit value of 6.8 (+1.4, −1.7) for the ^{7}Li/^{6}Li isotope ratio toward ζ Oph. It is clear from the data that the interstellar $^7Li/^6Li$ ratio toward both ζ Oph and ζ Per is less than the value of 12.6 representative of the early solar system. This result does not support the lower limit of 25 set for the $^7Li/^6Li$ ratio by Ferlet and Dennefeld (1984) from their Li I observations toward ζ Oph. However, it agrees much better with the recent $^7Li/^6Li$ measurement of 12.5 (+4.3, −3.4) toward ρ Oph by Lemoine et al. (1993). Since current models of Galactic chemical evolution predict that the $^7Li/^6Li$ ratio should increase with time to a present value near 25 (Abia and Canal 1988), these new results suggest that the production of ^6Li is being enhanced through cosmic ray spallation or some other process.

Detection of New Interstellar Molecules

Although only a few interstellar molecules have been identified in absorption at near-UV/optical/near-IR wavelengths, their usefulness as probes of interstellar abundances, densities, temperatures, and radiation fields makes them indispensable to studies of diffuse clouds. Until recently, this list of molecules consisted solely of the diatomic species CN, CH, CH^+, OH, and C_2 (with H_2 and CO at UV wavelengths) and had not expanded since the discovery of C_2 by Souza and Lutz (1977). The size of this list not only reflects the simplified chemistry that occurs in diffuse clouds relative to dark clouds due to their lower densities and UV extinction but also the difficulty of detecting the weak absorption lines of trace molecular constituents. In a recent effort to search for new interstellar molecules in the poorly-explored near-UV spectral region with the echelle spectrograph on the KPNO 4.0 m telescope, Meyer and Roth (1991a) succeeded in detecting the interstellar NH $A^3\Pi - X^3\Sigma$ (0,0) $R_1(0)$ line at 3358.053 Å toward both ζ Per ($W_\lambda = 0.37 \pm 0.08$ mÅ) and HD 27778 ($W_\lambda = 1.1 \pm 0.3$ mÅ) through high S/N observations at a spectral resolution of 0.08 Å. A control spectrum of γ Cas, a star with little interstellar material in its line of sight, was also obtained to make sure that these NH features were indeed interstellar and not some instrumental or atmospheric artifact. Such precautions are necessary to establish convincing detections of very weak absorption lines. The most striking aspect of the Meyer and Roth discovery of interstellar NH is that the column density of this molecule toward ζ Per is much greater than that predicted by the steady-state gas-phase models (van Dishoeck 1992) while the chemically similar species OH and CH agree quite well in such a comparison. Consequently, this result may signify a new aspect of the ion-molecule chemistry in diffuse clouds or the production of NH from grain surface reactions (Mann and Williams 1984). Also, it is now clear that CCDs have the near-UV sensitivity to make searches for other important interstellar molecules such as C_3, SH, SH^+, and OH^+ a much more viable enterprise.

The Galactic Halo and High-Velocity Clouds

Although high-velocity, metal-absorbing gas appears to be quite sparse in the Galactic halo, most of the sightlines sampled have involved high-latitude stars rather than extragalactic sources because few of these have been bright enough for measurement of the strong UV lines with *IUE* or the weak Ca II lines with a 4 m class telescope. Consequently, the existing absorption line data on the velocity distribution of gas over pathlengths which cover the full extent of the Galactic halo is very limited in terms of sky coverage and sensitivity. In this regard, the recent explosion of the Type Ia supernova SN 1991T in the Virgo Cluster galaxy NGC 4527 presented an interesting opportunity. Not only did the coordinates of SN 1991T place it 1.4° on the sky away from the bright ($V = 12.8$), well-studied QSO 3C 273, SN 1991T was also brighter than 3C 273 at optical wavelengths for well over a month near its maximum ($V = 11.5$). Thus, the stage was set for a sensitive absorption line probe of the interstellar and extragalactic gas along two nearly parallel sightlines out to the Virgo Cluster. Using the echelle spectrograph on the KPNO 4.0 m telescope, Meyer and Roth (1991b) took advantage of this situation by obtaining high S/N observations of the Ca II absorption toward SN 1991T ($S/N \approx 220$) and 3C 273 ($S/N \approx 120$) at a spectral resolution of 0.18 Å in respective net exposures of 13 and 10 hours. In addition to large line strength differences at low velocities, they detect weak ($W_\lambda < 20$ mÅ) Ca II components at LSR velocities of −87, +215, and +263 km s^{-1} toward SN 1991T that are absent in the 3C 273 spectrum. Given the velocity (+1733 km s^{-1}) of NGC 4527, these features are most likely due to the Galactic high-velocity cloud (HVC) phenomenon primarily observed in H I 21 cm emission (Wakker 1991). Although the metal-line counterparts to HVCs have rarely been observed, the ability to detect these weak Ca II components is well beyond the sensitivity of almost all previous absorption line work on extragalactic sources outside the vicinity of the Magellanic Clouds. If low column density HVCs are a common constituent of the Galactic halo, it would have important implications for the intervening halo interpretation of QSO absorption line systems (Bahcall and Spitzer 1969). Indeed, the complex Ca II profile spanning 350 km s^{-1} toward SN 1991T is very reminiscent of a QSO absorber. In any case, the ability to obtain high S/N spectra of interstellar absorption lines toward faint objects has the potential to expand the study of diffuse clouds to other galaxies.

CONCLUSION

With *HST* reaching its full potential and the advent of 8 m class telescopes with new high resolution spectrographs over the next few years, it is clear that interstellar absorption line studies will continue to enrich our understanding of diffuse clouds in the Galaxy and elsewhere.

ACKNOWLEDGMENTS

It is a pleasure to thank Jason Cardelli, Isabel Hawkins, Lew Hobbs, Katherine Roth, and Dan Welty for helpful discussions. Much of the author's work discussed in this paper was supported by the National Science Foundation through grants AST 90-21641 and AST 90-23372.

REFERENCES

Abia, C., and Canal, R. 1988, *A&A*, **189**, 55

Audouze, J., Boulade, O., Malinie, G., and Poilane, Y. 1983, *A&A*, **127**, 164

Bahcall, J. N., and Spitzer, L. 1969, *ApJ*, **156**, L63

Cardelli, J. A., Mathis, J. S., Ebbets, D. C., and Savage, B. D. 1993, *ApJ*, **402**, L17

Cardelli, J. A., Savage, B. D., Bruhweiler, F. C., Smith, A. M., Ebbets, D. C., Sembach, K. R., and Sofia, U. J. 1991, *ApJ*, **377**, L57

Cardelli, J. A., Savage, B. D., and Ebbets, D. C. 1991, *ApJ*, **383**, L23

Cowie, L. L., and Songaila, A. 1986, *ARA&A*, **24**, 499

Diego, F., Fish, A., Spyromilio, J., and Shortridge, K. 1993, *AAO Newsletter*, **64**, 1

Federman, S. R., Sheffer, Y., Lambert, D. L., and Gilliland, R. L. 1993, *ApJ*, in press

Ferlet, R., and Dennefeld, M. 1984, *A&A*, **138**, 303

Gilliland, R. L., Morris, S. L., Weymann, R. J., Ebbets, D. C., and Lindler, D. J. 1992, *PASP*, **104**, 367

Hobbs, L. M., and Welty, D. E. 1991, *ApJ*, **368**, 426

Hobbs, L. M., Welty, D. E., Morton, D. C., Spitzer, L., and York, D. G. 1993, *ApJ*, in press

Lambert, D. L., Sheffer, Y., and Crane, P. 1990, *ApJ*, **359**, L19

Lemoine, M., Ferlet, R., Vidal-Madjar, A., Emerich, C., and Bertin, P. 1993, *A&A*, **269**, 469

Mann, A. P. C., and Williams, D. A. 1984, *MNRAS*, **209**, 33

Meyer, D. M. 1990a, in *CCDs in Astronomy*, G. H. Jacoby, San Francisco: ASP, 345

Meyer, D. M. 1990b, *ApJ*, **364**, L5

Meyer, D. M. 1993, in preparation

Meyer, D. M., Hawkins, I., and Wright, E. L. 1993, *ApJ*, **409**, L61

Meyer, D. M., and Roth, K. C. 1991a, *ApJ*, **376**, L49

Meyer, D. M., and Roth, K. C. 1991b, *ApJ*, **383**, L41

O'Dell, C. R., Valk, J. H., Wen, Z., and Meyer, D. M. 1993, *ApJ*, **403**, 678

Savage, B. D., Cardelli, J. A., and Sofia, U. J. 1992, *ApJ*, **401**, 706

Souza, S. P., and Lutz, B. L. 1977, *ApJ*, **216**, L49

van Dishoeck, E. F. 1992, in *Astrochemistry of Cosmic Phenomena*, P. D. Singh, Dordrecht: Kluwer, 143

Wakker, B. P. 1991, *A&A*, **250**, 499

White, R. E. 1986, *ApJ*, **307**, 777

The First Symposium on the Infrared Cirrus and Diffuse Interstellar Clouds
ASP Conference Series, Vol. 58, 1994
Roc M. Cutri and William B. Latter (eds.)

VERY HIGH RESOLUTION, HIGH SIGNAL-TO-NOISE SPECTROSCOPY OF DIFFUSE INTERSTELLAR BANDS

J. KRELOWSKI[1], C. SNEDEN
Department of Astronomy and McDonald Observatory, University of Texas, Austin TX 78712

ABSTRACT

This paper reviews briefly the history of diffuse interstellar bands (DIBs) and describes the current state-of-art of their spectroscopic investigations. DIB intensity ratios vary greatly from cloud to cloud in the ISM, strongly suggesting that several carriers must be identified to account for the whole DIB spectrum. Some of these unidentified species apparently do not require any protection against the penetration of energetic, far-UV photons. High resolution profiles of DIBs show Doppler splitting, even in some modestly reddened stars, indicating the presence of multiple ISM components. In single cloud (non-Doppler split) DIB cases, the profile shapes do not change with the varying intensity ratios which in turn correlate with the shapes of vacuum UV extinction curves, as well as with absorption strengths of simple interstellar molecules. Thus DIBs can become very powerful diagnostic tools for probing physical conditions inside interstellar clouds.

INTRODUCTION

The discovery by Heger (1922) of quite broad, constant wavelength absorptions centered around 5780 and 5797 Å in the spectra of some early type binary stars opened the era of the investigations of the diffuse interstellar bands, hereafter called DIBs. The interstellar nature of a few DIBs was established by Merrill (1934, 1936) and by Beals & Blanchet (1937). The list of DIBs grew slowly until the publications of York (1971) and Herbig (1975), which boosted the number of certain or very probable DIB members to 39, with a few additional suspects.

To the present day the identification of the carrier(s) of DIBs has eluded spectroscopists. Almost all conceivable forms of matter have been proposed as the possible carriers, ranging from dust grains, to free molecules of very different sizes and structures, to even the hydrogen negative ion. None of the published hypotheses has succeeded in reproducing the whole set of the known diffuse bands or even a substantial subset. The numerous theoretical hypotheses are beyond the scope of this review; here we concentrate on the observational literature.

---

[1]FULBRIGHT FELLOW ON LEAVE FROM NICHOLAS COPERNICUS UNIVERSITY, TORUŃ, POLAND.

Many investigators have catalogued the overall absorption characteristics of DIBs, and in particular the 5780 and 5797[2] bands have been subject to several extensive analyses. A comprehensive survey of the properties of these bands from photographic spectra was conducted by Herbig (1975). In that paper he drew first attention to the possible existence of multiple overlapping features comprising each major band. For example, Herbig found that the prominent DIB at 5780 is inextricably intertwined with a very broad ($\simeq$ 20 Å) but extremely shallow DIB usually called the 5778 band. Herbig drew attention to the possibility of further structure within the general profile of the 5778 band, especially noting a feature at about 5774.0 Å, easily seen in his Figure 3. The DIB features discovered most recently (usually in the near infrared) are typically very weak (Herbig 1988, Herbig & Leka 1991).

Snell & Vanden Bout (1981) first used very high resolution, high signal-to-noise spectroscopy to search for the possible structures of molecular origin inside the broad profile of the 5780 DIB. They demonstrated some differences between profiles seen in spectra of different targets, but no convincing fine structure was observed. Herbig & Soderblom (1982) showed that the extremely narrow 6196 DIB has a complex, Doppler-split profile toward some lines-of-sight. Since that paper it has been clear that the DIBs observed in spectra of heavily reddened, distant stars may be composed of contributions from multiple clouds with different radial velocities.

The advent of high quantum efficiency, low noise solid-state detectors in stellar spectroscopy led to detection of DIBs in stars with small color excesses, where the line-of-sight may intersect only one cloud. This enhanced observational capacity has led to greatly expanded descriptions of the DIB properties during the last decade. In the next 6 sections we summarize the current state-of-the-art in DIB observations and discuss the correlations of DIBs with other interstellar medium features.

FAMILIES OF DIFFUSE BANDS

It is now clear that a single carrier probably cannot account for all DIBs. The publications of Krełowski & Walker (1987) and of Krełowski & Westerlund (1988) revealed the varying strength ratios of the major diffuse bands (5780 and 5797) from object to object. This phenomenon, confirmed in other papers (e.g. Josafatsson & Snow 1987; Herbig 1993), is illustrated in Figure I. The ratio of 5780 to 5797 varies greatly in different lines-of-sight, as does the ratio of 6284 to 6270. This fact suggests that the well-known DIBs are not of the same origin. Moreover, the different ratios of the DIBs are related to the extinction characteristics of interstellar clouds. Krełowski & Westerlund showed that the DIB strength ratios in σ Sco (HD147165) and ζ Oph (HD149757), which resemble closely those of HD2905 and HD24398 shown in Figure I, correlate well with the ultraviolet extinction curve peculiarities (Bless & Savage 1972) in these lines-of-sight. This effect is documented widely by Krełowski et al. (1992).

The division of the various DIBs into "families" (defined as sets of diffuse

---

[2]In this review we will label the major DIB features by their central wavelengths in Å.

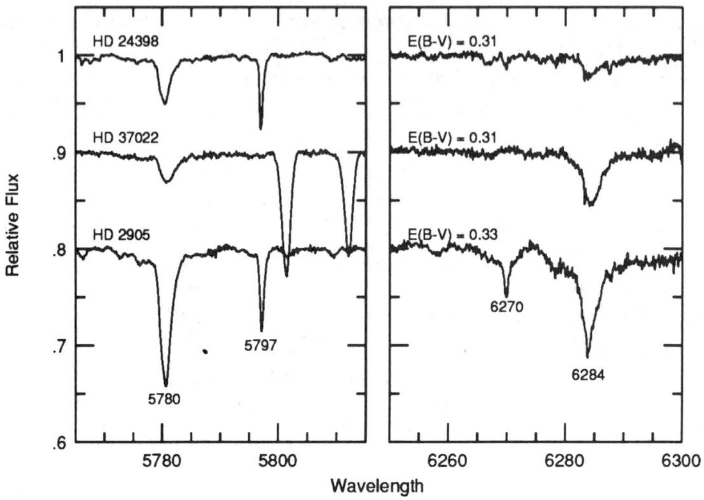

FIGURE I Two pairs of DIBs observed in spectra of equally reddened stars. Note the varying intensity ratios of these bands

features characterized by constant intensity ratios in spectra of every target) was first addressed by Krełowski & Walker (1987) and Josfatsson & Snow (1987). More recent attempts at DIB genealogy include the papers of Krełowski (1989), Porceddu, Benvenuti & Krełowski (1991), Westerlund (1989). The division into families has proven to be an enormeous challenge. The limited amount of high-quality (adequate resolution and S/N) spectra makes the lists of "family members" very unstable. New discoveries usually force observers to conclude that well-known, strong DIBs very rarely belong to the same "family". We illustrate this point with two examples.

As a first case, consider the neighbor DIBs at 5844 and 5850 Å. The 5850 band is a fairly strong and sharp feature attributed by Krełowski & Walker (1987) to the same family of DIBs that contains the strong 5797 band. The 5844 DIB is a broad and shallow one. The recent spectra of Krełowski et al (1993) show several examples in which pairs of stars show the same intensity of 5850 but 5844 is sometimes strong and sometimes not detectable (Figure II). This is but one example; see Herbig (1993) for spectra which include these top stars and several others.

Second, consider the alleged DIB at 6203 Å. Porceddu, Benvenuti & Krełowski (1991) have shown that its structure, often reported as asymmetric, actually consists of two partially overlapping bands. Moreover their intensity ratio varies from object to object. These obviously are separate bands from two distinct families. The moral is clear: much care must be taken with a statistically meaningful sample of spectra to produce correct lists of family members as well as to define all the existing DIB families.

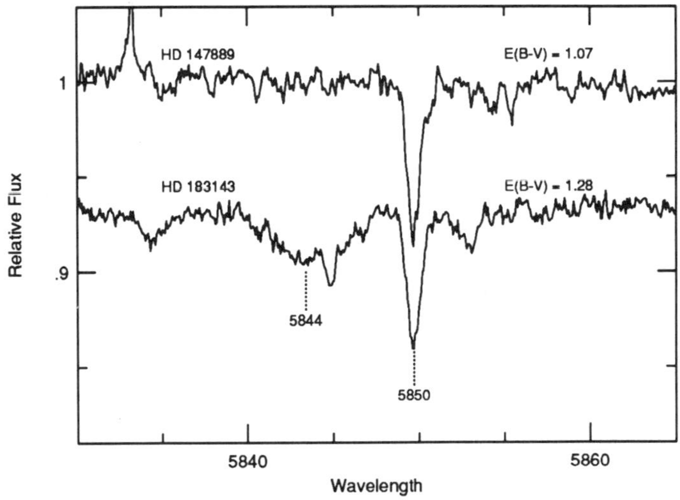

FIGURE II The 5844 and 5850 DIBs in spectra of two heavily reddened stars. Note the extreme weakness of 5844 in HD147889. Data: CFHT high resolution spectra of Krełowski et al., (1993).

DIFFUSE BAND VS. OTHER INTERSTELLAR FEATURES

The spectra of bright, early type, OB stars, usually contain a variety of interstellar features: a) continuous extinction by dust grains, which affects the broad-band colors of stars; b) polarization of starlight, which requires alignment of the grains; c) spectral lines of rarefied atomic gases, especially in far-UV; d) absorption features of simple molecular species such as CH, CH^+, CN,C_2, H_2; e) infrared spectral features (observed usually in opaque clouds), whose relations to other absorptions remain uncertain; and of course e) diffuse interstellar bands (DIBs).

The traditional reddening parameter E_{B-V} is of little help in classifying the properties of diffuse interstellar clouds; the detailed dust extinction properties, in the UV and IR regimes, and the atomic and molecular abundances can vary significantly from cloud to cloud (e.g. Bless & Savage 1972; Cardelli 1988; Joseph et al. 1986, Crawford 1989 and 1992, Cowie et al. 1979, Joseph, Snow and Seab 1989); see also the summary by Snow 1992. Some correlations among these quantities have been suggested. For example, gas phase depletions, implied by neutral atomic abundances, have been linked to the shape of the UV extinction curve (Joseph, Snow, & Seab 1989; Cardelli, Federman, & Smith 1991). A further link between depletions and molecular abundances, probably indicative of an underlying relationship between depletion and UV extinction curve shape, was uncovered by Joseph et al. (1986).

The existence of these tentative relationships suggests that all of the interstellar cloud properties are somehow linked; it is difficult to change one without also affecting the others. The question then turns to a search for the common denominator, the parameter on which the other quantities depend.

More directly, one wants to find those correlations with ISM cloud param-

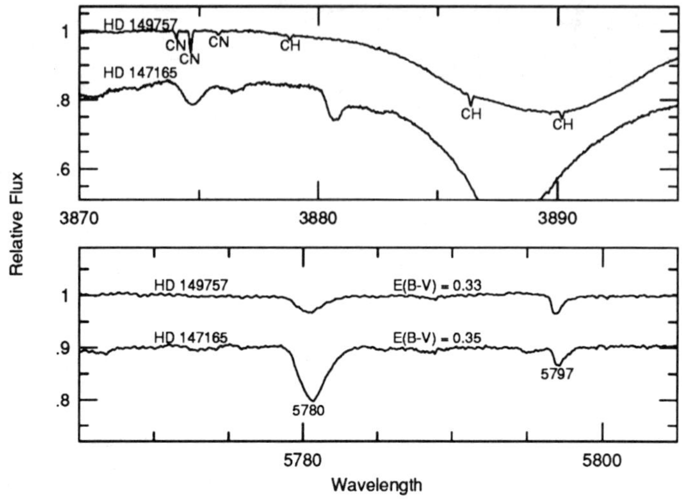

FIGURE III Spectra of two stars that suggest a strong relation between DIB ratio and the strength of molecular features (Snow et al. (1993).

eters that will help identification of the DIB carriers. For example, Krełowski et al. (1992) suggest that the strengths of the CN features around 3875 Å as well as those of the CH between 3878 and 3890 Å may be positively correlated with the 5797/5780 DIB intensity ratio (Figure III). Krełowski et al. (1987) showed that the ratio of famous 4430 DIB absorption to E_{B-V} was found to be inter-related with that of the prominent 5797 and 5780 DIBs, as well as to a parameter characterizing the far-UV rise of the extinction curve. And, as Sneden, Woszczyk & Krełowski (1991) have demonstrated a clear correlation with the 5797/5780 strength ratio with the UV extinction properties, we expect this ratio to be a valuable parameter characterizing interstellar clouds. The ratio of two optical features can be much more easily (and precisely!) determined than the details of the whole extinction curve.

INTRINSIC PROFILES AND DOPPLER SPLITTING

Herbig & Soderblom (1982) demonstrated the existence of Doppler splitting in the profile of the narrowest known DIB – 6196. This has heightened awareness that DIBs often contain components originating in different clouds along any line of sight. In many cases it is not possible to detect the separate components clearly. DIBs are intrinsically so broad that only some modification of profiles can be expected instead of clear splitting (Figure IV).

Such profile modifications were demonstrated by Westerlund & Krełowski (1988a) in the spectra of distant stars of the Carina–Crux complex. This makes the description of "intrinsic" DIB profiles (arising from single cloud with one velocity) quite difficult: all DIB profiles in heavily reddened lines of sight arise from multiple cloud absorptions.

Westerlund & Krełowski (1988b) used high resolution (R $\simeq$ 100,000) spectra of nearby, lightly reddened stars (mostly from the Sco OB2 association) to

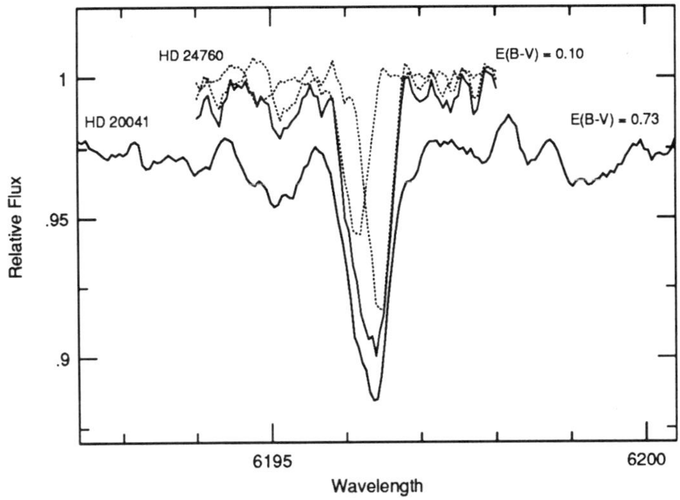

FIGURE IV The addition of two Doppler–shifted 6196 "intrinsic" DIBs (dotted–line spectra) from lightly reddened HD24760 yield a composite profile (upper solid line) that is quite comparable to the complex 6196 DIB in HD20041 (lower solid line).

determine the intrinsic profiles of 5 DIBs. They also showed that single cloud profile shapes of DIBs are the same regardless of their intensity ratios. Thus the varying DIB intensities depend on the abundance variations of the DIBs carriers, not on any modifications of their structure due to physical parameter differences (T, ρ) of their cloud environments. Such (unobserved) profile differences would have been likely to happen if DIBs originate in solid grains or very complicated molecules (see e.g. Shapiro & Holcomb 1986).

DIBs have been observed in quite modestly reddened stars. Many such lines–of–sight have multiple ISM Na D line components in spite of their small E_{B-V} values (*e.g.* Krełowski & Westerlund 1988 – HD102997) and complex DIB profiles can also be seen. The clouds contributing to these DIBs must be of very low opacity, and so their carrier(s) must be resistant against destruction by far-UV photons. Even the 5780 DIB in γAra ($E_{B-V} = 0.05$) has a profile nearly identical to the "intrinsic" one (Westerlund & Krełowski 1989). The recent paper of Herbig (1993) gives other examples of DIBs observed in spectra of lightly reddened stars.

Westerlund & Krełowski (1988a) demonstrated that composite profiles of DIBs observed in distant objects may be reproduced using the "intrinsic" profiles and the Doppler shifts determined from the sodium lines (see example in Figure IV). They suggested, however, that some clouds may not produce any DIB components. This fact was confirmed later by Porceddu, Benvenuti & Krełowski (1992) who discovered quite reddened stars whose spectra entirely lack the strong 5797 DIB (like HD37022 in Figure I).

Intrinsic profiles have not been determined for many known DIBs. Usually this occurs because the extant spectra are of too low S/N or resolution to determine details of the profiles. Even the published central wavelengths are not

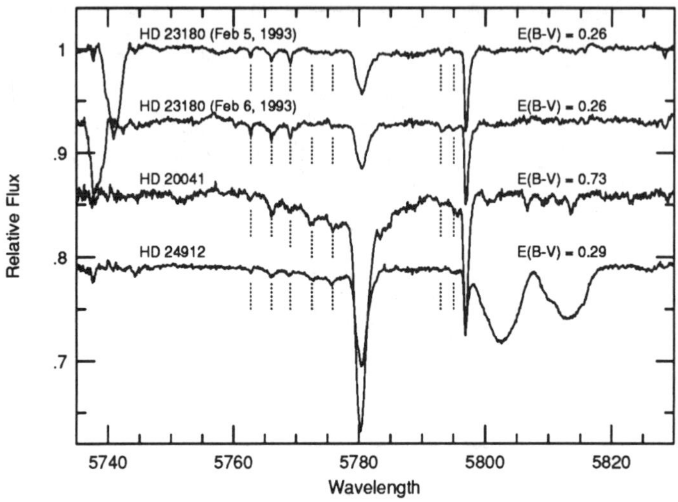

FIGURE V Four spectra of 3 stars showing the pattern of weak features which apparently do not participate in the Doppler motion easily observable in the SiIII 5740 Å line. The weak features are observed in HD24912 (O7.5III), where they are not broadened by stellar rotation, as well as in HD23180 (B1III) and slowly rotating HD20041 (A0I). Their pattern varies, however, from object to object.

always consistent, often due to the asymmetric shapes of profiles. The most striking example is the 5797 DIB. This feature is flat–bottomed and asymmetric and so it is difficult to decide which wavelength is "central". It is always possible that an asymmetric feature is simply a composite of two overlaping ones (e.g. the 6203 and 6205 DIBs (Porceddu et al. 1991).

VERY WEAK INTERSTELLAR FEATURES

Herbig (1975) and later Chlewicki et al. (1987) pioneered the investigation of DIB substructures through their studies of the 5778 band. Herbig tentatively identified a feature at approximately 5774.0 Å. Chlewicki et al. mentioned two absorption peaks at 5772.5 and 5775.9 Å whose strengths varied with those of the 5797 DIB rather than those of 5780 one. The possible existence of other, weak features inside the broad and shallow profile was not excluded. The same paper confirmed also the existence of the shallow 5795 DIB, partially overlapping the strong 5797 feature. Both Herbig and Chlewicki et al. assumed this feature to be broad and thus 5797 to be completely superimposed onto it. Jenniskens & Xavièr (private communication) have obtained spectra of heavily reddened O type stars. Their stars differ in rotational velocities, creating a wide range of FWHM of the stellar CIV features longward of 5800 Å, but the alleged DIB substructures in this region show approximately invariant profiles. This supports attribution of these features to the interstellar medium (see HD24912 in Figure V).

Are the differences exhibited by strong DIB features along different lines of sight repeated in the weaker substructures as suggested by Chlewicki et al.? The answer to this question lies in the detection of the substructures in spectra of bright stars, known to be obscured by one cloud only. To identify the carrier of any of the DIB families one first needs a list of family members. To this end we have gathered new high resolution, high S/N, echelle spectra of several stars with well-known properties of major DIB structures and of ultraviolet absorption characteristics (Krełowski & Sneden, in preparation). The spectra in vicinity of 5780 and 5797 DIBs in some of our program stars are depicted in Figure V. There are several indications that the very weak features marked in this figure are of interstellar origin:

- they are not instrument effects, for they appear in spectra of the same target star observed with several different spectrographs
- they are stationary – their wavelengths do not change as the stellar SiIII 5740 Å feature does in two spectra of the short period binary HD 23180;
- they are clearly seen in several reddened stars strongly differing in their spectral types and luminosity classes;
- they are invariant in shape (e.g. in their FWHM), irrespective of the *vsini* of the star being observed;
- they do not appear in model stellar spectra synthesized for a wide range of temperatures and gravities;
- they are not seen in any of the unreddened stars with spectral types and luminosities similar to the reddened targets.

We conclude that with the resolution (R $\simeq$ 60,000) and S/N values (> 500) achieved in our observations, the weak spectral features (especially if repeated in a couple of objects) are real spectral lines or bands of interstellar origin.

The weak substructure absorptions observed in this spectral region are centered at: 5760.37, 5762.80, 5766.08, 5769.09, 5772.63, 5775.70, 5793.04, 5795.20, 5809.13, 5818.69, and 5828.62 Å, with wavelengths referenced to a rest wavelength scale determined from Na D features seen in another echelle order of the same observations. Our inspection of Herbig's (1975) spectrum of HD 183143 reveals definite features at 5766.1 and 5772.6 Å, and possibly at 5775.7 Å. The latter two features probably make up Herbig's "5774" detection. We also confirm the existence of three DIB substructures tentatively identified by Chlewicki et al. (1987; they listed them at 5766.0, 5772.5, and 5775.9 Å). Many of the features that we have identified coincide with those Jenniskens & Xavièr (private communication); a detailed comparison will be given in a forthcoming paper. A few features are given by Snow et al. (1993), but apparently not at a rest ISM velocity scale. The 5795 feature, mentioned by Herbig (1975) seems to be a composite of two weak DIBs. Thus the broad features upon which the major DIBs are superimposed appear themselves to be complicated structures, composed of many barely detectable absorption features. Moreover, the pattern of detectable substructures seems to change significantly from cloud to cloud characterized by the different 5780/5797 DIB ratios (compare the substructure pattern between HD23180 and HD24912 in Figure V). There is some hope in the near future that the substructures may themselves be assigned definite family memberships.

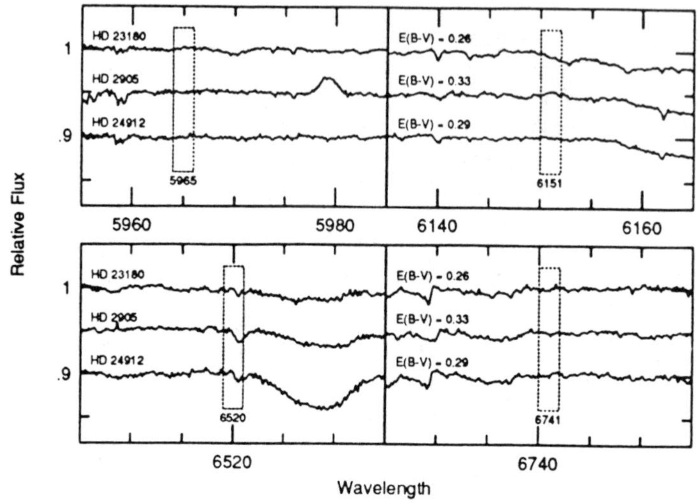

FIGURE VI The results of the search for four naphtalene cation features expected to be strong by Salama & Allamandola (1992). The broken–line boxes cover the predicted wavelengths ± 1 Å. Only the feature centered near 6520.7 Å matches approximately the predicted value

POLYCYCLIC AROMATIC HYDROCARBONS AS DIB CARRIER

Recently, polycyclic aromatic hydrocarbons, hereafter PAHs, have been proposed as a source of the observed DIBs (Léger & d'Hendecourt 1985, Crawford, Tielens & Allamandola 1985). PAHs form a broad class of molecular species that are resistant destruction by far-UV photons. Their presence in certain infrared sources seems to be evident. However, small neutral PAHs absorb only in the ultraviolet and cannot produce the visual wavelength DIBs. However, in the rarefied ISM clouds a large fraction of the PAHs are expected to be ionized and to have absorption features in the visible.

A recent paper of Salama & Allamandola (1992) presents the naphtalene ion $C_{10}H_8^+$ as a possible DIB carrier. The laboratory results predict several spectral features of different strengths. Some well-known DIBs (6376, 5848) seem to match the naphtalene wavelength pattern. These features are, however reported as very weak in the laboratory experiments. Are the strong lab features observable in spectra of reddened stars?

Figure VI shows the results of searching for 4 such features in the spectra of the reddened stars observed with the McDonald Observatory 2.1m telescope and Sandiford echelle spectrograph. We do not see the strongest of the predicted features (6741) in any of our spectra. A feature matching approximately this wavelength was suggested by Herbig (1975). Thus the feature may appear only in some spectra like *e.g.* 5844 (see Figure II) or 5780 and 5797 in HD29647 (Snow & Seab 1991), but it is never very strong.

We detect a weak and narrow interstellar absorption near the predicted 6520 Å strong naphtalene absorption. It closely resembles the weak features observed in the vicinity of 5780 and 5797 DIBs (Figure V). The feature clearly shares the Doppler velocities of the ISM clouds. It is also not observed in any of the unreddened stars. If it is the predicted naphtalene ion absorption, the profiles of other strong lab features (Salama & Allamandola 1992) should resemble it closely.

We do not see any absorption matching exactly the predicted 6151 naphtalene feature. However the well-known, very broad 6177 DIB (described by Herbig 1975) seems also to be populated with many weak features. Two of the possible weak absorptions are centered near 6146 and 6152 Å. Their profiles are uncertain for they are blended with weak, apparently stellar lines. We cannot decide whether any of these features matches the predicted one.

We do not detect any ISM feature in the close vicinity of the last strong naphtalene absorption, 5965. Our survey thus creates doubts that the naphtalene ion is responsible for the creation of any DIBs. The well-known DIBs which coincide with the predicted wavelengths also cannot be considered as support of the naphtalene hypothesis. The 6376 DIB is always observed in close vicinity of its much stronger twin: 6379. The latter does not belong to the naphtalene ion spectrum. The 5850 DIB, close to the predicted weak 5848 feature is, in fact, one of the strongest DIBs. It is much stronger in all astronomical sources than any of the "strong" features we have searched for (Figure VI).

DISCUSSION

The diffuse interstellar bands have remained unidentified for more than a half of century – lack of high S/N, high resolution spectroscopic data has precluded a complete description of all members of numerous "families" of DIBs, including very weak features. The identification of their carriers will depend on having a full set of DIB attributes: central wavelengths, relative intensities, profile details and behavior in different types of dark clouds.

The fact that so many separate features (strong and weak) are located in certain narrow spectral ranges (like around 5780/5797 or 6177) may suggest a physical relationship between the observed bands. Data concerning only strong features are not sufficient for a reliable carrier identification. The pattern of strong and weak absorptions, being possibly of the same origin, may suggest a molecular carrier (if a strong feature is a "head" of some band).

The intrinsic DIB profiles are seemingly insensitive to dust-related changes in the physical conditions of ISM clouds. Thus no modifications of the DIB carrier(s) internal structure is allowed; DIBs simply appear or disappear, suggesting a molecular rather than a grain origin.

The relations between different absorption features originating in the interstellar clouds are never very tight. It is quite difficult to decide what is the cause of the observed scatter. This problem was discussed for DIBs by Herbig (1993). One frustrating observational difficulty is that equivalent width measurements of DIBs by different authors still differ by larger factors than would seem justified by the S/N and resolutions of the spectroscopic data. Observers must exercise greater care in DIB measurement techniques; clear understanding of assumptions when measuring overlapping DIBs is essential.

The correlations between interstellar absorption features are usually quite tight when our sightlines probe the general interstellar medium; the strongest discrepancies are observed in cases of opaque individual clouds. We emphasize the necessity of observing lines–of–sight obscured by only one cloud wherever possible. Intensities of molecular features as well as of diffuse bands do not obey the general relations toward every target and so average relations cannot be blindly applied to an individual case. They also cannot be extrapolated – the behavior of the spectral features may change outside of the observed range(s) of their intensities.

The spectra shown in Figure VI do not vindicate the naphtalene ion as the carrier of any set of DIBs. This does not mean, however, that the PAH hypothesis is totally incorrect – naphtalene is only one of many PAHs! Determination of the absorption spectra of other PAHs and comparison with DIB features is well–advised. We also await theoretical modelling which may provide us with expectations what physical conditions are to be expected inside dark interstellar clouds and what species can form and survive.

The first DIB spectra of J.K. were obtained by Dr. Bengt Westerlund at ESO. That collaboration provided the impetus for his work in the last decade in this field, and is gratefully acknowledged. This project has been supported by the Fulbright Foundation grant No. 16288 to J.K., and by National Science Foundation grant AST91-15026 and Texas Advanced Research Program grant 003658-585 to C.S.

REFERENCES

Beals, C. S. and Blanchet, G. H. 1937, PASP, 49, 224.

Bless, R. C. and Savage, B. D. 1972, ApJ, 171, 293.

Cardelli, J. A. 1988, ApJ, 335, 177.

Cardelli, J. A., Federman, S. R., and Smith, V. V. 1991, ApJL, 381, L17.

Chlewicki, G., de Groot, M. S., van der Zwet, G. P., Greenberg, J. M., Alvarez, P. P., and Mampaso, A. 1987, A&A, 173, 131.

Cowie, L. L., Songaila, A., and York, D. G. 1979, ApJ, 230, 469.

Crawford, I. A. 1989, MNRAS, 241, 575.

Crawford, I. A. 1992, MNRAS, 254, 264.

Crawford, M. K., Tielens, A. G. G. M., and Allamandola, L. J. 1985, ApJL, 293, L45.

Heger, M. L. 1922, Lick Obs. Bull., 10, 146.

Herbig, G. H. 1975, ApJ, 196, 129.

Herbig, G. H. 1988, ApJ, 331, 999.

Herbig, G. H. 1993, ApJ, in press.

Herbig, G. H. and Leka, K. D. 1991, ApJ, 382, 193.

Herbig, G. H. and Soderblom, D. R. 1982, ApJ, 252, 610.

Josafatsson, K. and Snow, T. P. 1987, ApJ, 319, 436.

Joseph, C. L., Snow, T. P., and Seab, C. G. 1989, ApJ, 340, 314.

Joseph, C. L., Snow, T. P., Seab, C. G., and Crutcher, R. M. 1986, ApJ, 309, 771.

Krełowski, J. 1989, in: *IAU Symp. 135, Interstellar Dust*, eds Allamandola, L. and Tielens, A. (Dordrecht: Kluwer Academic) p. 67.

Krełowski, J., Snow, T. P., Papaj, J., Seab, C. G., and Wszołek, B. 1993, PASP, in press.

Krełowski, J., Snow, T. P., Seab, C. G., and Papaj, J. 1992, MNRAS, 258, 693.

Krełowski, J. and Walker, G. A. H. 1987, ApJ, 312, 860.

Krełowski, J., Walker, G. A. H., Grieve, G. R., and Hill, G. M. 1987, ApJ, 316, 449.

Krełowski, J. and Westerlund, B. 1988, A&A, 190, 339.

Léger, A. and d'Hendecourt, L. B. 1985, A&A, 146, 81.

Merrill, P. W. 1934, PASP, 46, 206.

Merrill, P. W. 1936, PASP, 48, 179.

Porceddu, I., Benvenuti, P., and Krełowski, J. 1991, A&A, 248, 188.

Porceddu, I., Benvenuti, P., and Krełowski, J. 1992, A&A, 260, 391.

Salama, F. and Allamandola, L. J. 1992, ApJ, 395, 301.

Shapiro, P. R. and Holcomb, K. A. 1986, ApJ, 305, 433.

Sneden, C., Woszczyk, A., and Krełowski, J. 1991, PASP, 103, 1005.

Snell, R. L. and Vanden Bout, P. A. 1981, ApJ, 244, 844.

Snow, T. P. 1992, Australian J. Phys., 45, 543.

Snow, T. P., Krełowski, J., Seab, C. G., Wszołek, B., and Papaj, J. 1993, PASP, in press.

Snow, T. P. and Seab, C. G. 1991, ApJ, 382, 189.

Westerlund, B. 1989, Rev. Mexic. Astron. Astrof., 19, 39.

Westerlund, B. and Krełowski, J. 1988a, A&A, 189, 221.

Westerlund, B. and Krełowski, J. 1988b, A&A, 203, 134.

Westerlund, B. and Krełowski, J. 1989, A&A, 218, 216.

York, D. G. 1971, ApJ, 166, 65.

The First Symposium on the Infrared Cirrus and Diffuse Interstellar Clouds
ASP Conference Series, Vol. 58, 1994
Roc M. Cutri and William B. Latter (eds.)

VARIABILITY OF INTERSTELLAR EXTINCTION AND ITS RELATIONSHIP TO ENVIRONMENT

JASON A. CARDELLI
Department of Astronomy, University of Wisconsin, Madison, WI 53706

ABSTRACT This paper reviews the current status of interstellar dust extinction studies. These studies have opened up new opportunities for a better understanding of the nature of interstellar grains, gas-grain interactions, and the role that dust plays in influencing the structure and evolution of interstellar clouds.

INTRODUCTION

Interstellar dust plays a direct and vital role in the chemical evolution of interstellar clouds. For example, the absorption and scattering of starlight by dust determines the radiation field within interstellar clouds and thereby influences the atomic and molecular processes that occur there. In turn, the properties of the gas-phase strongly influence the dust through grain surface chemistry. This includes processes associated with the depletion of atomic species as well as the formation of simple molecules like H_2, a precursor to the formation of other molecular species (e.g. CH, CN, etc.). Thus, detailed study of the extinction law, or more importantly the *variations of extinction with wavelength*, can provide vital clues as to the nature of this gas-grain interaction, particularly when associated with specific tracers of environmental conditions.

Although the subject of this conference centers on diffuse clouds, the characteristics of dust in such environments are strongly influenced by the effects of processing in other environments through which the gas and dust are cycled as shown below. What is interesting about diffuse clouds is that they seem to represent an 'environmental way-station' into which material is fed from all other

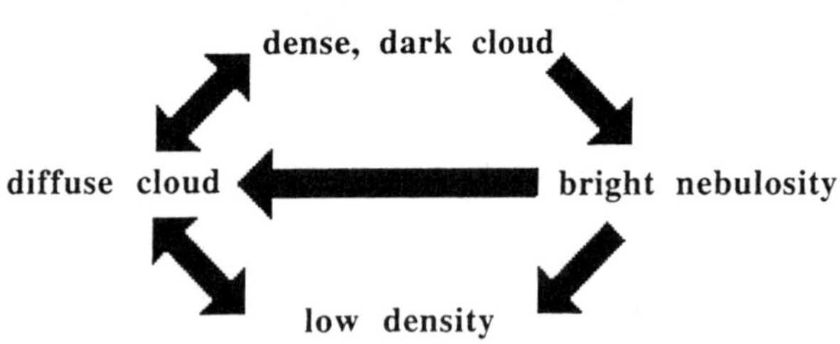

other environments. Consequently, understanding the nature of dust and its relationship to and impact on the gas-phase requires comprehensive studies involving a broad range of environmental characteristics.

This paper reviews the current status of extinction studies and the impact they have had on our understanding of the nature of gas-grain interactions and the role that dust plays in influencing the structure and evolution of interstellar clouds. An important aspect of many of these studies has been that they have attempted to merge the results from both multi-wavelength (IR, optical, UV) and interdisciplinary (IR emission, atomic and molecular spectroscopy, molecular chemistry, dust theory) analyses. This approach is important in that it establishes 'boundary conditions' which characterize the limits of dust processing. In turn, use of these boundary conditions to constrain theoretical dust studies offers a unique opportunity to explore the detailed nature of interstellar dust and the various physical processes associated with gas-grain chemistry.

BASIC DEFINITIONS

As employed here, interstellar extinction is strictly defined as the result of the absorption and single scattering of photons, emitted from some background source, off of dust grains. [In this regard, we do not include analyses involving either reflection nebulae or circumstellar envelopes. However, such studies do provide important additional constraints on the nature of dust grains.] Absolute extinction, measured in magnitudes, follows from the equation of radiative transfer and is defined as $A(\lambda) = 2.5\log[I(\lambda)_0/I(\lambda)] = 1.086\ \tau(\lambda)_{ext}$, where $I(\lambda)$ and $I(\lambda)_0$ are the observed and intrinsic (unreddened) source intensity, respectively, and $\tau(\lambda)_{ext}$ is the extinction optical depth $[\tau(\lambda)_{ext} = \tau(\lambda)_{abs} + \tau(\lambda)_{sca}]$.

Because it is directly related to the grain abundance, size distribution, and optical properties, we are most interested in exploring variations of $A(\lambda)$. Unfortunately, $A(\lambda)$ cannot generally be derived directly. Instead, measured extinction is typically expressed as the difference between the extinction at a pair

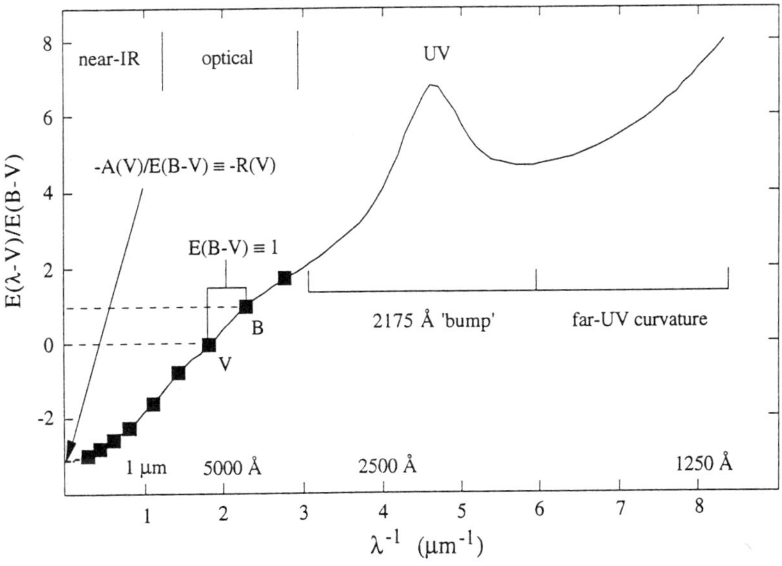

FIGURE 1 Typical extinction curve plotted as $E(\lambda\text{-}V)/E(B\text{-}V)$ vs. $1/\lambda$ (μm^{-1})

of wavelengths, or color excess, where $A(\lambda)$ - $A(\lambda_{std})$ = $E(\lambda\text{-}\lambda_{std})$. Traditionally, $\lambda_{std} = \lambda_V \approx 5500$ Å, the central wavelength of the Johnson V filter. In order to compare measured results for different sight lines, E(λ-V) is typically normalized by E(B-V), where $\lambda_B \approx 4400$ Å, the central wavelength of the Johnson B filter. In addition, E(λ-V)/E(B-V) is typically expressed as a function of wavenumber, $x \equiv 1/\lambda$, in units of μm^{-1}. Finally, A(λ) can be obtained from E(λ-V)/E(B-V) via R(V) = A(V)/E(B-V), the ratio of total-to-selective extinction [i.e. since $A(\lambda) \rightarrow 0$ as $x \rightarrow 0$, E(λ-V)/E(B-V) → -A(V)/E(B-V) = -R(V)].

A representative example of an extinction curve in the form E(λ-V)/E(B-V) is shown in Figure 1. This curve, which covers the wavelength range 3.4 μm $\leq \lambda \leq$ 0.12 μm (1200 Å), is essentially the same as the one presented by Savage & Mathis (1979, hereafter SM79; near-IR, optical, and UV) and Seaton (1979, hereafter S79; UV). The filled symbols in the optical/near-IR portion correspond to extinction derived from broad-band photometric data (e.g. Johnson U, B, V, R, I, J, H, K, and L). In this example, the curve corresponds to R(V) = 3.1.

The most prominent features in the extinction curve occur in the UV and include the feature centered at about 2175 Å and the steep rise toward short wavelengths. The general consensus is that these features arise from a small grain population with the 2175 Å bump being an absorption feature produced by some form of graphitic carbon (Draine & Lee 1984, Draine & Malhotra 1993, Mathis 1993, and references therein) and/or possibly polycyclic aromatic hydrocarbons (PAH's; Puget & Léger 1989, Joblin, Léger, & Martin 1992).

ANALYSIS OF EXTINCTION CURVE VARIABILITY

Limits of Observed Variability

Since the late 1970's, the extinction curve in Figure 1 (i.e. SM79; S79) has been adopted as representative of the typical or 'average' extinction in the Galaxy. As such, it has been extensively used in a variety of applications including chemical modeling of interstellar clouds, analyses of interstellar dust, and the dereddening of spectra. However, observed extinction actually shows a wide range of variation at all wavelengths, especially in the UV (Massa, Savage, & Fitzpatrick 1983; Savage et al. 1985). In the optical, this variability is seen as a change in slope of the curve on either side of the B and V normalization wavelengths and more notably as a change in R(V). Examples of the range of variability in UV extinction are shown in Figure 2. Also plotted is the 'average' curve from Figure 1. These data were taken from the sample of Fitzpatrick & Massa (1986, hereafter FM; 1988; 1990) which includes sight lines sampling a wide range of environments (i.e.diffuse clouds, star forming regions, and dense clouds). Also included are extinction curves arising from dense, dark cloud environments toward HD 29647 in Taurus, HD 62542 in the Gum Nebula region (Cardelli & Savage 1988), and VCT 10 in CrA (Cardelli & Clayton 1991).

As seen in the figure, the deviations of individual extinction curves from the traditional 'average' curve can be quite dramatic and have historically been referred to as being peculiar. However, does the 'average' curve represent the normal or expected state of interstellar dust? Are the extinction curves in Figure 2 truly peculiar in that they represent *unique and independent* grain characteristics?

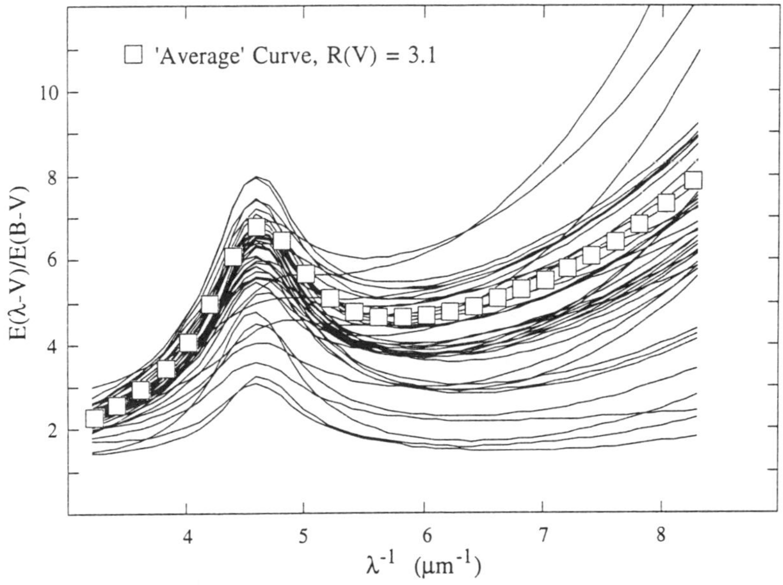

FIGURE 2 Range of extinction variability observed in the ultraviolet. Open symbols correspond to the curve of Seaton (1979) and Savage & Mathis (1979)

UV Extinction Curve Components: ***Universality***

A major step in the understanding of extinction curve variability has come through a technique which utilizes a decomposition of the observed curve into selected mathematical functions. This technique, first explored by Savage (1975) and later verified by S79 and Massa & Fitzpatrick (1986), involves fitting the observed extinction curve with a combination of a linear (in $x \equiv \lambda^{-1}$ in units of μm^{-1}) background and a bump profile such as a Lorentzian function. FM expanded on this earlier work by adopting a "Drude" function for the bump, which can be understood as an absorption cross section under the assumption of the Drude theory of metals (Bohren & Huffman 1983), and a cubic polynomial far-UV curvature function. Although this decomposition may be more mathematical than physical (the possible exception being the bump profile), it does provide an opportunity to quantitatively explore specific aspects of the extinction curve between different sight lines. A representation of this decomposition of the UV extinction curve is shown in Figure 3.

Massa & Fitzpatrick (1986) and FM applied this fitting technique to numerous extinction curves, including a fair number exhibiting a large range of deviations from the 'average' curve (see Figure 1). In every case, they found that all these curves were *equally well fitted with the* ***same*** *set of functions*. This technique has also been shown to apply equally well to very extreme cases where the extinction curve appears rather flat and featureless in the region of the bump and has a very steep far-UV rise (Fitzpatrick 1986; Cardelli & Savage 1988). The ability to fit a diverse range of observed extinction curves with the same functions underscores the ubiquitous nature of interstellar grain components.

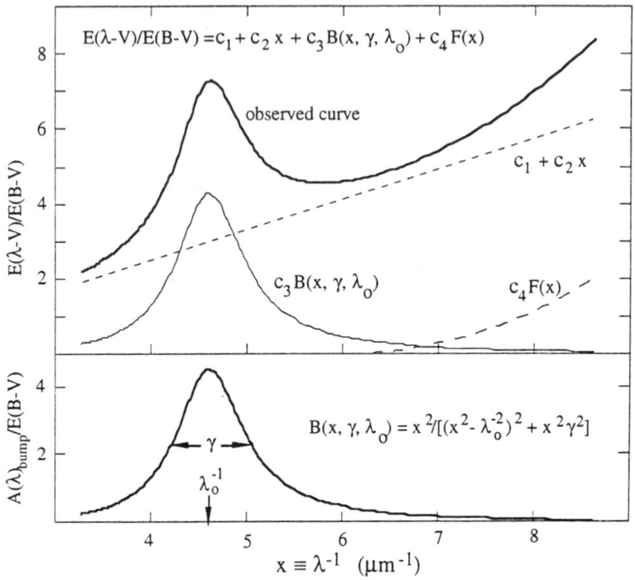

FIGURE 3 Extinction curve 'decomposition' functions developed by Fitzpatrick & Massa (1986, 1988, 1990)

The R(V)-Dependent Extinction Law

Does the universality of extinction curve components observed in a wide range of environments suggest that the basic characteristics of the different UV grain components are more or less the same every where? Are the observed differences related in some simple way and are they the result of some identifiable environmental process? What is the relationship between extinction deviations seen in the UV and those observed in the optical and near-IR?

Adding the near-IR and optical portions to the UV curves of FM, Cardelli, Clayton, & Mathis (1988; 1989, hereafter CCM) explored the deviations among these curves. Specifically, CCM compared extinction, expressed in the form A(λ)/A(V), to $R(V)^{-1}$ and found that a strong linear correlation exists at all wavelengths from the near-IR (3.4 μm) down to 0.1 μm (1000 Å). Examples of their findings are shown in Figure 4a for selected wavelengths. From a linear fit to this data, CCM found that the mean extinction at all wavelengths is well represented by $\langle A(\lambda)/A(V)\rangle = a(x) + b(x)/R(V)$. An example of the application of this R(V)-dependent extinction law is shown in Figure 4b for three stars representing a wide range of R(V) values. The wavelength dependent UV standard deviations of the CCM sample are shown in the inset and represent the relative accuracy to which this parametrization can be applied for any R(V) value.

The results of CCM show that the extinction curves shown in Figure 1 are in fact not peculiar, but rather represent a continuum of shapes related through variations in R(V). Specifically, curves like those seen in Ophiuchus [R(V) ≈ 4] and Orion [R(V) ≈ 5] *are not peculiar, but rather represent normal extinction for their respective R(V) values.* In addition, the 'average' curve of S79 and SM79 is

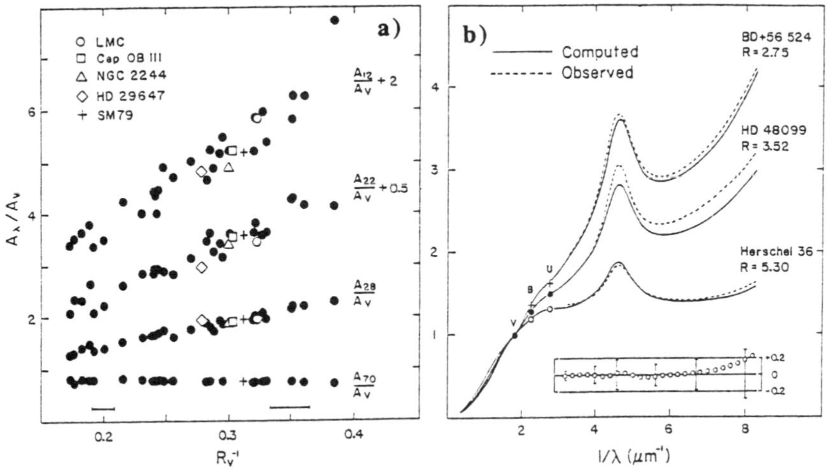

FIGURE 4 **a**) A(λ)/A(V) vs. R(V)$^{-1}$ for selected wavelengths; **b**) observed vs. computed curves from fit to data in **a**); from Cardelli, Clayton, & Mathis (1989)

neither a global average nor a 'normal' curve, but simply the expected shape for R(V) = 3.1 **only** (the open symbols in the inset in Figure 4b show the difference between the CCM curve for R(V) = 3.1 and this 'average' curve. The deviations at the shortest wavelengths are a result of the way the 'average' curve was constructed: see CCM). Use of the R(V)-dependent extinction law for the purpose of dereddening and chemical modeling of interstellar clouds is preferred over the 'average' curve, especially if R(V) is different from 3.1. This is particularly true in the case of chemical modeling of molecular clouds when photodestruction dominates and where R(V) can often differ from 3.1 (Cardelli 1988; Cardelli & Wallerstein 1989; Cardelli et al. 1990; Mathis 1990; Jenniskens et al. 1992).

The existence of the R(V)-dependence suggests that the processing of dust between different environments is systematic and rather efficient. For example, sight lines with similar R(V) values may have different processing histories and yet they all generally conform to the same extinction characteristics. CCM argued that the existence of R(V)-dependent extinction variations is the result systematic changes in the grain size distribution through grain-grain interactions (coagulation), with larger R(V) values corresponding to larger mean grain sizes.

EXTINCTION VARIABILITY AND ENVIRONMENT

General Deviations from the R(V)-Dependence

The R(V)-dependence extinction law provides a basis for understanding the general nature of extinction curve variability. In the near-IR/optical, the CCM extinction law is universal in that we have found no example (in over 100 sight lines) where the observed extinction exhibits deviations that larger than the uncertainties in the measurements. However, in the UV, the standard deviations shown in Figure 4b are larger than the measurement errors and there are also examples of extinction curves (e.g. HD 62542; Cardelli & Savage 1988) that show large deviations from the R(V)-dependence, particularly at the shorter UV wavelengths. Thus, while the R(V)-dependence may reflect the primary source of grain modification (size changes/coagulation), there are clearly secondary processes at work. Consequently, *deviations from the R(V)-dependence* offer the best opportunity to explore the nature of interstellar grains and their processing, *especially when linked to specific environmental characteristics.*

Cardelli & Clayton (1991; hereafter CC) explored deviations from the R(V)-dependence by considering extinction arising from two extreme environmental groups, namely sight lines associated with very bright nebulosity and dense, dark clouds. Relative to the extinction 'expected' for their R(V) values, CC found that the dark cloud sight lines exhibit unique and systematic deviations. An example is shown in Figure 5 where we plot the difference between the observed and expected extinction at two wavelengths, λ = 2200 Å and 1200 Å (a more generalized analysis of the wavelength dependence of these deviations can be found in Mathis & Cardelli 1992). The 'error bars' represent the CCM standard deviations at these two wavelengths.

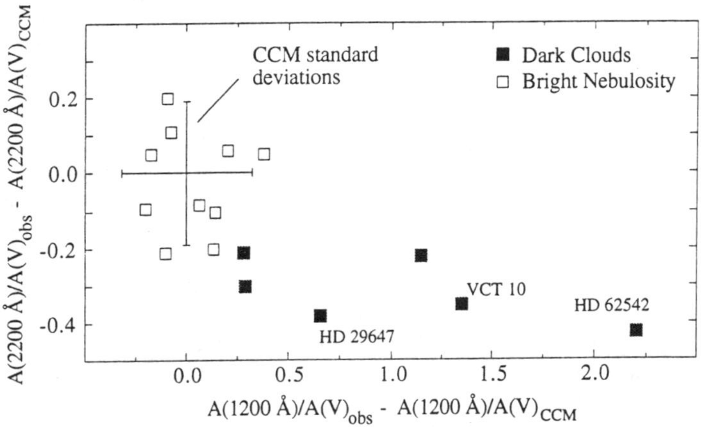

FIGURE 5 Deviations from the R(V)-dependent extinction law for sight lines sampling dark clouds and regions of bright nebulosity

CC concluded that the *deviations* from the R(V)-dependence for dark clouds are a result of *secondary grain processing* (i.e. in addition to the R(V)-dependent processing) via coatings and that the absence of such deviations outside of these environments is the result of their removal or modification by UV radiation. This conclusion has been quantified by Boulanger et al. (1993) who found that large deviations are present in sight lines with low mid-IR emission (12 and 25 μm) and that these deviations dramatically decrease as the mid-IR emission increases.

The 2175 Å Bump

The UV bump is particularly useful to study because one can characterize it by its measured strength, width, and central position, parameters that are useful for describing intrinsic variations in absorption profiles (see Figure 3). CCM found that the strength (or area) of the fitted bump normalized to A(V) shows a modest correlation with $R(V)^{-1}$ (i.e. $A_{bump}/A(V)$ decreases as R(V) increases). However, they found no relationship between $R(V)^{-1}$ and γ or x_o indicating that variations intrinsic to the bump carrier occur independently of the processes which give rise to the R(V)-dependence. Furthermore, in dense *and* dark environments, Cardelli & Savage (1988), CC, and Boulanger et al. (1993) have found examples of extremely broad, weak bumps which also exhibit shifts in the bump central position to wavelengths as short as 2110 Å (HD 62542).

The relationship between bump width (γ in μm^{-1}) and central position ($x_o = 1/\lambda$ in μm^{-1}) for various environmental groups is shown in Figure 6. Like the deviations from the R(V)-dependence, dark clouds and regions of bright nebulosity exhibit extreme behavior, particularly in bump width, which again suggests processing by UV radiation. The sensitivity of γ to UV radiation intensity, I_{uv}, is supported by the dark cloud/nebulosity group which tends to exhibit broad bumps although not as extreme as the dark cloud group. It seems that even modest increases of I_{uv} above the diffuse radiation field are sufficient to *reduce* or perhaps *inhibit* the formation of larger values of γ. However, since these dark cloud groups also tend toward larger values of spatial density, n, relative to the more diffuse cloud groups, differences in γ must also be contributed to by variations in density. The simplest notion is that regions of dense gas create broad bumps through the mantling of coatings of some sort (Mathis 1993) while regions with large values of I_{uv} tend to reduce such coatings. Specific values of γ are therefore the result of a balance between the effects of density and radiation which can be represented through variations in the parameter n/I_{uv}. The interplay between density and radiation is seen in the comparison between the bright nebulosity and low density (Cardelli, Sembach, & Mathis 1992; Cardelli & Sembach 1993) groups whose bump characteristics are similar even though their environments are radically different. We conclude that these bump similarities are the result of similar and relatively small values of n/I_{uv}. The low density group is typically characterized by low density gas and the diffuse I_{uv} while the bright

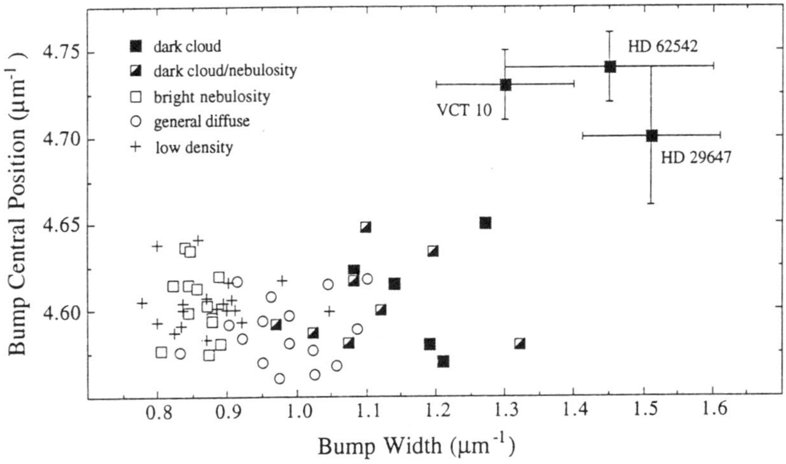

FIGURE 6 Bump central position ($x_o = 1/\lambda$ in μm^{-1}) vs. bump width (γ in μm^{-1}) for a sample of different environmental groups (from Cardelli 1993)

nebulosity group sight lines tend toward relatively large densities and *very* large I_{uv}. Thus, bumps in bright nebulosity sight lines are narrow because I_{uv} is very large ('destructive' processes dominate) while bumps in low density sight lines are narrow because n is relatively small ('growth' processes are suppressed).

Relationship to Gas-Phase Chemistry

Figures 5 and 6 provide important clues to the underlying nature grain processing by suggesting that variations in n/I_{uv} play a direct role in determining specific extinction characteristics. However, while this may suggest in part *why* these variations occur, it does little to tell us *how* or *where* such variations occur. It is therefore useful to explore this variability with respect to specific gas-phase characteristics which can serve as direct probes of n and I_{uv}.

Figure 7a (Cardelli 1993) shows a plot of γ vs. N(CH)/A(I), the column density of the CH molecule normalized to the absolute extinction at the I passband, $\lambda_I \approx 9000$ Å. A(I) is used instead of $N(H_{total}) = 2N(H_2) + N(H\ I)$ because $N(H_{total})$ data are not available for most of the sight lines. However, under the assumption of a constant dust-to-gas ratio, $A(I)/N(H_{total})$, this plot is analogous to the gas-phase abundance of CH, or $N(CH)/N(H_{total})$. In addition, A(I) was used rather than A(V) because A(I) is beyond the influence of R(V)-dependent extinction variations (CCM). Figure 7b (Cardelli 1993) shows a plot of γ vs. $f(H_2) \equiv 2N(H_2)/N(H_{total})$ for the same sample of sight lines. The points with error bars represent the only data for which both $N(H_2)$ and N(H I) were available. For the open symbols, $f(H_2)$ was 'calibrated' from the correlation ($r^2 \approx 0.85$) between N(CH)/A(I) and $f(H_2)$ for an independent sample of stars. This correlation arises from the fact that CH formation is initiated by the reaction $C^+ + H_2$ and that N(CH) and $N(H_2)$ are known to be highly correlated.

The results shown in Figure 7a yield the first tangible evidence that specific grain modification processes, as measured by extinction variations, are directly associated with specific *in situ* gas-phase processes. Since the gas-phase chemistry of species such as CH are a function of variations of n/I_{uv}, at least in cases where

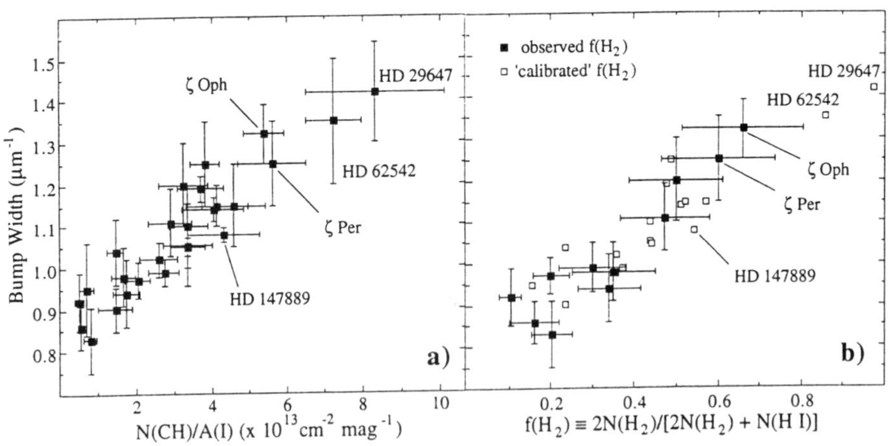

FIGURE 7 Bump width, γ, vs. **a**) N(CH)/A(I) and **b**) $f(H_2) \equiv 2N(H_2)/N(H_{total})$

photochemical processes dominate, it seems plausible to describe the processes responsible for the observed grain variations in similar terms as postulated earlier. The correlation between γ and $f(H_2)$ is particularly interesting since H_2 is postulated to form on grain surfaces.

These results argue that bump width variations must be the result of coatings (possibly hydrogen or hydrogen bearing material - PAHs?), *the processing of which is tied directly to the same physical conditions specifically controlling the chemistry of CH.* This identification of the primary site of bump grain modification is supported by a plot of γ vs. N(CN)/A(I), which shows a non-linear dependence with larger scatter (Cardelli 1993). Since CN is produced only after substantial amounts of the precursor molecules (CH, NH, C_2) are available, CN absorption samples much more restricted (denser) portions of the cloud. If bump width variations were restricted to this denser gas, we would have expected γ to correlated more with N(CN)/A(I) than either N(CH)/A(I) or $f(H_2)$.

REFERENCES

Bohren, C. F., & Huffman, D. R. 1983, *Absorption and Scattering of Light by Small Particles*, (Wiley and Sons: New York)
Boulanger, F., Prévot, M. L., & Gry, C. 1993, *AA*, in press
Cardelli, J. A. 1988, *ApJ*, **335**, 177
Cardelli, J. A. 1993, submitted to *ApJ*
Cardelli, J. A., & Clayton, G. C. 1991, *AJ*, **101**, 1021 (CC)
Cardelli, J. A., Clayton, G. C., & Mathis, J. S. 1988, *ApJ*, **329**, L33
Cardelli, J. A., Clayton, G. C., & Mathis, J. S. 1989, *ApJ*, **345**, 245 (CCM)
Cardelli, J. A., & Savage, B. D. 1988, *ApJ*, **325**, 864
Cardelli, J. A., & Sembach, K. R. 1993, in preparation
Cardelli, J. A., Sembach, K. R., & Mathis, J. S. 1992, *AJ*, **104**, 1916
Cardelli, J. A., Suntzeff, N. B., Edgar, R. J., & Savage, B. D. 1990, *ApJ*, **362**, 551
Cardelli, J. A., & Wallerstein, G. 1989, *AJ*, **97**, 1099
Draine, B. T. & Lee, H. M. 1984, *ApJ*, **285**, 89
Draine, B. T. & Malhotra, S. 1993, *ApJ*, in press
Fitzpatrick, E. L. 1986, *AJ*, **92**, 1068
Fitzpatrick, E. L., & Massa, D. 1986, *ApJ*, **307**, 286 (FM)
Fitzpatrick, E. L., & Massa, D. 1988, *ApJ*, **328**, 734
Fitzpatrick, E. L., & Massa, D. 1990, *ApJS*, **72**, 163
Jenniskens, P., Ehrenfreund, P., & Désert, F. X., 1992, *AA*, **265**, L1
Joblin, C., Léger, A., & Martin, P. 1992, *ApJ*, **393**, L79
Massa, D., & Fitzpatrick, E. L. 1986, *ApJS*, **60**, 305
Massa, D., Savage, B. D., & Fitzpatrick, E. L. 1983, *ApJ*, **266**, 662
Mathis, J. S. 1990, *ARAA*, **28**, 37
Mathis, J. S. 1993, *ApJ*, in press
Mathis, J. S., & Cardelli, J. A. 1992, *ApJ*, **398**, 610
Puget, J.-L., & Léger, A. 1989, *ARAA*, **27**, 161
Savage, B. D. 1975, *ApJ*, **199**, 92
Savage, B. D., & Mathis, J. S. 1979, *ARAA*, **17**, 73 (SM79)
Savage, B. D., Massa, D., Meade, M., & Wesselius, P. R. 1985, *ApJS*, **59**, 397
Seaton, M. J. 1979, *MNRAS*, **187**, 73p (S79)

The First Symposium on the Infrared Cirrus and Diffuse Interstellar Clouds
ASP Conference Series, Vol. 58, 1994
Roc M. Cutri and William B. Latter (eds.)

OPTICAL STUDIES OF HIGH-LATITUDE DUST

P. GUHATHAKURTA[1,2]
Department of Astrophysical Sciences, Princeton University, Princeton, NJ 08544-1001

R. M. CUTRI[1]
Steward Observatory, University of Arizona, Tucson, AZ 85721

ABSTRACT We present a review of an extensive set of optical imaging data, obtained using CCD images and digitized photographic plates, of high-latitude interstellar dust clouds. A variety of dust clouds, ranging from very thin cirrus to dense nebulae, have been studied. Quantitative comparisons are made between the surface brightness distributions of clouds at far infrared (*IRAS*: 12, 25, 60, and 100 μm) and optical (*BVRI*) wavelengths. For most clouds, there is good agreement between the overall morphology seen at different wavelengths. The *IRAS* colors show evidence for optically-thick cloud cores in a few cases. The typical ratio of blue to 100 μm brightness for thin cirrus is lower than would be expected for isotropically-scattering interstellar dust grains illuminated by ambient starlight, but is consistent with the grains having a moderately forward-scattering phase function. The optical $B - R$ and $R - I$ colors of the cirrus are too red to be explained in terms of scattering of starlight; a large fraction of the R and I radiation probably comes from luminescence in very small hydrogenated carbon grains (as is the case for bright reflection nebulae). Atomic hydrogen and molecular gas in the form of the $^{12}CO(1 \rightarrow 0)$ line is detected in even the most tenuous of cirrus clouds. Despite the general morphological similarities, there are striking differences in the spatial distributions of the various components on arcminute scales or smaller, implying variations in the grain properties (*e.g.* size distribution, gas-to-dust ratio, chemical composition) within each cloud. The regions of most extreme optical–far infrared color are occasional patches of optical emission that have no *IRAS* counterpart, and these may contain unusually cold dust grains. The optical reradiation from high-latitude cirrus exhibits a great deal of low-level structure on scales down to a few arcseconds and this is a potential hindrance for future deep imaging studies of extragalactic objects.

---

[1]Visiting Astronomers at Kitt Peak National Observatory and Cerro Tololo Inter-American Observatory, National Optical Astronomy Observatories, operated by Association of Universities for Research in Astronomy, Inc., under cooperative agreement with the National Science Foundation.

[2]Hubble Fellow.

INTRODUCTION

The optical brightness of the night sky is the sum of several diverse components, including atmospheric airglow, scattering of sunlight by dust in the Solar System (zodiacal light), reradiation of starlight by interstellar dust (smoothly distributed along the Galactic disk and contained in discrete clouds), and distant galaxies. It is difficult, if not impossible, to disentangle the contributions of these sources with the help of absolute measurements. One must instead rely on differential optical brightness measurements between various sky locations, thereby making use of the specific spatial (angular) dependence of each component. The two most common sources of (time-independent) structure on scales of a few arcminutes or less are: (1) clustering of distant galaxies (Schectman 1973, 1974; Tyson 1988; Efstathiou et al. 1991), and (2) reprocessing of Galactic starlight by filamentary dust clouds ("cirrus").

Optical nebulosity from diffuse interstellar dust at high latitudes was first noticed almost forty years ago (de Vaucouleurs 1955, 1960). Several high-contrast photographic plates dating back to the early 1970s show evidence for faint filamentary nebulosity, but the significance of this was not fully explored until the study by Sandage (1976) who showed that scattering of the general interstellar radiation field (ISRF) by dust grains could produce such nebulosity. The *Infrared Astronomical Satellite* (*IRAS*, Neugebauer et al. 1984) produced spectacular confirmation of this: the intricate optical filaments seen in the direction of M81/M82 on Sandage's photographic plate coincide with bright strands of 100 μm thermal emission from dust (Low, private communication). Early *IRAS* maps in the 60 and 100 μm bands clearly demonstrated that cold dust in the form of cirrus is an important constituent of the high-latitude sky (Low et al. 1984).

The first systematic comparison between the *IRAS* 100 μm and optical morphologies of dust clouds was conducted by de Vries and le Poole (1985). This and a few other studies (Paley 1990; Stark 1993), all based on photographic data, reveal a good correspondence between optical and infrared cirrus. Quantitative optical measurements of cirrus clouds using Charge-Coupled Device (CCD) images have been carried out by Guhathakurta and Tyson (1989, hereafter GT) and by Paley et al. (1991). Despite the fact that the average cirrus cloud has a very low optical surface brightness and is diffuse in nature, deep CCD images are more sensitive than *IRAS* at detecting small patches of faint cirrus, especially patches that are significantly smaller than the $3'$ resolution of the 100 μm image. The superior angular resolution of optical images (compared to the far infrared) make them good probes of the fine structure of dust clouds.

Over the last five years, we have collected a large body of CCD and digitized photographic images of dust clouds and these data are currently being used to carry out an extensive investigation of the optical properties of interstellar dust grains (Cutri and Guhathakurta 1993). A small set of *IRAS*-selected clouds have been mapped in the H I 21-cm line and/or the ^{12}CO(1$\rightarrow$ 0) line in order to study their gas-to-dust ratio and kinematics (Guhathakurta et al. 1991). We are in the process of extending our study into near infrared wavelengths using J- and K-band images of a few optically-bright nebulae obtained with the Steward Observatory 1.5-m telescope.

"CIRRUS" CLOUDS

The targets of our optical study are at Galactic latitudes $|b| \gtrsim 30°$ as this ensures relatively low line-of-sight extinction and avoids excessive line-of-sight superposition of dust clouds and crowding of the image by disk stars. The *IRAS* 100 μm all-sky maps show widespread wispy, filamentary features. It is this aspect of high-latitude interstellar dust (*i.e.* its morphology) that has earned it the name "cirrus" (Low et al. 1984). Most of cirrus is in large inter-connected complexes spanning up to tens of degrees on the sky. However, we have selected our targets to be small, relatively isolated cirrus knots less than about 1° in angular extent. These are loosely referred to as "clouds" in this paper.

The principal reason for selecting small cirrus clouds is the limited field of view of optical CCD cameras. The first attempt at optical CCD surface photometry of cirrus (GT), along with most of the subsequent deep optical studies, relied on images smaller than about 10′. In order to be able to make differential measurements of the cirrus, there must be a subtantial change in the dust column density (and hence of its surface brightness) within the optical field of view. The target fields have been selected to have a high contrast in far infrared brightness within them ($\Delta S_{100\mu m}$ is typically a few MJy sr^{-1}). In fact, all of the deep optical images have deliberately been centered on the sharpest possible cloud edges. These edges are mostly unresolved at the $\sim 3'$ in-scan resolution of the *IRAS* 100 μm band.

There has been rapid progress in CCD technology during the five years since they were first used to study cirrus. The most significant development has been the availability of relatively wide fields of view (up to 1°) through a combination of: (1) large format CCDs containing as many as 2048 × 2048 pixels, and (2) their use on small telescopes with short focal lengths. Modern CCDs have smaller pixel-to-pixel quantum efficiency variations than their precursors and are capable of producing more accurately flat-fielded images, which is particularly important for the study of faint extended objects such as cirrus. All of these features have, over the last two years, made it possible to obtain optical CCD images of entire cirrus clouds that are tens of arcminutes in size.

Clouds are selected on the basis of either: (1) *IRAS* 100 μm maps, or (2) Palomar Observatory Sky Survey (POSS) plates. Wherever possible, the far infrared selection has been done using cleaned and coadded *IRAS* Skyflux plates. Prior to the availability of these data, *IRAS* HCON-3 maps were used. The 100 μm maps are preferred over 60 μm maps as they provide higher signal to noise measurements of the cirrus, despite the superior angular resolution of the latter (1.′5 at 60 μm versus 3′ at 100 μm). The optically-selected clouds are a subset of Lynds' catalog of bright nebulae (LBN), a sample of diffuse interstellar reflection nebulae that were identified by Lynds (1965) on the the POSS O (blue) and E (red) photographs. The LBNs understandably have a higher mean optical surface brightness than the infrared-selected clouds.

The LBNs often tend to be translucent clouds ($A_V \sim 1$–2 mag) whereas the *IRAS*-selected clouds are usually optically thin. Magnani, Blitz, and Mundy (1985) have detected CO in a third of the LBNs indicating that they contain dense condensations. Optically-thin clouds have the advantage that their optical colors are easier to interpret since these are not affected by reddening within the cloud and the entire cloud is subject to the same (ambient) radiation field. In many LBNs on the other hand, the illuminating star is identifiable and it may be used to measure the distance to the nebula and the spectrum of the radiation incident on the cloud.

OPTICAL OBSERVATIONS

The three categories of data described below—deep CCD images, Schmidt observations, and digitized photographic plates—represent a sequence of measurements of decreasing depth and of increasingly wide fields of view. While deep CCD images are well suited for the detection of subtle surface brightness variations in the image and for making accurate measurements, the importance of a large field of view cannot be overstated. The wider the extent of the optical image the better one is able to measure the "sky" background off the cloud and thereby determine its absolute brightness and color (instead of having to make *relative* measurements between differents parts of the cloud).

Deep CCD Images

CCD images straddling sharp edges of four infrared-selected clouds (ir1–ir4) were obtained by GT during 1987 and 1988 using the Lowell Observatory 1.1-m telescope and an RCA 320 × 512 CCD. The CCD was binned 2 × 2 yielding a pixel scale of 2$''\!.$7 pixel^{-1} and an image size of 7′ × 9′. Short, disregistered exposures in the B_J (4500 Å), R (6500 Å), and I (9000 Å) bands have been coadded to yield effective exposure times of 1–2 hr in each band. An out-of-field chopping technique has been used to construct accurate sky-flats and to build up mosaic images up to 20′ on a side (see GT for details). The surface brightness detection thresholds in B_J, R, and I are about 27.5, 26.0, and 24.5 mag arcsec^{-2}, respectively, averaged over 10″ × 10″.

Another dozen or so targets, consisting of both optically-bright LBNs and *IRAS* 100 μm-selected clouds, have been studied between 1990 and 1992 using the Steward Observatory 2.3-m telescope on Kitt Peak. Most of the images discussed in this paper were acquired using a Texas Instruments 800 × 800 CCD binned 2 × 2 pixels with a scale of 1$''\!.$2 pixel^{-1} (after binning) and an 8′ × 8′ field of view. A set of standard Johnson *BVRI* filters were used. The mosaicing/flat-fielding technique of GT has been employed with typical effective exposure times of about 1.5–2 hr per cloud per filter. The limiting surface brightnesses are comparable to those of GT, perhaps slightly fainter.

Schmidt Telescope Observations

In April 1991, we used the Curtis-Schmidt telescope[1] (0.6-m corrector, 0.9-m primary mirror) at Cerro Tololo Inter-American Observatory (CTIO) with a Texas Instruments 800 × 800 CCD camera to study two LBNs (LBN 8 and LBN 1069) and two infrared-selected dust clouds (ir 1130–04 and ir 1147–08). The CCD was read out with 2 × 2 binning and each binned pixel covers roughly 3″ with a square field almost 20′ on a side. The effective integration time was about 75 min (consisting of several coadded 300 s exposures) per cloud through

[1]The Curtis-Schmidt telescope is jointly operated by the University of Michigan and CTIO.

each of the (Johnson) *BVRI* filters. No mosaicing was done but we used the usual technique of coadding data frames to generate accurate dark sky-flats in each of the four bands.

We have also used the 0.6-m/0.9-m (corrector, primary mirror) Burrell Schmidt telescope[2] at Kitt Peak National Observatory on three observing runs in 1991 and 1992. A Tektronix 2048 $\times$ 2048 CCD camera, with a scale of about $2''$ pixel^{-1}, was used. After accounting for vignetting by the filter wheel, the usable field of view is a circle of diameter $\sim 1°$. CCD images were obtained through standard *BVRI* filters with many 10 min frames in a given band combined to yield effective exposure times ranging from 0.5–1.5 hr per field per filter. Stray light from bright stars (presumably scattered off the filter glass) sometimes affects the edges of the field so that only the central $30'$ portion is uncontaminated. Images of more than a dozen clouds, mostly optically-bright LBNs, have been obtained, though only partial data exist in a few cases.

Digitization and Calibration of POSS Plates

Cutri, Low, and Guhathakurta (1993) have developed a method to calibrate surface photometry of faint extended sources (such as diffuse interstellar nebulae) using the National Optical Astronomy Observatories PDS scanning engine (Monet 1984) to digitize the POSS and European Southern Observatory/Science Research Council (ESO/SRC) Sky Survey photographic plates. Since this technique relies on a linear approximation to the characteristic curve, it is only appropriate for sources with surface brightnesses fainter than about μ_B, $\mu_R = 23$, 22 mag arcsec^{-2}. Thus, it is ideal for studying LBNs and we have made B and R measurements of a few optically-bright clouds. A detailed comparison between photographic scanning measurements of LBN 917 and CCD photometry of the same source shows that the former technique yields accurate photometry (1σ = 0.1–0.15 mag) down to brightness levels of $\mu_B = 26$ mag arcsec^{-2} and $\mu_R = 25$ mag arcsec^{-2}. The size of the scanned image is determined by the degree of uniformity of the background-sky fog on the photographic plate and is typically limited to $\lesssim 1°$.

PROPERTIES OF INTERSTELLAR DUST

While there is rough similarity in the morphology of a given cloud as seen at various wavelengths, there are spatially coherent variations in the optical, infrared, and optical–infrared colors within most clouds which shed light on the physical properties of the dust grains that make up the cirrus. Some of these are outlined below:

Warm Dust/Dense Cloud Cores

The peak far infrared brightness of a typical *IRAS*-selected cloud (on-off contrast) is $\Delta S_{100\mu\text{m}}$ = 5–10 MJy sr^{-1}. A far infrared surface brightness of $S_{100\mu\text{m}} =$ 10 MJy sr^{-1} corresponds to a neutral hydrogen column density $N_\text{H} = 10^{21}$ cm^{-2}

---

[2]The Burrell Schmidt telescope of the Warner and Swasey Observatory, Case Western Reserve University (CWRU) is jointly operated by CWRU and KPNO.

for a typical line of sight within the Galaxy (Boulanger and Perault 1988), which in turn implies a visual extinction $A_V = 0.6$ mag (Savage and Mathis 1979; Mathis 1990). The far infrared color of diffuse interstellar clouds is similar to that seen along average Galactic lines of sight, $S_{60\mu m}/S_{100\mu m} = 0.2$, consistent with that expected for warm interstellar dust grains (T $\sim$ 20 K) heated by the ambient ISRF (Mathis, Mezger, and Panagia 1983, hereafter MMP). A portion of the 60 μm flux is produced by stochastically heated small grains (*cf.* Draine and Anderson 1985), which occasionally reach temperatures of $\sim$ 200 K and radiate in the 12 μm and 25 μm bands. The observed mid-to-far infrared colors of clouds vary significantly around a typical value of $S_{12\mu m}/S_{100\mu m} \sim S_{25\mu m}/S_{100\mu m} \sim 0.05$.

A few clouds, most notably LBN 627 and LBN 917, show evidence for dense clumps, as indicated by bright 100 μm knots with unusually low 60-to-100 μm flux ratios. The 25 μm and particularly the 12 μm emission tends to avoid these clumps. This is most easily explained in terms of an optically-thick cloud core in which the large grains are colder than average (T $<$ 20 K) and there is little, if any, transient heating of small grains, due to the fact that this region is shielded from the ambient optical and ultraviolet photons most of which are absorbed in the outer parts of the cloud.

Reradiation of Optical (Blue) Starlight

Each of the forty-odd clouds that we have targeted for study so far is detected at optical wavelengths as faint, extended structure that is correlated with the far infrared emission. Of the optical bands, the B-band cloud morphology agrees most closely with that at 100 μm, but the agreement is not perfect. There are substantial variations in the blue–100 μm color within clouds and from cloud to cloud. The mean color $\langle S_B/S_{100\mu m}\rangle$, averaged over the cloud, is between 4×10^{-4} and 3×10^{-3} for four infrared-selected cirrus clouds, and is in the range 1–3×10^{-3} for four optically-bright LBNs.

The blue to far infrared brightness ratio expected for an optically-thin dust cloud illuminated by starlight is:

$$\frac{S_B}{S_{100\mu m}} = \frac{c\, u_B\, \omega_B^{sca}}{4\pi} \left[\frac{\tau_B}{N_{\rm H}}\right] \left[\frac{N_{\rm H}}{S_{100\mu m}}\right] = 4.1 \times 10^{-3} \qquad (1)$$

where c is the speed of light, u_B the B-band energy density of the standard MMP ISRF, ω_B^{sca} the dust grain albedo for blue light, and $\tau_B/N_{\rm H}$ the ratio of B-band extinction optical depth to hydrogen column density in the cloud (given by Mathis 1990). The albedo is computed on the basis of the interstellar dust grain model of Draine and Lee (1984), consisting of a mixture of silicate and graphite grains with a power-law distribution of grain diameters (a): $dn/da \propto a^{-3.5}$ (Mathis, Rumpl, and Nordsieck 1977). The above brightness ratio represents an average over all possible scattering angles and is exact only for dust particles that scatter isotropically.

The measured values of $\langle S_B/S_{100\mu m}\rangle$ range from being comparable to that expected for an optically-thin dust cloud (Eqn. 1) to being an order of magnitude smaller. The flux ratio is expected to be suppressed in regions of increased optical depth, as may be the case in the dense cores of LBNs. However, most of the cirrus clouds studied are likely to be optically thin (based on their 100 μm brightness). The low blue-to-100 μm brightness ratios of these clouds, compared to the isotropic value of 4×10^{-3}, is probably due to the forward-scattering phase function asymmetry in grains and the fact that we are looking at (mostly) back-scattering of disk starlight by high-latitude clouds (*cf.* Witt 1977).

Evidence for Luminescence in Hydrogenated Carbon

In general, there is better agreement between the cirrus structure seen in the different optical bands than the correlation between any of the optical images and the *IRAS* 100 μm image. This may be due, in part, to the vastly different angular resolutions of the optical and far infrared data, even though we have tried to account for the difference by smoothing the optical images. There are distinct variations in optical *BVRI* colors within clouds and between different clouds, though these are smaller than the scatter in optical–infrared colors. The mean $B - R$ colors of three infrared-selected clouds ir1–3 are 1.7, 1.2, and 1.0 mag, respectively, while they range from 0.8–2.4 mag for four LBNs; Lynds (1965) had noted that the optically-bright nebulae were often more prominent on the POSS E (red) plate than on the O (blue) plate. The average $R - I$ color has been measured in a few cases and it is about 1.4 mag.

By analogy with Eqn. 1, the ratio of the brightness of scattered starlight in two optical bands (i and j) is given by:

$$\frac{S_i}{S_j} = \frac{u_i\,\omega_i^{sca}\,\tau_i}{u_j\,\omega_j^{sca}\,\tau_j} \quad . \tag{2}$$

The color of the incident radiation (u_i/u_j) is taken to be that of the MMP ISRF which is in agreement with *Pioneer 10* measurements (Toller, Tanabe, and Weinberg 1987) of diffuse stellar light at high Galactic latitudes ($B - R \sim 1.3$). The Draine and Lee (1984) dust models yield similar albedos (and phase function asymmetry factors) across *BVRI*, with $\omega_R^{sca}/\omega_B^{sca} = 1$ and $\omega_I^{sca}/\omega_R^{sca} = 0.8$. The ratio of optical depths, $\tau_R/\tau_B = 0.63$ and $\tau_I/\tau_R = 0.57$, are taken from measurements of the extinction law in the Galaxy (Cardelli, Clayton, and Mathis 1989; Mathis 1990). The expected colors of an optically-thin dust cloud are (assuming the phase function asymmetry is wavelength-independent):

$$B - R = 0.77, \qquad R - I = 0.16 \quad . \tag{3}$$

The measured $B-R$ and $R-I$ colors of the cirrus are *redder* than those predicted by the simple scattering model outlined above by up to 1–2 mag.

The red color of the cirrus is widespread and is characteristic of even the most tenuous clouds. It is therefore unlikely to arise from optical depth effects (although a few dense LBN cores have red $B - R$ colors for this reason). Simple tests indicate that the radiation incident on a typical *IRAS*-selected cloud is unlikely to be dominated by a single star near the cloud (GT), confirming that the appropriate color to use in Eqn. 2 is that of the ambient ISRF. GT have searched unsuccessfully for Hα line emission from cirrus and conclude that it probably has little effect on the broad band R flux.

The amount and nature of the excess R and I flux in cirrus (over that expected from scattering alone) is similar to that seen in bright reflection nebulae

(Witt, Schild, and Kraiman 1984; Witt and Schild 1985, 1986). Spectra of reflection nebulae such as NGC 7023 and NGC 2023 display a broad bump in the 6500 Å–1 μm range that is thought to be caused by luminescence in small hydrogenated amorphous carbon (HAC) grains (Witt and Schild 1988). Such a bump has also been observed in the Red Rectangle (Schmidt, Cohen, and Margon 1980), which is a bipolar outflow excited by a central star, and in dense clouds illuminated by ambient starlight: *e.g.* Lynds 1780 (Mattila 1979). Photoluminescence in HACs has been studied in the laboratory (Watanabe, Hasegawa, and Kurata 1982; Furton and Witt 1993). By analogy with interstellar dust nebulae (and partly by elimination of some of the other possibilities), it appears that the extreme red optical colors of the cirrus are caused by luminescence in HACs, or possibly in polycyclic aromatic hydrocarbons.

Phantom Optical Emission?

Earlier in this section, we have tried to interpret the *mean* blue-to-100 μm color of clouds in terms of a simple scattering model. There are patches of optical emission in certain clouds (ir1, ir2, LBN 574, and LBN 627) that have no *IRAS* 100 μm counterpart so that $S_B/S_{100\mu\mathrm{m}} > 10^{-2}$, which a factor of 10 or more larger than the average value for the cloud.

This unusual optical emission appears to be reproducible in images taken through different filters over a time scale of a few months, arguing against time-dependent phenomena such as airglow or zodiacal light. The amount of stray light (due to scattering in the camera optics, for example) is probably small as judged from the excellent agreement in the surface photometry of an LBN using digitized POSS plates compared to using CCD images taken with the Burrell Schmidt telescope. The "phantom" optical patches may be scattered starlight off an inhomogenously mixed population of unusually cold dust grains ($T \lesssim 15$ K), that are either poor absorbers of ultraviolet/optical radiation or efficient emitters at far infrared wavelengths.

Fine Structure of Cirrus

Small scale variations in the optical and infrared colors of cirrus clouds, even in optically-thin portions, indicate the presence of compositional inhomogeneities (*e.g.* in the mixing ratio of different grain sizes/species). There are several mechanisms that might be responsible for such variations, including chance superposition of clouds with different compositions, processing of grains during the passage of a shock, segregation produced by a magnetic field, etc.. Some clouds do indeed show multiple kinematical components (see below).

Cirrus has a two-dimensional spatial power spectrum of the form: $P(k) \propto k^{\alpha}$, with $\alpha \sim -3$, where k is the reciprocal of angular scale θ (Gautier 1993). Power spectra derived from high signal-to-noise *IRAS* 100 μm maps (for θ ranging from a few arcminutes to degree scales) have the same power law index as those derived from optical CCD surface photometry ($\theta \sim 15''$–$3'$). High-contrast far infrared maps show that a large fraction of the sky is covered by faint cirrus (Low et al. 1984; Beichman 1987). Low level arcsecond-scale fluctuations in the surface brightness of scattered light ($B \sim 30$ mag arcsec^{-2}), caused by subtle variations in the dust column density, could pose a problem for ultra-deep extragalactic imaging studies.

ATOMIC AND MOLECULAR GAS ASSOCIATED WITH CIRRUS

Knapp (private communication), using the Bell Laboratories 7-m radio dish, has detected $^{12}CO(1\rightarrow 0)$ line emission in all of the five infrared-selected cirrus clouds searched. The molecular gas appears to be localized at the peaks in the 100 μm brightness. VLA 'D' array 21-cm observations of three of these clouds show H I associated with them. There appear to be spatial variations in the atomic-to-molecular gas ratio and in the H I-to-100 μm flux ratio, but accurate measurements are hampered by the missing 'zero spacing' in the interferometric observations which cause HI on large angular scales to go undetected. The atomic hydrogen shows multiple kinematical components, each with $\Delta v \sim 10\,\mathrm{km\,s^{-1}}$ and all within $\pm 20\,\mathrm{km\,s^{-1}}$ of the LSR velocity. The $^{12}CO(1\rightarrow 0)$ line is typically narrower ($\Delta v \sim 2$–$5\,\mathrm{km\,s^{-1}}$). The properties of atomic and molecular gas associated with cirrus clouds are currently under investigation.

SUMMARY

Our studies of the optical emission associated with infrared cirrus have produced the following results:

1. Cirrus clouds are detected in the optical *BVRI* bands as faint extended nebulosity that has the same overall structure as the 100 μm image.

2. The mean blue-to-100 μm brightness ratio observed in cirrus is lower than that expected for isotropic scattering of the general Galactic radiation field by interstellar dust grains, but may be explained in terms of forward-scattering grains.

3. Optical nebular emission is occasionally seen in regions with no detectable *IRAS* 100 μm flux (but is apparently associated with a cirrus cloud), and may indicate the presence of cold, efficiently-scattering grains.

4. The optical colors ($B-R$, $R-I$) of the cirrus are significantly redder than those expected for scattering of ambient starlight, and this is probably caused by luminescence in very small hydrogenated carbon grains (similar to what is seen bright reflection nebulae and some dense interstellar clouds).

5. The spatial distribution of optical–infrared and optical colors suggests intracloud (and cloud-to-cloud) variations in the relative fractions of different grain species (and therefore in the chemical composition, optical properties, etc.).

6. Molecular gas, in the form of the $^{12}CO(1\rightarrow 0)$ line, and atomic hydrogen has been searched for in a few clouds and the detection rate is 100%. The CO line width is smaller than that of the H I.

ACKNOWLEDGMENTS

We are grateful to Jacqueline van Gorkom for helping with the VLA observations and data reduction, and to Jill Knapp for carrying out the $^{12}CO(1\rightarrow 0)$ observations. We would also like to thank Ed Carder, Bruce Draine, Craig Gullixson, Frank Low, Bill Schoening, and Tony Tyson for assistance with various aspects of the optical and *IRAS* data and for helpful suggestions. The staff at CTIO, KPNO, Lowell Observatory, Steward Observatory, and NRAO are thanked for their help and hospitality. This work is supported in part under the ADP program (NASA grant NAG 5-1252 to RMC), by Mission Research Corporation contracts SC-0014-88-002 and SC-0028-93-0001 (RMC), and by NASA grant number HF-1033.01-93A (PG) from the Space Telescope Science Institute, which is operated by AURA, Inc., under NASA contract NAS 5-26555.

REFERENCES

Beichman, C. A. 1987, *ARA&A*, **25**, 521

Boulanger, F., and Perault, M. 1988, *ApJ*, **330**, 964

Cardelli, J. A., Clayton, G. C., and Mathis, J. S. 1989, *ApJ*, **345**, 245

Cutri, R. M., and Guhathakurta, P. 1993, *BAAS*, **24**, 1198

Cutri, R. M., Low, F. J., and Guhathakurta, P. 1993, *PASP*, **105**, 106

Draine, B. T., and Anderson, N. 1985, *ApJ*, **292**, 494

Draine, B. T., and Lee, H. M. 1984, *ApJ*, **285**, 89

Efstathiou, G., Bernstein, G., Katz, N., Tyson, J. A., and Guhathakurta, P. 1991, *ApJ*, **380**, L47

Furton, D. G., and Witt, A. N. 1993, *ApJ*, **415**, L51

Gautier, T. N. 1993, this volume

Guhathakurta, P., Cutri, R. M., van Gorkom, J. H., and Knapp, G. R. 1991, *BAAS*, **22**, 1240

Guhathakurta, P., and Tyson, J. A. 1989, *ApJ*, **346**, 773 (GT)

Low, F. J. et al. 1984, *ApJ*, **278**, L19

Lynds, B. T. 1965, *ApJS*, **12**, 163

Magnani, L., Blitz, L., and Mundy, L. 1985, *ApJ*, **295**, 402

Mathis, J. S. 1990, *ARA&A*, **28**, 37

Mathis, J. S., Mezger, P. G., and Panagia, N. 1983, *A&A*, **128**, 212 (MMP)

Mathis, J. S., Rumpl, W., and Nordsieck, K. H. 1977, *ApJ*, **217**, 425

Mattila, K. 1979, *A&A*, **78**, 253

Monet, D. 1984, in *Astronomical Microdensitometry Conference*, D. A. Klinglesmith, NASA Conf. Publ. 2317, 291

Neugebauer, G. et al. 1984, *ApJ*, **278**, L1

Paley, E. S. 1990, M.S. thesis, University of Arizona

Paley, E. S., Low, F. J., McGraw, J. T., Cutri, R. M., Rix, H.-W. 1991, *ApJ*, **376**, 335

Sandage, A. 1976, *AJ*, **81**, 954

Savage, B. D., and Mathis, J. S. 1979, *ARA&A*, **17**, 73

Schectman, S. A. 1973, *ApJ*, **179**, 681

Schectman, S. A. 1974, *ApJ*, **188**, 233

Schmidt, G. D., Cohen, M., and Margon, B. 1980, *ApJ*, **239**, L133

Stark, R. 1993, Ph.D. thesis, University of Leiden

Toller, G., Tanabe, H., and Weinberg, J. L. 1987, *A&A*, **188**, 24

Tyson, J. A. 1988, *AJ*, **96**, 1

de Vaucouleurs, G. 1955, *Observatory*, **75**, 129

de Vaucouleurs, G. 1960, *Observatory*, **80**, 106

de Vries, C. P., and le Poole, R. S. 1985, *A&A*, **145**, L7

Watanabe, I., Hasegawa, S., and Kurata, Y. 1982, *Japanese J. Phys.*, **21**, 856

Witt, A. N. 1977, *ApJS*, **35**, 7

Witt, A. N., and Schild, R. E. 1985, *ApJ*, **294**, 225

Witt, A. N., and Schild, R. E. 1986, *ApJS*, **62**, 839

Witt, A. N., and Schild, R. E. 1988, *ApJ*, **325**, 837

Witt, A. N., Schild, R. E., and Kraiman, J. B. 1984, *ApJ*, **281**, 708

The First Symposium on the Infrared Cirrus and Diffuse Interstellar Clouds
ASP Conference Series, Vol. 58, 1994
Roc M. Cutri and William B. Latter (eds.)

X-RAY SHADOWS OF CIRRUS CLOUDS

U. MEBOLD, J. KERP, P. MORITZ, J. ENGELMANN AND U. HERBSTMEIER
Radioastronomisches Institut der Universität Bonn, 53121 Bonn, FRG

ABSTRACT We review the utility of soft X-ray observations with the Röntgen Satellite (*ROSAT*) X-ray telescope (XRT) for studies of Cirrus clouds. A number of Cirrus clouds have been detected as X-ray shadows against the diffuse soft X-ray background (SXRB). These shadows can be used to derive the amount of X-ray absorbing matter so that total masses as well as chemical and physical properties can be determined. We discuss three Cirrus clouds that are seen as X-ray shadows: the local H I filament LVC 88+36–2 which is embedded in the hot plasma of the Local Hot Bubble (LHB), the X-ray shadows of the Draco nebula and of IVC 86+38–44 which are located outside the galactic disk. In addition we discuss X-ray emission associated with halo type objects like the high velocity clouds (HVC) M I and M II. The cirrus clouds and the HVCs populate the distance range from about 60 pc to more than a few kpc. The soft X-ray shadows imply the presence of a pervasive X-ray emitting plasma between the sun and and the lower galactic halo in which the neutral Cirrus clouds are embedded. The soft X-ray emission close to the HVCs possibly indicates interaction of these clouds with the galactic disk medium. We presents a new technique to derive total column densities of hydrogen nuclei for molecular clouds, by the comparision of the X-ray shadows, H I and IR emission. This technique is applied to the molecular clumps of the Draco nebula.

INTRODUCTION

The discovery of the galactic infrared Cirrus by the IRAS (Infra-Red Astronomical Satellite, Low et al. 1984) came as a surprise to the astronomical community because the sensitivity of the detectors for diffuse distributed matter was better than expected. The discovery of Cirrus clouds as shadows against the diffuse SXRB came as a surprise to most astronomers, too. But this discovery was unexpected because the leading experts in the field did not accept the presence of large amounts of X-ray emitting gas outside the LHB. Although a distant origin of the SXRB was predicted (e.g. Marshall & Clark 1984) on the basis of a negative correlation between the SXRB and the galactic H I 21-cm line brightness distribution, proposals to observe the shadows cast by Cirrus clouds against that SXRB with the *ROSAT* XRT were turned down on the basis of the forecasts of the LHB model for the X-ray emitting plasma in our Galaxy (see Snowden et al. 1990 and references therein). As a result, the first discovery of such a soft X-ray shadow at high galactic latitude by Snowden and the X-ray group in Bonn (Snowden et al. 1991) was made accidentally during the performance tests of the XRT. – In the mean time it has become clear that Cirrus clouds are associated with X-ray emission not only with the signature of shadows, i.e., absorption, but also with positive positional correlations.

After the original discovery of the SXRB by Bowyer et al. (1968) observations giving more details about its distribution became available with the Wisconsin Soft X-Ray Survey (McCammon et al. 1983) and the SAS-3 observations (Marshall & Clark 1984). Global absorption-model fits to the 1/4 keV diffuse background (Marshall & Clark 1984) predicted the existence of a ubiquitous hot plasma outside the Galaxy which is absorbed by the galactic layer of gas and dust. However, this model was challenged based on of the fact that the observed column densities of neutral material will let only 5 to 10% of the observed X-ray photons pass to the observer from a distant X-ray emitting layer. The bulk of the photons are emitted by the LHB (Snowden et al. 1990, McCammon & Sanders 1990, see also Frisch & York 1983). Based on the Wisconsin data Hirth et al. (1991) deduced the existence of a hot plasma intersperced with other - in particular neutral - constituents of the ISM throughout and outside the galatic disk. Finally, the existence of a galactic soft X-ray corona has been proved by images obtained by the *ROSAT* (Trümper 1983) of X-ray shadows of Cirrus clouds (Snowden et al. 1991; Burrows & Mendenhall 1991) in conjunction with the known distances of these clouds (Mebold et al. 1985; Lilienthal et al. 1991; Lilienthal et al. 1992; Wennmacher et al. 1993). The existence of a hot galactic corona has opened up the possibility to study Cirrus clouds by means of standard X-ray diagnostics.

In the next section we review the most important instrumental parameters of the *ROSAT* mission. After a discussion of the radiation transport of soft X-rays through the interstellar medium (ISM) we review the present knowledge about X-ray shadows of Cirrus clouds and discuss the shadows of LVC 88+36–2, the Draco nebula and IVC 86+38–44 in some detail. As a brief aside we present evidence for a spatial association of HVCs in the complexes M and C with soft X-ray emission.

THE *ROSAT* MISSION

The results from the *ROSAT* mission (Trümper 1983) which are presented below demonstrate the utility of this X-ray telescope (Aschenbach 1988) in combination with the position sensitive proportional counter (PSPC; Pfeffermann et al. 1987) for studies of the SXRB. The instrumental characteristics of this telescope system and its modes of operation (survey, pointed) will be summarized in the present section.

The XRT mirror is a fourfold nested Wolter type I system with 83 cm aperture and 240 cm focal length. The unabsorbed collecting area is 1141 cm^2 which is almost a factor of three larger than that of the Einstein telescope. The PSPC provides a circular field of view of about 2° in diameter and an angular resolution of 34" (FWHM) at 1/4 keV. The raw images have to be corrected for vignetting, and the influence of a wagon-wheel like detector support structure in addition to meshes at the detector entrance window. The effective area of the telescope and the detector efficiency vary across the field of view of the PSPC. Therefore exposure corrections of the raw X-ray map are required for observations of extended (> 10 arcmin) objects in the pointed mode. Further, the exposure corrections vary from one energy band to the other. Standard software for the X-ray data reduction is provided by the EXSAS package (Zimmermann et al. 1993).

The *ROSAT* PSPC has relatively limited spectral resolution ($E/\Delta E \approx 1$ at 1/4 keV and $E/\Delta E \approx 2$ at 1 keV). The usable energy range extends from 0.1 to 2.4 keV. The PSPC can distinguish up to 7 energy bands, frequently called R1 to R7, which are, however, partially overlapping. Their energy ranges defined as 10% of the peak response are given in Table 1:

TABLE 1 *ROSAT* Energy Bands

Band Name		R1	R2	R3	R4	R5	R6	R7
Energy range,	from	110	140	200	440	560	730	1050
10% of peak (eV)	to	284	284	830	1010	1210	1560	2040

The R1 and the R2 band are somewhat analogous to the B (Boron) and C (Carbon) bands of the Wisconsin all-sky survey. The combined R1 and R2 band is called the *ROSAT* 1/4 keV band. Bands R4 and R5 cover roughly the energy ranges of the Wisconsin M1 and M2 bands and together form the 3/4 keV band which is similar to the Wisconsin M band. Similarly, bands R6 and R7 cover the Wisconsin I and J bands and together can be used as a 1.5 kev band. More details can be found in Snowden et al. (1993).

In the present paper we use data wich were obtained in the survey mode of operation (e.g., the data for the Draco nebula and the HVCs) and in the pointed mode (the data for LVC 88+36–2 and for IVC 86+38–44). During survey phase (Snowden & Schmitt 1990; Voges 1992), *ROSAT* scans in such a way that one great circle is swept out on the sky each orbit which takes about 95 minutes. The plane of the great circle is rotated by 4 arcmin each orbit around a line connecting the ecliptic poles. With this geometry the exposure time rises from appr. 600 seconds near the ecliptic equator to appr. 30 000 seconds close to the ecliptic poles.

In order to permit quantitative studies of the diffuse cosmic X-ray background and of extended X-ray sources the contamination by non-cosmic particle events has to be corrected for. Any such contamination is difficult to distinguish from extended galactic emission e.g. from the soft X-ray emission close to the sun in the LHB. Obviously a correction for non-cosmic events is vital for quantitative comparision of pointed observations. Especially it is critical for a proper treatment of the radiation transport. As we will see below additive contamination terms to the cosmic count rate affects not only solutions for the LHB count rate but may also invalidate the decomposition of soft X-ray spectra. Four different components are contributing to this contamination (Snowden et al. 1993) of the *ROSAT* X-ray data. These are the particle induced instrumental background (Snowden et al. 1992), the background from solar X-rays scattered into the field of view by the residual Earth atmosphere along the line of sight towards the observed source (Snowden & Freyberg 1993), and two empirically identified components (Snowden et al. 1993). These are distinguished by their temporal properties as short term enhancements with a tyical duration of 1 to 2 orbits and as long term enhancements with a duration of typically 1 to 2 days during the survey operation. While the former two components are well understood and can be corrected for and short-term enhancements are easily recognized in plots of count rate versus time and can be excluded from further analysis, long-term enhancements can only be corrected for by normalization relative to the *ROSAT* all-sky survey (Snowden et al. 1993).

RADIATION TRANSPORT AND THE X-RAY ABSORPTION CROSS SECTION

Soft X-rays can be used for diagnostics of the interstellar medium through which they pass to the observer. This is of particular interest for galactic Cirrus clouds because recent observations have established that a significant fraction of the observed soft X-ray photons originate outside the galactic gas layer coming from the galactic halo and from extragalctic sources. Passing through the galactic gas the X-ray photons are partially absorbed. The observed count rate, $C_{\rm obs}$, can be used to determine the count rate, C_1, produced in the galactic halo and that, C_0, produced in the LHB, if the column density, $N(\rm H)$ of the absorbing medium and the absorption cross section, $\sigma[N(\rm H)]$ are known:

$$C_{\rm obs} = C_0 + C_1 e^{-\overline{\sigma} N(\rm H)}. \tag{1}$$

For most regions in the galactic Cirrus, i.e., if molecular hydrogen, H_2, can be neglected, the column density $N(\rm H)$ is sufficiently well approximated by the column density N(H I) of neutral hydrogen. The contribution of the heavy elements to the absorption is taken into account by using absorption cross sections that are representative of the standard mixture of the chemical elements in the interstellar medium. The dominant absorption mechanism at the energies considered here is photoelectric absorption, which has a $E^{-8/3}$ (Morrison & McCammon 1983) between absorption edges. In regions with a significant amount of molecular hydrogen its absorption must be accounted for. Between 150 and 280 eV Denne (1970) finds cross sections for one H_2 molecule which are between 2.2 to 3 times that of atomic hydrogen, respectively. In equation (1) band-averaged cross sections are used. Because of the steep energy dependence, band-averaged cross sections are dependent on the incident X-ray spectrum. Absorption by intervening gas selectively reduces the intensity from the lowest energies in the bands. This "hardens" the spectrum and reduces the band-averaged cross section of the next cloud of absorbing material. As a result, the band-averaged cross sections depend both on the incident X-ray spectrum and on the absorbing column density $\overline{\sigma} = \sigma[N(\rm H)]$. For the observations discussed below we have used cross sections which are consistent with the ones given by Snowden et al. (1993). The cross section per particle is $\overline{\sigma} \approx 10^{-20}$ cm^2 for the 1/4 keV band and $\overline{\sigma} \approx 10^{-21}$ cm^2 for the 3/4 keV band if a thermal plasma of $T = 10^6$ K is adopted as background source. These numbers happen to fall in a range where typical Cirrus clouds with column densities between 10^{20} cm^{-2} and 10^{21} cm^{-2} can be observed as X-ray shadows. Of course, shadows can only be identified if the column density in the surroundings of the absorbing clouds is small enough that the SXRB is not completely blocked. As a result, the first soft X-ray shadows have been found in high latitude areas (Snowden et al 1991; Burrows & Mendenhall 1991; Snowden 1993).

In low latitude regions 1/4 and even 3/4 keV photons from the galactic background will mostly be blocked out completely. Therefore shadows can only be expected from high column density molecular clouds against the 1.5 keV X-ray background. These shadows are of particular interest for studies of massive molecular clouds with column densities larger than 10^{22} cm^{-2}. They provide an independent means of determining the total masses of star forming regions in our Galaxy.

X-RAY SHADOWS OF CIRRUS CLOUDS

The most spectacular result from *ROSAT* relating to the SXRB has been the observation of shadows cast by Cirrus clouds (Snowden et al. 1991; Burrows & Mendenhall 1991). In the mean time a dozen clouds have been investigated for possible shadowing with only 4 of them showing a 1/4 keV shadow and 5 of them showing a 3/4 keV shadow. Table 2 lists the clouds for which X-ray shadowing has been observed.

TABLE 2 Cirrus clouds with X-ray shadows

Name	$l(°)$	$b(°)$	dist (pc)	1/4 keV shadow	3/4 keV shadow
*MBM*36[a]	4	36		no	yes
*MBM*40[a]	38	45	70	no	
LVC 88+36–2	88	36	60	yes	no
IVC 86+39-44	88	39	>300	yes	no
Draco	91	40	>300	yes	
Cepheus[a]	114	16	450	no	
UMa[a]	150	50		yes	
*MBM*12[a]	155	-35	65	no	yes
*MBM*20[a]	210	-35	<125	no	yes
β, CMa[a]	226	-15		no	
Chamaeleon[a]	289	-15	140	no	yes
Ophiuchus[a]	350	20	150	no	yes

[a]Adopted from Snowden (1993)

The clouds showing a shadow against the 3/4 keV background but showing no 1/4 keV shadow are obviously located at the distance of the gas forming the boundary of the LHB. This gas does not allow the softer X-rays to penetrate from the background into the LHB and to the observer. The harder X-rays can obviously penetrate this gas but are strongly blocked by the clouds themselves thus producing the shadows. The clouds called LVC 88+36–2, Draco, and IVC 86+39–44 are discussed in some detail below. More details about MBM 12 are published by Snowden et al. (1993). Details for the rest of these clouds are not available.

Detailed analyses of X-ray shadows provide powerfull tools for investigations of both the neutral, cold component of the ISM, i.e., the interstellar clouds, and

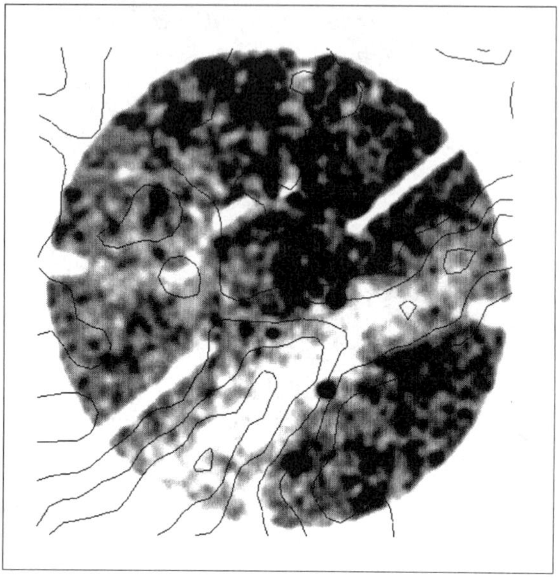

FIGURE 1 *ROSAT* image of LVC 88+36–2 in the energy band defined in the text. The count rates range from $2.9 \cdot 10^{-4}$ (black) to $6.6 \cdot 10^{-4}$ counts $\cdot$ $s^{-1} \cdot$ $arcmin^{-2}$ (white). The H I column density distribution in the velocity interval -90 to $+36$ kms^{-1} is overlayed as contours (lowest contour: $1.8 \cdot 10^{20} cm^{-2}$, contour step: $0.23 \cdot 10^{20}$ cm^{-2}).

of the fully ionized hot plasma in our Galaxy. They can e.g. be used for
- distance determinations,
- determination of the total absorbing column density,
- separation of the emission in front and behind the absorber, and
- estimates of the heating and ionization of the absorbing cloud by the incident X-ray radiation.

In the present paper we are mainly concerned with the first two items. But we will also give emission measures of the hot plasma in the LHB and outside the Galactic gas layer and as a brief aside discuss associations of X- ray emission and H I high velocity clouds (HVC's).

A DENSE H I FILAMENT IN THE LOCAL HOT BUBBLE

To start with the most local one of the X-ray shadows, we discuss the evidence for a shadow cast by the low velocity H I and cirrus filament LVC 88+36–2 first. Wennmacher et al. (1992) point out that this cold ($T < 150$ K) and dense ($n = 75\,cm^{-3}$) H I filament at $(l, b) = (88°, 36°)$ for which interstellar absorption line studies give a distance of $< 60 \pm 20$ pc, is visible as a shadow in the X-ray data of Burrows & Mendenhall (1991). This leads us to observe the position $(l, b) = (88.7°, 36.5°)$, with the *ROSAT* PSPC with the aim to verify and quantify this shadow (see Kerp et al. (1993) for more details). Figure 1 shows the X-ray images which Kerp et al. (1993) found after $\sim$ 27000 sec of intergration time (field diameter 2°) for the 1/4 keV band. Overlayed are H I column density contours obtained from Effelsberg 21-cm line spectra. The anticorrelation of the soft X-ray count rates and the H I column densities is clearly visible.

The absorption cross section in the lowest part of the R1 band is about $2.25 \cdot 10^{-20}\,\mathrm{cm}^2$ and the column density of the filament and the diffuse H I layer is $\sim 2.8 \cdot 10^{20}\,\mathrm{cm}^{-2}$. Therefore, nearly all coronal X-ray photons are blocked. The shadow cast by the filament is therefore due to absorption by the H I filament of X-ray photons which come from the hot plasma which is located between the back side of the filament and the diffuse H I layer, i.e., coming from distant regions of the LHB. This implies that the H I filament is located within the LHB.

From the depth of the shadow, which is $\sim$ 30 % of the off-filament count rate, the distance to the filament is $\sim$ 2/3 of the radius of the LHB in this direction. Assuming a radius of $\sim$ 100 pc (Snowden et al. 1991), the filament is at a distance of 65 ± 20 pc. This is consistent with the upper limit of 60 ± 20 pc derived from interstellar absorption line studies.

LVC 88+36–2 is the first example of a dense filament of neutral gas and dust that is embedded in the hot X-ray emitting plasma of the LHB. From recent *ROSAT* data Kerp (this volume) finds a temperature of about $8 \cdot 10^5$ K for the plasma in the LHB. This raises a number of problems. Some of these are: What is the life time and the origin of LVC 88+36–2? Is edge brightening observable in the EUV? Is the filament confined by magnetic fields? How many more filaments of this kind do exist in the LHB? Is the filament LVC 270-30–10 (McGee et al. 1986) which has an IR and H I morphology similar to that of LVC 88+36–2 also embedded in the LHB? What is the effect of these neutral enclosures on the temperature distribution of the hot plasma in the LHB?

THE X-RAY SHADOW OF THE DRACO NEBULA

Next along the line of sight in a region rather close to the H I filament discussed above we find the X-ray shadow cast by the Draco nebula. The Draco nebula is a molecular cloud (Mebold et al. 1985; Herbstmeier et al. 1993) at high galactic latitudes $(l, b) = (91°, 38°)$. The distance has been estimated by various methods. A lower distance limit of $d > 300$ pc has been derived from interstellar absorption line studies (Lilienthal et al. 1991). From star counts and estimates of the extinction (Mebold et al. 1985; Lilienthal et al. 1991) and from UBV photometry of stars in the most obscured regions of the nebula (Georigk & Mebold 1986) a distance of about 600 pc has been estimated. During the verification phase of the *ROSAT* mission two pieces of the shadow were found (Snowden et al. 1991; Burrows & Mendenhall 1991). As the depth of these shadows was about 50 %, it was obvious that large amounts of diffuse X-ray photons are emitted behind the Draco nebula, i.e., at distances > 600 pc most probably in the lower galactic halo. It appeared that our Galaxy has an X-ray corona.

We have compared *ROSAT* 1/4 keV data obtained during the survey phase with the IRAS 100 μm data for the entire Draco nebula. Figure 2a shows the 1/4 keV count rate distribution and Fig. 2b the 100 μm emission of the nebula. We see that all parts of the Draco nebula show up as X-ray shadows.

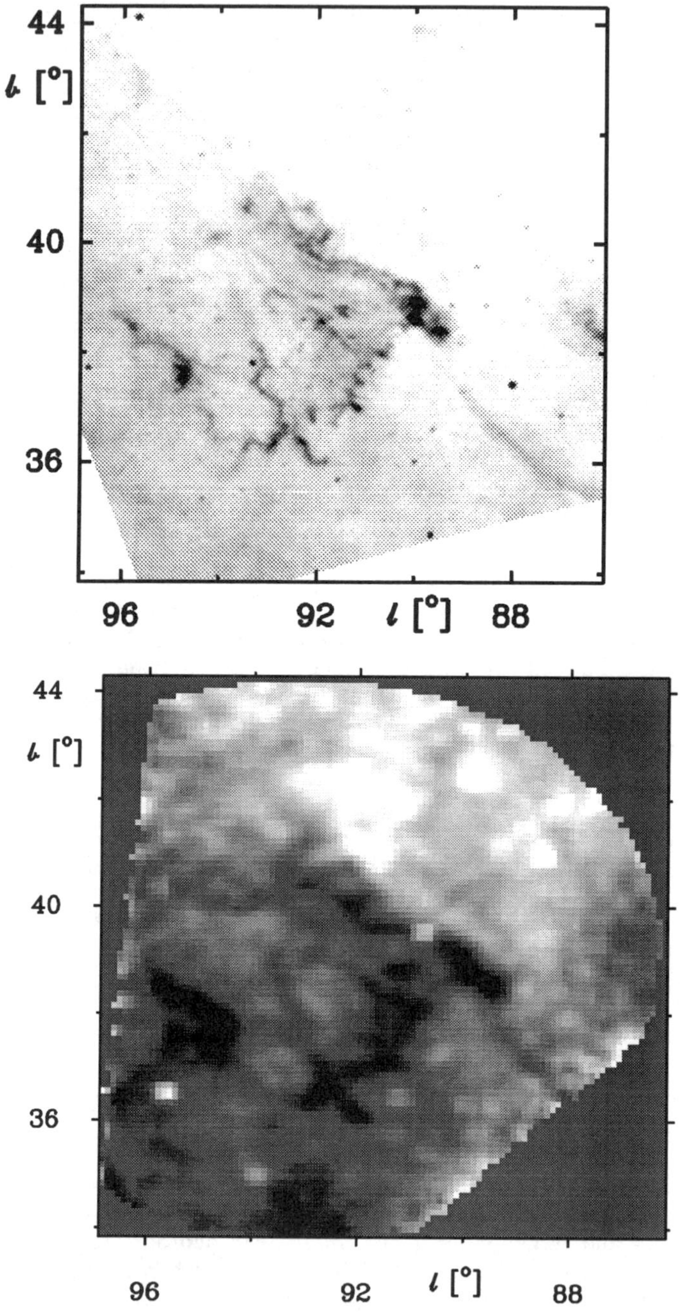

FIGURE 2 Top: The *ROSAT* 1/4 keV survey image of the region of the Draco nebula (provided by S. Snowden). The count rates range from $12 \cdot 10^{-4}$ (white) to $7 \cdot 10^{-4}$ counts $\cdot$ s^{-1} $\cdot$ arcmin^{-2} (black). Contours indicate the range from $8 \cdot 10^{-4}$ to $18 \cdot 10^{-4}$ counts $\cdot$ s^{-1} $\cdot$ arcmin^{-2}. Bottom: The same region at 100 μm wavelength from the IRAS Sky Survey Atlas. A constant background is subtracted. The intensities range from 1 (white) to 5.5 MJy ster^{-1} (black).

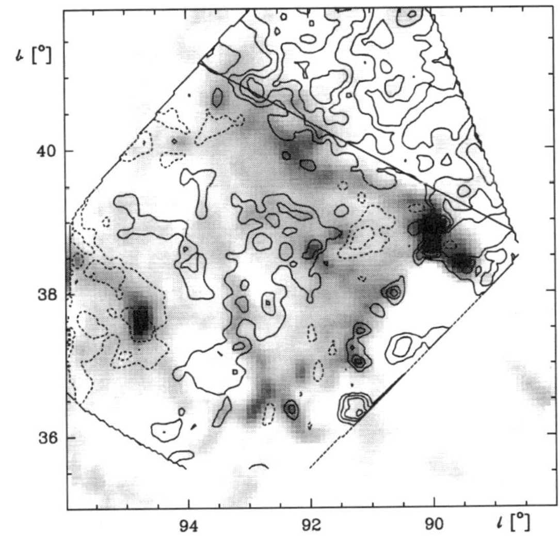

FIGURE 3 Draco nebula: The relative deviation of the observed soft X-ray count rates from the model count rates [$(C_{\rm obs} - C_{\rm model})/C_{\rm model}$, see text] overlayed on the IRAS 100 μm image. The model fitted to the data includes the X-ray emission of the LHB and a coronal X-ray component which is attenuated by the gas of the Draco nebula and that of the Galactic disk. Negative values are represented by dashed contours. The contour interval is ± 0.1.

The shadow is getting as deep as 60 % of the off-nebula count rate. Since the material in the nebula is optically thick for background soft X-rays in the deepest part of the shadow, only the X-ray photons of the foreground are observed here. Therefore only 40 % of the SXRB observed in the region around the Draco nebula is contributed by the foreground. The other 60 % are coronal emission. Since this emission is attenuated by an ubiquitous diffuse H I gas of known column density (see above) we can correct for this attenuation and derive the coronal emission. We find an emission measure of the coronal X-ray emission of ($EM_{\rm halo} = 3.2 \cdot 10^{-2}\,{\rm cm}^{-6}{\rm pc}$) which is about a factor of four larger than that of the LHB ($EM_{\rm LHB} = 0.8 \cdot 10^{-2}\,{\rm cm}^{-6}{\rm pc}$).

The depth distribution of the shadow in Fig. 2a has been used to derive the column density distribution in units of hydrogen nuclei, N(H), of the absorbing material (Moritz 1993). Here a smooth surface brightness has been assumed for the X-ray background of the Draco nebula. Further, absorption cross sections consistent with those of Snowden et al. (1993) have been used. The derived column density distribution has been compared to the 100 μm emission, I_{100}. Assuming a constant conversion factor for the whole nebula, the best fit factor becomes

$$a_{100} = I_{100}/N(H) = 1.1 \cdot 10^{-20}\,{\rm MJy\ sr^{-1}\ cm^{2}}. \qquad (2)$$

This conversion factor is consistent with that ($a_{100} \approx 1.3 \pm 0.3 \cdot 10^{-20}$ MJy sr^{-1} cm^{2}) derived for the same nebula from a comparison of I_{100} with the column density of atomic hydrogen N(H I) and the velocity integral of the ^{12}CO J=1$\rightarrow$0 line (Herbstmeier et al. 1993). This is remarkable because a_{100} for the Draco nebula is larger than typical values derived for the galactic plane near the solar circle (Bloemen et al 1990).

The conversion factor derived above does not fit, however, equally well for all parts of the nebula. Calculating the relative deviation $[(C_{\rm obs} - C_{\rm model}) / C_{\rm model}]$ from the observed and the modelled count rates, the most striking deviations are found for the boundary towards low l and low b (galactic south-west (SW)) (see Figure 3). Here Herbstmeier et al (1993) find a chain of molecular clumps. The observed depth of the shadow is smaller than that expected from the observed IR emission and the best fit value for a_{100}. To explain the depth of the shadows of the molecular clumps, less $N(\rm H)$ is required than implied by the value of a_{100} given above. Moritz (1993) finds

$$a_{100} > 1.3 \cdot 10^{-20}\,\mathrm{MJy\ sr^{-1}\ cm^{2}}. \tag{3}$$

This value is even farther from that found for molecular clouds in the galactic plane near the solar circle ($0.8 \cdot 10^{-20}\,\mathrm{MJy\ sr^{-1} cm^{2}}$) than the best fit value given above. Various interpretations are possible. It can for instance be explained by dust temperatures which are larger than for the bulk of the Draco nebula or by a different chemistry in the clumps along the SW boundary. This would be consistent with an interpretation of this boundary as a low-velocity shockfront (Herbstmeier et al. 1993).

This analysis of X-ray shadows is a new technique to derive the total column density of hydrogen nuclei for massive molecular clouds. If the H I content of a molecular cloud is known from 21-cm emission or absorption lines, the molecular content can be derived from the difference of total column density of the X-ray absorbing material, $N(\rm H)$, and the column density, $N(\rm H\,I)$ of the atomic gas: $N(H_2) = [N(H) - N(HI)]/2$. In this case one can also derive the molecular fraction.

Another region of strong deviation of a_{100} from the best fit value for the whole nebula is found to the galactic north-west (NW) of the Draco nebula. Here we find an excess of X-ray emission which can hardly be explained by a lack of absorbing material. The excess of X-ray emission is positionally coincident with the HVCs observed in this area. This coincidence is discussed below.

X-RAY EDGE-BRIGHTENING OF IVC 86+38–44

The intermediate velocity cloud IVC 86+38–44 is a dust and atomic gas cloud located at $(l, b) = (86°, 38°)$ which is only about 5° towards the (galactic) West from the Draco nebula and about 3° towards the North of LVC 88+36–2. It was discovered in the 100 μm IRAS band and is found in the 21-cm line at a radial velocity of $-44\,\mathrm{km\,s^{-1}}$. Morphologically the IVC appears to be located at the low-l, low-b tip of an H I HVC stream. This coincidence suggests that IVC 86+38–44 is the interface of an inelastic collision between the HVC stream and gas at zero velocity (Herbstmeier 1989; Mebold et al. 1989). Although no detailed information about the distance is available so far, the lack of any noticeable star count deficit on the Palomar Observatory Sky Survey plates implies that the IVC is probably located outside the galactic plane.

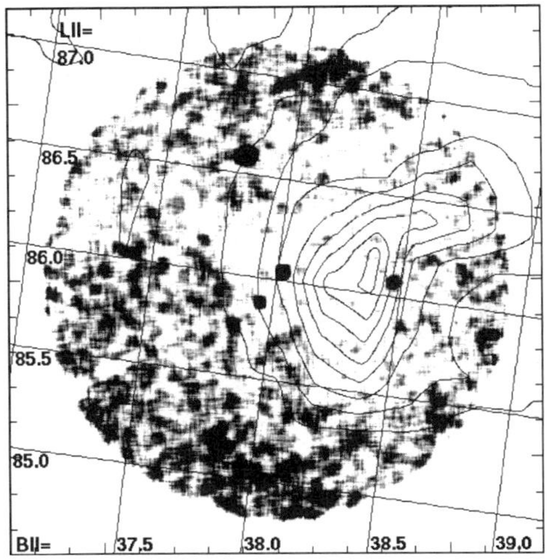

FIGURE 4 *ROSAT* X-ray image of IVC 86+38–44 in the R2 band. The grey scale indicates count rates between 7.3 (dark) and 4.8 · 10^{-4} counts · s^{-1} · $arcmin^{-2}$. The isophotes indicate H I column densities between 0.3 and 2.75·10^{20} cm^{-2}. The step between contours is: 0.35 · 10^{20} cm^{-2}. A clear anti-corrrelation between X-ray intensity and H I column density is indicated.

In Figure 4 we show a grey-scale representation of the 1/4 keV X-ray count rate distribution which was derived from a pointed *ROSAT* observation with about 7000 seconds of integration time (Herbstmeier et al. 1993). The overlay represents the H I column density of IVC 86+38–44 integrated between LSR velocities of -73 and -35 km s^{-1}. The peak column density is 3.5·10^{20} cm^{-2}. The average column density in the area around the IVC is 2 · 10^{20} cm^{-2}.

A clear anti-correlation of the X-ray count rates and the H I column densities is seen. This anti-correlation can only be associated with IVC 86+39–44 because the foreground H I emission appears rather smooth. The observed count rate depression of about 35 % of the average count rate (7 · 10^{-4} counts $s^{-1}$$arcmin^{-2}$) in the area around the IVC can be well fitted by absorption of an soft X-ray background by the H I gas of the IVC (Engelmann 1993). Here the fitted background (10 · 10^{-4} counts $s^{-1}$$arcmin^{-2}$ at $T \simeq 10^6$ K) and foreground count rates (4.5·10^{-4} counts $s^{-1}$$arcmin^{-2}$) are consistent with the corresponding count rates found for the Draco nebula and LVC 88+36–2. This implies that the emission measures for the LHB and the galactic corona are also similar to those given above for the Draco nebula.

No detailed radiation transport solution has been modelled so far because the situation is complicated by enhancements of the X-ray count rates close to the edge of the IVC above the general off-cloud count rates. These enhancements are statistically significant (Engelmann 1993) and may be interpreted as rim-brightening of IVC 86+38–44. Although such an interpretation is rather ambiguous it is very attractive because similar phenomena are observed for the H I high velocity clouds discussed in the next section.

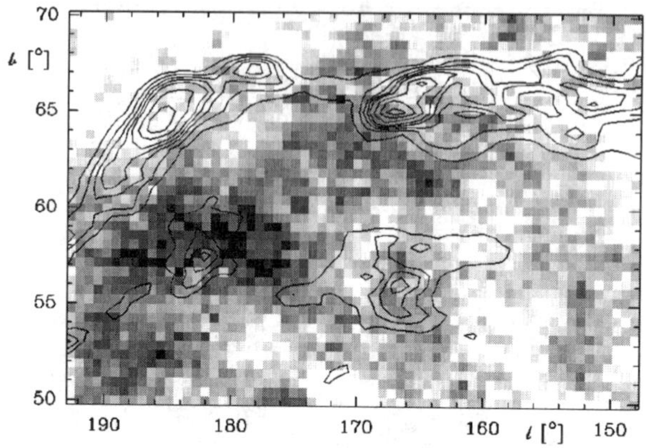

FIGURE 5 The *ROSAT* 1/4 keV survey data around the HVCs M I and M II (provided by S. Snowden) are represented as a grey scale image. The grey scale indicates values between $4.5 \cdot 10^{-4}$ (white) and $8.0 \cdot 10^{-4}$ counts^{-1}arcmin^{-2} (dark). The H I column density of the HVCs in the velocity range -145 to - 67 kms^{-1} is presented as contour plot. This H I image of Giovanelli et al. (1973) has been re-digitized and projected to the same coordinate system as the *ROSAT* image by Westphalen (1993). The column densities range from $\sim$ 2 (lowest contour) to $\sim 17 \cdot 10^{19}$ cm^{-2} (highest contour).

X-RAY EMISSION ASSOCIATED WITH HVC'S

A positional association of the chains C and A of H I high velocity clouds (HVC's) (cf., e.g., Wakker 1991) with spurs in the data of the Madison soft X-ray survey (McCammon et al. 1983) has already been pointed out by Hirth et al. (1985). Distances to HVC's are still a matter of much controversy. But most authors agree that they belong to the halo population of our Galaxy or are even members of the local group of galaxies (Wakker & van Woerden 1991; Kaelble et al. 1985).

HVC's form the most distant group of objects discussed in the present paper. The positional association of some HVC's and diffuse soft X-ray spurs found by Hirth et al. (1985) is confirmed by *ROSAT* survey data. The area to the NW of the Draco nebula (see above) is one example of such an association. The HVC's observed in this area (Giovanelli et al. 1973) are part of the HVC complex C whose association with an X-ray spurs was already found by Hirth et al. (1985).

Two further examples are the HVC's called M I and M II located at $(l, b, v) \sim (165°, 65°, -110\,\mathrm{kms}^{-1})$ and $(l, b, v) \sim (185°, 60°, -100\,\mathrm{kms}^{-1})$, respectively. Figure 5 shows the distribution of the soft X- ray count rates in the area of the two HVCs; the corresponding H I column density of the HVCs is shown as overlay (Giovanelli et al. 1973; Westphalen 1993). A comparison of the two distributions suggests that the X-ray emission looks like a combination of edge-brightening and shadowing in the case of M I. In the region of HVC M II the H I gas is associated with a patch of fairly strong soft X-ray emission $[8 \cdot 10^{-4}\mathrm{countss}^{-1}(\mathrm{arcmin})^{-2}]$.

If the suggestion of rim-brightening is supported by further observations, it holds important clues to the understanding of the nature of soft X-ray emission in the galactic halo. An anticorrelation of X-ray and H I emission for M I can be interpreted in various ways. Attractive hypotheses are: absorption of background emission by the neutral gas in the HVC, the HVC as the densest inner part of a larger complex of gas whose outer parts are hot enough to emit soft X-ray photons, or even magnetic reconnection at the boundaries of the clouds which move at rather large veolcities. However, a quantitative analysis of the data displayed in Fig. 5 cannot distinguish between these alternatives. Instead higher resolved X-ray and H I data with higher angular resolution are required (Wakker 1991b). From data of this kind we hope to get the required clues to the understanding of the nature of soft X-ray emission in the galactic halo.

CONCLUSIONS

In the previuos sections we have presented evidence that the diffuse soft X-ray distribution observed with the *ROSAT* PSPC instrument is very useful for the analysis of both the cold Galactic Cirrus clouds and the interspersed hot plasma. Indeed, the observations are consistent with an ubiquitous X-ray emitting plasma in which the neutral cirrus clouds are embedded.

First and closest to the observer, the Cirrus H I filament LVC 88+36–2 is found to be embedded in the plasma of the Local Hot Bubble (LHB). This finding raises questions about the origin and the lifetime of a dense neutral filament that is immersed in an X-ray emitting plasma and about the homogenity of the temperature and density distribution of the plasma in the LHB. It also opens up the opportunity to study the transition region between neutral gas at about 100 K and a fully ionized plasma at about 10^6 K, e.g. by searching for rim-brightening in the EUV range.

Next along the line of sight, we find the X-ray shadows of the Draco nebula and IVC 86+38–44 which imply the presence of large amounts of X-ray emitting gas in the background of the nebulae, i.e., in the galactic corona above $|z| \simeq$ 400 pc. Assuming that this background radiation is optically thin, the emission measure of the background gas is about 4 times larger than that of the LBH. However, analysis of other sources from Table 1 indicates that this corona is obviously not homogeneous but is rather patchy (Burrows & Mendenhall 1993). This is consistent with our finding of an association of IVC and HVC clouds with filaments of X-ray emission. The correlation of X-ray shadows with H I and IR images of molecular clouds is a new technique to derive the total column density of hydrogen nuclei for these clouds. We present results of an application of this technique to the molecular clumps of the Draco nebula. These results are consistent with those derived for the densest part of this nebula by Herbstmeier (this volume).

Farthest along the line of sight, we find evidence for an association of soft X-ray emission with the IVC 86+39–44, the H I high velocity clouds (HVCs) M I, M II and with the HVC complexes A and C in the northern sky. However, the nature of this association is not clear yet. The apparent anticorrelation may indicate either absorption of background emission or edge brightening of the H I clouds. A careful treatment of the attenuation of the background emission by all of the intervening material, i.e., not only the high or intermediate velocity gas, is not available yet. Also better angular resolution of both X-ray and H I data is required before the nature of this association can be established with a reasonable degree of reliability.

ACKNOWLEDGMENTS

We acknowledge the support for this work by Steve Snowden and J. Trümper from the Max Planck Institut für Extraterrestrische Physik in Garching. This work was supported by the German space agency DARA under project No. 50 OR 9203.

REFERENCES

Aschenbach, B. 1988, *Appl. Opt.*, **27(8)**, 1404

Blitz, L, Bazell, D., Désert, F.X. 1990, *ApJ*, **352**, L13

Bloemen, J.B.G.M., Deul, E.R., Thaddeus, P. 1990 *A&A*, **233**, 437

Bowyer, C.S., Field, G.B., Mack, J.E. 1968, *Nature*, **217**, 32

Burrows, D.N., Mendenhall, J.A. 1993, *Adv. Sp.Res.*, in press

Burrows, D.N., Mendenhall, J.A. 1991, *Nature*, **351**, 629

Denne, D.R. 1970, *J. of Phys.*, **3**, 1392

Désert, F.X., Bazell, D., Boulanger, F. 1988, *ApJ*, **334**, 815

Engelmann, J. 1993, *Diploma Thesis, University of Bonn*, in prep.

Frisch, P.L., York, D.G. 1983, *ApJ*, **271**, L59

Giovanelli, R., Verschuur, G.L., Cram, T.R. 1973, *A&A Suppl*, **12**, 209

Goerigk, W., Mebold, U., 1986, *A&A*, **162**, 279

Herbstmeier, U. 1989, in Structure and Dynamics of the Interstellar Medium, IAU Colloquium No. **120**, Lecture Notes in Physics, **350**, Ed. Tenorio-Tagle et al., Springer Berlin etc, 458

Herbstmeier, U., Moritz, P., Engelmann, J., Kerp, J., Westphalen, G. 1993, in *Back to the Galaxy*, Proc. of the 3rd Annual October Astrophysics Conference in Maryland, in press

Herbstmeier, U., Heithausen, A., Mebold, U. 1993, *A&A*, **272**, 514

Hirth, W., Mebold, U., Müller, P. 1985, *A&A*, **153**, 249

Hirth, W., Mebold, U., Dahlem, M., Müller, P. 1993, *AP&SS*, **186**, 211

Kaelble, A., de Boer, K.S., Grewing, M. 1985 *A&A*, **143**, 408

Kerp, J., Herbstmeier, U., Mebold, U. *A&A*, **268**, L21

Kramer, C. 1992, PhD Thesis, Universität zu Köln

Lilienthal, D., Hirth, W., Mebold, U., de Boer, K.S. 1992, *A&A*, **255**, 323

Lilienthal, D., Wennmacher, A., Herbstmeier, U., Mebold, U. 1991, *A&A*, **250**, 150

Low, F.J. et al. 1984, *ApJ*, **278**, L19

Marshall, F.J., Clark, G.W. 1984, *ApJ*, **287**, 633

Martin, C., Hurwitz, M., Bowyer, S. 1990, *ApJ*, **354**, 220

McCammon, D., Sanders, W.T. 1990, *Ann. Rev. Astr. Ap.*, **28**, 657

McCammon, D., Burrows, D.N., Sanders, W.T., Kraushaar, W.L. 1983, *ApJ*, **269**, 107

McGee, R.X., Haynes, R.F., Grognard, R.J.-M., Malin, D. 1986, *MNRAS*, 543

Mebold, U., Cernicharo, J., Velden, L., Reif, K., Crezelius, C., Goerigk, W. 1985, *A&A*, **151**, 427

Mebold, U., Heithausen, A., Reif, K. 1987, *A&A*, **180**, 213

Mebold, U., Herbstmeier, U., Kalberla, P.W.M., Souvatzis. I. 1989, in Structure and Dynamics of the Interstellar Medium, IAU Colloquium No. **120**, Lecture Notes in Physics, **350**, ed. Tenorio-Tagle et al., (Berlin: Springer), 458

Moritz, P. 1993, *PhD Thesis, University of Bonn*, in prep.

Morrison, R., McCammon, D. 1983, *ApJ*, **270**, 119

Pfeffermann E., et al. 1987, in *Soft X-ray Otics and Technology, Proc. SPIE*, **733**, ed. E.-E. Koch & G. Schmahl, 519

Snowden, S.L. 1993, in *Recent Results in X-ray and EUV Astronomy*, COSPAR-Symposium 1992 in Washington D.C. in press

Snowden, S.L., Burrows, D.N., Mendenhall, J.A. 1993, *ApJ*, in press

Snowden, S.L., Cox, D.P., McCammon, D., Sanders, W.T. 1990, *ApJ*, **354**, 211

Snowden, S.L., Freyberg, M.J. 1993, *ApJ*, **404**, 403

Snowden, S.L., Mebold, U., Hirth, W., et al. 1991, *Sci*, **252**, 1529

Snowden S.L., Plucinsky, P.P., Briel, U., Hasinger, G., Pfeffermann, E. 1992, *ApJ*, **393**, 819

Snowden, S.L., Schmitt, J.H.M. 1990, *AP&SS*, **171**, 207

Trümper, J., 1983, *Adv. Sp. Res.*, **2(4)**, 241

Voges, W. 1992, in *Proc. Satellite Symp.*, **3**, ed. T.D. Guyenne & J.J. Hunt (Noordwijk: ESA Publ. Div.), 9

Wakker, B.P. 1991, *A&A*, **250**, 499

Wakker, B.P. 1991b, *A&A Supl*, **90**, 495

Wakker, B.P., van Woerden, H. 1991, *A&A*, **250**, 509

Wennmacher, A., Lilienthal, D., Herbstmeier, U. 1992 *A&A*, **261**, L9

Westphalen, G. 1993 *Diploma Thesis, University of Bonn*

Zimmermann , H.U. et al. 1992, in *Data Analysis in Astronomy*, **IV**, ed. V. Di Gesù et al. (New York: Plenum), 141

The First Symposium on the Infrared Cirrus and Diffuse Interstellar Clouds
ASP Conference Series, Vol. 58, 1994
Roc M. Cutri and William B. Latter (eds.)

PHYSICAL CONDITIONS IN CLOUD ENVELOPES FROM MEASUREMENTS OF OPTICAL ABSORPTION LINES

S.R. FEDERMAN, C.J. STROM
Department of Physics and Astronomy, University of Toledo, Toledo, OH 43606

D.L. LAMBERT, V.V. SMITH
Department of Astronomy, University of Texas, Austin, TX 78712

JASON A. CARDELLI, C.L. JOSEPH
Department of Astronomy, University of Wisconsin, Madison, WI 53706

ABSTRACT Observations of absorption from CH, C_2, and CN were made toward moderately reddened stars lying behind the envelopes of the ρ Ophiuchus Molecular Cloud, the Cep A Molecular Cloud, and the dark, filamentary clouds in Taurus and Auriga. Our measurements were supplemented with available data for sight lines passing through the envelopes. The combined survey provides the basis for a detailed analysis of the physical conditions prevailing in the gas from considerations of molecular excitation and molecular chemistry. Of particular interest are gas density and temperature as well as the strength of the ultraviolet radiation field permeating the medium. Here, we emphasize our results for stars in the Sco OB2 association, which probe the envelope of the ρ Oph Molecular Cloud. Our analysis reveals that the regions probed by C_2 and CN absorption have a higher pressure than more extended portions of the envelope.

INTRODUCTION

Physical conditions control the processes occurring in interstellar clouds. In particular, density, temperature, and radiative flux determine which routes are effective for molecular production and how individual atomic and molecular levels are populated. Observations of atoms and molecules provide a way to determine the processes that are taking place and to extract the physical conditions responsible for these processes. In this study, knowledge about the physical conditions is obtained from analysis of abundances for CH, C_2, and CN and of the distribution among rotational levels in C_2 and CN. The focus is on the ρ Oph Molecular Cloud because it is one of the closest molecular clouds and because its envelope is sampled by absorption from both lightly and moderately reddened stars. The combined study of several envelopes appears in Federman et al. (1993b).

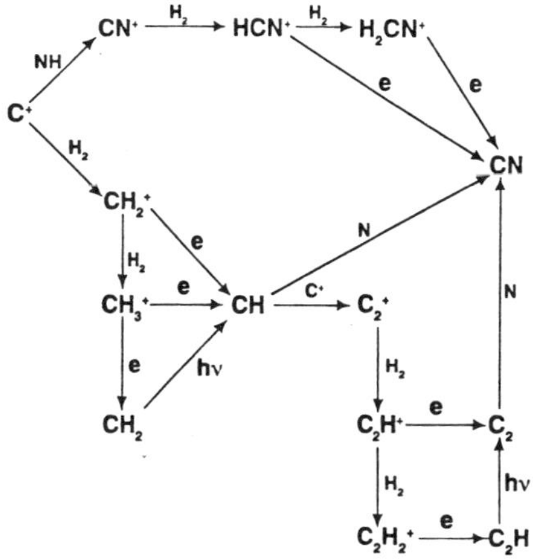

FIGURE I Chemical network

OBSERVATIONS

Moderately reddened stars in the Sco OB2 association with A_V of 2 - 4 mag were observed with the coudé spectrograph of the 2.1 m telescope at McDonald Observatory. There were separate setups for CH at 4300 Å, for C_2 at 8760 Å, and for CN and CH at 3880 Å. The stellar spectra were imaged onto either an RCA CCD with 30 μm pixels or a Tektronics CCD with 27 μm pixels. The nominal resolution was 10 km s^{-1}; it must be noted that the aim of this study was not to obtain accurate velocities, but instead to seek absorption for more opaque sight lines than were obtained in previous work. Whenever absorption was detected, the velocity at line center was consistent with measurements with higher spectral resolution. The observed equivalent widths were converted into column densities with a curve of growth based on a Doppler parameter of 1 km s^{-1}.

DISCUSSION

Initial values for density and temperature for the chemical analysis were taken from available information on C_2 and CO excitation (e.g., Danks & Lambert 1983; van Dishoeck & de Zeeuw 1984; van Dishoeck et al. 1991). Refinements to these values were obtained in the chemical analysis, where a comparison was made between observations and theoretical models based rate equations in the form of column densities (see Federman et al. 1984; Federman & Huntress 1989 for earlier versions). In these calculations the *observed* column densities of progenitor species was used. The chemical network for this analysis appears in Fig. 1.

The results of our analysis for the envelope of the ρ Oph Molecular Cloud appear in the following Table. The respective columns list the star, density

Star	n	T	τ_{uv}	I_{uv}	N(CH) obs	$N(C_2)$ obs	$N(C_2)$ pred	$\frac{obs}{pred}$	N(CN) obs	N(CN) pred	$\frac{obs}{pred}$
	cm^{-3}	K			cm^{-2}	cm^{-2}	cm^{-2}		cm^{-2}	cm^{-2}	
HD147889	700	40	4.73	1	1.20(14)[a]	1.20(14)	1.17(14)	1.03	3.50(13)	4.13(13)	0.85
HD147343	700	40	2.78	1	6.40(13)	...	6.26(13)	...	9.30(12)	9.92(12)	0.94
HD147648	600	50	3.86	5	6.50(13)	...	1.86(13)	...	<3.90(12)	3.62(12)	1.08
HD147701	1000	50	3.17	1	5.90(13)	5.40(13)	6.67(13)	0.81	1.50(13)	1.40(13)	1.08
o Sco	600	50	3.21	1	5.00(13)	5.50(13)	4.15(13)	1.32	6.00(12)	9.01(12)	0.67
ρ Oph A	400	50	2.04	1	2.40(13)	2.60(13)	2.25(13)	1.15	2.00(12)	2.34(12)	0.86
ρ Oph C	600	40	2.09	1	2.10(13)	...	2.66(13)	...	2.70(12)	2.86(12)	0.94
χ Oph	400	60	2.30	1	3.40(13)	3.50(13)	2.39(13)	1.47	1.30(12)	3.12(12)	0.42
ζ Oph	400	60	1.98	1	2.50(13)	2.20(13)	2.44(13)	0.90	2.60(12)	2.46(12)	1.06

Notes: [a] $1.20(14) = 1.20 \times 10^{14}$.

(n), kinetic temperature (T), optical depth for grains at 1000 Å (τ_{uv}), a factor to account for enhancements in the ultraviolet flux over the average interstellar value (I_{uv}), observed and predicted column densities and their ratios. The observational data are a compilation of our measurements and those available from previous studies (Danks & Lambert 1983; Danks et al. 1984; Federman et al. 1984; van Dishoeck & de Zeeuw 1984; Cardelli & Wallerstein 1986; van Dishoeck & Black 1989; Gredel et al. 1991). For the most reddened stars, the optical depth, τ_{uv}, is reduced to take into account the larger-than-average grain size deduced from photometric studies (Carrasco et al. 1971; Whittet 1974; Vrba et al. 1975) and observations of lower ultraviolet extinction (Bohlin & Savage 1981; Cardelli et al. 1989; Fitzpatrick & Massa 1990; Snow et al. 1990; Green et al. 1992).

For the envelope of the ρ Oph Molecular Cloud, the pressure [n_H T, where $n_H = n(H) + 2\,n(H_2)$] appears to be constant in the regions sampled by C_2 and CN, regardless of the visual extinction, A_V. The density is somewhat higher and the temperature is correspondingly lower in the more opaque sight lines (e.g., HD147889) relative to the conditions derived for sight lines sampling material permeated by a stronger radiation field (e.g., ζ Oph). The variation in density and temperature under pressure equilibrium is the result of a variation in photoelectron heating caused by the attentuation of the radiation. The pressure found for the regions sampled by C_2 and CN is about a factor of 2 larger than that obtained from studies of CO isotopes and excitation, as well as C **I** excitation (Sheffer et al. 1992; Lambert et al. 1993). Since CO shields itself from dissociating radiation, it occupies a more extended volume than either C_2 or CN, whose chemistries are very sensitive to density and ultraviolet radiative flux. The atomic gas probed by C **I** also appears more widely distributed (Federman et al. 1993a). These facts suggest that the outer regions of the cloud envelope have lower pressure than the denser inner clumps and thus, the envelope might be expanding. Based on 21 cm observations, Andersson et al. (1992) found a similar situation for the atomic envelope around the molecular cloud B5.

REFERENCES

Andersson, B.-G., Roger, R.S., & Wannier, P.G. 1992, *A&A*, **260**, 355

Bohlin, R.C., & Savage, B.D. 1981, *ApJ*, **249**, 109

Cardelli, J.A., Clayton, G.C., & Mathis, J.S. 1989, *ApJ*, **345**, 245

Cardelli, J.A., & Wallerstein, G. 1986, *ApJ*, **302**, 492

Carrasco, L., Strom, S.E., & Strom, K.M. 1973, *ApJ*, **182**, 95

Danks, A.C., Federman, S.R., & Lambert, D.L. 1984, *A&A*, **130**, 62

Danks, A.C., & Lambert, D.L. 1984, *A&A*, **124**, 188

Federman, S.R., Danks, A.C., & Lambert, D.L. 1984, *ApJ*, **287**, 219

Federman, S.R., & Huntress, W.T. 1989, *ApJ*, **338**, 140

Federman, S.R., Sheffer, Y., Lambert, D.L., & Gilliland, R.L. 1993a, *ApJ*, in press

Federman, S.R., Strom, C.J., Lambert, D.L., Cardelli, J.A., Smith, V.V., & Joseph, C.L. 1993b, *ApJ*, in press

Fitzpatrick, E.L., & Massa, D. 1990, *ApJS*, **72**, 163

Gredel, R., van Dishoeck, E.F., & Black, J.H. 1991, *A&A*, **251**, 625

Green, J.C., Snow, T.P., Cook, T.A., Cash, W.C., & Poplawski, D. 1992, *ApJ*, **395**, 289

Lambert, D.L., Sheffer, Y., Gilliland, R.L., & Federman, S.R. 1993, *ApJ*, submitted

Sheffer, Y., Federman, S.R., Lambert, D.L., & Cardelli, J.A. 1992, *ApJ*, **397**, 482

Snow, T.P., Allen, M.M., & Polidan, R.S. 1990, *ApJ*, **359**, L23

van Dishoeck, E.F., & Black, J.H. 1989, *ApJ*, **340**, 273

van Dishoeck, E.F., Black, J.H., Phillips, T.G., & Gredel, R. 1991, *ApJ*, **366**, 141

van Dishoeck, E.F., & de Zeeuw, T. 1984, *MNRAS*, **206**, 383

Vrba, F.J., Strom, K.M., Strom, S.E., & Grasdalen, G.L. 1975, *ApJ*, **197**, 77

Whittet, D.C.B. 1974, *MNRAS*, **168**, 371

The First Symposium on the Infrared Cirrus and Diffuse Interstellar Clouds
ASP Conference Series, Vol. 58, 1994
Roc M. Cutri and William B. Latter (eds.)

A COMPARISON OF THE DIFFUSE CLOUDS IN FRONT OF ζ OPHIUCHI AND σ SCORPII

THEODORE P. SNOW
University of Colorado

JACEK KREŁOWSKI
Nicolaus Copernicus University

ABSTRACT The well-studied stars ζ Ophiuchi and σ Scorpii have gross similarities in visual interstellar extinction (color excesses $E_{B-V} = 0.32$ and 0.40, respectively), but significant contrasts in many other interstellar parameters. The extinction curve for σ Sco is much flatter than that for ζ Oph, so that the former has a ratio of total to selective extinction of $R_V = 3.80$ while the latter has $R_V = 3.09$, close to the canonical diffuse ISM value of 3.1. In the ultraviolet the contrast is especially severe; ζ Oph has a very steep far-UV extinction rise and a somewhat broadened 2175 Å bump, while σ Sco has very low far-UV extinction and a narrower bump. One result is that even though σ Sco has a significantly greater total gas column density ($N_H = 2.3 \times 10^{21}$ cm^{-2} versus 1.4×10^{21} cm^{-2} for ζ Oph), the internal cloud radiation field intensity is more intense for σ Sco. The molecular abundances for the two clouds are very different, with ζ Oph exhibiting much higher column densities of species such as H_2, CH, and CN (and probably CO). There are marked contrasts between the two clouds in terms of diffuse interstellar band behavior as well. DIBs known to belong to different families (e.g., the bands at 5780 and 5797 Å) show very different relative strengths; this information, combined with the contrasts in dust properties and molecular abundances, may prove helpful in the search for the carriers of the DIBs.

In this paper, we present new observational results on optical interstellar lines toward ζ Oph and σ Sco, including data on the DIBs. We also bring together existing data on atomic and molecular abundances, and re-analyse the depletions of the two lines of sight with the goal of specifying how the differing optical properties of the grains may be related to the depletion process. We include some discussion of the possible evolutionary histories of the two clouds, and comment on how the properties and histories of these clouds may be related to the properties of the infrared cirrus clouds.

INTRODUCTION

The diffuse clouds toward ζ Ophiuchi and σ Scorpii are among the most thoroughly studied interstellar lines of sight, and are often cited as prototypes of standard diffuse cloud properties. Many assumptions about ultraviolet extinction curves, depletions, and chemical processes that are based on these stars (particularly ζ Oph) are applied to other clouds. Yet there can be very significant contrasts in physical and optical properties of diffuse clouds which may appear superficially similar, and it turns out that σ Sco and ζ Oph represent extreme examples. Thus it is important to understand the contrasts between these two clouds, as an aide toward gaining a better understanding of the processes that govern the evolution of diffuse clouds and the interstellar dust. Thus despite the large body of data and the extensive literature available for both σ Sco and ζ Oph, we find it worthwile to reexamine them in a comparative study.

Table 1 summarizes the comparison between the material observed toward ζ Ophiuchi and σ Scorpii. Here we see that the two are at similar distances and have comparable reddenings, although σ Sco has a significantly higher ratio of total to selective extinction ($R = 3.80$ versus 3.09 for ζ Oph) and hence a larger total extinction A_V.

TABLE 1 Comparison of ζ Ophiuchi and σ Scorpii

	ζ Oph	σ Sco
Spectral Type	O9.5Vn	B1III
Distance	138 pc	142 pc
Visual Magnitude	2.56	2.90
E_{B-V}	0.29	0.34
A_V	0.99	1.50
R_V	3.09	3.80
N(H I)	5.2×10^{20}	(2.2×10^{21})
$N(H_2)$	4.4×10^{20}	(6.1×10^{19})
N_H (Total)	1.4×10^{21}	(2.3×10^{21})
$f = \frac{2N(H_2)}{N(HI)+2N(H_2)}$	0.63	0.053
Average n_H	3.3×10^{-3}	5.3×10^{-3}
N(CN)	4.8×10^{12}	$\leq 1.3 \times 10^{11}$
N(CH)	3.4×10^{13}	4.5×10^{12}
$N(CH^+)$	9.3×10^{12}	5.0×10^{11}

THE DUST

One of the most striking contrasts between the two clouds arises in connection with the optical properties of the dust. In Fig. 1 we show the full extinction curves (from near-IR through far-UV) of the two stars (the IR and broad-band UV *TD-1* data are from Wegner, Papaj, and Krełowski (1991); the continuous far- UV extinction curves are from Bless and Savage (1972). In this figure we see that there are major differences, in that ζ Oph has a much steeper far-UV extinction rise and a somewhat broader 2175 Å bump, and there is a difference of slope in the visual and near-IR regions. These differences are further indicated by Fig. 2, in which the polarization as a function of wavelength is shown for both

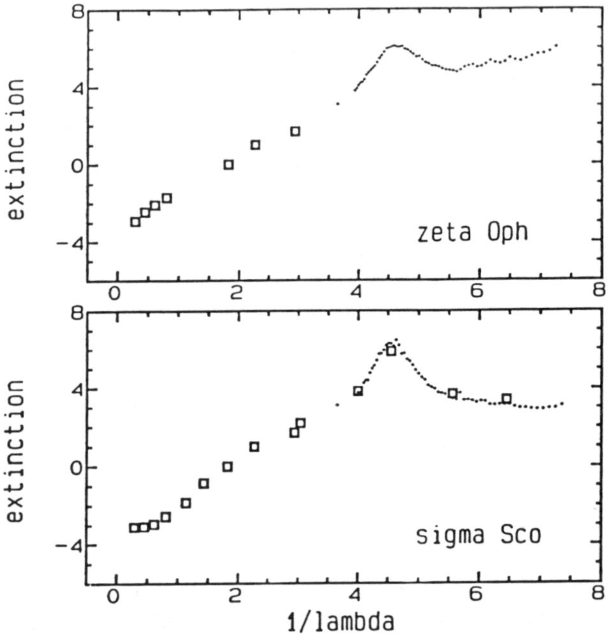

FIGURE 1 Extinction curves for ζ Oph and σ Sco. These represent a combination of ground-based optical and near-IR spectra, along with broad-band UV photometry obtainted by the *TD-1* satellite.

stars, based on data from Gehrels and Sylvester (1965), Coyne and Gehrels (1966), Gehrels (1974), Serkowski, Mathewson, and Ford (1975), Wilking et al. (1980), and Bailey and Hough (1982). Here we see that the wavelength of maximum polarization for σ Sco is displaced longward of the peak for ζ Oph.

Both the shape of the extinction curve and the location of the peak polarization suggest a contrast in dust grain size distribution, with the σ Sco cloud containing relatively fewer small grains than the ζ Oph line of sight. Comments on how this may have arisen will be found below.

The *IRAS* broad-band IR data have been examined, with results shown in Fig. 3. Here we see curves showing the best fit to the grain temperatures (provided through the courtesy of J. Saken) at points close to each star. In both cases there is visible nebulosity (cirrus) near the star, and in both cases we have derived the dust temperature for a two-pixel by two-pixel region (i.e., about 4′ by 4′) located within the cirrus (about 6′ away from ζ Oph and 8′ away from σ Sco). A two-component fit was made to the data, showing similar results: in both cases there is evidence for a warm component ($T = 168\ K$ for ζ Oph and $T = 164\ K$ for σ Sco) and a cool component ($T = 44.5\ K$ and $38.2\ K$, respectively). The warm dust is derived largely from the 12/25 μm ratio, while the cool dust is probed primarily by the 60/100 μm ratio. In most interpretations, the warm dust is thought to consist of very small grains, small enough to be heated by

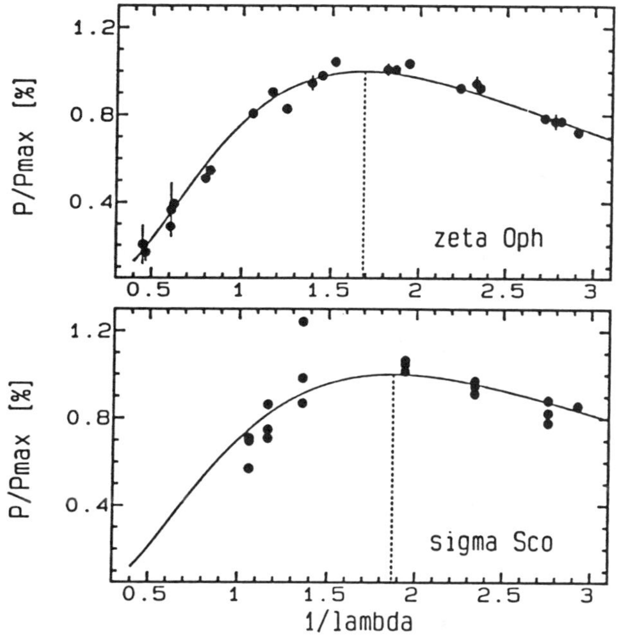

FIGURE 2 Polarization curves for ζ Oph and σ Sco. These are based on compilations from several authors, , and show that λ_{max}, the wavelength of maximum polarization, is longer in the line of sight toward σ Sco than toward ζ Oph.

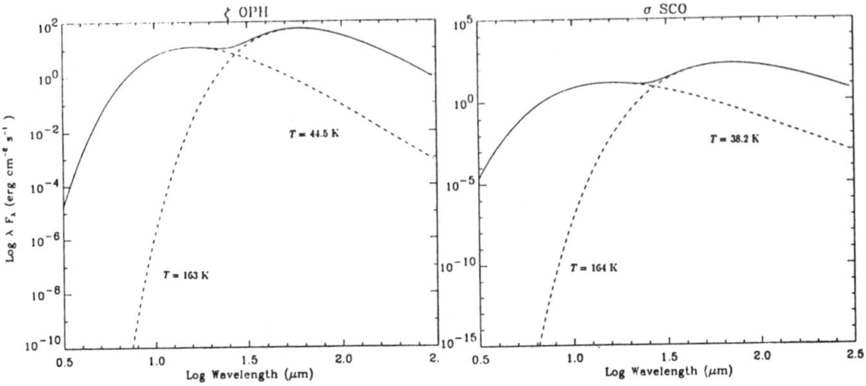

FIGURE 3 Two-component fits to the dust temperature near ζ Oph and σ Sco. The cooler of the two temperatures is based on the ratio of 60μm to 100μm flux, while the warmer temperature is based on the ratio of 12μm to 25μm fluxes. It is thought that the cooler temperature represents the relatively large grains which have constant temperature, while the warmer temperature represents the fraction of very small grains that are temprarily heated by single-photon absorption.

single-photon absorption. If this is correct, it is interesting and perhaps a bit surprising that the σ Sco vicinity shows strong evidence for the presence of such grains, in view of the foregoing discussion of the extinction and polarization curves. This may imply that the small grains that produce the 12-μm emission are not the same as those which produce the far-UV extinction, or that the region we sampled in the *IRAS* survey is physically distinct from the region probed by absorption spectroscopy of the sightline toward σ Sco itself.

ATOMIC ABUNDANCES

Extensive and detailed measurements of interstellar absorption lines toward both stars have been published (e.g., Herbig 1968; Shulman, Bortolot, and Thaddeus 1969; Morton 1974, 1975; Snow and Meyers 1979 for ζ Oph, and Allen, Snow, and Jenkins 1991 for σ Sco). From measurements of the dominant ionization state, which require ultraviolet observations (done by the *Copernicus* satellite for both stars), a pattern of abundances and depletions is derived that shows a general similarity between the two clouds, except that for some elements the depletion appears to be greater for ζ Oph.

This tendency becomes more clear-cut when lines of neutral atoms are examined, as was done by Snow (1984) and by Snow, Joseph, and Meyers (1986). In these studies, ratios of neutral atomic column densities were compared in order to infer relative column densities of the more abundant ions; this involves some approximations, but much of the uncertainty in the ionization balance cancels out. The abundance ratios and depletions that are inferred differ from those derived from direct measurements of the ions in that they refer specifically to the portion of the line of sight where the neutral atoms are most abundant; i.e., the densest portion. Thus from the observations of neutral atoms it is possible to gain information on depletions in diffuse cloud cores, as opposed to depletions integrated over the entire line of sight.

These studies show greater depletions of calcium and iron toward ζ Oph than toward σ Sco; the typical difference is nearly a full order of magnitude. This, combined with the tendency for some ions to show greater depletions toward ζ Oph, indicates that significantly greater depletion has occurred in the ζ Oph cloud. This would be consistent with a greater relative abundance of small grains in this cloud, since small grains provide more surface area (per unit of visual extinction) for the accretion of gas particles.

MOLECULAR ABUNDANCES

Another very striking contrast between the two clouds is found in their relative column densities of molecular species. First, the molecular fraction of hydrogen is far smaller for σ Sco than for ζ Oph (only 0.053 of the hydrogen nuclei are bound into H_2 molecules toward σ Sco, as compared with 0.63 toward ζ Oph). We find also that other species, notably CN and CH, are less abundant toward σ Sco. Clearly the chemical conditions are very different.

The explanation of this contrast may invoke aspects of the dust and of the environment. It is believed that H_2 forms on the surfaces of dust grains, and it was pointed out by Snow (1983) that the low abundance of H_2 in the ρ Oph cloud (where σ Sco resides) may be explained by the relatively small grain surface area (per unit of extinction) that is available there.

The enhanced far-ultraviolet radiation field intensity must also play a role. The contrast in far-UV extinction between the two clouds shows that far-UV photons will penetrate the σ Sco cloud much more readily, thus bathing its interior with radiation capable of dissociating molecules.

In addition, CH, CN, and most other molecular species are linked to the abundance of H_2 through the ion-molecule reaction sequence that is responsible for most of the chemistry in diffuse clouds (e.g., van Dishoeck and Black 1986). Therefore a reduced H_2 abundance in the σ Sco cloud must play a role in keeping the column densities of the other species low as well.

Finally, we note that the CH^+ abundances in the two lines of sight are are also very different, again with the higher abundance in the ζ Oph cloud. In this case the chemistry is very different than that which produces the other molecules discussed, so the reason for the enhancement toward ζ Oph is not clear. There is an extensive literature on the problem of CH^+ formation, which variously invokes shocks, turbulence, and radiative formation mechanisms (for recent reviews, see Gredel, van Dishoeck, and Black 1993; Snow 1993; and Duley et al. 1992). It is noteworthy that despite the presumed higher UV radiation field intensity in the σ Sco cloud than in the ζ Oph line of sight, σ Sco does not show an enhanced CH^+ abundance, as might be expected if radiative effects dominated its formation. Instead the CH^+ column density toward σ Sco is an order of magnitude smaller than toward ζ Oph. On the other hand, it is possible that the CH^+ exists on the outer boundaries of the clouds, where the internal shielding due to dust is insignificant, so that the contrasts in extinction curve properties may not affect the formation of CH^+.

DIFFUSE BANDS

Modern studies of the unidentified diffuse interstellar bands (DIBs) have shown that there are subgroups, or families, of bands which exhibit similar behavior and are therefore assumed to arise from a common carrier (Krełowski and Westerlund 1988; Josafatsson and Snow 1987). The strong bands at 5780 Å and 5797 Å belong to two different families, and thus are probably formed by distinct carriers. They show very different behavior in the ζ Oph and σ Sco clouds, the 5780 Å band being deeper than λ5797 in the spectrum of σ Sco and shallower than λ5797 in the spectrum of ζ Oph. If all the underlying causes of the contrasts between the two clouds can be understood, this may aide in the quest to identify the diffuse band carriers.

DO OTHER CLOUDS EXHIBIT SIMILAR CONTRASTS?

The foregoing discussion has been focused entirely on the two clouds toward ζ Oph and σ Sco, but in fact there are many other examples similar to each. Thus any explanation of the contrasts may have broader applicability, and it is therefore important to attempt to develop such an explanation.

The work of Bless and Savage (1972) and, more recently, of Fitzpatrick and Massa (1986, 1988, 1990) has shown that ultraviolet extinction curves exhibit a wide range of behavior. Most stars are intermediate between the characteristics represented by ζ Oph and σ Sco, but other examples are found that are equally or even more extreme. Parameterization of the extinction curve shape shows that in most cases the entire curve can be derived from a single property, the ratio of total to selective extinction A_V (Cardelli, Clayton, and Mathis 1988, 1989). But σ Sco is an exception, in that the low curvature of the far-UV portion of its extinction curve is not predicted by its value of R_V. This reinforces the notion that something physically different has occurred to modify the dust in this cloud, and that it cannot be viewed simply as having a different mixture of the same grain types as are found in other clouds. The ζ Oph extinction curve is more closely fit by the Cardelli et al. formalism, although some deviation may be present.

On the generality of the contrast in depletions between the two clouds, little quantitative work has been done. But the study by Joseph et al. (1986) did show a tendency for stars having steep far-UV extinction to have greater depletions, and the more recent work by Cardelli, Federman, and Smith (1991) argued that in cloud cores where neutral atoms predominate, the is a greatly enhanced level of depletion.

The general trend toward enhanced molecular abundances in clouds with steep far-UV extinction is very clear, and is linked to the value of R_V (Cardelli 1988). This is not surprising, in view of the relatively low UV radiation intensity in such clouds, but may also be due in part to the preponderance of small grains and the correspondingly high surface area available for H_2 formation.

Lastly, the DIBs exhibit consistent behavior similar to that seen in the ζ Oph and σ Sco clouds. It has been shown by Krełowski et al. (1992) that in other clouds the 5780 Å band appears deeper than λ5797 when the far-UV extinction is low, and shallower when the far-UV extinction is steep.

Thus the ζ Oph and σ Sco clouds are not isolated cases of extreme and contrasting properties, but instead may be viewed as prototypes for general classes of behavior in diffuse clouds. This enhances the interest in finding an explanation for the contrasts.

A POSSIBLE SCENARIO

It is suggested that the principal reason for the differences between ζ Oph-like and σ Sco-like clouds is their differing prior histories.

We suggest that a ζ Oph-like cloud is an undisturbed, relatively isolated cloud which has not been the site of recent star formation. Previous shock processing has fragmented the dust grains so that a significant population of very small grains is mingled with the larger dust, thus producing an extinction curve with a steep far-UV rise and relatively small values of R_V and λ_{max} for polarization. The steep UV extinction, coupled with the presence of small grains, provides a relatively high available surface area for the formation of H_2 and for the depletion of elements from the gas phase onto the grains. The low UV radiation field, along with the high production rate for H_2, favors the production of other molecular species. These conditions, by mechanisms as yet unknown, also favor the λ5797 DIB over the band at 5780 Å.

On the converse, in a σ Sco-like cloud, we envision a prior (or concurrent) phase of star formation occurring within the cloud. During the phase of cloud collapse and star formation, dust grains coagulated, removing tiny grains at the expense of larger ones; this has been specifically proposed for the ρ Oph cloud by Jura (1980). This alters the overall grain size distribution in such a way that the grains are not effective scatterers at far-UV wavelengths, accounting for the shallow far-UV extinction. At the same time, the presence of young stars within the cloud, coupled with the low far-UV extinction, strongly enhances the radiation field intensity. This, in combination with the relatively low grain surface area available for H_2 formation, inhibits the formation of molecules. Depletion is also inhibited due to the low grain surface area; it may be that the bulk of the observed depletions in these clouds is attributable to the formation process, or some other prior stage in the evolution of the dust. Finally, these conditions, for reasons not yet understood, favor the DIB at 5780 Å over the one at 5797 Å.

Thus the prior history of the diffuse clouds may account entirely for the observed contrasts, both by creating initial differences in grain properties and by creating environmental factors which drive the evolution of the gas and dust in different directions. To further test this hypothesis, we are continuing a large-scale program to assemble as wide a range of data on as many stars as possible, to see whether correlations among cloud properties continue to follow the trends seen so far. This work will include a search for evidence of grain mantle formation, which we expect will be favored in the ζ Oph-like clouds, where the high abundance of small grains provides the necessary surface are for the accretion of mantles.

There may well be additional factors that complicate the situation. For example, the shock history of a cloud should come into play in modifying the grain size distribution; this could explain how separate clouds within the same complex could have very different properties (e.g., the contrasts between o Persei and ξ Persei; the former lies behind a ζ Oph-like cloud while the latter resembles the sightline toward σ Sco).

ACKNOWLEDGMENTS

We are grateful for the assistance of J. Saken (Colorado) in the analysis of the *IRAS* data for the regions near ζ Oph and σ Sco. TPS is also thankful for the hospitality at the Canada-France-Hawaii Telescope, where the ground-based spectra were obtained, and for the hospitality of the Research Centre for Theoretical Astrophysics of the University of Sydney, where he visited on sabbatical while some of this work was done.

REFERENCES

Allen, M. M., Snow, T. P. and Jenkins, E. B. 1991, *ApJ*, **355**, 130

Bailey, J. and Hough, J. H. 1982, *PASP*, **94**, 618

Bless, R. C. and Savage, B. D. 1972, *ApJ*, **171**, 293

Cardelli, J. A. 1988, *ApJ*, **335**, 177

Cardelli, J. A., Clayton, G. C., and Mathis, J. S. 1988, *ApJ*, **329**, L33

Cardelli, J. A., Clayton, G. C., and Mathis, J. S. 1989, *ApJ*, **345**, 245

Cardelli, J. A., Federman, S. R., and Smith, V. V. 1991, *ApJ*, **381**, L17

Coyne, G. V. and Gehrels, T. 1966, *AJ*, **71**, 355

Duley, W. W., Hartquist, T. W., Sternberg, A., Wagenblast, R., and Williams, D. A. 1992, *MNRAS*, **255**, 463

Gehrels, T, 1974, *AJ*, **79**, 590

Gehrels, T. and Sylvester, A. B. 1965, *AJ*, **70**, 579

Herbig, G. H. 1968, *Zs. f. Ap.*, **68**, 243

Fitzpatrick, E. L. and Massa, D. 1986, *ApJ*, **307**, 286

Fitzpatrick, E. L. and Massa, D. 1988, *ApJ*, **328**, 734

Fitzpatrick, E. L. and Massa, D. 1990, *ApJS*, **72**,163

Gredel, R., van Dishoeck, E. F., and Black, J. H. 1993, *A&A*, *in press*

Josafatsson, K. and Snow, T. P. 1987, *ApJ*, **319**, 436

Joseph, C. L., Snow, T. P., Seab, C. G., and Crutcher, R. M. 1986, *ApJ*, **309**, 771

Krełowski, J. 1992, *Acta. Astr.*, **42**, 211

Krełowski, J., Snow, T. P., Seab, C. G., and Papaj. J. 1992, *MNRAS*, **258**, 693

Krełowski, J. and Westerlund, *A&A*, **190**, 339

Morton, D. C. 1974, *ApJ*, **193**, L35

Morton, D. C. 1975, *ApJ*, **197**, 85

Shulman, S., Bortolot, and Thaddeus 1969

Snow, T. P. 1983, *ApJ*, **269**, L57

Snow, T. P. 1984, *ApJ*, **287**, 238

Snow, T. P. 1993, *ApJ*, **403**, L73

Snow, T. P., Joseph, C. L., and Meyer, D. M. 1986, *ApJ*, **303**, 433

Snow, T. P. and Meyers, K. A. 1979, *ApJ*, **229**, 545

van Dishoeck, E. F. and Black, J. H. 1986, *ApJS*, **62**, 109

Wegner, W., Papaj, J., and Krełowski, J. 1991, *Acta. Astr.*, **41**, 149

The First Symposium on the Infrared Cirrus and Diffuse Interstellar Clouds
ASP Conference Series, Vol. 58, 1994
Roc M. Cutri and William B. Latter (eds.)

THE SIZE DISTRIBUTION OF INTERSTELLAR DUST PARTICLES AS DETERMINED FROM EXTINCTION

SANG-HEE KIM[1]
Department of Astronomy, University of Toronto, Toronto, ON, Canada M5S 1A7

P. G. MARTIN[1]
Canadian Institute for Theoretical Astrophysics, University of Toronto, Toronto, ON, Canada M5S 1A7

PAUL D. HENDRY
Department of Astronomy, University of Toronto, Toronto, ON, Canada M5S 1A7

ABSTRACT Variations in the shape of the interstellar extinction curve from place to place in the Galaxy indicate that the size distribution of interstellar dust particles is also changing. We have investigated the character of these changes and illustrate them with two contrasting cases: the diffuse interstellar medium ($R_V = 3.1$), and a dense cloud region ($R_V = 5.3$). For this exploratory study we adopted spherical bare silicate and bare graphite particles as in the modeling by Mathis, Rumpl, & Nordsieck (MRN). To extract the size distributions of the two compositional components as objectively as possible we used the Maximum Entropy Method, with an observational abundance constraint implemented.

For the diffuse interstellar medium we confirm the qualitative features of the MRN distribution, a simple power law out to a sharp cutoff at $a_+ = 0.25$ μm. However, we now show how the size distribution falls off smoothly beyond a_+, perhaps compatible with an exponential cutoff. Compared to the case of the diffuse interstellar medium, the size distribution of a denser cloud region has a significant reduction in the number of smaller particles (< 0.1 μm) and a more modest increase at larger sizes.

INTRODUCTION

Changes in the wavelength dependence of interstellar extinction with environment have been placed on a systematic basis by Cardelli, Clayton, & Mathis (1989), leading to the hope that corresponding systematic changes in the size distribution of interstellar dust particles might be quantified. They showed that the entire extinction curve could be parameterized by the ratio of total to selective extinction, R_V, and we investigate two representative cases, 3.1 (diffuse

---

[1]Also Theoretical Astrophysics, California Institute of Technology, Pasadena, CA 91106

clouds) and 5.3 (dense clouds).

FITTING THE EXTINCTION

The extinction produced along any line of sight is given by

$$A_\nu/1.086 = \int dl \int_0^\infty \pi a^2 Q_{e,\nu}(a) n(a)\, da \;=\; \int dl \,\frac{3}{4\rho}\int_0^\infty \frac{Q_{e,\nu}(a)}{a} m(a)\, da$$
$$\simeq \int dl\, \frac{3}{4\rho} \sum_i m_i \int_{a_i}^{a_{i+1}} \frac{Q_{e,\nu}(a)}{a}\, da\ .$$

Usually, the size distribution has been expressed in terms of $n(a)$, the number of particles per unit volume in the size interval a to $a + da$, but there are important technical and astrophysical reasons to prefer $m(a)$, the corresponding mass. The unknowns m_i are to be determined by fitting the predicted extinction to that observed at a number of different frequencies; we used the range $0.2 - 10\ \mu m^{-1}$. The goodness of fit is measured by the usual (weighted) χ^2.

In the fitting process within MEM the smoothest possible solution near some prescribed default and consistent within some target χ^2 is obtained by maximizing the Shannon-Jaynes form of entropy. In practice the entropy is written:

$$S_M = \sum_i w_i m_i \left[1 - \ln(f_i/d_i)\right]\ .$$

Here m_i is the discretized function to be obtained. The default (or template) d_i is some smooth function containing minimal information about the desired solution. The w_i are simply weights.

Constraints were applied to the total mass of each material so as not to use more than the cosmically available abundances; this requires knowing the amount of extinction per unit hydrogen column density, A_ν/N_H.

INTERSTELLAR SIZE DISTRIBUTIONS

Diffuse Cloud: $R_V = 3.1$

The size distributions derived for silicate and graphite components are displayed in Figure 1. In this logarithmic representation we plot $m(a)a$ (dimension mass), normalized to the mass of hydrogen in the same volume. Also shown for comparison are the MRN (Mathis, Rumpl, & Nordsieck, 1977) power laws. To a first approximation our MEM size distributions resemble power laws with an exponential cutoff; the overall slope found by MRN is roughly reproduced.

At intermediate sizes ($0.02 - 0.2\ \mu m$), the MEM size distributions have more structure than a simple power law. This structure is robust and necessary to achieve a good fit to the data at U, B, and V where the extinction curve changes slope.

The sharp cutoff in the MRN distribution at a_+ is clearly unphysical and is of concern since $m(a)a$ is still rising. Our MEM result shows a smooth decrease beginning at about $0.2\ \mu m$; this decrease out to $\sim 1.0\ \mu m$ is forced on the solution by the extinction data.

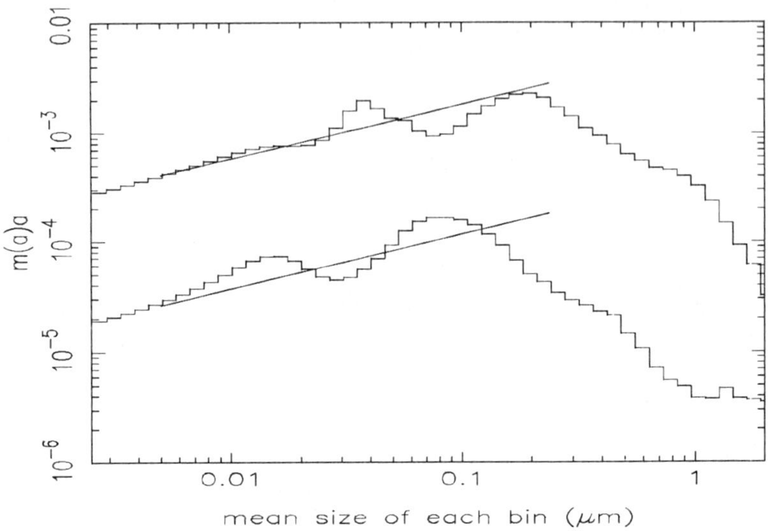

FIGURE 1 The resultant mass distribution for the case of $R_V = 3.1$. The upper bin is for silicate and the lower bin is for graphite (plotted 10 times lower for clarity). The two solid lines are the corresponding MRN results.

For small grains ($< 0.02\ \mu$m), determining the shape of the size distribution from the extinction is not possible, basically because such small particles only absorb radiation and $Q_{e,\nu}(a)/a$ is constant with a. Their total mass is constrained but the size distribution is just a replica of the shape of the default. This was chosen to extrapolate the size distribution smoothly to the smallest particles (Fig. 1); while cosmetically pleasing, this is only suggestive.

Dense Cloud: $R_V = 5.3$

The extinction curve is substantially different for $R_V = 5.3$. It has the same shape in the infrared, but much flatter behaviour through the ultraviolet. MEM two component solutions are shown in Figure 2. The most striking change is for intermediate sizes: the relative numbers of particles with $a < 0.1\ \mu$m are much reduced for both silicate and graphite components; this now quantifies our understanding of the difference in ultraviolet extinction.

Small graphite particles are required to produce the 2200 Å extinction bump which is still present in dense clouds. This accounts for the obvious structure in the graphite size distribution, which is not as smooth as that for silicate.

The total mass in this model is about the same as for $R_V = 3.1$. Since mass in the form of small and intermediate sized particles has been removed, there is a corresponding increase for $a > 0.2\ \mu$m. But because the observed shape of the infrared extinction curve is rather invariant with R_V, so too is the shape of the derived size distribution between 0.2 and 1.0 μm.

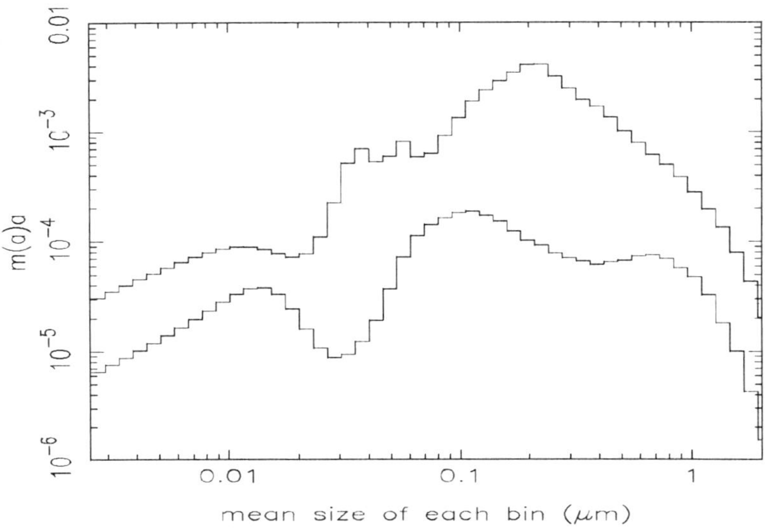

FIGURE 2 Same as Fig. 1, except for the case of $R_V = 5.3$

Scattering

From the MEM size distributions we can calculate the scattering properties of the interstellar grains. Two commonly used diagnostics are the albedo and the asymmetry parameter of the phase function, g. For $R_V = 5.3$, both are increased because of the relative shift in importance from smaller to larger particles.

Within the frequency range 1.0 to 10 μm^{-1} their values for $R_V = 3.1$ are quite similar to those that we calculate from the MRN distribution (see also Draine & Lee 1984). Because the MRN power laws are truncated abruptly at $a_+ = 0.25\,\mu m$, while the derived MEM size distributions are found to fall off more smoothly through larger particles, there are significant differences in the infrared ($< 1\ \mu m^{-1}$). Compared to the MRN model, the scattering is somewhat less isotropic, but the more important difference is in the albedo, which is markedly increased toward lower frequencies (by a factor 6.5 by 0.2 μm^{-1}).

ACKNOWLEDGMENTS

This research has been supported by NSERC of Canada.

REFERENCES

Cardelli, J. A., Clayton, G. C., & Mathis, J. S. 1989, *ApJ*, **345**, 245

Draine, B. T., & Lee, H. M. 1984, *ApJ*, **285**, 89

Mathis, J. S., Rumpl, W., & Nordsieck, K. H. 1977, *ApJ*, **217**, 425

The First Symposium on the Infrared Cirrus and Diffuse Interstellar Clouds
ASP Conference Series, Vol. 58, 1994
Roc M. Cutri and William B. Latter (eds.)

A MULTI-WAVELENGTH STUDY OF THE PLEIADES REGION IN CONJUNCTION WITH WISP

S. GIBSON, K. H. NORDSIECK, A. AFFLERBACH,
C. M. ANDERSON, K. P. JAEHNIG, AND D. E. MICHALSKI
Space Astronomy Lab, University of Wisconsin, Madison, WI 53706

MULTI-WAVELENGTH STUDY

Reflection nebulae (RN) consist of OB stars illuminating nearby clouds, and as such are ideal objects for studying the distribution of Galactic dust. The Pleiades represent the closest bright RN, an optimal target for a multi-wavelength investigation of diffuse interstellar matter. The Wide-Field Imaging Survey Polarimeter (WISP), scheduled for launch this fall, will observe this area in polarized ultraviolet (1300-2800 Å) to provide data on UV dust scattering properties and the three-dimensional geometry of the region. To complement the WISP data, we are assembling information on the region in other parts of the spectrum, as summarized below.

TABLE I Multi-Wavelength Data Applications

Wavelength Regime	Contribution
WISP UV polarimetry	UV scattering function and 3-D nebular structure
Schmidt CCD frames	visible structure and optical scattering function
IRAS images	dust grain temperature and albedo
H-I 21-cm maps	neutral gas distribution
Hα scans	ionized gas distribution

All of these data provide information on the properties and distribution of various phases of diffuse interstellar matter in the direction of the Pleiades. Taken together, the multi-wavelength image data should give a coherent picture of the structure and behavior of the nebular region. Ultimately we wish to build an all-sky database to select future WISP targets. We will apply what we learn from this study to facilitate that selection.

WISP

WISP is a sounding-rocket payload designed to take 15 arcsec/pixel, $5^\circ \times 1.7^\circ$ field polarimetric images in the UV, with broadband filter peaks at 1640 and 2180 Å. Dust scatters strongly at these wavelengths, while the diffuse ultraviolet background is quite low. The small number of stars hot enough to provide illumination makes scattering paths tracable for single, outlying clouds. With reasonable distance data, we can then reconstruct the 3-D distribution of the nebulosity. Since the UV scattering properties of dust depend strongly on composition as well as geometry, WISP will also provide useful constraints on theoretical models, such as the determination of dust grain albedo A and scattering phase function asymmetry g. The latter of these has inferred values ranging from nearly isotropic ($g < 0.2$; Hurwitz *et al.* 1991) to very forward-throwing ($g \sim 0.75$; Witt *et al.* 1992). These conflicting values depend on assumed scattering geometry, which WISP should be able to determine.

WISP is scheduled for first flight in the fall of 1993, to observe the Pleiades. Several other targets are under consideration. The target list is given below. We would welcome any suggestions of additional objects.

TABLE II Currently Planned WISP Targets

Name	RA	Dec	ℓ^{II}	b^{II}	Comment
Pleiades	3.7	24	167	-24	closest (126 pc) bright RN[a]
NGC 7023	21.0	68	104	14	brightest RN; Witt study
N America Nebula	21.0	44	85	-1	large RN
M81/M82	9.9	70	141	40	Sandage DGL[b] area; M82 knots
Draco Cloud	16.8	60	90	39	DGL; X-ray shadow
Lockman Window	10.7	58	151	52	DGL; extragalactic background
LMC	5.7	-69	279	-32	EG[c]; large angular size
M31	0.7	41	121	-22	EG; DGL

[a]Reflection Nebula: individual OB stars illuminating nearby clouds within the field of view. Use to study UV optical properties of dust.

[b]Diffuse Galactic Light: diffuse clouds illuminated by distant stars. In the UV a small number of OB association illuminators will be identifiable by polarization position angle. Use to study 3-D structure of galactic dust.

[c]External Galaxy: offers a chance to view DGL from the outside. Use to study dust properties and 3-D structure of other galaxies.

PLEIADES REGION IMAGE DATA

The images in Figure 1 give a brief overview of the region's appearance at various wavelengths. We are especially interested in deep CCD mapping of the WISP field for comparison with detected UV features. We have applied for Schmidt time to map the field in the B and R filter bands. Two trial R-band images (courtesy P. Marcum) with field diameters of $\sim 1^\circ$ are shown which illustrate (1a) faint extended nebulosity and (1b) the central region including Merope, which exhibits 3-12% visible polarization. These were taken with 800 and 600 second exposures, respectively, using a Tek 2048^2 CCD, then binned 3x3. Despite some uncertainty in flatfielding, structure is visible at a few percent of sky. The nebulosity should appear brighter in B.

A complete deep-CCD map should detect more of the outer structure which is just visible in the Palomar Sky Survey (1c). These extended wisps are probably part of one sheet of matter in front of the cluster, evidenced by the organized large-scale structure and the ubiquitous 7 km/s interstellar absorption line in cluster star spectra. Figure 1c shows the $5^\circ \times 1.7^\circ$ rectangular WISP field, with the Schmidt field locations marked. The 5° length of the WISP field is equivalent to 10 pc for the assumed cluster distance.

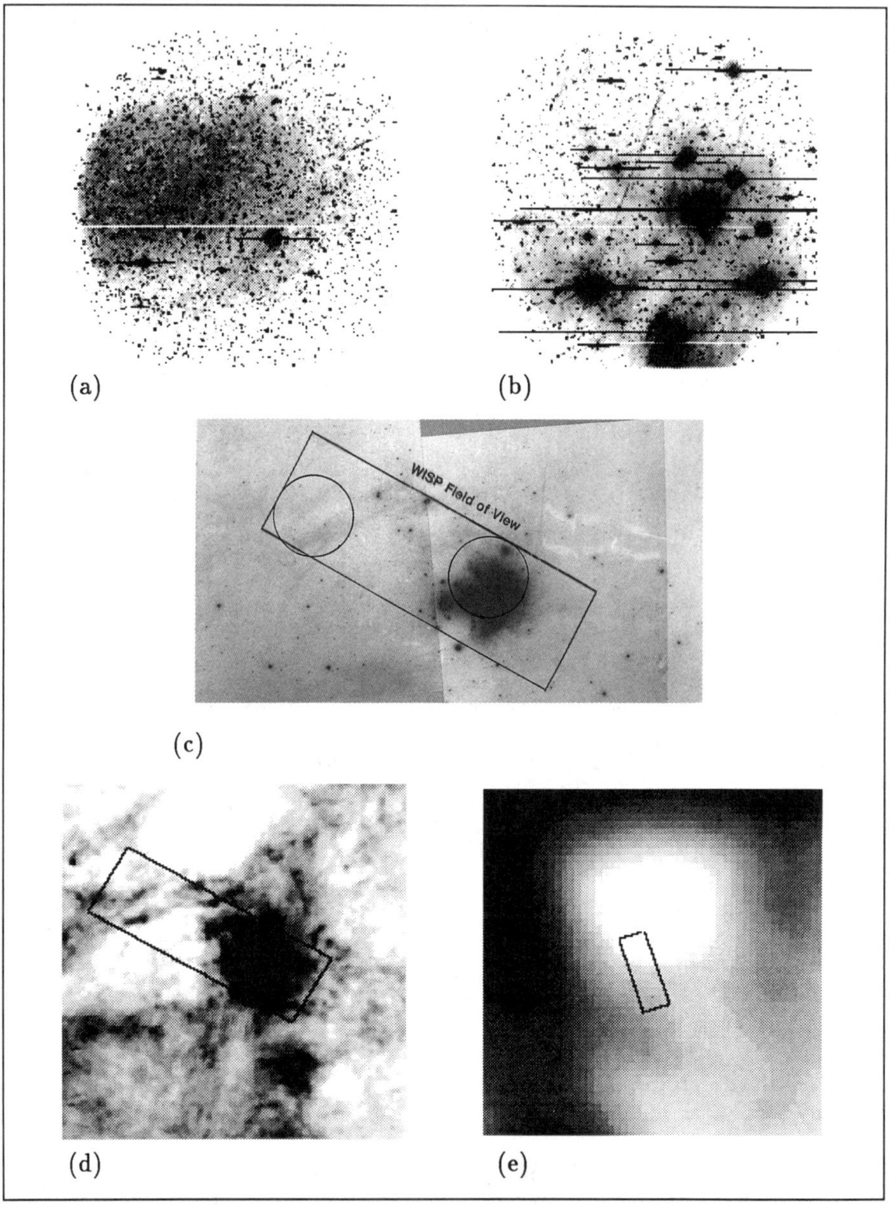

FIGURE I Pleiades multi-wavelength images: all are shown with north up, east left, except (e), which is oriented in galactic coordinates and thus tipped clockwise; (a) R-band CCD image of extended nebulosity; (b) R-band CCD image of central nebula; (c) Palomar Sky Survey O print (blue) mosaic; (d) IRAS 100 μm; (e) H-I column density (Stark *et al.* 1992)

The outer reflection nebulosity seen in the Sky Survey and Schmidt maps appears to have associated infrared emission, expecially in the 100 μm IRAS band (1d). The radiation responsible for heating the nebula is primarily the same as that which is seen in reflection in the UV. The WISP field is again marked.

Finally, the WISP field is marked on an H-I column density map (1e) taken from the Bell Labs Survey data of Stark *et al.* (1992). These data are oriented in the galactic coordinate frame, which is tilted clockwise $\sim 35^\circ$ here. The Pleiades region seems to exists between two H-I column "spurs", though this may only be a line of sight effect. Several measured Hα fluxes (courtesy R. Reynolds & S. Tufte) appear to anticorrelate with the local H-I column density map, but the data are too sparse to decide whether the H-I hole may be "filled" by ionized gas. We plan to examine the velocity distributions of these components as part of the overall study of the region.

ACKNOWLEDGMENTS

We would very much like to thank to P. Marcum for taking test Schmidt images and R. Reynolds and S. Tufte for their Hα scan data. All image data were processed using the IRAF software package. WISP is supported by NASA grant NAG5-647.

REFERENCES

Hurwitz, M., Bowyer, S., & Martin, C., 1991. *ApJ* **372**, 167

Stark, A. A., Gammie, C. F., Wilson, R. W., Bally, J., Linke, R. A., Heiles, C., & Hurwitz, M., 1992. *ApJS* **79**, 77-104

Witt, A. N., Petersohn, J. K., Bohlin, R. C., O'Connell, R. W., Roberts, M. S., Smith, A. M., & Stecher, T. P., 1992, *ApJ* **395**, L5

The First Symposium on the Infrared Cirrus and Diffuse Interstellar Clouds
ASP Conference Series, Vol. 58, 1994
Roc M. Cutri and William B. Latter (eds.)

SCATTERING FROM INTERSTELLAR DUST IN THE COALSACK: THE OPTICAL CONSTANTS IN THE FUV

JAYANT MURTHY

R. C. HENRY
The Johns Hopkins University

J. B. HOLBERG
Univ. of Arizona

INTRODUCTION

The Coalsack nebula is one of the most prominent dark nebulae in the southern Milky Way and is easily visible with the naked eye as a dark patch against the bright background of early type stars. We have obtained three observations of dust scattering in this region using the ultraviolet spectrometers (UVS) aboard the two *Voyager* spacecraft and have used these observations to constrain the optical properties of the interstellar dust grains.

OBSERVATIONS

The *Voyager* UVS instruments are compact Wadsworth spectrometers with a field of view of $0.87° \times 0.1°$. The spectral range of both spectrometers is about 500 — 1700 Å, although the sensitivity drops drastically at wavelengths above Ly α, with a spectral resolution of 38 Å (FWHM) for diffuse sources which fill the slit. There are three components to the signal observed by the UVS instruments: an instrumental dark count induced by radiation from the spacecraft's radioactive thermoelectric generator (RTG); emission lines from resonantly scattered solar lines; and the "true" astrophysical signal. The three sources of signal are independent of each other and thus both the dark noise and the interplanetary lines can be subtracted using templates. A more comprehensive discussion of the instruments and the subtraction procedure is given in Murthy *et al.* (1993).

A strong continuum with a peak level of about 4×10^4 photons cm^{-2} s^{-1} sr^{-1} Å^{-1} is observed in two of our observations (Targets A and B; which look at the same direction but with different instruments and orientation). The peak level in the third observation is 3×10^4 photons cm^{-2} s^{-1} sr^{-1} Å^{-1} (Target C). While the continuum in Targets A and B is due entirely to dust scattered starlight, about half the astrophysical emission in Target C is due to a B6 star in the field of view.

DISCUSSION

The dense mass of the Coalsack acts as a curtain keeping out light from any star not in the foreground. In addition, three of the five brightest stars in the FUV (as seen from the Earth) are located at distances between 100 and 160 pc directly in from of the nebula. Hence the radiation field at any point between the Sun and teh Coalsack is dominated by these three early B stars (α Cru, β Cru, and β Cen) and can be modelled by the spectrum of a B0 star. While we had originally expected to see only back scattered light from dust in the nebula, detailed extinction studies show a significant amount of dust in the foreground (Franco 1989; Seidensticker 1989) and it is this dust which is responsible for most of the observed emission.

Although we have attempted to model the scattered light in a manner similar to our previous observations of the diffuse FUV radiation field (Murthy et al. 1991, 1993), there is too much uncertainty in the distribution of the dust along the line of sight for us to place useful limits on the optical constants of the interstellar grains. Nevertheless, if the phase function is constant over the spectral range of concern and every photon observed has been scattered once and once only, we can derive the simple relationship $I_O/I_\star \propto a\sigma \exp^{-n\sigma}$ where I_O and $I_\star$ are the observed and incident intensities, respectively, a is the grain albedo, σ is the extinction cross-section and n is the total HI column density along the line of sight. The ratio $I_O/I_\star$ is virtually constant between 900 and 1400 Å in our data and, assuming a lower limit of 10^{20} cm^{-2} on the H I column density (correponding to an A_V of 0.3 from Franco 1993) we can show that the albedo does not decrease over this spectral range. If a new population of zero albedo small grains were responsible for the FUV rise in the extinction curve (*c.f.* Witt *et al.* 1993), we would expect the albeod to drop significantly at short wavelengths. No such trend is seen in our data, perhaps implying an increase in the number density of the small grains rather than a dramatic change in their optical constants.

REFERENCES

Franco, G. A. P. 1989, *A&A*, **215**, 119

Franco, G. A. P. 1993, (in preparation)

Murthy, J., Henry, R. C., & Holberg, J. B. 1991, *ApJ*, **383**, 198

Murthy, J., Im, M., Henry, R. C., & Holberg, J. B. 1993, *ApJ*, (in press)

Witt, A. N., Petersohn, J. K., Holberg, J. B., Murthy, J., Dring, A., & Henry, R. C. 1993, *ApJ*, **410**, 714

The First Symposium on the Infrared Cirrus and Diffuse Interstellar Clouds
ASP Conference Series, Vol. 58, 1994
Roc M. Cutri and William B. Latter (eds.)

CARBON CHEMISTRY IN THE POLARIZED 2175 Å BUMP SIGHTLINE TOWARD HD 197770

MARGARET MURRAY HANSON AND GEOFFREY C. CLAYTON
Center for Astrophysics & Space Astronomy, University of Colorado, Boulder, CO 80309-0389

CHRISTOPHER M. ANDERSON
Space Astronomy Laboratory, University of Wisconsin, 475 N. Charter St., Madison, WI, 53706

KIM A. VENN
Dept. of Astronomy, University of Texas, RLM 15.308, Austin, TX 78712

ABSTRACT In December 1990, the Wisconsin Ultraviolet Photo-Polarimeter Experiment (WUPPE) made the first spectropolarimetric observations of interstellar polarization in the UV. WUPPE provided data on the wavelength dependence of interstellar polarization along 8 lines of sight (Clayton et al. 1993). HD 197770 (a B2 III star) was the only line of sight to show evidence for a polarization bump at 2175 Å. Is there some unique environment of the dust along this line of sight which allows the bump grains to be aligned? We have obtained very high resolution, high signal-to-noise spectra of the star HD 197770 to study the atomic and molecular gas in the cloud. We present here a preliminary analysis.

We have serendipitously discovered temporal variations in the stellar lines of HD 197770 and believe the star to be a candidate β Ceph.

INTRODUCTION

Polarization data allow us to take a new look at dust grain models which have been used to fit the extinction and polarization in the visible and infrared. The polarization bump seen toward HD 197770 can be well fit with small aligned graphite disks. The apparent necessity of aligning both small silicate and graphite grains seems to raise questions about theories of grain alignment. For a full discussion see Wolff et al. (1993). HD 197770 seems to be a classic diffuse dust line of sight, with a "normal" UV extinction curve, though its molecular and atomic abundance has never been studied. It is not known to be associated with a cluster or association and lies at a rather high galactic latitude ($b = 9°$). It is only about 400 pc distant and reddened by 0.57 mag. This is a high value of reddening per distance and since it is so nearby, most of the extinction may be caused by starlight passing through the outer edge of a single dark cloud.

OBSERVATIONS

The data set presented here has come from 3 different observing runs. In June 1992, MMH obtained CN, CaII, CaI and CH+ at the KPNO Coudé Feed, obtaining a resolution of ~120,000. In October of 1992, MMH & KAV obtained C_2 at the McDonald 2.7m telescope at coudé focus, obtaining a resolution of ~80,000. Also in October of 1992, CMA obtained all features other than the C_2 lines at the Coudé Feed at KPNO. The resolution is ~80,000 for these data.

RESULTS

Two velocity components have been detected (see Table 1). The main feature, at $-3\pm$ 1 km s^{-1} contains (nearly) all the molecules in the sightline. The second feature at $+7 \pm 1$ km s^{-1} is traced in CaI, CaII, FeI, and possibly a 1-σ detection of CH^+. The star has a radial velocity of -15 km s^{-1}.

Table 1
Molecular and Atomic Column Densities

LogN (cm^{-2})	Feature 1 -3 km s^{-1}	Feature 2 +7 km s^{-1}
LogN(CaI)	10.3	~10
LogN(FeI)	12.1	~11.4
LogN(CH) total	13.4	...
parity − (upper)	13.2	
parity + (lower)	12.9	
LogN(CN) total	12.8	...
J=0	12.7	
J=1	12.3	
LogN(C_2) total	13.5[a]	...
J=0	12.1	
J=2	12.8	
J=4	12.9	
J=6	12.8	
J=8	12.6	
LogN(CH^+)	12.6	~12.0

[a]T_{kin}=50K was assumed. See the discussion on C_2.

CN

Without question, the R(0) line is saturated. The column density shown in Table 1 assumes a Doppler b-value of 1.0 km s^{-1}. From the column densities quoted there, the excitation temperature of the molecule is T_{10}=2.70 K. Within the errors, this is explained solely by thermal excitation from the cosmic background radiation (T_{bg}=2.730 K). As a smaller Doppler *b*-value is assumed, the T_{10} goes *below* 2.7 K. While the error in our column density determination may allow for slightly smaller *b*-values, our most reasonable assumption is that *b* is not less than 0.8 km s^{-1}. The lack of evidence for collisional excitation above the cosmic background temperature suggests the density is not high (< 1000 cm^{-3}), however our analysis does not take into consideration the spin-rotation fine-structure components (probably a small effect compared to the error in our equivalent width measurements).

C_2

The relative population of the rotational J states in C_2 are dependent on the rate of collisions with H and H_2 and through radiative excitation. These effects are equally important in the low density range of diffuse clouds ($25 < n < 500$ cm^{-3}). In clouds with densities above this, the population distribution becomes thermal ($T_{rot} \rightarrow T_{kin}$), and no information about the cloud environment (other than the cloud kinetic temperature) can be derived. Because n, T_{kin}, and ionization field all play a role in creating the population distribution seen, usually T_{kin}, is approximated using other methods, leaving the density, as a function of ionizing radiation to be determined from the population ratios (given as: $\frac{n_c \sigma_o}{I}$, where n_c is density of collision partners, σ_o is the collisional cross section for de-excitation (which is not well determined), and I is a scaling factor for the radiation field, van Dishoeck 1984). The values given in Table 2 below were taken from van Dishoeck (1984).

Table 2

Rotational Populations of C_2

J	$n_x(0,J)/n_x(0,2)$ observed	$n_x(0,J)/n_x(0,2)$ T=50K, $\frac{n_c \sigma_o}{I} = 4 \times 10^{-14}$	$n_x(0,J)/n_x(0,2)$ T=100K, $\frac{n_c \sigma_o}{I}$ =Thermal
0	0.20	0.262	0.234
2	1.0	1.0	1.0
4	1.25	1.23	1.25
6	1.0	0.916	1.018
8	0.63	0.66	0.609

The low ratio of J=0/J=2, indicates T_{kin} is at least greater than 20 K. The population ratios for T_{kin}=50 that match our observed ratios is shown ($\frac{n_c \sigma_o}{I} = 4 \times 10^{-14}$ is rather average). However, for a T_{kin} of 100 K, the distribution is best fit by a thermal one (and is a very close match to the observed ratios). The higher J values, which we do not have, are the most sensitive to deviations between T_{kin} and T_{rot}. In order for the total C_2 column density to be determined, the distribution of the rotational population must be known. For T=50 K, 25% of the molecules are in states above $J=8$. For a thermal distribution of T=100 K, only 10% of the molecules exist above $J=8$.

CH

The CH in this sightline is strong enough and our signal-to-noise high enough, to detect the weak 3886 and 3890 Å lines. Generally, the stronger 4300 Å line is used, but that feature is actually a blend of two lines, arising from different ground states. The interaction of the nuclei and the rotational angular momentum ($\Lambda=1$ for the $X^2\Pi$ ground state) splits each J–level into two parity components, + and – as shown in Table 1 (this is called Λ-doubling). The 3886 Å feature comes from a blend of two lines, which both originate from the –parity state, while the 3890 Å feature is a single line, originating from the +parity state. Their combined column density should match that obtained from the 4300 Å feature. We have used this fact to estimate the Doppler b-value. Using a b-value of 1 km s^{-1} for the 4300 Å CH line, and summing the oscillator strengths from its two unresolved features, we obtain a column density of Log(N)=13.4, consistent with that obtained from the weaker features used in Table 1. Note we do not get an equal population in the two ground states $[N(-) \sim 2N(+)]$.

DISCUSSION

While our numbers are still preliminary, our data indicate that the cloud is quiescent, harboring a large column density of CN, CH and C_2, with a small column density of CH^+. There appears to be a single dominant cloud in which the molecules exist. The physical conditions in the cloud, as determined by our 'first run' analysis of the rotational populations of CN and C_2 are consistent with a density from several hundred cm^{-3} to 1000 cm^{-3}. Perhaps this provides the conditions for the bump grains to align but since the alignment mechanism is mysterious, little more can be said to explain the polarized 2175 Å bump. However, it is clear that HD 197770 is an excellent sightline for the testing of interstellar chemical models because the column densities are large and the sightline appears clean of interfering clouds.

Finally, with our high quality spectra, we have made a serendipitous discovery about the star HD 197770. Peculiar changes in the stellar lines over our observing period suggest that HD 197770 is perhaps a β Cep variable star. A recent paper by Jerzykiewicz (1993) identified HD 197770 as having photometric variations.

REFERENCES

Clayton, G.C. et al. 1992, *ApJ*, 385, L53

Jerzykiewicz, M. 1993, *A&AS*, 97, 421

Van Dishoeck, E.F. 1984, PhD Thesis, University of Leiden

Wolff, M. et al. 1993, *ApJ*, 403, 722

The First Symposium on the Infrared Cirrus and Diffuse Interstellar Clouds
ASP Conference Series, Vol. 58, 1994
Roc M. Cutri and William B. Latter (eds.)

ULTRAVIOLET INTERSTELLAR POLARIZATION OBSERVED WITH THE HUBBLE SPACE TELESCOPE

D. C. B. WHITTET
Department of Physics, Rensselaer Polytechnic Institute, Troy, NY 12180

W. B. SOMERVILLE
Department of Physics & Astronomy, University College London, United Kingdom

P. G. MARTIN[1]
Canadian Institute for Theoretical Astrophysics, University of Toronto, Toronto, Canada

ABSTRACT We have used the Faint Object Spectrograph of the Hubble Space Telescope to observe interstellar linear polarization from 1300 to 3200 Å in three stars with well-studied interstellar polarization at optical wavelengths. In each case, the wavelength dependence of polarization declines smoothly with decreasing wavelength and is devoid of structure associated with the prominent 2175 Å absorption bump in the interstellar extinction curve. The position angle of polarization is independent of wavelength and consistent with that measured in the visible. Out of a total of 11 stars now studied by means of spectropolarimetry in the satellite ultraviolet (including 8 observed with WUPPE), only one (HD 197770) shows any evidence for polarization associated with the 2175 Å bump.

INTRODUCTION

The availability of spectropolarimetric data in the satellite ultraviolet (UV) provides a unique opportunity to investigate the nature and alignment properties of the interstellar dust grains responsible for extinction at these wavelengths. The UV extinction curve is dominated by the dramatic 'bump' feature centered at 2175 Å, superposed upon continuum extinction arising, according to current models, in at least two distinct populations of dust. The bump is often attributed to graphite, but there are significant problems with this identification: indeed, the nature of the particles responsible for the bump is amongst the most significant unsolved problems in astronomical spectroscopy (Draine 1989).

The wavelength dependence of polarization across a dust-related absorption feature is a vital diagnostic of the shape, size distribution, refractive index

[1]Currently on leave at the Department of Astronomy, California Institute of Technology, Pasadena, CA 91125

and alignment properties of the carrier: excess polarization is expected for significant alignment of aspherical carrier grains. For example, such an excess has been observed at infrared wavelengths corresponding to the principal vibrational resonance in silicates near 10 μm, indicating that silicate particles are alignable and may contribute to the general continuum polarization. Whether or not this property is shared by the carrier of the bump is a key question with important implications for grain models: Draine pointed out that non-spherical graphite grains, if appreciably aligned, would produce distinctive linear polarization at UV wavelengths and that the predicted polarization is an important test of graphite as a viable component of grain models. Spectropolarimetric results available to date are intriguing. An early experiment (Gehrels 1974) suggested a lack of significant excess above the continuum predicted by an extrapolation of the optical polarization curve to ultraviolet wavelengths, a result which seems to have prompted widespread adoption of graphite 'spheres' in model calculations. Much more recent data from the Wisconsin Ultraviolet Photopolarimeter Experiment (WUPPE; Clayton et al. 1992) suggest the presence of a 2175 Å excess in one star out of several studied with this instrument.

In this paper, we briefly describe observations of UV interstellar polarization made with the Hubble Space Telescope (HST). A fuller account of these results and their significance will be given elsewhere (Somerville et al. 1993).

OBSERVATIONS AND RESULTS

Observations of three stars with previously known optical interstellar extinction and polarization curves (HD 7252, HD 98695 and HD 161056) were made in cycle 1 of the HST Guest Observer program in September, 1992. The Faint Object Spectrograph was used in spectropolarimetric mode (see Allen et al. 1993 for a description). The total integration time was 410 minutes. The spectral range 1250–3200 Å was covered in three segments using the G130H, G190H and G270H gratings at a nominal spectral resolution of 2 Å. (The G130H and G270H data for HD 98695 were subsequently rejected). Results were binned to 10 data points per grating segment in order to improve signal to noise. In all cases, the position angle of polarization in the UV was found to be consistent with that measured in the optical, with no evidence for systematic dependence on wavelength.

Optical and UV data were combined for the three stars to produce polarization curves extending in wavenumber from 0.4 to 8.0 μm^{-1}. Standard empirical curves represented by the Serkowski Law (Whittet et al. 1992 and references therein) were fitted to the optical data, and extrapolated shortward to examine the extent to which such a formulation is applicable in the UV. From an analysis of WUPPE data, Wolff et al. (1993) recognised three types of behavior: (i) stars with UV data well fitted by an extrapolation of the Serkowski Law; (ii) stars exhibiting excess continuum polarization with respect to an extrapolation of the Serkowski Law; and (iii) one star with excess polarization apparently associated with the 2175 Å bump. Preliminary analysis of HST data suggests that two of the three stars (HD 98695 and HD 161056) fall into category (i), and the third (HD 7252) falls into category (ii). Significantly, none exhibits category (iii) behavior: out of a total of 11 stars now studied in the UV with WUPPE or HST, only HD 197770 shows evidence for polarization associated with the

2175 Å bump. Our data for HD 161056 set a particularly stringent upper limit on the amplitude of any such polarization excess in that line of sight.

THEORETICAL MODELING

Observations of UV interstellar polarization described above are providing impetus for current theoretical work designed to test existing models for the wavelength dependence of extinction and to extend models for grain alignment to very small particles (see Wolff et al. 1993 for a discussion of the applicability of existing grain models to WUPPE data). A goal for future research will be to perform a quantitative test of models which attribute the bump to free graphite particles, using the observed wavelength dependence of extinction and polarization near 2175 Å to delimit the allowed grain size, shape and composition along each line of sight observed. The presence or absence of polarization associated with the far UV extinction rise will also be investigated. It has been argued (Mathis 1986) that the very small grains responsible for this component of the extinction should not be aligned, in qualitative agreement with the lack of obvious far UV excess in the polarization data available to date.

ACKNOWLEDGMENTS

Support for this work was provided by NASA through grant number GO–2245.01–87A from the Space Telescope Science Institute, which is operated by the Association of Universities for Research in Astronomy, Inc., under NASA contract NAS5–26555.

REFERENCES

Allen, R. G., Smith, P. S., Angel, J. R. P., Miller, B. W., Anderson, S. F. & Margon, B., 1993, *ApJ*, **403**, 610

Clayton, G. C., Anderson, C. M., Magalhães, A. M., Code, A. D., Nordsieck, K. H., Meade, M. R., Wolff, M. J., Babler, B., Bjorkman, K. S., Schulte–Ladbeck, R., Taylor, M. J. & Whitney, B. A., 1992, *ApJ*, **385**, L53

Draine, B. T. 1989, in *IAU Symp. no. 135, Interstellar Dust*, eds. L. J. Allamandola & A. G. G. M. Tielens (Kluwer, Dordrecht), p. 313

Gehrels, T., 1974, *AJ*, **79**, 590

Mathis, J. S., 1986, *ApJ*, **308**, 281

Somerville, W. B., Allen, R. G., Carnochan, D. J., He, L., McNally, D., Martin, P. G., Morgan, D. H., Nandy, K., Walsh, J. R., Whittet, D. C. B., Wilson, R. & Wolff, M. J., 1993, *ApJ*, preprint

Whittet, D. C. B., Martin, P. G., Hough, J. H., Rouse, M. F., Bailey, J. A. & Axon, D. J., 1992, *ApJ*, **386**, 562

Wolff, M. J., Clayton, G. C. & Meade, M. R., 1993, *ApJ*, **403**, 722

The First Symposium on the Infrared Cirrus and Diffuse Interstellar Clouds
ASP Conference Series, Vol. 58, 1994
Roc M. Cutri and William B. Latter (eds.)

SCATTERING BY DIFFUSE CLOUDS IN THE GALAXY: SEARCHING FOR THE COSMIC BACKGROUND

ADOLF N. WITT, JENS K. PETERSOHN
The University of Toledo, Ritter Observatory, 2801 Bancroft, Toledo, OH 43606, USA

ABSTRACT

We have developed a Monte Carlo code for the simulation of diffuse galactic light (DGL) in our Galaxy, based on multiple scattering in an interstellar medium consisting of individual clouds exposed to the local anisotropic radiation field. This model has been used for the analysis of existing sets of measurements of the diffuse galactic background radiation in the UV, with two main purposes:

1. the determination of scattering properties of dust grains in the diffuse medium;
2. the determination of the extragalactic cosmic background in the UV, possible only after correctly accounting for the scattered foreground radiation originating in the Galaxy.

In this paper, we discuss systematic errors arising in the commonly employed practice of extrapolating high-$|b|$ measurements of the DGL to $N(HI) = 0$ or to $\csc(|b|) = 0$ to derive the level of the "cosmic" extragalactic background.

INTRODUCTION

The diffuse background radiation at high galactic latitudes at visible and UV wavelengths, after correction for airglow, faint star and zodiacal light contributions is dominated by two components:

1. scattered light from galactic dust in the diffuse interstellar medium;
2. a "cosmic" isotropic component, mainly extragalactic in nature.

Here, we examine the question of how to separate these two components, given a set of measurements of the total background intensity over a wide range of galactic latitudes.

APPROACH

Commonly used methods for separating the cosmic background from the galactic scattered light include extrapolations of the measured intensities to $\csc(|b|) = 0$

or to $N(HI) = 0$. The underlying assumptions are a plane-parallel galactic dust layer with the sun in the central plane, in the first case, and a constant dust-to-gas ratio, with HI a tracer of the gas distribution, in the second case. When these two extrapolations are applied to the same data set, they generally produce significantly different results for the derived extragalactic component (e.g. Wright 1992). In an effort to decide which approach is more likely to yield reliable results, we have simulated the production of diffuse galactic light with a model involving multiple scattering in diffuse clouds, which are illuminated by the actual (observed) non-isotropic interstellar radiation field. The following characteristics describe the model:

1. Effective model wavelength: $\lambda = 1565$Å;
2. Three-cloud spectrum, with optical depth radii of the clouds of $\tau = 0.25$, 1.25, and 5.0, with relative spatial frequencies of 1728:27:1;
3. Dust scattering properties defined by the single-scattering albedo a, and the phase function asymmetry g;
4. Line-of-sight optical depth and the average number of clouds along a given sightline derived from $N(HI)$, taken from the Bell Lab $N(HI)$ survey (Stark et al. 1992);
5. Radiation field based on positions and fluxes of 58,000 stars contained in the TD-1 catalog;
6. Complete multiple scattering within each cloud, leading to detailed surface brightness profiles;
7. Intensity integration along a given line-of-sight by Monte Carlo method.

The observational data on the diffuse UV background which we subjected to an analysis are those of Fix, Craven, & Frank (1989). These data were obtained with the Dynamics Explorer 1 (DE 1) spacecraft.

RESULTS

Since our model does not include an a-priori cosmic light component, extrapolations of our predicted intensities to $\csc(|b|) = 0$ and $N(HI) = 0$ should produce a zero intercept. We find, however, that extrapolations of predicted DGL intensities to $\csc(|b|) = 0$ produces negative values. Two separate effects are responsible. One is related to the concentration of the non-isotropic radiation field at low-$|b|$ or high $\csc(|b|)$ values. Identical clouds composed of forward scattering grains will appear with higher surface brightness when located at low $|b|$ than at high $|b|$, this steepening the slope of $I(DGL)$ vs. $\csc(|b|)$ and leading to a negative intercept. The second effect is related to the fact that the HI layer in the Galaxy is *not* plane-parallel. Plots of $N(HI)$ vs. $\csc(|b|)$ generally lead to *negative* values for $N(HI)$ at $\csc(|b|) = 0$. This is the case, in particular, when the $N(HI)$ for the DE 1 target areas are plotted against the $\csc(|b|)$

values for these areas. Since we assume the dust column density to be proportional to $N(HI)$, this leads to a further depressions in the value of the derived extragalactic component.

The two effects described produce a negative bias to the extrapolated cosmic background at 1565Åof about -300 photons cm^{-2} s^{-1} Å^{-1} sr^{-1}. Wright's (1992) result for the upper limit on the cosmic UV background of 4 ± 80 photons cm^{-2} s^{-1} Å^{-1} sr^{-1} derived from an extrapolation to $\csc(|b|) = 0$ needs to be corrected for this bias, leading to a more acceptable value of 300 ± 80 photons cm^{-2} s^{-1} Å^{-1} sr^{-1} for the cosmic background in the UV.

The derivation of the cosmic background from an extrapolation of $I(DGL)$ to $N(HI) = 0$ is more likely to yield a reliable result, because the assumption that the interstellar medium is a plane-parallel layer is no longer made. Nevertheless, systematic errors arise due to two effects. At low-$N(HI)$ sightlines, additional dust associated with ionized or molecular gas may produce higher intensities than expected (Wright 1992), and at high-$N(HI)$ values there is a greater statistical likelyhood that the sightline is closer to peak radiation field intensities, thus increasing the predicted surface brightness in the case of strongly forward scattering grains. The combination of these two effects is that the cosmic light derived from an extrapolation to $N(HI) = 0$ is systematically higher than that found from an extrapolation to $\csc(|b|) = 0$.

Our analysis of the DE 1 data has yielded a best estimate of the dust scattering properties at $\sim$ 1500Å of $a = 0.5$ and $g = 0.9$, with a cosmic (isotropic) background of 300 ± 80 photons cm^{-2} s^{-1} Å^{-1} sr^{-1}. (Witt and Petersohn 1993).

REFERENCES

Fix, J.D., Cravens, J.D., & Frank, L.A. 1989, *ApJ*, **345**, 203.

Stark, A.A., et. al. 1992, *ApJS*, **79**, 77.

Witt, A.N., & Petersohn, J.K. 1993, in preparation.

Wright, E.L., 1992, *ApJ*, **391**, 34.

The First Symposium on the Infrared Cirrus and Diffuse Interstellar Clouds
ASP Conference Series, Vol. 58, 1994
Roc M. Cutri and William B. Latter (eds.)

THE ORIGIN OF THE SOFT X-RAY BACKGROUND

JÜRGEN KERP
Radioastronomisches Institut der Universität Bonn, Auf dem Hügel 71, 53121 Bonn, Federal Republik of Germany

ABSTRACT ROSAT observations of atomic and molecular clouds at high galactic latitudes reveal a wealth of new information about the physical conditions in which these clouds are embedded. The coexistence of cold neutral matter with a hot X-ray emitting plasma on one hand and the significant contribution of a very large extended X-ray source located behind the galactic H I layer on the other hand questions our understanding of origin and evolution of neutral matter.

INTRODUCTION

Soft X-ray observations performed with sounding rocket flights revealed that the Sun is embedded within a hot X-ray emitting environment. These observations suggested that a hot ($\log(T) \simeq 6$ K) plasma partly fills the local cavity of neutral matter (Local Hot Bubble (LHB); Snowden et al. 1990a), the amount of residual neutral material was found to be in the order of $7 \cdot 10^{18}\,\mathrm{cm}^{-2}$ (Juda et al. 1991). In direct opposition to these observational results is the detection of cold ($T \simeq 100$ K) dense ($n_{HI} \simeq 75\,\mathrm{cm}^{-3}$) matter within the local cavity. Wennmacher et al. (1992) suggested the location of a H I 21-cm filament (LVC 88+36−2) within the LHB. Based on NaID absorption line measurements they derived an upper distance limit of 60 ± 20 pc taking into account a LHB radius of 100 pc into the filament direction; this upper distance limit implies its location within the LHB. Kerp et al. (1993) found indeed a soft X-ray shadow of this filament derived from a deep ($t_{int} \cong 28$ ksec) pointed ROSAT PSPC observation. The observed soft X-ray intensity distribution is in quantitative agreement with the measured H I column density distribution of the cloud, therefore this confirms the coexistence of the two different temperature states of matter. Recently Snowden et al. (1993) found soft X-ray shadows of the near (~ 65 pc; Hobbs et al. 1986) molecular cloud MBM12 (Magnani et al. 1985) from a ROSAT observation. The content of neutral matter within the LHB is therefore obviously not negligible and more than a factor of 100 over the amount derived from sounding rocket flights. The reason for this disagreement between the first X-ray observations and the ROSAT results is caused by the very different angular resolution of the used optics. With ROSAT the angular resolution is usually better than some arcmin while the sounding rocket experiments were restricted to a resolution of some degrees! Therefore we can conclude that the sounding rocket flights gave an average view of the soft X-ray sky while ROSAT or imaging X-ray telescopes offer a sharp detailed view of all X-ray sources.
In 1991, some months after the launch of ROSAT, a soft X-ray shadow caused

by the high latitude molecular cloud IVC 91+38-21, called Draco nebula, was detected (Snowden et al. 1991). The Draco nebula is more distant than 300 pc (Lilienthal et al. 1991) and located high above the galactic plane (~180 pc). This observed shadow is only explainable by a second soft X-ray source being located behind the cloud. The LHB as the only source of soft X-rays was overcomed.

ANALYSIS

I studied soft X-ray spectra derived from the ROSAT soft X-ray observations of LVC 88+36–2 and of the socalled Draco–nose. The spectra gave direct evidence for the contribution of two soft X-ray sources to the observed spectrum (Fig. 1). One component is always observable also on the densest parts of both clouds where the distant X-ray background source is very efficiently blocked by the high H I amount of the clouds. This component must be associated with the LHB. Its temperature is in the order of $\log(T) = 5.85 \pm 0.05$ K and in agreement with the sounding rocket results which sample the integral properties of the LHB (Snowden et al. 1990b). The second component has to be associated with a distant source located behind the total H I column density along the line of sight. Its temperature is in the order of $\log(T) = 6.5 \pm 0.05$ K. A two component fit of the soft X-ray spectra assuming a plasma in collision equilibrium after Raymond & Smith (1977) fits the data very well. The emission measure (1:5; LHB:distant source) is very constant which implies the quantitative agreement of X-ray intensities with the amount of absorbing matter (Kerp 1993). Burrows & Mendenhall (1991) suggested from their ROSAT observation of the Draco–nose that this background source is only associated with an isolated source located behind the Draco cloud. Latest results of Garmire et al. (1992), Wang (1992) and Wang & McCray (1993) show the existence of a pervasive hot plasma intermixed and also behind the galactic H I -layer with a temperature of about $\log(T) \sim 6.4$ K. Therefore it is very likely that the galaxy has a soft X-ray emitting halo.

CONCLUSION AND FURTHER PERSPECTIVES

ROSAT soft X-ray observations revealed the coexistence of neutral atomic and also molecular gas with the hot soft X-ray emitting plasma of the LHB. These results are striking for both, the understanding of the lifetime of the hot plasma and of the clouds. One important question is obviously weather or not the cold cloud boundaries are visible as emission in other wavelengths due to the interaction with the hot plasma, perhaps in the UV? What is the cooling function of the hot gas and what are the stabilization mechanisms of the clouds in such an aggressive environment? There are a lot of hints for the existence of a hot gas behind the galactic H I layer, therefore it is possible that the galaxy has a soft X-ray emitting halo with a temperature of $\log(T) \sim 6.4$ K. The gas content of the halo (HVCs and IVCs) and their interaction with this plasma have to be further investigated as well as the heating mechanism of the halo gas.

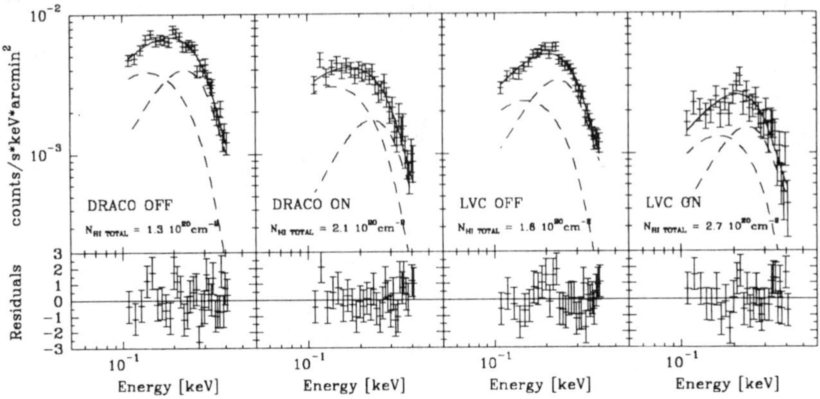

FIGURE 1 Soft X-ray spectra obtained with the ROSAT X-ray telescope. Obviously the spectral shape is different between the ON and OFF positions. At the ON positions the count rate is nearly constant up to 0.15 keV. This observational fact can directly be linked to the contribution of the LHB. The integral fit of the data is indicated by the solid line while the dashed lines display the influence of each temperature component to the spectrum. I used a LHB temperature of $T_{LHB} = 5.85\,\mathrm{K}$ and $T_{DISTANT} = 6.5\,\mathrm{K}$ for the fit of a Raymond & Smith model to the data. The integral H I column density within the line of sight is also noted.

REFERENCES

Burrows D.N., Mendenhall J.A., 1991, *Nat*, **351**, 629

Garmire G.P., Nousek J.A., Apparao K.M.V. et al., 1992, *ApJ*, **399**, 694

Hobbs L.M., Blitz L., Magnani L., 1986, *ApJ*, **306**, L109

Juda M., Bloch J.J., Edwards B.C. et al., 1991, *ApJ*, **367**, 182

Kerp J., 1993 submitted to *A&A*

Kerp J., Herbstmeier U., Mebold U., 1993, *A&A*, **268**, L21

Lilienthal D., Wennmacher A., Herbstmeier U., Mebold U., 1991, *A&A*, **250**, 150

Magnani L., Blitz L., Mundy L., 1985, *ApJ*, **295**, 402

Raymond J.C., Smith B.W., 1977, *ApJS*, **35**, 419

Snowden S.L., Cox D.P., McCammon D., Sanders W.T., 1990a, *ApJ*, **354**, 211

Snowden S.L., Schmitt J.H.M.M., Edwards B.C., 1990b, *ApJ*, **364**, 118

Snowden S.L., Mebold U., Hirth W. et al., 1991, *Sci*, **252**, 1529

Snowden S.L., McCammon D., Verter F., 1993, *ApJ*, **409**, L21

Wang Q., 1992, *ApJ*, **392**, 509

Wang Q., McCray R., 1993, *ApJ*, **409**, L37

Wennmacher A., Lilienthal D., Herbstmeier U., 1992, *A&A*, **261**, L9

16

14

CHAPTER II

INFRARED SPATIAL AND SPECTRAL STUDIES

The First Symposium on the Infrared Cirrus and Diffuse Interstellar Clouds
ASP Conference Series, Vol. 58, 1994
Roc M. Cutri and William B. Latter (eds.)

DUST AND GAS IN THE INFRARED CIRRUS

F. BOULANGER
Institut d'Astrophysique Spatiale Bât. 121, Université Paris XI, 91405 Orsay Cedex France

ABSTRACT This paper discusses infrared properties of cirrus derived from the IRAS sky images in relation with complementary observations of the associated dust and gas. We go into the use of the 100 μm emission of cirrus as a tracer of interstellar gas irrespective of it being atomic or molecular. The IRAS data shows that dust evolves within cirrus. We summarize the evidence for variations of the size distribution of dust particles and in particular the abundance of mid-IR emitters. We show that this abundance correlates with changes in the 2200 Å bump and far-UV section of the extinction curve. The mixed nature, atomic-molecular, of interstellar gas in cirrus is discussed. We point out correlations between the dust emission and extinction properties with the abundance of CO and CH^+.

INTRODUCTION

When Low et al. (1984) in the first series of papers on IRAS reported the discovery of extended sources with a filamentary structure reminiscent of cirrus clouds in our atmosphere, one wondered if they comprised some new class of celestial objects. Already in this first paper Low et al. pointed at the association of several cirrus features with known interstellar clouds detected in the 21 cm line of H I, but the correlation was not one to one. The idea that some of them could be in the outer parts of the solar system was around, but no systematic search for parallaxes was ever carried out so that this idea could not be dismissed. The difficulty of carrying such a search illustrates the complexity of the IR sky revealed by the IRAS images. Structures are interwoven from the smallest resolved scales to large scale features such as shells covering several tens of degrees. Structures are present at all brightness levels with a wide range of contrast and it is only over a few pieces of the high latitude sky that no real structure is observed in the 100 μm images. Obviously the information in these images may not be summarized in a catalog of objects with tabulated properties. Since one lacks mathematical tools to globally characterize the observed structures, studies of cirrus have so far been limited to a small number of localized features. Global studies have been carried out only at the $1/2°$ resolution of the H I sky surveys. Ten years after the IRAS launch much work still remains to be done. Compared to early studies new investigations will be made easier by the high quality of the images in the recently completed IRAS Sky Survey Atlas.

The low resolution comparison between far-IR brightness and the 21 cm H I emission has demonstrated that most of the infrared cirrus is the emission of interstellar dust associated with interstellar clouds outside of the Galactic plane, and heated by the Galactic interstellar radiation field. Numerous features in the high latitude sky images have been identified with interstellar dust locally heated by an early type star (van Buren 1989; Murthy et al. 1992). These features, which have been used to estimate the filling factor of interstellar gas versus density (see van Buren, this volume), are not discussed further in this paper. Away from local heat sources, the mean 100 μm emission per H nucleon, 0.85 MJy/sr for 10^{20} H cm^{-2}, measured from the correlation between the IRAS and H I low resolution all-sky map coincides almost exactly with the value predicted by Mathis et al. (1983) and Draine and Lee (1984) using the Mathis, Rumpl and Nordsieck (MRN; 1977) dust model. The discovery of infrared cirrus was thus theoretically expected and the discovery was made thanks to an unexpected stability of the detectors which made the mapping of extended emission possible.

The presence of gas at high latitudes was already known from studies of absorption lines towards stars, extinction and reflected light in optical plates (e.g., Sandage 1976), 21 cm emission from H I (see Kulkarni and Heiles 1987) and also line emission from CO (Magnani et al. 1985). Compared to these earlier observations the big step forward made by IRAS was to provide a spectacular view at the intricate structure of this matter and this over the whole sky and over a large range of brightness. Combined with other data they may be used to study the composition and structure of interstellar matter. In the following, we list specific topics which will be discussed in this paper. The far-IR brightness is widely used as a tracer of gas within cirrus irrespective of the chemical composition and physical state of the gas. Observations of cirrus have greatly contributed to change our view at interstellar dust. The energy distribution of the cirrus emission between the four IRAS wavelengths is related to the abundance and the size distribution of dust particles. Variations of this energy distribution among and within cirrus clouds indicate that this size distribution changes on small scales within cirrus clouds. The IRAS colors are correlated with the shape of the UV extinction curve. There is already some observational evidence suggesting that the changes in dust composition are coupled to the internal structure of cirrus and to the gas chemistry. The cirrus seen at high Galactic latitudes is a widespread component of interstelar matter which accounts for a dominant fraction of the diffuse IR emission from the Galaxy. Images of cirrus in the inner parts of the Galaxy have been obtained by filtering IRAS images of the Galactic plane (see also Waller and Boulanger, this volume).

CORRELATION BETWEEN FAR-IR EMISSION AND GAS DISTRIBUTION

The 100 μm – H I correlation

The correlation between far-IR emission and gas distribution was first tested and established from the correlation of the IR brightness with the 21 cm emission from atomic gas (see Boulanger and Pérault 1987; Désert et al. 1988 and references therein). These comparisons, initially done with the low resolution sky surveys, were later extended to H I observations with resolution comparable or better than IRAS (Herter et al. 1990; Joncas et al. 1992). At low resolution there is a general correlation between the spatial distribution of the 21 cm and far-IR emission which implies that the bulk of the far-IR emission at high latitude comes from dust associated with atomic gas. There are regions of excess or deficit IR emission with respect to the mean correlation, but they account for only $\sim 2\%$ of the total emission. The slope of the correlation is observed to vary from place to place over the sky, but by less than 50% over the sky area sampling matter where the heating radiation may be assumed to be roughly uniform. In practice this applies to regions at $b > 20°$, away from the Sco-Cen and Orion OB associations. This result suggests that large variations of the far-IR emission per nucleon, such as those measured from the molecular ring to the outer parts of the Galaxy (Bloemen et al. 1990) result from changes in the radiation field and not in the dust emissivity temperature or abundance.

At higher angular resolution one may clearly identify clouds for which the IR–H I correlation extends down to the IRAS resolution and others for which there is no correlation (e.g., de Vries et al. 1987). Much of the latter cirrus coincides with high latitude molecular clouds seen in emission in CO (e.g., Weiland et al. 1986) but regions of excess far-IR emission with respect to the mean correlation with H I emission have been identified with sizes up to several degrees (Blitz et al. 1990; Boulanger et al. 1994). In the few atomic couds which were observed at 21 cm at high resolution, the correlation is never perfect and regions of excess and deficient IR emission exist. The regions of excess IR emission may indicate places where the H I gas is cold and the 21 cm emission is optically thick or where some molecular hydrogen is present. The regions of deficient IR emission do not have such a straightforward explanation. They raise the question of variability of the far-IR emission per nucleon among different velocity components. The H I observations show that cirrus clouds have a complex velocity structure (e.g., Deul and Burton 1990; Joncas et al. 1992). Some cirrus appears to correlate spatially with gas in distinct velocity components extending into the Intermediate Velocity Cloud (IVCs) range. Some cirrus coincide with IVC clouds (Désert et al. 1990; Herbstmeier et al. 1993) which show that dust is present in these clouds but no systematic investigation of the coincidence of cirrus features with high and intermediate velocity gas, possibly interacting with gas at the mean Solar Neighborhood velocity, has yet been carried out. The new Leiden-Dwingeloo survey of Galactic H I (Hartmann and Burton 1993) is very much suited to such an investigation. The questions calling for such an investigation are to what extent the morphology of cirrus results from shocks in between merging gas flows, and what is the effect of shocks (those able to throw gas out of the galactic disk) on the abundance and composition of interstellar dust.

The 100 μm Emission per Nucleon in Molecular Clouds

The CO emission is not a good tracer of the distribution of gas within nearby molecular clouds. The reliability of the far-IR emission as a tracer of gas distribution within molecular clouds is therefore better investigated by comparing the 100 μm brightness with the visible extinction, rather than with the CO emission (e.g., Laureijs et al. 1987, 1991). However, this is not easy to do for most of the high latitude cirrus where the extinction is small and difficult to measure. One must rely on nearby clouds at intermediate latitudes with moderate extinctions, A_V up to 2 – 3 mag. In Figure 1 we show two examples of $I_\nu(100\ \mu m)$–A_V diagram selected from a study of the Chamaeleon region (Boulanger et al. 1994). The clouds, Cham III and G300-17 are close to each other, have a comparable range of opacities but differ in IRAS colors. G300-17 has a strong 12 and 25 μm halo while the mid-IR brightness of Cham III is small. In both diagrams the data points are compared with model calculations taken from Bernard et al. (1992). These calculations are based on the dust model of Désert et al. (1990). They correspond to spherical clouds with uniform dust properties and a gas density law following a r^{-2} law. The observed correlation is significantly different between the two clouds but the mean slope measured between 0 and 2 mag are close, 10.5 MJy/sr per mag for G300-17 compared to 7 for Cham III. These values are somewhat lower than the mean value of 15.9 MJy/sr per mag for atomic clouds at high latitude. The model calculations are useful in predicting how the IR emission per nucleon is lowered by the attenuation of the radiation field within non-transparent clouds. They have also been used to estimate the effect of cloud clumpiness on the IR emission of clouds (Bernard et al. 1993). The ratio of IR luminosity to gas mass is not very dependent on the clumpiness as long as the surface filling factor of dense sub-structures is close to 1, or in other words as long as the opacity of the dense sub-structures is not larger than the overall opacity of the cloud

In Figure 1, the results of the model do not fit the data but are close to it. The difference of slope between atomic and molecular clouds with moderate opacity is thus roughly accounted for by what is expected from the attenuation of the radiation field. The peculiar shape of the diagram for G300-17, with a marked increase of the slope of the correlation above 1 mag where a flattening of the relation is expected from the attenuation of the heating radiation, coincides with a sudden change in the IRAS colors in between the halo and the core of the cloud (Bernard et al. 1993). We consider this as possible evidence for an effect of the abundance of mid-IR emitters onto the far-IR emission per nucleon of interstellar matter. The two scatter diagrams in Figure 1 look qualitatively alike when we plot the total IR emission summed over the four IRAS bands, instead of the 100 μm brightness alone, but quantitatively the difference in the ratio between IR emission and H column density, is much larger in this case: a factor 2.5 higher for G300-17 compared to Cham III.

From the existing data, we conclude that for extinctions up to the order of 2 mag the 100 μm brightness may be considered as a good tracer of the column density of gas within a factor of 2. The far-IR and sub-mm spectrum of the Galaxy measured by COBE (Wright et al. 1991) has answered an outstanding question about the use of IRAS data as a tracer of interstellar matter by demonstrating that any dust component, too cold to have been detected by IRAS, accounts for at most a negligible fraction of the emission from interstellar dust. Of course this result applies on average in the Galaxy, it may not be true locally on small scales.

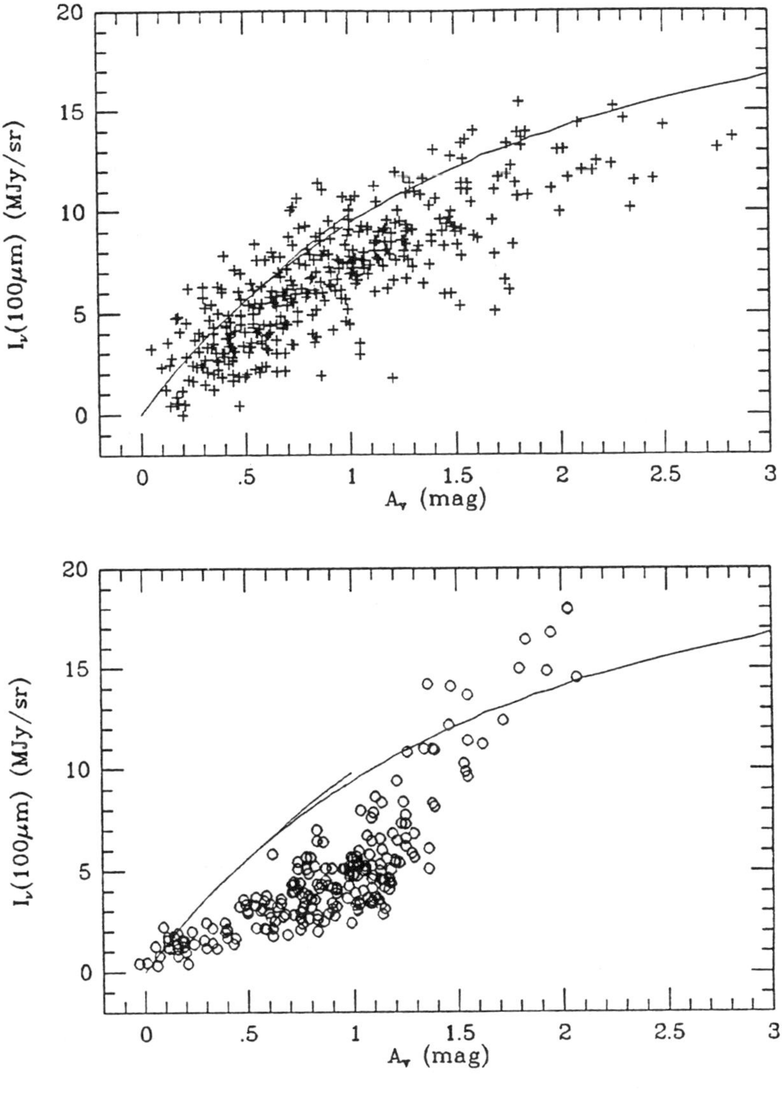

FIGURE 1 Comparison of the 100 μm brightness versus A_V measured from star counts for two adjacent clouds in Chamaeleon: Cham III (plus signs) and G300-17 (circles) (Boulanger et al. 1994). The clouds differ in their IRAS colors. Cham III has little mid-IR emission and a low 60/100 μm color; G300-17 has a halo of $A_V = 1$ mag with high 12, 25 and 60/100 μm colors. The solid lines represent model calculations taken from Bernard et al. (1992) for spherical clouds with total A_V of 1 and 4 mag, a r^{-2} density law and uniform dust properties.

DUST IN CIRRUS

The Near and Mid-IR Emission and the Need for Three Dust Components

The power radiated by interstellar dust in cirrus is distributed over a range of wavelengths much wider than the emission of grains at any given temperature. The emission must therefore originate from particles emitting over a wide range of temperatures. By detecting mid-IR emission from cirrus (Boulanger et al. 1985), the IRAS observations confirmed spectacularly the discovery, first made in reflection nebulae (Sellgren et al. 1983), of the conspicuous presence in the ISM of small particles heated by the absorption of photons to temperatures much higher than the equilibrium temperature of large grains ($T_{eq} \sim 20$ K). Such particles were not included in pre-IRAS dust models (e.g., MRN model) based only on extinction and scattering data. The IRAS and near-IR data on the emission of dust imply that the size distribution of dust particles extends without any gap down to molecules with a few tens of atoms. These molecules are identified with Polycyclic Aromatic Hydrocarbons (PAHs) account for the emission bands in the near-IR (see review by Puget and Léger 1989). The 3.3 μm emission of cirrus associated with the molecular ring (see ection Looking Further Away) has been detected by Giard et al. (1989). In the dust model of Désert et al. (1990), PAHs account for the near-IR and 12 μm IRAS emission, very small grains (VSGs) in the size range 15 to 50 Å account for the 25 μm and about half of the 60 μm emission. The rest of the emission is due to larger particles at equilibrium temperature. PAHs and VSGs each account for $\sim$ 20% of the emission.

Variations in the IRAS Colors

The energy distribution of the IR emission of interstellar matter within the IRAS bands has been shown to vary within nearby molecular cloud complexes and among cirrus clouds (Weiland et al. 1986; Heiles et al. 1988; Penprase and Helou, this volume). Relative to the 100 μm emission, the 12 and 25 μm emission vary more than one order of magnitude around the mean colors, R(12,100) [$= I_\nu(12\ \mu\text{m})/I_\nu(100\ \mu m)$] = 0.04 and R(25,100) = 0.05. Boulanger et al. (1990) have argued that these variations trace changes in the abundance of mid–IR emitters. Laureijs et al. (1991) have shown that the 100 μm emission from molecular clouds may be separated in warm and cold components on the basis of the R(60,100) color. The warm component has a R(60,100) color $\sim$ 0.2 comparable to the mean color ratio of cirrus. The cold one, with R(60,100) $\sim$ 0.1 or smaller, is often not detectable in the 60 μm images. This cold component is absent or weak for most cirrus clouds except the most massive clouds like MBM 12.

Bernard et al. (1992, 1993) have coupled radiation transfer calculations to the dust model of Désert et al. to quantify the variations in IRAS colors in terms of PAHs and VSGs abundance changes. In a second paper they have applied the model calculations to specific brightness profiles across clouds in Chamaeleon and Ophiuchus. For haloes of clouds with high mid-IR brightness [R(12,100) and R(25,100) $\sim$ 0.1] they find a PAH abundance enhanced by a factor of five with respect to the mean value in the Solar Neighborhood. For

VSGs the enhancement is only a factor of 2. For the cold component with a low R(60,100) ($\sim$ 0.1) and no measured mid-IR emission the PAHs and VSGs abundances are depleted by a factor 4 and 2 with respect to the mean Solar Neighborhood value.

Correlation between Dust Emission and Extinction

To further characterize the changes in dust composition within the ISM, Boulanger et al. (1993a) obtained extinction curves for 8 stars in Chamaeleon behind clouds with mid-IR to 100 μm colors differing by one order of magnitude. The extinction curves are shown in Figure 2. The 12/100 μm color of the absorbing matter increases from the top to the bottom of the figure by an order of magnitude. With decreasing R(12,100) the 2200 Å bump is observed to become weaker (smaller integrated area) and wider; the far-UV extinction and in particular the non-linear far-UV rise increases. The extinction curve measured for the high latitude star HD 210221 fits in this trend. It has a weak bump and a strong far-UV extinction (Welty and Flower 1992) which fits with the low mid-IR brighness of the foreground cirrus [$I_\nu(12\mu m) \sim 0.15$ MJy/sr]. The fact that the extinction curve is observed to change with the mid-IR brightness of the absorbing clouds supports the idea that the IRAS colors indicate changes in the composition of dust. Evidence for the evolution of dust particles within cirrus also also comes from differences in optical colors and in between IR and optical images (Guhathakurta and Tyson 1989, Paley et al. 1991, Guhathakurta and Cutri, this volume).

An important result of the extinction study is that the amplitude of none of the spectral features in the extinction curve scales with the abundance of the mid-IR emitters. This means that none of these spectral features is specific to the mid-IR emitters which is not necessarily surprising since the size range of particles susceptible to contribute to spectral features in the UV extintion, up to sizes of a few 100 Å, is much larger than that of mid-IR emitters (size < 30 Å). Boulanger et al. (1993a) conclude from this result that the abundance of mid-IR emitters depends on a process which changes the size distribution of small particles but not their overall abundance. Where their abundance is low these particles have been transformed by accretion into particles too large to emit in the mid-IR but still small enough to be optically thin to UV radiation. This interpretation implies that small particles grow in size by mutual coagulation or accretion of atoms and molecules but do not condense on large grains. The size distribution of small particles may also be affected by shocks.

GAS CHEMICAL COMPOSITION VERSUS INFRARED EMISSION

Gas in Cirrus is dominantly atomic but the comparison between IRAS, CO, and H I observations suggest that a large fraction if not all of the cirrus features contain some molecular gas. This is in particular demonstrated by the large success rate in detecting CO emission in cirrus selected on the basis of the IRAS images alone (e.g., Stark 1993; Reach et al. 1993; Guhathakurta and Cutri, this volume). The molecular fraction of gas is observed to be highly variable from one cirrus feature to another.

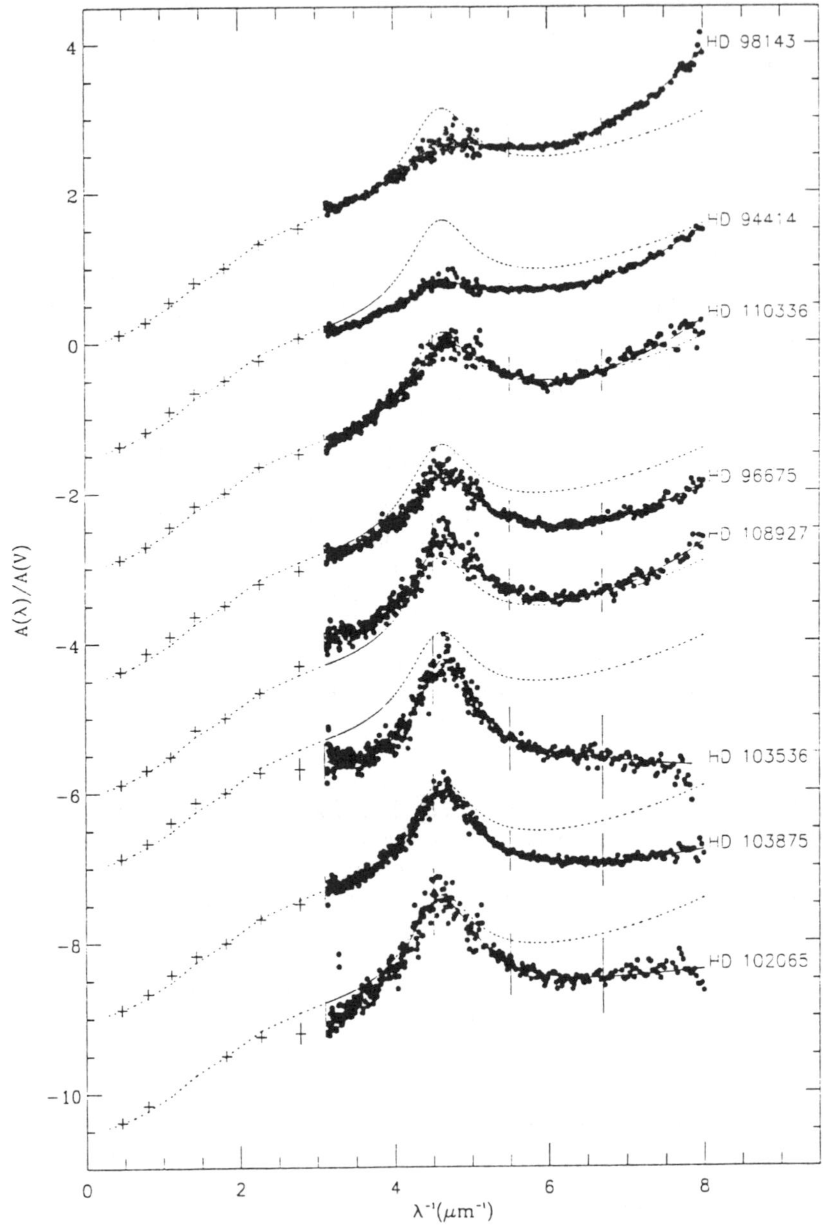

FIGURE 2 Extinction curves for a set of stars in Chamaeleon behind clouds sampling one order of magnitude in mid-IR emissivity as measured by the ratio $I_\nu(12\mu m)/A_V$ (Boulanger et al. 1993a). The curves are presented in order of increasing mid-IR emissivity from top to bottom. For each curve, the dotted line is the mean Solar Neighborhood extinction curve and the solid line an analytical fit to the data.

Dust Properties versus Molecular Content

The detection of molecular clouds at high latitude through CO observations and the discovery of IRAS cirrus have prompted numerous studies of absorption lines towards high latitude stars. One motivation was to estimate the distance to cirrus clouds (Hobbs et al. 1986, 1988; Welty et al. 1989). Another important motivation is to study the gas chemistry within these clouds. If some molecular cores have been detected in emission lines of NH_3, H_2CO, C_3H_2 and HC_3N (e.g., Turner et al. 1989; Turner, this volume), over the bulk of the infrared cirrus, even where CO emission is detected, such lines cannot presently be detected. With maps of H I and CO emission, absorption studies provide the basic information we have about the chemical composition of cirrus clouds. The IRAS data brings morphological information about clouds detected through absorption line studies. They also provide a mean to correlate the chemical composition and the physical state of the gas with observed emission and extinction properties of dust.

In Figure 3 we present preliminary results of a comparison of molecular observations with IRAS data for a set of stars in Chamaeleon (the same as that used in the extiction study) complemented with other stars behind high latitude molecular clouds for which CH has been detected (de Vries and van Dishoeck 1988; Penprase 1993). CH has been shown to be well correlated with the H_2 column density in diffuse clouds (Danks et al. 1984) and in dark clouds (Mattila 1986). The top diagram in Figure 3 shows that the CH abundance, CH/H_2 [N(H_2) is derived from A_V using the standard N(H)/A_V ratio, and assuming that all of the gas is molecular], is within a factor of 2, $\sim 2.5 \times 10^{-8}$, for all of the stars. This value is within the range of values measured for diffuse and dark clouds. For four of the Chamaeleon stars with CH detections, the J=1–0 emission line of CO was not detected with a 1σ sensitivity of 0.1 K. This is in particular the case for the three stars behind a cloud in Chamaeleon with bright mid-IR emission which was surveyed for CO emission over an extended area without any detection (Boulanger et al. 1994). This is a clear instance of a cloud with molecular gas with no detected CO emission.

For the same lines of sight as in the top diagram, the bottom diagram in Figure 3 is a plot of the CH^+ to CH column density ratio versus the 12 μm per unit extinction $I_\nu(12\,\mu m)/A_V$. Unlike CH, the abundance of CH^+ is known to be highly variable within diffuse and translucent clouds (e.g., Gredel et al. 1993). The measured abundance ratios differ by more than one order of magnitude and values of the ratio of the order of 1 are observed only for lines of sights crossing clouds with high mid-IR emission as measured by $I_\nu(12\,\mu m)/A_V$. The nature of this correlation between the abundance of mid-IR emitters and the abundance of a molecule which may be formed only in shocks or in high temperature gas is unclear.

CO Emission versus Far-IR Emission

The comparison of CO and the far-infrared emission of cirrus shows little correlation. Based on a large and deep CO survey of cirrus clouds around the North Celestial Pole, Heithausen et al. (1993) concluded that there is not a one-to-one correlation between the two intensities and that there is no threshold of I_ν(100 μm) for the detection of CO. For the cloud G300-17 in Chamaeleon, 90% of the

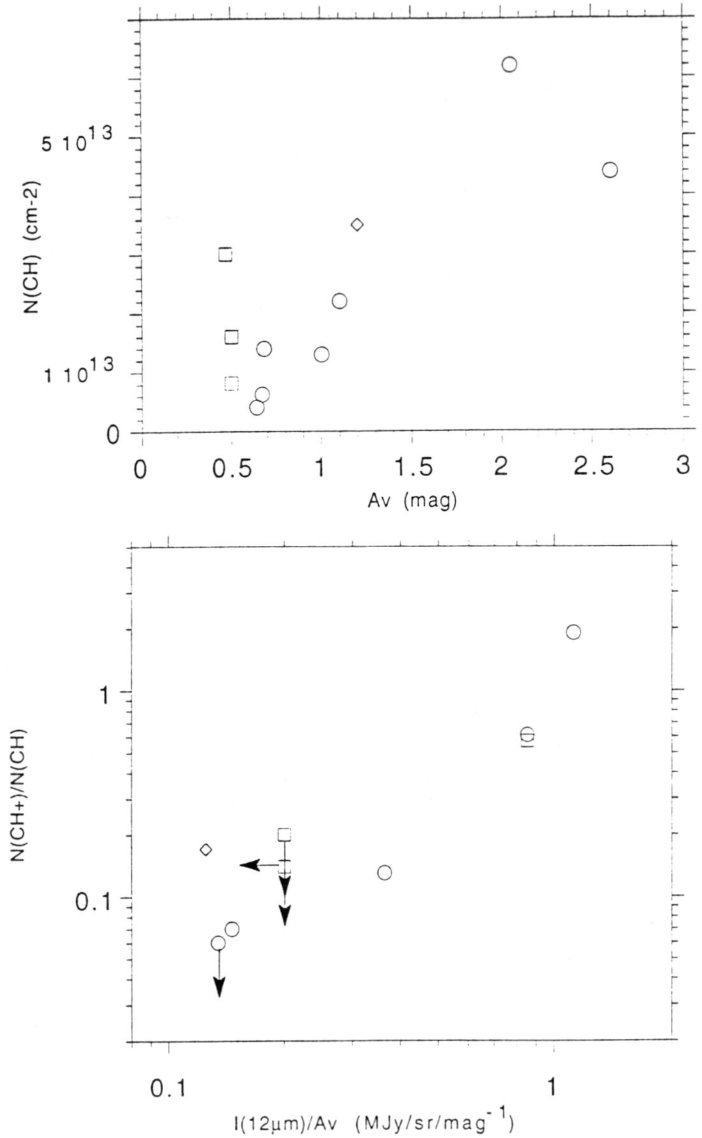

FIGURE 3 CH and CH+ observations in absorption towards background stars versus the extinction toward the star and the 12 μm brightness from the cloud at the line of sight to the star. The plotting symbols are circles for the Chamaeleon stars (Boulanger et al. 1993b), squares for three stars at high ecliptic and galactic latitude (HD 57336, HD 77005 and HD 146010) taken from the sample of Penprase (1993) and a diamond for HD 210221 (de Vries and van Dishoeck 1988).

matter seen at 100 μm is not seen in emission in CO (J=1–0) at a sensitivity level of 0.25 K (1σ) (Boulanger et al. 1994). Lines of sight in Chamaeleon with I_ν(100 μm) up to 10 MJy/sr, and A_V values of 1 mag have no CO (J=1–0) emission at a sensitivity level of 0.1 K (Boulanger et al. 1993b), while integrated CO emission up to 8 K km/s are detected where $I_\nu(100\ \mu m) = 3$ MJy/sr [N(H)$\sim 5 \times 10^{20}$; Heithausen et al. 1993]. For molecular fragments within the Draco Nebula, Herbstmeier et al. (1993) have measured values of the X_{CO} ratio between NH_2 and CO (J=1–0) integrated emission one to two orders of magnitude smaller than the mean Galactic value of 2.3×10^{20} cm^{-2} (K km/s)$^{-1}$. The CO emission is clearly far from being simply related to the column density of molecular hydrogen in cirrus.

Measurements of the CO column density, together with investigations of physical conditions indicate that cirrus clouds span the critical regime where the abundance of CO starts to become significant but in which most of the carbon is still in atomic form (van Dishoeck et al. 1991, Gredel et al. 1992). Theoretical models (van Dishoeck and Black 1988; Viala et al. 1988) show that in this regime, small fluctuations in involved physical parameters can lead to large variations in observable CO line intensity. The abundance of CO depends not only on the total column density but also on the gas density, the far-UV radiation field and the dust extinction in the far-UV. Once the molecule is present, the measured brightness also depends on excitation conditions and the radiation transfer which in a much structured medium may be more complex than inferred from what available models tell us. It is in any case not clear what explains the formation of molecular condensations in some cirrus features and not others.

The Chamaeleon data (Figure 2) suggest that there is a difference in far-UV extinction between clouds with and without CO emission. Other observations have pointed out a strong far-UV extinction for lines of sight with molecular gas (Cardelli and Clayton 1991; Jenniskens et al. 1992). Dust may facilitate the onset of formation of self-shielded CO by attenuating radiation coinciding with photo-dissociative transitions. Is this alone sufficient to explain the observed differences in CO abundance? This question remains to be tested.

The CO emitting gas has been shown to be structured on scales much smaller than the IRAS resolution (Falgarone et al. 1991). Small scale clumping is also necessary to explain the observed excitation of the CO rotational transitions. Gas densities in CO emitting pieces of cirrus are estimated to be at least 10^3 H_2 cm^{-3} (Falgarone et al. 1991; van Dishoeck et al. 1991). This lower value is much larger than the mean densities one gets from the mean column densities and structure sizes derived from the IRAS images, especially where CO is detected and the mean column density of gas as estimated from I_ν(100 μm) is low. The density and velocity stucture of clouds (see Falgarone, this volume), could be the key point explaining the differences in the gas and dust composition observed among and within cirrus clouds. The good correlation observed between the cold 100 μm emission (see section on dust) and the ^{13}CO emission in nearby molecular clouds (Laureijs et al. 1991; Abergel et al. 1993) suggests that the evolution of dust in clouds is indeed related to their density structure. This correlation is expected if the variations in dust composition result from coagulation of grains and/or the accretion of gas species on dust particles.

LOOKING FURTHER AWAY

Up to several degrees in latitude, Galactic longitude cuts across the IRAS Galactic plane images show a clear emission enhancement over the longitude range ±30° of the molecular ring (Pérault 1987). Since the distribution of gas does not show a similar enhancement, the excess emission within the molecular ring longitude range is interpreted as coming from high z gas above the molecular ring heated by young stars in the ring, or in other words of cirrus clouds associated with the molecular ring. The molecular ring cirrus are brighter than the nearby ones; the mean far-IR emission per nucleon in the molecular ring has been estimated to be a factor $\sim$ 6 higher than the Solar Neighborhood value (Bloemen et al. 1990). Studies of the origin of the infrared emission form the Galaxy agree to conclude that 50% of the total IR emission (70% of the diffuse component) comes from the atomic component of the ISM heated by the Galactic interstellar radiation field, i.e., from Galactic cirrus. For an extended review on the origin of the IR emission from the Galaxy the reader is referred to Cox and Mezger (1989) and references therein.

More recently Waller and Boulanger (1993) have filtered galactic plane images to produce spectacular images of cirrus features extending from intermediate latitudes down to the Galactic plane. The mean brightness of these features scales like the mean Galactic emission profile versus Galactic latitudes. Cirrus observed within a few degress from the plane are too bright to be local. They must be associated with the molecular ring which must be confirmed by associations with features in H I and CO maps. This will allow an estimate of their distance from the measured velocities.

CONCLUSION

The discovery of infrared cirrus by IRAS has opened a new perspective on interstellar matter. Matter was known to exist in between the largest and densest molecular clouds, but little was known about its structure, dynamical state and chemical composition at the time of the IRAS launch. The spectacular images revealed by IRAS have brought many people to work on this matter. CO and H I observations have broadened the IRAS perspective by providing images on smaller scales and velocity information. They show that the filamentary morphology extends to much smaller scales than the IRAS resolution and goes together with a complex velocity structure (Falgarone, this volume).

Cirrus clouds are the most widespread component of dense interstellar matter. They comprise the nearest clouds to us for which we can get the most detailed spatial information about the structure of interstellar matter. Observations of cirrus have provided new insights about many key processes in the evolution of interstellar matter. In particular, the chemical transition between atomic and molecular gas, the formation of CO, the composition of dust and its evolution within the interstellar medium, and the link between dust evolution and the physical state and chemistry of gas. These questions are probably all related to the understanding of the labyrinthine structure of cirrus.

ACKNOWLEDGMENTS

We are grateful to Edith Falgarone for comments on the manuscript.

REFERENCES

Abergel, A., Boulanger, F., Mizuno, A. and Fukui, Y. 1993, *ApJ*, in press

Bernard, J. P., Boulanger, F., Désert , F. X., and Puget, J. L. 1992, *A&A*, 263, 258

Bernard, J.P., Boulanger, F., and Puget, J. L. 1993, *A&A*, in press

Blitz, L., Bazell, D., and Désert , F. X. 1990, *ApJ*, 352, L13

Bloemen, J.B.G.M., Deul, E. R., and Thaddeus, P. 1990, *A&A*, 233, 437

Boulanger, F., Baud, B., and van Albada, G. D. 1985, *A&A*, 144, L9

Boulanger, F. and Pérault, M. 1988, *ApJ*, 330, 964

Boulanger, F., Falgarone, E., Puget, J.L., and Helou, G. 1990, *ApJ*, 364, 136

Boulanger, F., Prévot, M. L., and Gry, C. 1993a, *A&A*, in press

Boulanger, F., et al. 1993b, in preparation

Boulanger, F., Bronfman, L. and Thaddeus, P. 1994, *A&A*, submitted

Cardelli, J. A. and Clayton, G. C. 1991, *AJ*, 101, 1021

Cox, P. and Mezger, P. G. 1989, *A&A*Rev. 1, 49

Danks, A. C., Federman, S. R. and Lambert, D. L. 1984, *A&A*, 130, 62

Désert , F. X., Bazell, D. and Boulanger, F. 1988, *ApJ*, 334, 815

Désert , F. X., Bazell, D. and Blitz, L. 1990, *ApJ*, 355, L51

Deul, E. R. and Burton, W. B. 1990, *A&A*, 230, 153

de Vries, C. P. and van Dishoeck, E. F. 1988, *A&A*, 203, L23

de Vries, H. W., Heithausen, A. and Thaddeus, P. 1987, *ApJ*, 319, 723

Draine, B. T., and Lee, H. M. 1984, *ApJ*, 285, 89

Falgarone, E., Phillips, T. G. and Walker, C. K. 1991, *ApJ*, 378, 186

Giard, M., Pajot, F., Lamarre, J.M., Serra, G., and Caux, E. 1989, *A&A*, 215, 92

Gredel, R., van Dishoeck, E. F., de Vries, C. P. and Black, J. H. 1992, *A&A*, 257, 245

Gredel, R., van Dishoeck, E. F., and Black, J. H. 1993, *A&A*, 269, 477

Guhathakurta, P. and Tyson, J. A. 1989, *ApJ*, 346, 773

Hartmann, D. and Burton, W. B. 1993, in *Back to the Galaxy*, S.S. Holt and F. Verter, American Institute of Physics, 279

Heiles, C., Reach, W. T. and Koo, B. C. 1988, *ApJ*, 332, 313

Heithausen, A., Stacy, J. G., de Vries, H. W., Mebold, U. and Thaddeus, P. 1993, *A&A*, 268, 265

Herter, T., Shupe, D. L. and Chernoff, D. F. 1990, *ApJ*, 352, 149

Hobbs, L. M., Blitz, L., and Magnani, L. 1986, *ApJ*, 306, L109

Hobbs, L. M., Blitz, L., Penprase, B. E., Magnani, L., Welty, D. E. 1988, *ApJ*, 327, 356

Jenniskens, P., Ehrenfreund, P. and Désert, F. X. 1992, *A&A*, 265, L1

Joncas, G., Boulanger, F., Dewdney, P. E. 1992, *ApJ*, 397, 165

Kulkarni, S. R. and Heiles, C. 1987, in *Interstellar Processes*, D. J. Hollenbach, H. A. Thronson Jr., Reidel, 87

Laureijs, R. J., Mattila, K. and Schnuur, G. 1987, *A&A*, 184, 269

Laureijs, R. J., Clark, F. O., and Prusti, T. 1991, *ApJ*, 372, 185

Low, F. J., et al. 1984, *ApJ*, 278, L19

Magnani, L., Blitz, L. and Mundy, L. 1985, *ApJ*, 295, 402

Mathis, J. S., Rumpl, W. and Nordsieck, K. H. 1977, *ApJ*, 217, 425

Mattila, K. 1986, *A&A*, 160, 157

Murthy, J., Walker, H. J., and Henry, R. C. 1992, *ApJ*, 401, 574

Paley, E. S., Low, F. J., McGraw, J. T., Cutri, R. M., and Rix, H. W. 1991, *ApJ*, 376, 335

Penprase, B. E. 1993, *ApJS*, in press.

Pérault , M. 1987, Thèse d'Etat, Université Paris VII

Puget, J.L. and Léger, A. 1989, *ARA&A*, 27, 161

Reach, W. T., Heiles, C. and Koo, B. C. 1993, *ApJ*, 412, 127

Sandage, A. 1976, *AJ*, 81, 954

Sellgren, K., Werner, M. W. and Dinerstein, H. L. 1983, *ApJ*, 271, L13

Stark, R. 1993, Ph.D. Thesis, Leiden University.

Turner, B. E., Rickard, L. J., and Xu, L. 1989, *ApJ*, 344, 292

van Buren, D. 1989, *ApJ*, 338, 147

van Dishoeck, E. F., and Black, J. H. 1988, *ApJ*, 334, 771

van Dishoeck, E. F., Black, J. H., Phillips, T. G., and Gredel, R. 1991, *ApJ*, 366, 141

Viala, Y. P., Letzelter, C., Eidelsberg, M. and Rostas, F. 1988, *A&A*, 193, 265

Waller, W.H. and Boulanger, F. 1993, in *Back to the Galaxy*, S.S. Holt and F. Verter, American Institute of Physics, 544

Weiland, J.L., Blitz, L., Dwek, E., Hauser, M. G., Magnani, L., and Rickard, L. J. 1986, *ApJ*, 306, L101

Welty, D. E., Hobbs, L. M., Blitz, L., Penprase, B. E. 1989, *ApJ*, 346, 232

Welty, D. E. and Fowler, J. R. 1992, *ApJ*, 393, 193

Wright, E.L. et al. 1991, *ApJ*, 381, 200

The First Symposium on the Infrared Cirrus and Diffuse Interstellar Clouds
ASP Conference Series, Vol. 58, 1994
Roc M. Cutri and William B. Latter (eds.)

CIRRUS AND FUTURE SPACE BASED ASTRONOMY

T. N. GAUTIER
Jet Propulsion Laboratory, Pasadena, CA

ABSTRACT

Astronomical observatories from space make possible observations of sensitivity and spatial resolution impossible in the past. This increase in sensitivity will both make possible the observation of new phenomena and will bring observations against limitations not encountered before. This paper discusses the effects that infrared cirrus and diffuse interstellar clouds will have on space based observations. Some special opportunities provided by space observations of cirrus are presented and a partial list of currently planned observation of cirrus by space telescopes is given.

INTRODUCTION

The IRAS sky maps show that the wispy emission structures of the infrared cirrus are found in all parts of the sky. The cirrus is densest and brightest a low galactic latitudes with fewer and fewer bright clouds appearing as one approached the galactic poles. The cirrus piles up and merges into the general galactic disk emission at very low latitudes so that individual cirrus clouds are no longer distinguishable. Maps from the IRAS Infrared Sky Survey Atlas show that cirrus is structured on all spatial scales down to the resolution limit at 6 arc minutes. This fine structure exists at all infrared wavelengths seen by IRAS. The cirrus presents a finely structured distribution of emitting and absorbing dust which has much to tell about the interstellar medium but through which which we must look to make observations of any celestial objects. In many respects going to space does not get telescopes out of the weather.

The same cirrus seen in the infrared is also detected in the visible. Detections of faint, highly structured visible reflection nebulae were reported by de Vaucouleurs and Freeman (1972) and Sandage (1976). These nebulae were later found to be correlated with 100 μm cirrus seen by IRAS (e.g. de Vries and le Poole, 1985). Deep ground based CCD images now routinely detect reflection nebulosity associated with infrared cirrus to surface brightnesses of 27 to 28 mag/arcsec^2, sufficient to see the large majority of 100 μm cirrus detected by IRAS. Fine structure in the visible nebulosity is seen at arc second scales.

Cirrus is also seen in absorption absorption measurements and in UV emission. Bates and Catney (1991) have seen spatial structure on the scale of a few arc minutes in interstellar Na D absorption in the direction of the globular cluster M22. Molaro, *et al.* report similar results in the direction of the large Magellanic cloud. Van Steenberg and Shull (1988) and Hackwell, *et al.* (1991) have found correlation between some kinds of IR cirrus and interstellar UV extinction using IUE spectra and the IRAS maps. Jakobsen, *et al.* (1987) found correlations at intermediate and high galactic latitudes between IRAS 100 μm and diffuse UV emission measured between 2100 angstroms and 1600 angstroms by sounding rockets.

EFFECTS OF CIRRUS

Cirrus clouds add noise to astronomical observations in three ways. Cirrus emission increases the background against which other sources must be measured, adding photon noise to measurements. Cirrus emission is not spatially smooth, thus presenting a variable background structure against which other sources must be extracted and measured. Finally, absorption by cirrus clouds is similarly mottled making correction for extinction in individual sources difficult.

The additional photon noise introduced by cirrus emission can be significant in observations from space where, particularly in the far infrared, a moderately bright cloud at mid to high galactic latitude can contribute the bulk of the background emission for lines of sight through that cloud. This increase in photon noise is inconvenient but does not present a fundamental problem since longer integration times can restore the signal to noise ratio.

The confusion noise caused by the spatial variations in surface brightness or extinction of cirrus clouds is a fundamental problem. Confusion noise from cirrus emission results from the inability to distinguish fluctuations in cirrus surface brightness from the source being measured. Longer integration times only produce an improved signal to noise ratio on the fixed fluctuation pattern which does not improve without limit the ability to sort out the cirrus from the desired source. A similar situation holds for spatial variations in extinction of background sources by cirrus clouds.

The magnitude of cirrus emission confusion noise can be calculated from distribution functions of cirrus brightness. Gautier, *et al.* (1992) discusses calculation of cirrus confusion noise from the spatial power spectrum of the cirrus surface brightness distribution. Cirrus clouds generally display a power law spectrum over the range of spatial scales observed by IRAS, as shown in figure 1 for a typical moderately bright cirrus cloud. Figure 2 shows the cirrus confusion noise calculated for an 85 cm space based telescope using a small photometer aperture and a closely spaced annular reference region with a line of sight through this cloud and through a very dim cirrus cloud . The cirrus spectrum power law was extrapolated to small spatial scales for this calculation. Point source confusion noise away from the galactic plane, calculated for the same aperture arrangement by the method of Helou and Beichman (1990), is also shown. Note that point source confusion may be expected to dominate cirrus confusion only for the dimmest cirrus clouds. The dim cirrus is from a region of low HI emission at $l = 151°$, $b = +50°$. Only several hundreds of square degrees of sky is no noisier than the dim cirrus. Some few hundred square degrees in smaller patches than 10×10 degrees may be dimmer yet and have less cirrus confusion noise.

Cutri (1993) has obtained the spatial power spectrum of a cirrus cloud in reflected V band light and finds that the spectrum is well approximated by a

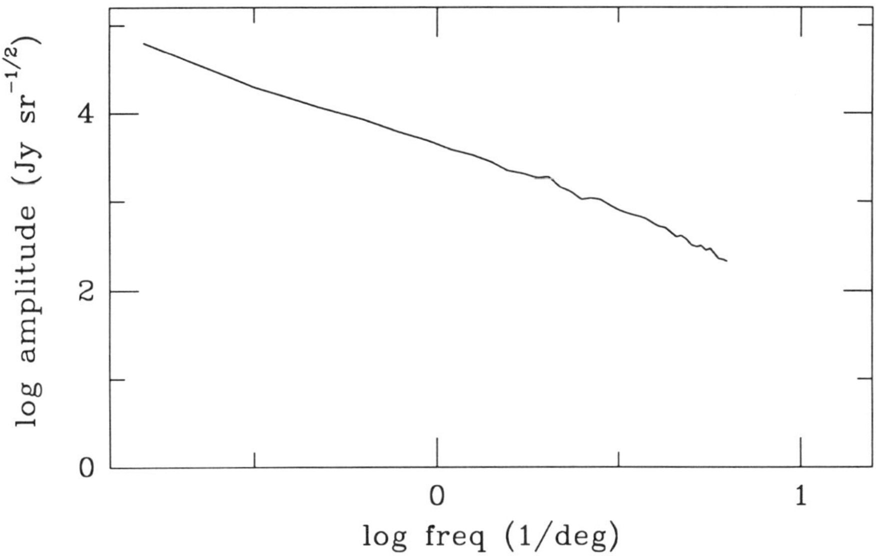

Figure 1. Spatial power spectrum of 100 μm emission from a moderately bright cirrus cloud near the north equatorial pole. This cloud is about 9 MJy/sr brighter than its surroundings. The spectrum is an azimuthal average of the amplitude of the 2 dimensional Fourier transform of the IRAS 100 μm map of the cloud.

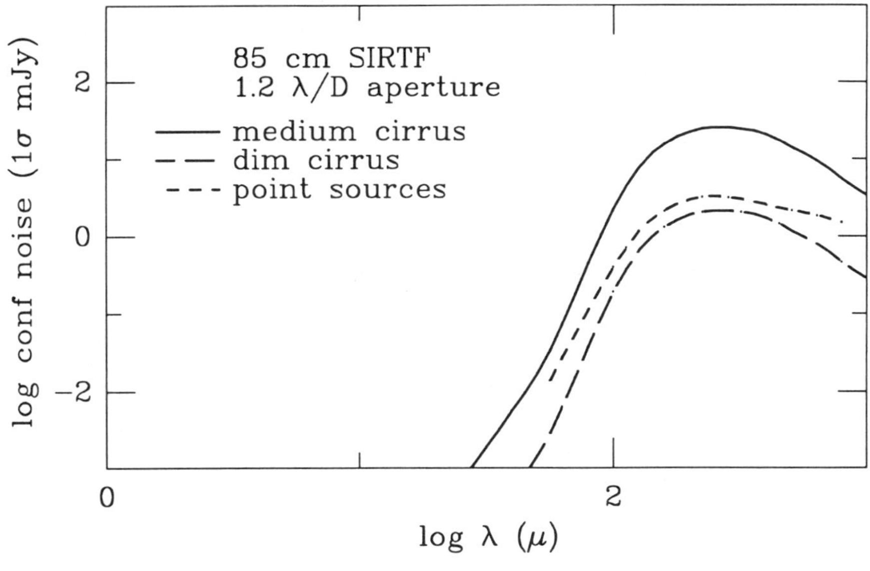

Figure 2. Predicted cirrus and point source confusion for the planned 85 cm SIRTF telescope. The "medium cirrus" is derived from the cirrus cloud used for the spectrum in figure 1. See text for explanation of point source and dim cirrus curves.

power law with the same index as found at 100 μm over spatial scales from 5 arc minutes to 1 arc second. No quantitative confusion noise calculation have been done for the reflection cirrus but there will clearly be confusion effects from this material when very sensitive observations are made. Guhathakurta and Tyson (1989) indicate that reflection cirrus confusion noise should become a problem in the visible at surface brightnesses of 30 to 31 B_J mag/arcsec^2. That is 15 times fainter than the V band cirrus referred to earlier.

Before leaving the subject of cirrus confusion I should point out that often one man's noise is another man's signal and the presence of spatial structure in the cirrus offers many opportunities to study the properties of the interstellar medium. Some of these opportunities will be discussed below.

Since the IR emission from cirrus is very often correlated with visible and UV extinction we can expect cirrus to produce a kind of absorption confusion noise in the visible and UV. The magnitude of this effect can be estimated using some representative values for the relation of A_V to the 100 μm surface brightness, S_{100}. Using $A_V(\text{mag}) = 5.3 \times 10^{-22}\ N_H$ and $N_H(\text{cm}^{-3}) = 10^{20}\ S_{100}(\text{MJy/sr})$ the variation in visible extinction with 100 μm brightness is $\delta A_V = 5.3 \times 10^{-2}\ \delta S_{100}$. Take, for example, the measurement of the 2 point correlation function for galaxies. The brightnesses of pairs of galaxies, measured through apertures of of diameter d, separated by distance Θ are compared. The double aperture form of the confusion noise formula in Gautier, *et al.* (1992), though not strictly applicable to this aperture geometry, can be used to estimate the cirrus noise from the cloud of figure 1. The power law index for 100 μm spatial spectra is generally near -2.9 which produces: $\delta S_{100}(\text{MJy/sr}) = 0.32\ A_0\ \Theta(\text{degrees})^{.63}\ d(\text{arcsec})^{-.18}$ where A_0 is the amplitude of the power spectrum at 1 arcmin^{-1}. $\Theta = 2°$ is interesting for galaxy correlation functions. Take $d = 1$ arcsec and use $A_0 = 11\ \text{Jy}/\sqrt{\text{sr}}$ from the spectrum in figure 1. Then $\delta S_{100} = 5.4$ MJy/sr and $\delta A_V = 2.9$ mag. De Lapparent *et al.* (1986) indicate that $\delta A_V < 0.04$ mag is needed to usefully measure the correlation function of galaxies at the depth of the Shane-Wirtanen counts at scales as large as 2.5 degrees. Smaller errors are needed for accuracy at larger spatial scales.

OBSERVATION OPPORTUNITIES FROM SPACE

Future spaced based astronomy promises many new opportunities to use cirrus to study structures and processes in the interstellar medium. Larger apertures and improved detectors will provide future infrared space telescopes with sensitivities in the IR hundreds to thousands of times greater than that of IRAS. Higher spatial resolution and higher spectral resolution will be available. In the visible and UV the higher spatial resolution, lower backgrounds and extended spectral coverage available from space make interesting cirrus studies possible.

An incomplete list of cirrus studies from space gives some idea of the possibilities:

1. Study of cirrus spatial structure at higher spatial resolution should eventually allow us to see structure characteristic of the processes which form and maintain the cirrus.

2. Spectral studies at moderate resolution should settle the question of what is emitting the mid IR flux from the cirrus.

3. Measurement and mapping of the CII(158 μm) line, an important coolant of cirrus, can study energy balance and kinematics.

4. The three dimension structure of the cirrus could be mapped by combining the methods of Gaustad and van Buren (1993) for finding heated dust around stars with Hipparcos data for star positions. The high sensitivity and spatial resolution of planned telescopes should be able to detect dust in the normal interstellar medium at densities of 1 H atom/cm^3 around A0 stars at distances greater than 200 pc.

5. Low surface brightness measurements in the visible and UV can be combined with IR emission measurements to yield composition information for the cirrus dust.

6. UV spectroscopic measurements of H_2 fluorescence may be possible and would allow determination of the H_2 density in cirrus clouds as has been done for reflection nebulae by Witt *et al.* (1989).

PLANNED CIRRUS OBSERVATIONS FROM SPACE

A partial summary of currently planned space observations of cirrus:

ISO (Europe)

A key project will map cirrus structure at high spatial resolution. ISO instrumentation can do spectral measurements of cirrus. The launch date is currently in late 1995.

IRTS (Japan)

Will mainly look at sky structure at low (8 arc minute) spatial resolution. About half of its two weeks of planned observations will look at the galactic plane and half at high latitudes. IRTS carries near and mid IR spectrometers and a far IR line mapper for 63 μm OI and 158 μm CII. a 1/2 degree resolution far IR photometer covers 4 bands from 120 to 1000 μm for cosmic background measurements. The flight is planned for 1995.

MSX (US Air Force)

Covers many photometric bands from 2 to 30 μm with spatial resolution greater than 10 to 20 times that of IRAS. Also carries a Fourier transform spectrometer. MSX will look at "everything". There are plans to release the data to the astronomical community some time after the planned flight in early 1994.

SIRTF (USA)

SIRTF will carry large format arrays, matched to the resolution of its 85 cm aperture, to perform photometry from 2 to 200 μm. Mid and high resolution spectrometers will cover wavelengths from the near IR to about 200 μm. No specific observing plans have been made. Launch is unlikely before 2001.

HST (USA)

Several on going observing programs are concerned with cirrus and several of the papers in this volume contain some of these results. Correction of the spherical aberration late this year should provide even greater opportunity for cirrus studies.

IUE (USA and Europe)

IUE will continue line absorption work relating to interstellar clouds until it dies.

Note that not all of the ideas above are well suited to planned space telescopes. In particular, planned and existing telescopes may not have large enough fields of view or small enough plate scales for optimum measurement of low surface brightness extended emission. Some consideration might usefully be given to intermediate sized telescopes, at all wavelengths, optimized for low surface brightness measurements.

REFERENCES

Bates, B. and Catney, M.G. 1991 *ApJ*, **371**, L37

Cutri, R. 1993 Private Communication

de Vaucouleurs, G. and Freeman, K.C. 1972 *Vistas in Astronomy*, **14**, 163

Gaustad, J.E. and van Buren, D. 1993 *PASP*, in press

Gautier, T.N., Boulanger, F., Puget, J.L. and Peŕault, M. 1992, *AJ*, **103**, 1313

Guhathakurta, P. and Tyson, J.A. 1989 *ApJ*, **346**, 773

Hackwell, J.A., Hecht, J.H. and Tapia, M. 1991 *ApJ*, **375**, 163

Helou, G. and Beichman, C.A. 1990 *Proc. 29th Liege International Astrophysical Colloquium: From Ground-Based to Space-Borne Sub-mm Astronomy*, Liege, Belgium, 3-5 July 1990, *ESA SP-314*, 117

Jakobsen, P., de Vries, J.S. and Paresce, F. 1987 *A&A*, **183**, 335

de Lapparent, V., Kurtz, M.J. and Geller, M.J. 1986 *ApJ*, **304**, 585

Molaro, P., Vladilo, G., Avila, G. and d'Odorico, S. 1989 *ApJ*, **339**, L63

Sandage, A. 1976 *AJ*, **81**, 954

van Steenberg, M.E. and Shull, J.M. 1988 *ApJ*, **335**, 197

de Vries, C.P. and le Poole, R.S. 1985 *A&A*, **145**, L7

Witt, A.N., Stecher, T.P., Boroson, T.A. and Bohlin, R.C. 1989 *ApJ*, **336**, L21

The First Symposium on the Infrared Cirrus and Diffuse Interstellar Clouds
ASP Conference Series, Vol. 58, 1994
Roc M. Cutri and William B. Latter (eds.)

THE 3 DIMENSIONAL DISTRIBUTION OF CIRRUS

D. VAN BUREN
Infrared Processing and Analysis Center, California Institute of Technology and Jet Propulsion Laboratory, Pasadena, CA 91125

J. E. GAUSTAD
Department of Physics and Astronomy, Swarthmore College, Swarthmore, PA 19081

ABSTRACT We just completed a large survey of the IRAS scanning data in the neighborhoods of Bright Star Catalog O and B stars with the purpose of mapping the distribution of the infrared cirrus. We find that within 400 pc the filling factor of interstellar gas with $n > 0.5$ cm^{-3} is approximately 15%. For densities greater than 1 cm^{-3} the density distribution function per logarithmic density interval follows a power law with index -1.25. When the detected cirrus "hotspots" are projected onto the galactic plane several large structures emerge: the local hole or trough, the Taurus-Auriga clouds and the Oph-Sco-Cen region of recent and ongoing star formation.

INTRODUCTION

We were motivated by the compelling correspondence between the infrared cirrus, the 21 cm emitting gas and the material responsible for the extinction of starlight to explore the morphology of the diffuse interstellar medium using the IRAS data. Traditionally the distribution of interstellar matter has been studied using line of sight methods such as the color excesses of stars, interstellar absorption lines in stellar spectra and the surface brightness distribution of emission lines arising in the gas. Often a kinematic model is combined with spectral data to determine the distance to various emitting or absorbing gas parcels. With the IRAS data we were presented with a new opportunity to find the 3 dimensional distribution of the diffuse ISM because bright stars heat dust associated with the gas in their neighborhoods enough to be detected as extended far infrared emission nebulae. We call such nebulae cirrus hotspots.

A simple calculation shows that stars earlier than A0 are luminous enough to cause cirrus hotspots in gas with $n = 1$ cm^{-3} at the IRAS detection level, approximately 5 Jy at 60 μm for an extended object against the galactic background. We thus chose all 1753 Bright Star Catalog stars located in regions surveyed by IRAS and not likely to be confused by IRAS with other BSC stars as our sample, and obtained 1-D coadds of the scanning data crossing each star's position. About 35% of these stars showed possible evidence for extended far infrared emission, so we obtained $4° \times 4°$ coadded images to see if the extended far infrared had the characteristics expected for a cirrus hotspot:

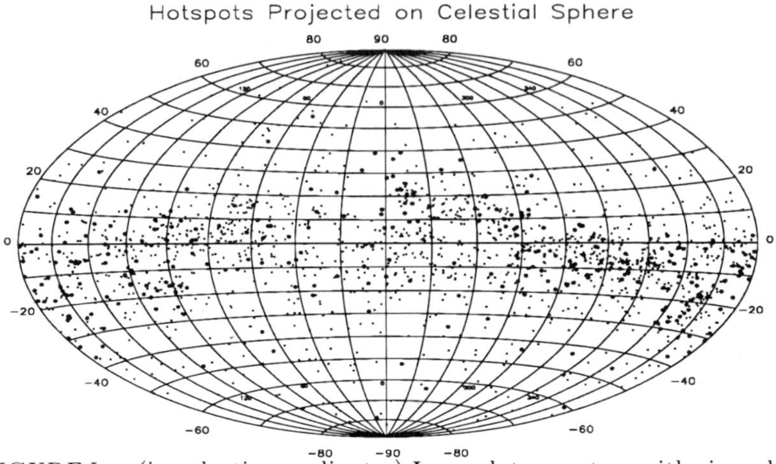

FIGURE I (in galactic coordinates) Large dots are stars with cirrus hotspots, small dots those without detections.

- in the direction of a known star,
- enhanced color temperature,
- temperature peaks at stellar position,
- extended on arcminute scales,
- $F_{FIR}/F_{Bol} < 0.1$ to eliminate HII regions.

Full details of the procedure are presented in our journal paper (Gaustad and Van Buren, 1993). The methodology follows closely that developed in the pilot study (Van Buren, 1989) and used by Buss, Gautier, Werner and Van Buren (1993), but differs in significant ways from the work of Murthy, Walker and Henry (1993).

MAPS

Using spectroscopic distances inferred from the BSC or SIMBAD, we can plot the 3-d distribution of stars with and without hotspots. These distances are not very accurate though, so we are awaiting publication of the Hipparcos results to produce a better set of maps than we present here. Figure I shows the positions of all program stars in the sky projected in galactic coordinates. Heavy dots represent stars with hotspots, small dots those with no detected FIR nebulae. In Figure II we show the distribution of hotspots projected onto the galactic plane. 302 hotspots are represented in these maps.

FILLING FACTOR

If the program stars are distributed in an unbiased manner with respect to the cirrus, a point discussed in some detail in our PASP paper, then the fraction of stars showing hotspots is equal to the fraction of space containing dust with sufficient density to yield an observable hotspot. By constructing complete

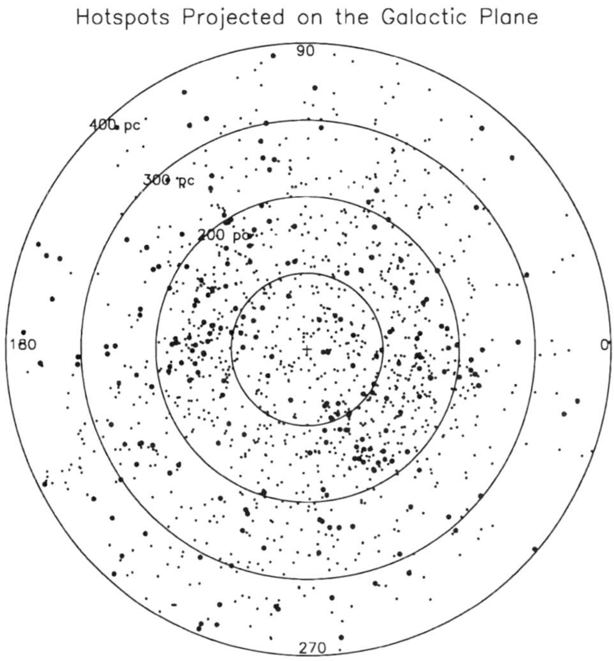

FIGURE II (projected on the galactic plane) Towards l=180 are the Taurus clouds at $d \approx 150pc$, the Oph-Sco-Cen region is near l=300, and in the close neighborhood of the sun there is a minimum in the hotspot distribution.

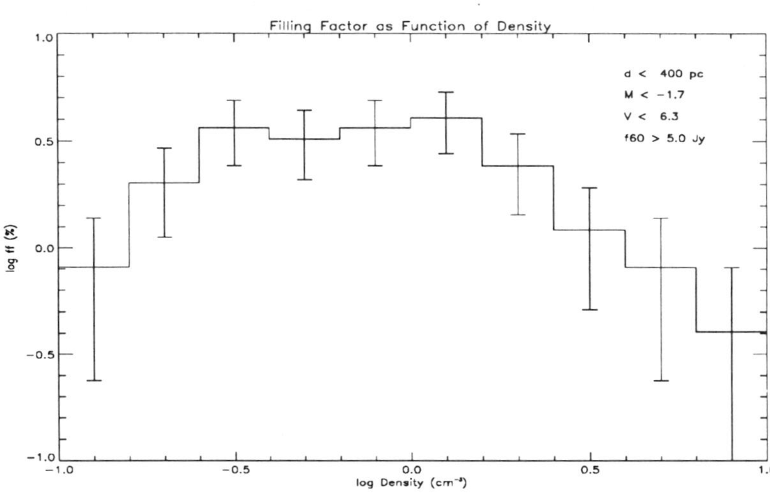

FIGURE III The density distribution function as determined from a complete subsample of our survey. At densities > 1 a power law with slope -1.25 is a good fit. We are not sensitive below densities of 0.5 cm^{-3}.

subsamples of the program stars on the basis of visual magnitude (governing inclusion in the BSC) and far infrared fluxes (governing detection by IRAS) we find that the filling factor of diffuse ISM with $n > 0.5$ cm^{-3} is $14.6 \pm 2.4\%$ within 400 pc.

Referring to Figure III, where the density distribution function of the dust-bearing medium is presented for a complete subsample of stars with $d < 400$ pc, note that a power law $d(\log f)/d(\log n) = -1.25$ is an adequate representation for $n > 1$ cm^{-3}. Because this is so steep, the sample size is not large enough to probe densities much above 10 cm^{-3}, but in any case no more than a few percent of interstellar space could contain material at higher densities, subject of course to the validity of our assumptions.

Combining our derived filling factor of 15% for the diffuse ISM, an extrapolation of the power law behavior at higher densities (which can only account for $\approx 1\%$ of the volume), and the estimates by Reynolds (1991 and references therein) that the medium with *electron* density ≈ 0.2 cm^{-3} comprises about 20% of the volume, we note that a volume fraction ≈ 0.64 appears to be void of any but the most tenuous gas or dust. The actual value of this fraction depends on to what extent the diffuse ionized gas and the low density cirrus overlap. With this overall picture in mind it is interesting to view figure 7 of Bregman, Kelson and Ashe (1993), where 21 cm channel maps show the best match for an ISM with a volume fraction 0.57 to 0.75 in holes.

ACKNOWLEDGMENTS

The following students assisted in the analysis of the data: Andrew Afflerbach, Matthew Alford, Fronefield Crawford, Gary Garber, Marshal Hampton, Lauren Jones, Meredith MacKenzie, Deborah Maraziti, Dawn Marshal, Anne McCormick, Joel Offenberg, Knut Olsen, Giray Pultar, Stephen Sample and Gregory Selvarian. We used SIMBAD, operated at CDS, Strasbourg, France. Funding for this research was obtained through the NASA ADP program (NAG 5-1152), the Keck Foundation, the Greenwall Foundation and the Howard Hughes Medical Institute. This work was done in part at JPL under contract for NASA. We thank IPAC staff for filling our many data requests and the people of the United States, United Kingdom and the Netherlands who collectively made the IRAS mission possible through their support of space astrophysics.

REFERENCES

Bregman, J. N., Kelson, D. D., and Ashe, G. A. 1993, *ApJ*, **409**, 682.

Buss, R. H. Jr., Gautier, T. N., Werner, M. W., and Van Buren, D., *in preparation.*

Gaustad, J. E., and Van Buren, D. 1993, *PASP*, *in press.*

Murthy, J., Walker, H. J., and Henry, R. C. 1992, *ApJ*, **401**, 574.

Reynolds, R. J. 1991, in IAU Symposium 144, *The Interstellar Disk-Halo Connection in Galaxies*, Dordrecht: Kluwer, p. 67.

Van Buren, D. 1989, *ApJ*, **338**, 147.

The First Symposium on the Infrared Cirrus and Diffuse Interstellar Clouds
ASP Conference Series, Vol. 58, 1994
Roc M. Cutri and William B. Latter (eds.)

IR PROPERTIES OF GALACTIC CIRRUS NEAR THE POLAR CAPS

BRYAN E. PENPRASE[1] AND GEORGE HELOU
IPAC, Caltech MS 100-22, Pasadena, CA 91125

ABSTRACT

We present a systematic study of the infrared properties of the high galactic latitude sky, by examining two large sections of the IRAS Sky Survey, which cover galactic latitudes from b=20 to b=90, and b=-20 to b=-90, at a reduced resolution of 7.5' / pixel. The *IRAS Sky Survey Atlas* (ISSA) plates are used to determine the variations in colors R(12,25),R(25,60) and R(60,100) of clouds with spatial scales of 1–5 degrees, as a function of galactic latitude, and also as a function of the brightness of the emitting material.

Our results use the improved sensitivity provided by the ISSA plates, to present images highlighting distinct clouds with atypical colors. The cirrus appears to consist of a mixture with typical colors at lower galactic latitudes, and several cirrus wisps with anomalous colors near the galactic poles. We also report colors of cometary nebulae seen near the Southern galactic pole which have values of R(25,60) $\geq$ 0.45. The colors of the high galactic latitude cirrus is comparable to weakly illuminated cirrus at the edges of M31 and in other galaxies (Xu and Helou 1993 ; Helou, 1986).

INTRODUCTION

Since the discovery of the IR cirrus, a number of authors have improved upon our understanding of the diffuse IR emission from the galaxy. Boulanger and Perault (1988) documented the average colors of galactic cirrus, the HI/I100μ emission ratio, and the variation of IR emission intensity throughout the high galactic latitude regions. They also reported the reduced emission at the polar caps which was apparent from about b=70 to b=80, which appeared to be the darkest patch in the IR sky. Desert, et al (1990) provided improved knowledge of the variations in color ratios for different regions in the galaxy, by relating them to variations in the relative abundance of three components: large grains, small grains and PAH's, with additional variations in ISRF. Recently, Helou (1986) and Xu and Helou (1993) have documented IR color variations in the galaxies M31 and M33, which appear to arise from regions in the galaxies with low levels of illumination, and a deficiency of large grains.

From the recently released Infrared Sky Survey Atlas (ISSA) a new oppor-

[1]present address: Pomona College Department of Physics and Astronomy, 610 N. College Ave., Pomona CA 91711-6348

tunity has arisen to reexamine the IR colors of the interstellar medium with the improved sensitivity and resolution of the ISSA. We report the first result of an examination of the properties of clouds with sizes of a few degrees, near the weakly emitting polar caps of the galaxy.

DATA AND DATA ANALYSIS TECHNIQUES

100 plates of 12x12 degree size were examined to determine the IR emission characteristics of the weakly emitting polar regions. The plates were selected to sample sightlines at high ecliptic and galactic latiitudes. Each plate was block averaged by a factor of 5 in both dimensions, to a reduced spatial resolution of 7', in order to improve the S/N for weakly emitting clouds. A background subtraction algorithm was next applied to each plate separately, and then the plates were mosaiced into 2 large images which were analyzed for color variations. The emission brightness from the polar regions were much weaker than the average emission from galactic cirrus, such as those reported in Boulanger and Perault (1988), with 100μ intensities mostly in the range of $0.1 < I(100\mu) < 1$ MJy Sr^{-1}, as opposed to typical cirrus for which $0.5 < I(100\mu) < 5$ MJy Sr^{-1}. The locations of the plates selected for the our study are presented in Figure 1a, which shows an Aitoff projection in galactic coordinates and the locations of the selected plates.

Background Subraction of the ISSA Plates

The ISSA plates are excellently suited for detecting the variations in IR emission on spatial scales of a few degrees. However, uncertain zodiacal background residuals do exist in all of the plates, which are a major obstacle in determining the actual emission from the regions studied. Since the backgrounds are a numerical residual, their spatial and spectral properties are not correlated in any consistent way with the original zodiacal background.

To address the problem of background removal, each plate was processed using newly developed routines which iteratively fit a planar background to the image, and subtract this background. The routine selects background points based on a consideration of the statistics of the image: points are labelled as background if they are within the range of the 0 to P percentile, and next the threshold P is increased until a set of points is derived which covers 3 out of 4 quandrants of the plate, with at least 2% of the background points coming from the least populated quadrant. The final value of P is typically around 5%, and varies with the covering fraction of IR-emitting material in the selected plate. A plane is then fit to the background points and subtracted from the plate. This process is repeated until each new iteration results in a residual at the edge of the plate of $R < 0.08$ for the 100μ images, and $R < 0.02$ for the other three bands.

The result is a uniform and repeatable background subtraction for the images. After background subtraction, histograms of the pixel intensities are examined for any remaining systematic offsets in either axis of the images. Figure 1b presents intensity histograms of the background subtracted images for the North Galactic Pole sample.

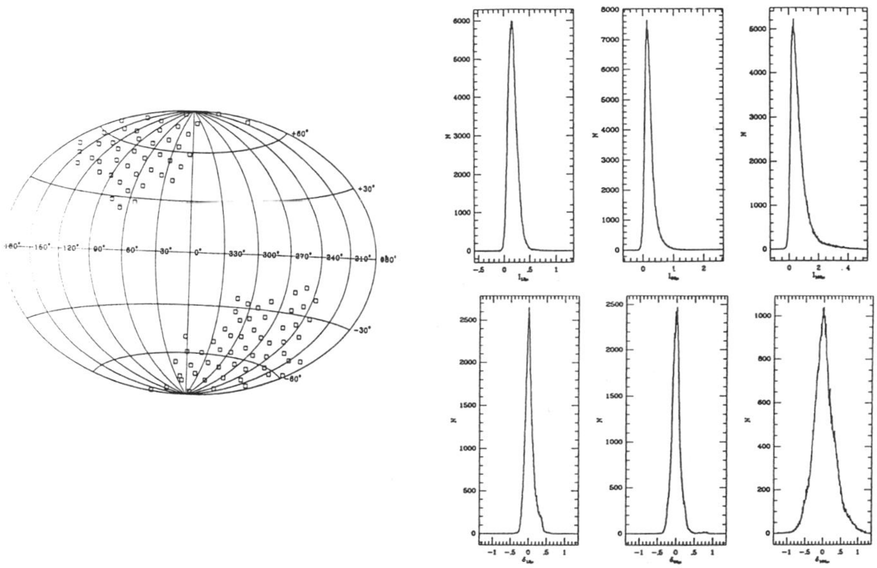

FIGURE I a.(left) Aitoff projection showing locations of plates used in study. b.(upper right) Histograms of pixel intensities for North Galactic Pole Region, at wavelengths (from left) 12μ,60μ and 100μ. c.(lower right) Histograms of differences in overlap regions (see text).

Measuring the Accuracy of Background Subtraction

One estimate of the uncertainties from the background subtraction may be obtained by measuring the width of the negative going tail in the histograms of Figure 1b, which for a correctly flattened plate will reflect only the instrumental noise of the IRAS survey, and will be free from systematic offsets.

An independent assessment of the quality of background removal may be obtained by considering the difference in pixel values for overlapping regions of neighboring plates. The overlap is largest near the north and south celestial poles, where it is as much as 40% of the plate pixels, and then decreases steadily at lower declinations to about 2% at the celestial equator. Each plate from the two groups was background subtracted, and the average value and variance of the difference values for the overlap pixels was evaluated. Histograms of the differences in the overlapping pixels are presented in Figure 1c for the North Galactic Pole sample. Gaussians were fit to the curves of Figures 1b and 1c and from both techniques we observe that the background uncertainties are on the order of 0.1 MJy Sr^{-1} for the 12μ and 60μ plates, and 0.25 for the 100μ plates.

RESULTS

An image of the background subtracted mosaic for the Northern Galactic Pole is shown in Figure 2a. From the figure, it is apparent that a number of structures with widely varying size scales exist in this region, suggesting different regimes of the ISM and/or different distances to the clouds. Visible at upper left in the image is the Draco cloud, and on the right is the Ursa Major molecular clouds,

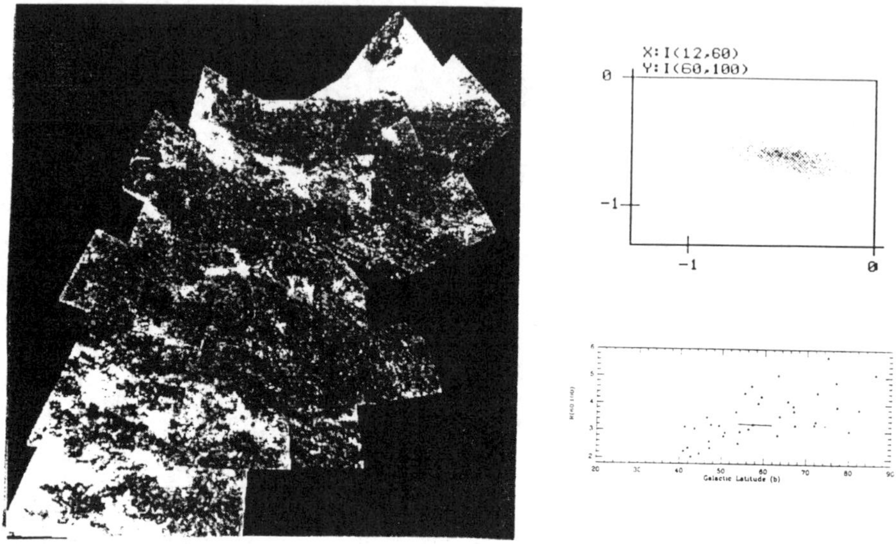

FIGURE II a.(left) 100μ mosaic for North Galactic Pole Region, showing anomalous wisp in center. b.(upper right) Color-color plot for same region;X axis=log[R(12,60)], Y axis=log[R(60,100)] c.(lower right) Plot of R(60,100) vs. galactic latitude.

and part of the 'North Polar Spur'. The wisp at the center of the image was observed to have larger R(60,100) than clouds at the edges, and a general trend of increasing R(60,100) was observed as a function of galactic latitude for the Northern hemisphere. Figure 2b presents color-color diagrams for the Northern hemisphere mosaic, and Figure 2c shows the trend of increasing R(60,100) as a function of galactic latitude. An additional result was the observation that the color R(25,60) for the cometary clouds near the Southern galactic pole is anomalously large, with R(60,100) $>$ 0.4.

Further studies are underway to determine possible causes of the anomalous colors seen in the ISSA plates, including difference in the grain populations due to shocks and other disruptive processes, variations in the ISRF, and effects of $||z||$ on the IRAS colors.

REFERENCES

Boulanger, F., and Perault, M. 1988, *Ap. J.* **330**, 964.
Desert, F.A., Boulanger, F., and Puget, J.L. 1990, *Astron. Astrophys.* **273**, 215.
Helou, G. 1986, *Ap. J.* **311**, L33.
Xu, C. and Helou, G. 1993, *preprint.*

The First Symposium on the Infrared Cirrus and Diffuse Interstellar Clouds
ASP Conference Series, Vol. 58, 1994
Roc M. Cutri and William B. Latter (eds.)

INTERSTELLAR FROTH: THE CIRRUS-LIKE FAR-INFRARED STRUCTURE NEAR THE GALACTIC PLANE

WILLIAM H. WALLER[1,2]
NASA Goddard Space Flight Center, Code 681, Greenbelt, MD 20771

FRANCOIS BOULANGER[3]
Institut d'Astrophysique Spatiale, B.P. 10, Route des Gatines, 91371 Verrières le Buisson Cedex, France

ABSTRACT IRAS/BIGMAP mosaics of the inner Galaxy have been accessed and analyzed in terms of the interstellar structures that are traced by the FIR emission. Through spatial filtering techniques, we have been able to eliminate the strong gradient in brightness towards the Galactic midplane thus revealing a "froth" of diaphanous structure in the FIR emission with size scales as small as the $4' \times 4'$ resolution of the images. *This cirrus-like structure is not merely of local origin, but consists of both nearby and more distant features in the disk.* Efforts to find interstellar counterparts to the FIR "froth" are discussed.

INTRODUCTION

Since the launch of the Infrared Astronomy Satellite (IRAS) in 1983, we have learned that far-infrared emission pervades every quadrant of the Galactic sky (cf. Beichman 1987). Moreover, the emission is rich with structure, thus providing important clues to the content and dynamics of the dusty ISM. At high Galactic latitudes, overlapping wisps, tendrils, and knots of FIR emission highlight the so-called Infrared Cirrus (cf. Boulanger 1990). Recent investigations of these presumably local features have revealed strong associations with HI features — often at anomolous (blueshifted) velocities (Deul and Burton 1990; Burton *et al.* 1993). Some of the Infrared Cirrus features are also evident in deep surveys of CO emission (Heithausen *et al.* 1993), thus indicating their multi-phase nature.

---

[1]National Research Council Resident Research Associate

[2]StarStuff Incorporated, Greenbelt, MD

[1]Ecole Normale Supérieure, Paris, France

FINE-SCALE FIR EMISSION

Using the IRAS/BIGMAP mosaics of the inner Galaxy, we have applied a form of spatial filtering which completely eliminates the strong gradient in brightness towards the Galactic midplane. The resulting images reveal a "froth" of diaphanous structure in the FIR emission with size scales as small as the $4' \times 4'$ resolution limit (see FIGURE 1). This cirrus-like structure extends all the way down to the Galactic midplane. Moreover, it scales in intensity with the smoothly varying background, independent of latitude, thus indicating that the fine-scale residual emission is co-extensive with the smooth background. If the *cosecant b* behavior of the smoothly varying background is the result of observing an edge-on disk of emitting dust, then the fine-scale residual emission originates from dust features that are similarly distributed. *Therefore, this cirrus-like structure is not merely of local origin, but consists of both nearby and more distant features in the disk.*

Although we had expected to find morphological evidence for supernova-driven "worms" or "chimneys" rooted in the Galactic plane (cf. Koo *et al.* 1992; Reach *al.* 1993), our processing shows the FIR fine-scale structure to be more complex (e.g. less coherent and less rooted) as viewed in projection. This may be due to the varying distances and sizes of the emitting features. The observed *froth of superposed filaments, voids, and shells* has yet to be reconciled with other tracers of the ISM (see below).

INTERSTELLAR COUNTERPARTS

In an effort to determine the interstellar counterparts to the FIR "froth," we have compared the fine-scale FIR structure with maps of CO emission, local molecular clouds, and radio supernova remnants. FIGURE 1 shows these comparisons. The CO emission (from Dame *et al.* 1987) has been spatially filtered to bring out the fine-scale structure. The resulting CO emission residuals are shown as contours overlaying the FIR images. Correspondences between the FIR enhancements and CO residuals are evident close to the Galactic plane but disappear beyond a few degrees from the midplane. This is at least partly due to the limited dynamic range of CO intensities being emitted and/or detected. Deeper CO maps may reveal more structure at higher latitudes (e.g. Heithausen *et al.* 1993). In the mid-latitude Ophiucus star-forming region (not shown), we have found strong correlations between the CO and FIR emission (Boulanger and Waller 1993). Therefore, CO counterparts to the FIR fine-scale structure appear to be restricted to regions where the CO emission is strong — independent of latitude. Close to the plane, the strongest CO emission comes from the molecular ring at R(gal) $\approx$ 5 kpc, implying a similar location for much of the midplane FIR fine-scale structure.

Comparisons with local molecular clouds (filled squares with error bars) show piecemeal correlation with the FIR enhancements. For example, the clouds comprising the Cygnus, Vulcan and Aquila complexes are poorly traced by the FIR residuals, perhaps because their dust is too cool for prominent emission at 100 μm. Comparisons with radio supernova remnants (open circles with error bars) yield poor correlation with both the FIR *enhancements* and the FIR *voids*.

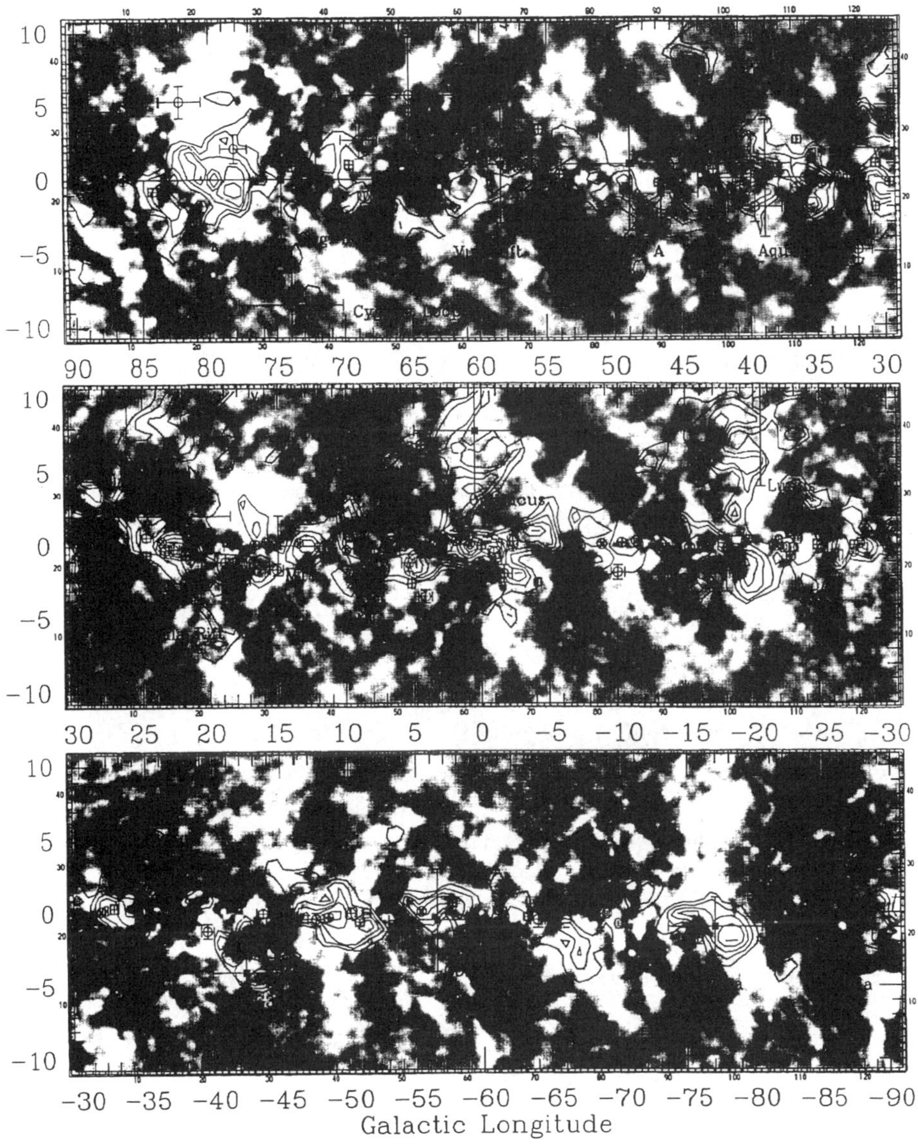

FIGURE 1. Far-infrared fine-scale structure in the inner Galaxy — from Cygnus (upper left), through the Galactic center (middle center), to Carina (lower right). Images are based on the IRAS/BIGMAP mosaics available from the IPAC. The *relative residual* emission at 100 μm is shown, after having spatially filtered the original mosaics to eliminate the strong gradient in brightness towards the Galactic midplane. Contours of residual CO emission are overlaid along with the positions and sizes of local molecular clouds (filled squares [from Dame *et al.* 1987] and radio supernova remnants (open circles [from Green 1990]).

We suspect that the closest counterparts to the filamentary and shell-like FIR structures will be found in well-resolved maps of the Galactic HI and radio-continuum emission. At high latitudes, the Infrared Cirrus is most closely traced by HI features at specific velocities (Deul and Burton 1990; Burton *et al.* 1993). At intermediate latitudes, the HI emission shows an abundance of filaments and loop-like structures (Heiles 1984; Hartmann 1993), some of which could correspond to FIR features. And at low latitudes, some of the FIR "worms" have radio-continuum counterparts (Koo *et al.* 1992; Reach *et al.* 1993). Detailed comparisons between our images of the fine-scale FIR emission near the Galactic plane and similarly processed maps of Galactic HI and radio continuum emission should reveal many correlated features, whose measured properties could then be organized into a more coherent picture of the diffuse interstellar medium.

ACKNOWLEDGMENTS

This research was funded in part by a NASA/ADP grant (NAG 5-1424) to StarStuff Inc. We also thank the folks at IPAC for their invaluable support.

REFERENCES

Beichman, C. A. 1987, *ARA&A*, **25**, 521

Boulanger, F. 1990, in *The Galactic and Extragalactic Background Radiation*, eds. S. Bowyer and C. Leinert (Dordrecht: Reidel), p. 139

Boulanger, F. and Waller, W. H. 1993, in *Back to the Galaxy*, eds. S. Holt and F. Verter (Woodbury, NY: AIP Publications), in press

Burton, W. B., Bania, T. M., Hartmann, D. and Yuan, T. 1993, in *Evolution of Interstellar Matter and Dynamics of Galaxies,* eds. J. Palous, W. B. Burton and P. O. Lindblad (Cambridge: Cambridge University Press), in press

Dame, T. M., Ungerechts, H., Cohen, R. S., de Geus, E., Grenier, I., May, J., Murphy, D. C., Nyman, L. -Å, and Thaddeus, P. 1987 *ApJ*, **322**, 706

Deul, E. R. and Burton, W. B. 1990, *A&A*, **230**, 153

Green, D. A. 1990, *PASP*, **103**, 209

Hartmann, D. 1993, in *Back to the Galaxy*, eds. S. Holt and F. Verter (Woodbury, NY: AIP Productions), in press

Heiles, C. 1984, *ApJS*, **14**, 1

Heithausen, A., Stacy, J. G., de Vries, H. W., Mebold, U., and Thaddeus, P. 1993, *A&A*, 268, 265

Koo, B. -C., Heiles, C., and Reach, W. T. 1992, *ApJ*, **390**, 108

Reach, W. T., Heiles, C., and Koo, B.-K. 1993, in *Back to the Galaxy*, eds. S. Holt and F. Verter (Woodbury, NY: AIP Productions), in press

The First Symposium on the Infrared Cirrus and Diffuse Interstellar Clouds
ASP Conference Series, Vol. 58, 1994
Roc M. Cutri and William B. Latter (eds.)

THE 100 MICRON EMISSION OF CLOUDS

R.J. LAUREIJS
Space Science Dept., ESTEC, POB 299, 2200 AG Noordwijk, The Netherlands

G. HELOU
IPAC, Caltech

F.O. CLARK
Geophysics Directorate, Phillips Laboratory

ABSTRACT We have analyzed the extinction and infrared properties of a sample of clouds selected from our own measurements and the literature. The mean ratio of 100 μm emission to blue extinction $< I_{100}/A_B >$ in the outer regions is tightly correlated with the cloud's central extinction. We infer that reasonably accurate cloud masses can be estimated on the basis of I_{100} images alone.

INTRODUCTION

The IR emission of clouds depends on radiative transfer effects, even for clouds with visual extinctions A_V <1 mag. We have analyzed the relationship between I_{100} and A_B for a sample of clouds with central opacities ranging from A_B <2 mag to A_B >8 mag. Both tracers are most likely due to the same component of the ISM, namely large (0.15 μm) dust grains. Since optical extinction is a column density tracer unbiased by radiative transfer effects, we are not only able to investigate radiative transfer effects on IR emission, but also able to gauge I_{100} as an cloud-mass estimator.

ANALYSIS OF $< I_{100}/A_B >$

The pixel to pixel correlation diagrams between I_{100} and A_B for all clouds show an increase of I_{100} with increasing A_B. Some clouds exhibit a non-linear behaviour: I_{100} appears to flatten at the higher extinction ends. In view of the scatter in the extinction measurements, the observed correlations are well approximated by a linear relationship for the data points constituting the largest fraction of the cloud area.

The observations show two noticeable trends:

1. the pixel-to-pixel correlation between I_{100} and A_B is remarkably linear for each cloud.
2. the mean ratio $< I_{100}/A_B >$ derived for each cloud from linear regression analysis (Fig 2) scales as a function of peak extinction $A_B(peak)$ in mag accord-

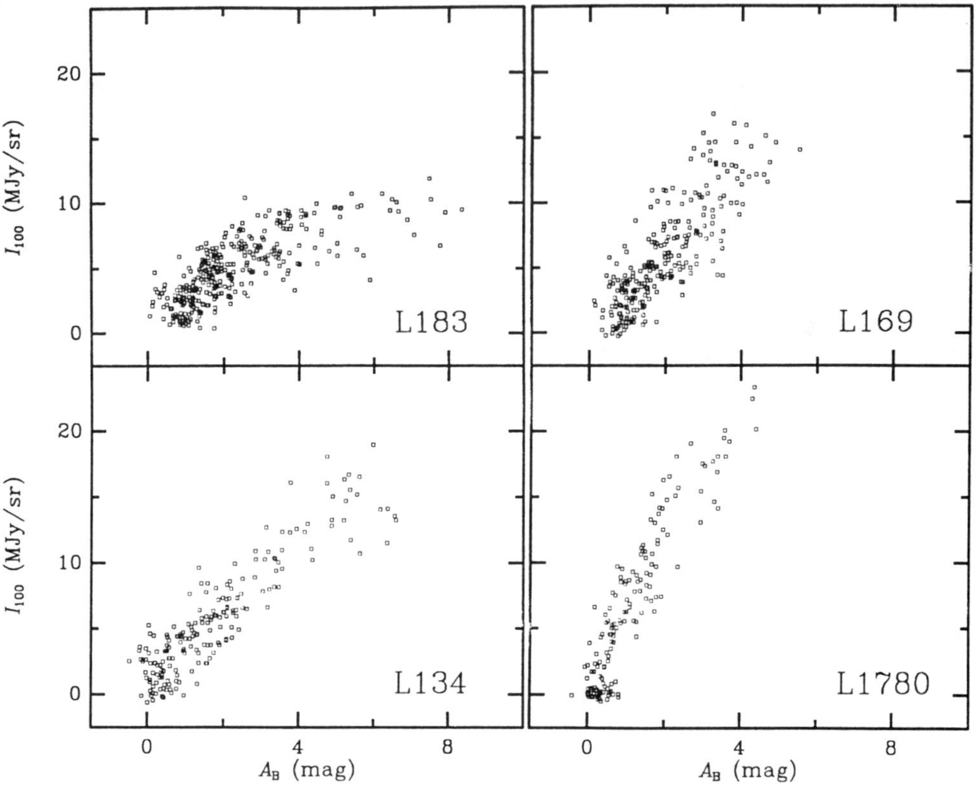

FIGURE I The pixel-to-pixel correlation diagrams for 4 clouds.

ing to

$$< I_{100}/A_B >= (14 \pm 2) \times \exp \frac{-A_B(peak)}{4.2 \pm 0.4} MJysr - 1mag - 1 \qquad (1)$$

IMPLICATIONS

Cloud structure

Fig. 2 suggests that $< I_{100}/A_B >$ is exclusively determined by radiative transfer in the clouds and that the ambient radiation density heating the dust at 100 μm is reasonably constant in the local solar neighborhood.

Model calculations assuming homogeneous spherical clouds predict a detectable drop in I_{100}/A_B as a function of AB in clouds with $A_B > 2$ mag. The empirical relationships show a less pronounced decline in I_{100}/A_B in the outer cloud regions ($A_B < 6$ mag). In fact, the relationship for the dark clouds L134 and L169 appear linear. Despite the lack of strong variations of I_{100}/A_B in individual clouds, Fig. 2 indicate that the mean infrared emission of a cloud is

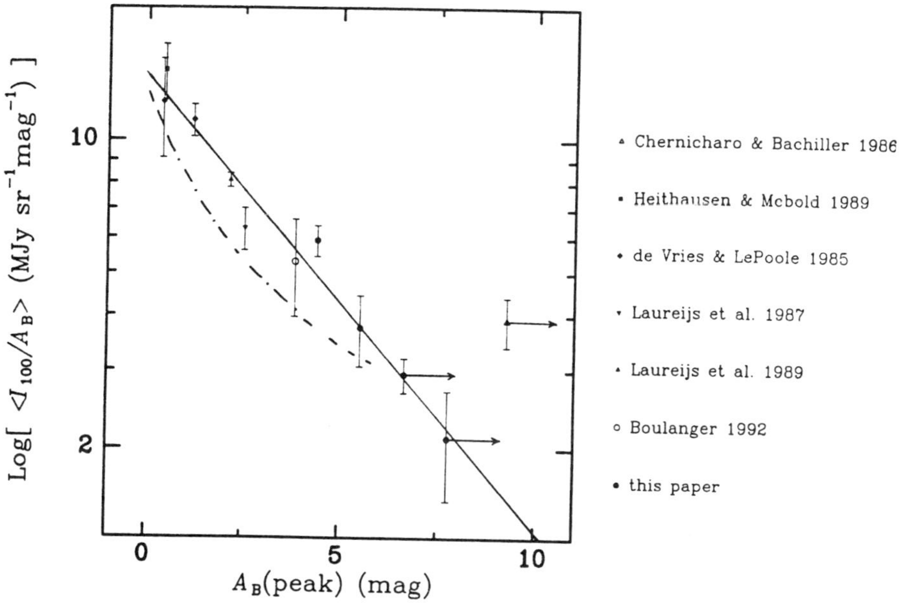

FIGURE II The relationship between the slopes of the regression lines, $< I_{100}/A_B >$, and A_B(peak) in clouds. The solid line is a regression fit excluding lower limits. The dashed line represents model calculations by Bernard et al. (1992) for clouds with an r^{-2} density distribution.

clearly a function of A_B(peak). Thus the outer cloud regions which constitute most of the cloud's area are affected by the presence of a higher opacity center.

An interpretation for the observations is that clouds are inhomo- geneous. In a clumpy medium the UV penetration is deeper if the volume filling factor and the interclump density are low (Boissé 1990). Our results suggest that A_B(peak) regulates the radiation field for the cloud as a whole, and that clumpiness could soften gradients in the radiation field as a function of cloud radius.

Emission properties at 100 μm

The value of $< I_{100}/A_B >$ for A_B=0 is remarkably consistent with what is known from the diffuse ISM. We obtain a value of 14±2 MJy/sr/mag which corresponds to I_{100}/N_H = 94±13 10^{-22} MJy $sr^{-1}cm^2$. This is close to 85 10^{-22} MJy $sr^{-1}cm^2$ which applies to the diffuse ISM in the local solar neighborhood (Boulanger and Perault 1988). Thus a linear extrapolation of data derived from dense and translucent clouds is adequate to predict the 100 μm emissivity in the diffuse ISM.

The observations offer a method to estimate masses of clouds for which A_B ¡ 10 mag purely on the basis of the 100 μm images. Assuming that equation 1 is valid and that the cloud is resolved, the following expression between I_{100}(peak) and A_B(peak) applies:

$$I_{100}(peak) = 14 \times A_B(peak) \times exp - \frac{A_B(peak)}{4.2} MJysr - 1. \qquad (2)$$

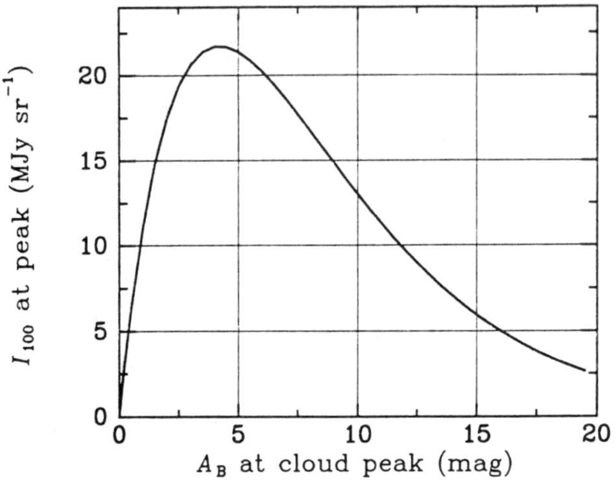

FIGURE III The relationship between peak 100 μm surface brightness and peak extinction predicted by the empirical relation (1).

The function has a maximum of I_{100}(peak)= 22 MJy/sr for A_B(peak) = 4.2 mag. Thus the central extinction of a cloud can be derived by measuring I_{100}(peak) see Fig 3. The ambiguity whether the cloud's opacity is higher or lower than 4.2 mag can easily be resolved.

Knowing the peak column density, the total mass of a cloud can be derived by first measuring the total 100 μm luminosity by integrating the surface brightness over the cloud's area, and subsequently by multiplying that value with the corresponding ratio for I_{100}/N_H derived from $< I_{100}/A_B >$. This method gives a reliable mass estimate since it applies to a large fraction of the cloud's volume, containing most mass. Moreover, I_{100} can be detected down to $A_B << 1$ mag, which is hard to achieve with molecular data for instance.

Equation 2 predicts that clouds with A_B(peak)=4.2 mag exhibit the highest 100 μm surface brightness of I_{100}(peak)=22 MJy/sr. Thus, as long as the ISRF is the typical one expected at high latitudes, genuine translucent clouds can be found by looking for the brightest cirrus features in the 100 μm images. Consequently, cirrus features that are brighter than 22 MJy/sr above the background have most likely nearby or embedded heating sources, or are a superposition of several clouds sufficiently far apart not to influence the radiation fields of each other.

We acknowledge K. Mattila and G. Schnur for kindly providing their extinction data of the L134 complex, and K. Mattila for many stimulating suggestions.

REFERENCES

Bernard, J.P, Boulanger, F., Désert, F.X., Puget, J.L., 1992, *A&A*, **263**, 258

Boissé, M. 1990, *A&A*, **228**, 483

Boulanger, F., Perault, M. 1988, *A&A*, **330**, 964

The First Symposium on the Infrared Cirrus and Diffuse Interstellar Clouds
ASP Conference Series, Vol. 58, 1994
Roc M. Cutri and William B. Latter (eds.)

INTERSTELLAR COLUMN DENSITIES USING IRAS

P. G. MARTIN[1]
Canadian Institute for Theoretical Astrophysics, University of Toronto, Toronto, ON, Canada M5S 1A7

ABSTRACT The potential of using the long wavelength (60 and 100 μm) IRAS Sky Survey Atlas to provide column densities is explored, obvious advantages being all-sky coverage and relatively good spatial resolution, about $4'$. It is demonstrated how a major challenge, converting the infrared surface brightness into a reliable column density, might be met by bootstrapping from larger beam 21-cm survey data.

INTRODUCTION

Extragalactic sources suffer energy-dependent photoelectric absorption by foreground material in our Galaxy. If the column density of absorbing material is not known independently it must be treated as a free parameter, which limits the precision with which other key properties such as the x-ray spectral index and flux can be determined. Uncertainties in the column density become increasingly detrimental as the observation energy decreases, being particularly acute for measurements with ROSAT and the EUVE observatory. Optical extinction can be estimated from the column density too.

21-cm Emission

At high latitudes, column densities can be probed using 21-cm HI emission; molecular emission is not all that prevelant. A preferred starting point is the "Bell Labs" survey (Stark et al. 1992), since the horn reflector antenna used has very much reduced sidelobe response away from the main beam; the main drawback is the relatively large beam (2°) since there is non-negligible structure in the column density at smaller angular scales (see below). Using $3'$ observations from Arecibo, Elvis et al. (1986) estimated that the typical error using Bell Labs data alone would be 0.6×10^{20} cm^{-2} (1 σ), which could be a significant proportion of the total column density in high latitude directions where one might hope to obtain low energy spectra. To obtain data at higher angular resolution, Lockman, Jahoda, & McCammon (1986) developed a technique of removing the stray radiation from the $21'$ beam of the 43 m telescope at NRAO, based on "bootstrapping" from the Bell Labs survey data. This produces an accuracy of about 0.1×10^{20} cm^{-2} and has been exploited by Elvis, Lockman, & Wilkes (1989) to obtain accurate column densities toward many quasars and

[1]Visiting Associate, Theoretical Astrophysics and IPAC, California Institute of Technology, Pasadena, CA 91106

AGN; they comment that an all-sky survey with similar beam size and accuracy would be of great utility for x-ray astronomy.

The IRAS Sky Survey Atlas

The main point of this investigation is to explore the potential of a complementary technique using the long wavelength (60 and 100 μm) IRAS database for such a purpose. Obvious advantages are (i) the all-sky coverage, (ii) the relatively good spatial resolution – about $4'$ in the IRAS Sky Survey Atlas (ISSA) recently released by IPAC, (iii) the archival nature and ready availability of the data (on CD-ROM), and (iv) a documented software executive with which to display and analyse it ("skyview"). These images certainly reveal spatial structure in the infrared cirrus down to the resolution limit. Power spectrum analysis of the structure (e.g., Gautier, this meeting) indicates that fluctuations are decreasing toward smaller scales, so that examination on finer scales is statistically of benefit in assessing a particular line of sight.

A major challenge is converting the infrared surface brightness into a reliable gas column density. However, it is known that at high latitudes the surface brightness of infrared emission from dust is correlated with gaseous 21 cm emission from the Bell Labs survey (Boulanger & Pérault 1988) and subsequent observations with the DRAO Synthesis Telescope show that there is a correlation at the higher spatial resolution of IRAS too (Joncas, Boulanger, & Dewdney 1992; Martin et al., this meeting). This suggests that bootstrapping from the Bell Labs survey data might again be possible, resulting in the desired extensive sky coverage (above $\delta = -40°$) from archival data. The emphasis here is on *general accessibility* and of course seeing if the technique even works. More precision might be achieved with a more sophisticated variant.

IMPLEMENTATION

The Bell Labs Survey

Details of the survey and data are given by Stark et al. (1992). The product used here was a file of emission spectra obtained from NSSDC (A8010_DATA.DAT). Spectra near the chosen line of sight were extracted and decoded using software by W. Reach, also available through NSSDC. Simple summation of the spectrum gives the column density, assuming the optically-thin limit holds. For this exploratory investigation, the horn beam was taken to have uniform illumination within a 2° diameter circle, a coarse approximation to the real beam.

ISSA

Before embarking on discussion of details, one of the real merits of using ISSA should be stressed, which is that simple inspection of the image (plate) will immediately reveal whether there is variation worth quantifying and the sign of the effect (many examples in the affirmative were shown in the poster version of this paper). The importance of the context or "bigger picture" provided by examination of the 12.5° square field should not be underestimated. Use of the ISSA plates does not require the heroic efforts characteristic of earlier analyses of IRAS cirrus. Nevertheless, one should read the Explanatory Supplement

(Wheelock et al. 1993) to be aware of potential pitfalls. The plates used were the "HCON= 0" versions (incorporating all scans of the region) in bands 4 and 3 (100 and 60 μm respectively). These have 1.5′ pixels and a resolution 4–5′.

Bootstrapping

The data in ISSA are linear in the infrared surface brightness I, but have an unknown offset. One first "observes" the ISSA plate with the radio antenna beam and calculates the effective average infrared surface brightness, I_B. For the simple adopted beams, this is readily measured within "skyview"; if there are bright sources in the beam, they can be avoided or eliminated. This procedure is most straightforward using the individual original pointings of the Bell Labs survey, so that the I_B are directly paired with $N_{\rm HI}(B)$ from these spectra without interpolation. Then I is measured in a much smaller region (say 5′ diameter) on or close to the desired line of sight. In the simplest implementation I is measured within the beam used to obtain I_B, though not necessarily at its centre. In this way a differential quantity ΔI is obtained, independent of the unknown offset.

HI gas and infrared emission by interstellar dust are correlated over large angular scales (Boulanger & Pérault 1988); the slope of this mean relation (in an appropriate latitude range) can be used to convert ΔI into $\Delta N_{\rm H}$, whence the desired $N_{\rm HI}(I)$ for the line of sight. There is some cosmic scatter about the mean relation and there are also changes in slope from region to region. It is possible to derive the local slope by "observing" the ISSA plate with many beams and studying the usually evident correlation between I_B and $N_{\rm HI}(B)$.

TESTS AND APPLICATIONS

As a first test $N_{\rm HI}(I)$ was deduced for lines of sight for which the discrepancy between Bell Labs and Arecibo (3′) measurements was large, namely six examples in Table 10 of Elvis et al. (1986) for which $|\Delta N_{\rm H}| > 5 \times 10^{20}$ cm^{-2}. Most of these are not high latitude sources, as was intended for this technique. Nevertheless, the new estimates of $N_{\rm HI}(I)$ are usually closer to the Arecibo results.

X-ray Spectra of Compact Radio Sources

$N_{\rm HI}(I)$ was measured for six sources for which Worral & Wilkes (1990) plot contoured confidence intervals for two parameter fits of $N_{\rm HI}$ and spectral index (their Figs. 1 and 2 and Table 3). Two of these sources have been measured by Elvis, Lockman, & Wilkes (1989) using the NRAO technique as well. The results were again encouraging, and can be taken as providing further constraint on the spectral index.

Worral & Wilkes also identified nine sources (including one of the above) for which they found a significant discrepancy between $N_{\rm HI}$ from the fit and that measured independently (see Table I). It is of interest to see whether the new $N_{\rm HI}(I)$ values are different in the sense of removing the discrepancy, or whether another explanation is needed (e.g., complex source spectrum). The results given in Table I are mixed. Six of these offer a further comparison of the new $N_{\rm HI}(I)$ values with the accurate NRAO values, which is favorable.

TABLE I COMPARISON OF HI COLUMN DENSITIES[a]

Name	b	NRAO	Bell	ISSA	discrepancy[b]
0332+320	−18.7	14.22		:	+
0923+392	46.2	1.69		1.46	−
1226+023	64.4		1.83	1.75	−
2126−158	−41.9	4.85		5.25	+
0219+428	−16.8		7.51	9.2	+
0235+164	−39.1	7.60		7.5	+
1219+285	83.3		1.81	1.91	+
2200+420	−10.4	20.15		19.7	+
2155−304	−52.2	1.36[c]	1.73	1.41	−

[a]In units of 10^{20} cm^{-2}
[b]+ means that $N_{\rm HI}$ from spectral fitting is greater than $N_{\rm HI}$ from NRAO or Bell Labs
[c]recent unpublished measurement by F.J. Lockman & B.D. Savage

EUV Photoelectric Absorption

Two EUVE approved targets have been examined. For NGC 5548 (an IRAS source), $N_{\rm HI}(I) = 1.57 \times 10^{20}$ cm^{-2}. Mrk 478 is in a relative hole in the cirrus and $N_{\rm HI}(I) = 0.94 \times 10^{20}$ cm^{-2}.

However, one of the more compelling cases is the BL Lac source PKS 2155−304 which was detected by the ROSAT WFC and EUVE deep (DSLex at ∼100Å) surveys. In fact, this was the first ISSA plate examined as this technique was being conceived, and it was immediately clear that there was a slight "cirrus hole" toward this object. This is quantified in Table I; the new independent estimates from NRAO and ISSA are consistently lower than that based on the lower resolution Bell Labs data. This change in column density might seem small but as can be seen from the EUV optical depth in Figure 1, it can have a dramatic effect on the spectral shape and intensity in this EUV window; the short wavelength spectrum which can be inspected in the EUVE archives (PKS2155_SW.spec) does indeed show a long wavelength cutoff near 100 Å (among other interesting things).

Some care has to be taken in using one measure of column density as a surrogate for another, or as a measure of another related quantity. The 21-cm emission measures the neutral hydrogen. IRAS surface brightness has been expressed in terms of neutral HI column density, but of course measures emitting dust present independent of the ionization state of the gas. This is also a different assessment of dust distinct from extinction. Photoelectric optical depth is also expressed in terms of $N_{\rm HI}$, but in this window is largely from He (assumed neutral in Fig. 1). Therefore, comparison of the column densities as measured by all of these independent physical processes potentially gives information on chemical abundance and ionization state.

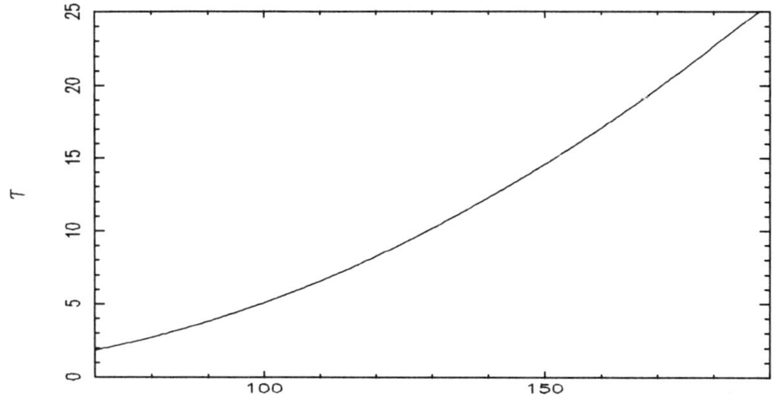

FIGURE 1 EUV photoelectric optical depth for $N_{\rm HI} = 1.4 \times 10^{20}$ cm^{-2}, computed using "ism.c" from the EUVE archive (Rumph, Bowyer, & Vennes 1993).

Optical and Ultraviolet Extinction

IRAS cirrus measures dust emission, which correlates with $N_{\rm HI}$. The extension to optical extinction anticipated by Boulanger, Baud, & van Albada (1985) requires a further calibration step, $N_H/E_{\rm B-V}$. Heiles (1989) gives an excellent introduction to the pitfalls of this step at high latitude. Nevertheless, it might be possible with differential cirrus measurements to improve on the Burstein & Heiles (1982) or other extinction estimates using a bootstrapping approach.

REFERENCES

Boulanger, F., Baud, B., & van Albada, G.D. 1985, *A&A*, **144**, L9

Boulanger, F. & Pérault, M. 1988, *ApJ*, **330**, 964

Burstein, D., & Heiles, C. 1982, *AJ*, **87**, 116

Elvis, M., et al. 1986, *ApJ*, **310**, 291

Elvis, M., Lockman, F.J., & Wilkes, B.J. 1989, *AJ*, **97**, 777

Heiles, C. 1989, in *Extreme Ultraviolet Astronomy*, ed. R. F. Malina and S. Bowyer (Oxford: Pergamon), p. 333

Joncas, G. , Dewdney, P.E. & Boulanger, F. 1992, *ApJ*, **397**, 165

Lockman, F.J., Jahoda, K., & McCammon, D. 1986, *ApJ*, **302**, 432

Rumph, T., Bowyer, C.S., & Vennes, S. 1993, *ApJ*, submitted

Stark, A.A., et al. 1992, *ApJS*, **79**, 77

Wheelock, S., et al. 1993, in preparation

Worral, D.M., & Wilkes, B.J. 1990, *ApJ*, **360**, 396

The First Symposium on the Infrared Cirrus and Diffuse Interstellar Clouds
ASP Conference Series, Vol. 58, 1994
Roc M. Cutri and William B. Latter (eds.)

IRAS LRS SPECTRA OF (ALMOST) CIRRUS

FRANK O. CLARK
Department of the Air Force, Phillips Laboratory, Geophysics Directorate (AFGL)

ROB ASSENDORP, PAUL WESSELIUS,PETER ROELFSEMA
Laboratory For Space Research, The Netherlands

R. F. SHIPMAN
U. Wyoming & USAF Phillips Lab

TOM KUCHAR
USAF Phillips Lab

ABSTRACT The Rosette nebula H II region is spatially resolved in IRAS infrared data. These IRAS data reveal new characteristics of the Rosette nebula which permit extraction of IRAS LRS slitless spectra from a well defined unresolved outer rim surrounding the Rosette. This rim consists of neutral material of very low density. IRAS infrared spectra from this rim reveal both a continuum component and an emission band component. These spectra are indicative of the processes contributing to the infrared emission of the diffuse interstellar medium, also known as the infrared cirrus, as these spectra are from a low density parcel of the diffuse interstellar medium with an enhanced but non-ionizing radiation field.

INTRODUCTION

The objective of the current work is to model the short wavelength infrared emission from H II regions, from 1 – 30 microns. The approach used has been to first model that component of the infrared emission which one can understand, the long wavelength infrared emission. New techniques were used for manipulating all of the data from the IRAS focal plane. Discriminating analysis was then used to confirm or select the appropriate model. Spectroscopy has been used to probe the relevant emission mechanisms.

LARGE H II REGIONS

Two large spatially resolved H II regions were initially chosen: the Rosette and Sharpless 252. Both are resolved spatially, and are regular in appearance. Both exhibit a central cavity at IRAS resolution, which is not the norm for H II regions. Both exhibit a rise in I(60)/I(100) towards the center. S252 is physically smaller, exhibits hotter dust emission and a steeper increase in the I(60)/I(100) ratio. Our initial concentration is on the Rosette because of its large angular size and regular appearance.

There are a number of observational characteristics in the resolved radio and infrared data which a model must reproduce (Botttinelli and Gouguenheim 1964, Celnick 1985, Menon 1962). All IRAS morphologies are more similar than dissimilar (12, 25, 60, 100 μm). The I(60)/I(100) ratio is not strongly peaked which implies that the temperature of the large grains, the dominant emitters at these wavelengths, is relatively flat within the H II region. The ratio I(12)/I(100) is nearly constant over the face of the H II region, which implies that some of the emitters at this wavelength reside within the ionized region. The ratio I(12)/I(100) is very similar to that of the diffuse ISM, perhaps implying similar emitters within the ionized region. The ratio I(25)/I(100) exhibits a slight rise in the central region, which again suggests that some type of small grain emitters must exist within the ionized region.

UNRESOLVED RIM

The Rosette IRAS emission exhibits an unresolved outer rim in all four IRAS bands. Small IRAS edge detectors were used to confirm that this outer rim is in fact unresolved at IRAS in-scan resolution at many locations surrounding the H II region. The unresolved outer rim is consistent with Spitzer's predicted weak expansion shock front.

INNER CAVITY

The inner stellar wind cavity is nearly unresolved in the IRAS detectors, a characteristic shared by the Effelsberg continuum data (Celnick 1985). Neither spherically symmetric models nor two dimensional models can reproduce such a steep inner cavity, implying that this cavity probably is cylindrical in character and aligned along the line of sight.

SPECTRA

The IRAS satellite carried with it a Low Resolution Spectrometer (LRS), which was of slitless design. The in-scan direction, or direction of spacecraft travel, was also the IRAS LRS dispersion direction. Although the LRS was intended for use on point sources, IRAS slitless LRS spectra may be extracted from extended objects which are unresolved in the dispersion or in-scan direction, and therefore for the unresolved regions in the outer rim of the Rosette. This operation was successfully performed using a new Laboratory for Space Research Groningen (ROG) LRS extraction package. Two stars were also extracted in order to verify the process. Of the spectra from the outer unresolved rim, about half are weak and flat exhibiting continuum like spectra. Emission features could be smoothed out in this set of spectra if these regions are partially resolved. Three locations for which spectra were extracted were cleanly unresolved at IRAS resolution (of order one arc minute) and show the unidentified infrared bands plus a continuum.

RADIATION FIELDS

At least three radiation fields are operable in the Rosette: direct stellar radiation, scattered radiation, and the diffuse interstellar radiation field. The large grain emission from the Rosette has been successfully modeled and we find that the radiation field is dominated by scattered photons.

Tom Kuchar has been studying the infrared characteristics of galactic H II regions. He has measured IRAS fluxes for cataloged H II regions in the first quadrant, identified by recombination lines, with known distances. He has identified average infrared emission features, and his data indicates that the Rosette nebula is **not** typical of galactic H II regions.

CONCLUSIONS (FOR ROSETTE ONLY)

The unresolved rim is consistent with Spitzer's predicted expansion shock. This is a neutral low density area outside of the ionized region which, while subject to an enhanced radiation field, is yet representative of the interstellar medium. The spectra from this unresolved outer rim are consistent with a combination of Sellgren small grain emission (continuum like) plus the unidentified infrared bands. The region from which these spectra were obtained is representative of low density infrared cirrus with an enhanced but non-ionizing radiation field.

REFERENCES

Bottinelli, L. and Gouguenheim L. 1964, Annales d'Astorphy., **27**, 685

Celnick, W. E. 1985, *A&A*, **144**, 171

Menon, T. K. 1985, *ApJ*, **135**, 394

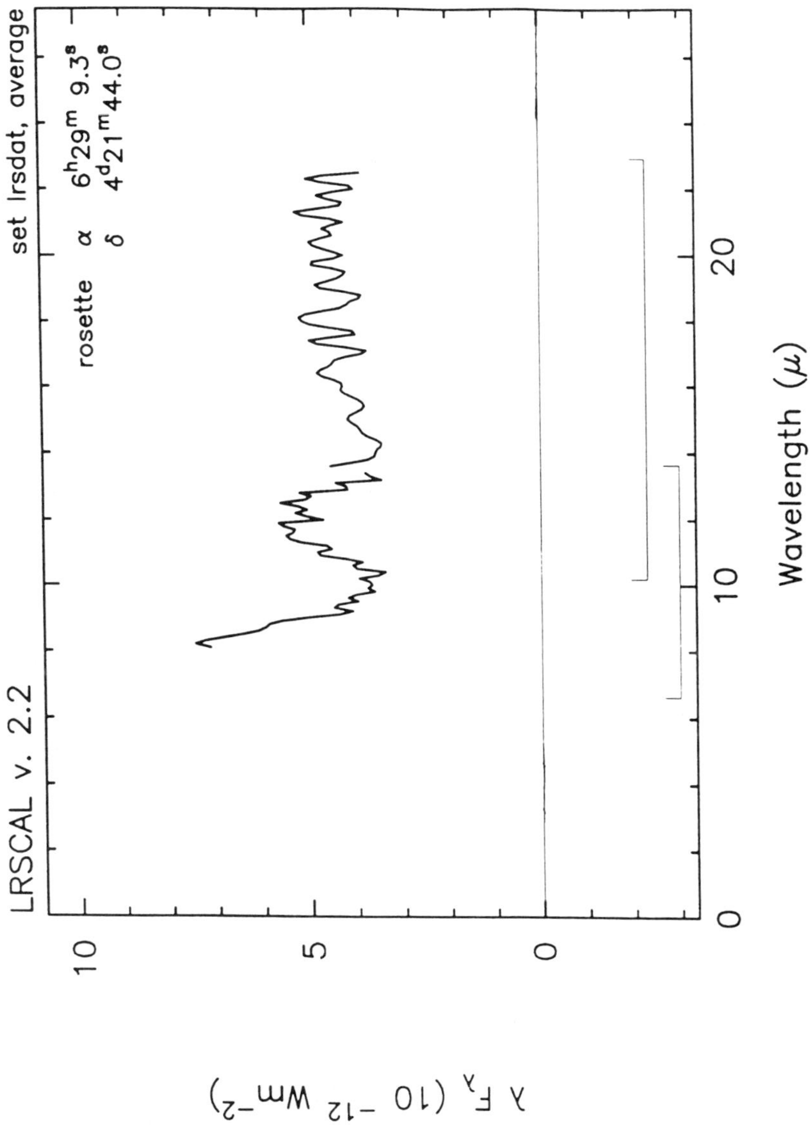
LRSCAL v. 2.2
set lrsdat, average
rosette α 6h29m 9.3s
δ 4d21m44.0s
10
5
0
0
10
20
Wavelength (μ)
λ F_λ (10 -12 Wm^-2)

The First Symposium on the Infrared Cirrus and Diffuse Interstellar Clouds
ASP Conference Series, Vol. 58, 1994
Roc M. Cutri and William B. Latter (eds.)

OBSERVATION OF [CII] 158μm EMISSION FROM THE DIFFUSE INTERSTELLAR MEDIUM AT HIGH GALACTIC LATITUDE

JAMES J. BOCK, VIKTOR V. HRISTOV, PHILIP D. MAUSKOPF, PAUL L. RICHARDS, ANDREW E. LANGE
Department of Physics, University of California, Berkeley, CA 94720

MITSUNOBU KAWADA, HIDEO MATSUHARA, TOSHIO MATSUMOTO, SYUUJI MATSUURA, MASAHIRO TANAKA
Department of Astrophysics, Nagoya University, Nagoya 464, Japan

ABSTRACT We report the first detection of [CII] 158μm line emission from the diffuse interstellar medium at high galactic latitude. We have measured the integrated line intensity in a 36' field-of-view along a triangular scan path in a 5° x 20° region in Ursa Major using a rocket-borne, liquid helium cooled spectrophotometer. The scan included high latitude infrared cirrus, molecular clouds, a bright external galaxy, M82, and the HI Hole, which is a region of uniquely low neutral hydrogen column density. Emission from [CII] is observed in all regions and, in the absence of appreciable CO emission, it is well correlated with neutral hydrogen column density . We observe a [CII] gas cooling rate which varies from (1.18 ± 0.4 to 3.25 ± 0.8) x 10^{-26} erg s^{-1} H-atom^{-1}, in good agreement with recent observations of UV absorption lines at high galactic latitude. Regions with CO emission have enhanced [CII] line emission over that expected from the correlation with neutral hydrogen column density. We measure a line-to-continuum ratio which varies from $I_{[CII]} / \lambda I_\lambda = 0.002$ to 0.008 in comparison with the all sky average of 0.0082 reported by FIRAS, which is heavily weighted towards the Galactic plane.

INTRODUCTION

Emission from the $^2P_{3/2}$ - $^2P_{1/2}$ fine structure transition of singly ionized carbon ([CII]) at 157.7μm is thought to be the dominant cooling mechanism for the gas phase of the diffuse interstellar medium (ISM) over a wide range of temperature and density (Dalgarno and McCray 1972). In statistical equilibrium, gas cooling by fine structure line emission from atoms and ions balances gas heating. The dominant heating mechanism of the gas in the diffuse ISM is thought to arise from photo-electrons ejected from the surface of dust grains by UV photons (Spitzer 1948, Watson 1972). Observations of [CII] emission probe the distribution and energetics of many phases of the diffuse ISM, including the cold medium (CM), the warm neutral medium (WNM) and the warm ionized medium (WIM).

Observations of [CII] 158μm line emission using balloon-borne and airborne telescopes have been limited to bright photodissociation regions (PDRs) in the plane of the Galaxy and in bright external galaxies (Shibai *et al.* 1991, Stacey *et al.* 1991). Since the gas densities and UV photon fields are orders of magnitude smaller in the diffuse ISM than in PDRs, emission from [CII] in the diffuse ISM is too faint to be detected by these techniques. The [CII] 158μm intensity from the diffuse ISM has been inferred from the population of the excited $^2P_{3/2}$ state of C^+ by ultraviolet absorption measurements (Pottasch *et al.* 1979, Savage *et al.* 1993). Recently, the FIRAS experiment aboard COBE produced a full sky map of [CII] 158μm emission in a 7° beam with sensitivity sufficient to detect the line in bright regions near the plane of the Galaxy (Wright *et al.* 1991). This paper reports the first detection of [CII] 158μm emission from the diffuse ISM at high galactic latitude.

OBSERVATIONS

The instrument and the results of the observations, including M82 and the HI Hole, are further described in Bock *et al.* (1993). Information about the rocket-borne payload and the flight can be found in Matsuhara *et al.* (1993). The scan path of the telescope, shown in Fig 1, includes high latitude infrared cirrus, molecular clouds, the galaxy M82, and the HI Hole. The scans between the vertices of the triangle were at a

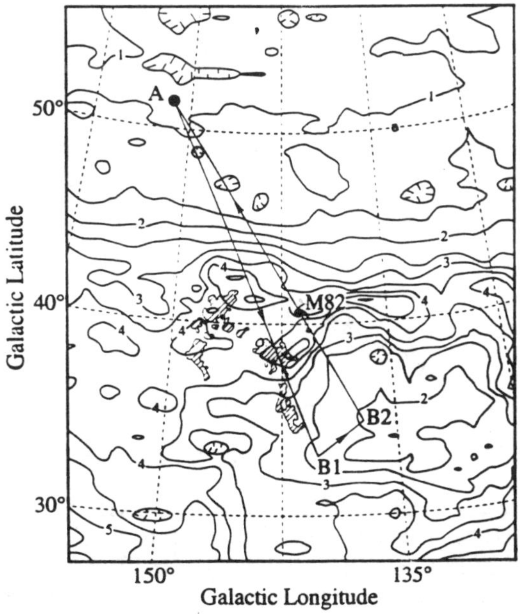

Fig. 1. The scan path of the telescope beam. The contours indicate neutral hydrogen column density (Heiles 1975) in units of 1 x 10^{20} cm^{-2}. The HI data used in the analysis comes from a separate measurement made in 1992. The shaded regions indicate CO emission from high latitude molecular clouds. The galaxy M82 is shown as the spiral. The size of the beam is shown as the solid circle at point A, which is the HI Hole.

constant rate of 36' (1 beam) per second. The telescope acquired point A, the HI Hole, at the end of the scan sequence for a 120s pointed observation.

The neutral hydrogen (N_{HI}) column density in a 36' (FWHM) beam was measured at 18' intervals along the scan path using the 85ft telescope at the Hat Creek Observatory (Heiles, private comm.). An offset was subtracted from the HI data to account for sidelobe contamination. Emission from the J=1 -> 0 transition of carbon monoxide (CO) is computed from a map obtained by deVries *et al.* (1987) using the Columbia 1.2m telescope with a 8.7' beam.

DISCUSSION

The measured [CII] 158μm intensity correlates with N_{HI} as shown in Fig 2. The gas cooling rate per hydrogen atom varies from $4\pi\, I_{[CII]} / N_{HI} = (1.18 \pm 0.4 \text{ to } 3.25 \pm 0.8) \times 10^{-26}$ erg s^{-1} H-atom^{-1}, excluding regions with CO emission where molecular gas may be present and regions where $N_{HI} < 10^{20}$ cm^{-2}, which are sensitive to statistical error. The correlation of [CII] intensity with N_{HI} outside the CO regions gives an average gas cooling rate of $(1.79 \pm 0.45) \times 10^{-26}$ erg s^{-1} H-atom^{-1}. Regions with CO emission have enhanced [CII] line intensity over that expected from the correlation with N_{HI}, consistent with emission from an increased C^+ column density associated with molecular

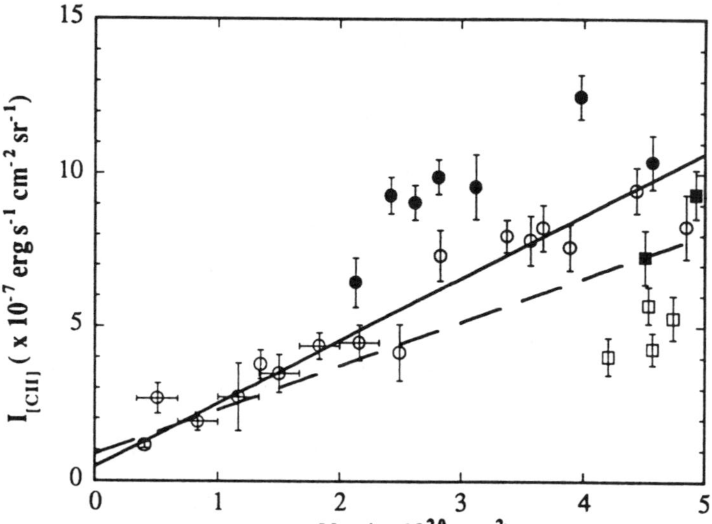

Fig. 2. The correlation of [CII] intensity, binned in 45' intervals along the scan, with HI column density. Horizontal error bars represent scanning data binned over intervals of 3.33×10^{19} cm^{-2} in N_{HI}, vertical error bars represent 1σ statistical uncertainty and do not include systematic errors. Regions with appreciable (> 0.1 K km s^{-1}) CO intensity are indicated with shaded symbols. A region with strong ($\lambda I_\lambda > 2 \times 10^{-4}$ erg s^{-1} cm^{-2} sr^{-1}) continuum intensity and anomalously low line-to-continuum ratio is indicated by square symbols. The best fit correlation given by the dashed line excludes data with CO emission. The solid line indicates the correlation if the region with low line-to-continuum is also excluded.

hydrogen. Further excluding the region with low line-to-continuum ratio results in a better statistical fit and gives a rate of $(2.56 \pm 0.57) \times 10^{-26}$ erg s^{-1} H-atom^{-1}. The low line intensity observed in this region indicates that the gas heating is reduced. This may be the result of shadowing of FUV photons responsible for photoelectric gas heating in this region by the CO cloud near (b = 36°, l = 142°).

The observed gas cooling rate agrees well with the rate of 2.1×10^{-26} erg s^{-1} H-atom^{-1} determined by Savage *et al* (1993) from a measurement of the UV absorption of the excited $^2P_{3/2}$ state of C^+ along a line of sight at high galactic latitude. It is much lower than the average rate of 1.3×10^{-25} erg s^{-1} H-atom^{-1} obtained by Pottasch *et al.* (1979), measured along several lines of sight to local stars.

We measure a range of line-to-continuum ratios from 0.002 to 0.008 in comparison to the all sky average of 0.0082 reported by FIRAS, which is heavily weighted towards the galactic plane. We have not included data with the lowest continuum signal levels ($\lambda I_\lambda < 10^{-4}$ erg cm^{-2} sr^{-1}) in this range because low intensity sources of continuum emission not associated with ISD emission will artificially depress the ratio.

CONCLUSIONS

We have measured [CII] intensity from the diffuse ISM at high galactic latitude and observe that, in the absence of CO emission, [CII] line intensity correlates with the column density of neutral hydrogen. The observed gas cooling rates agree with that derived from a recent UV measurement at high galactic latitude.

We would like to thank Carl Heiles, Jeff Beeman, Gordon Stacey, Andreas Heithausen, H. Okuda, the staff at ISAS, and the Daiko Science Foundation. This work was supported by NASA grants NAGW-1352 and NGT-50771, by a grant-in-aid from the Ministry of Education Science and Culture in Japan, and by the David and Lucile Packard Foundation.

REFERENCES

Bock, J. J. *et al.* 1993, *Ap. J.*, **410,** L115.
Dalgarno, A., and McCray, R. M. 1972, *Ann. Rev. Astr. Ap.*, **10,** 375.
deVries, H. W., Heithausen, A., and Thaddeus, P. 1987, *Ap. J.*, **319,** 723.
Heiles, C. 1975, *Astr. Ap. Suppl.*, **20,** 37.
Jahoda, K., Lockman, F. J., and McCammon, D. 1990, *Ap. J.*, **354,** 184.
Matsuhara, H. *et al.* 1993, in preparation.
Pottasch, S. R. *et al.* 1979, *Astr. Ap. (Letters)*, **74,** L15.
Savage, B. D. *et al.* 1993, *Ap. J.*, **404,** 124.
Shibai, H. *et al.* 1991, *Ap. J.*, **374,** 522.
Spitzer, L., Jr. 1948, *Ap. J.*, **107,** 6.
Stacey, G. J., Geis, N., Genzel, R., Lugten, J. B., Poglitsch, A., Sternberg, A., and Townes, C. H. 1991, *Ap. J.*, **373,** 423.
Watson, W. D. 1972, *Ap. J.*, **176,** 103.
Wright, E. L. *et al.* 1991, *Ap. J.*, **381,** 200.

The First Symposium on the Infrared Cirrus and Diffuse Interstellar Clouds
ASP Conference Series, Vol. 58, 1994
Roc M. Cutri and William B. Latter (eds.)

ISOPHOT: A UNIQUE INSTRUMENT FOR CIRRUS OBSERVATIONS WITH THE ISO SATELLITE

R.J. LAUREIJS
Space Science Dept., ESTEC, POB 299, 2200 AG Noordwijk, The Netherlands

D. LEMKE
Max Planck Inst. für Astronomie, Heidelberg, Germany

T. LILJESTRÖM
Observatory and Astrophysics Lab., Helsinki, Finland

ABSTRACT ISOPHOT, one of the 4 instruments on board the Infrared Space Observatory (ISO) satellite, has dedicated capabilities for undertaking a thorough study of the structure and composition of infrared cirrus. The photometric, spectroscopic and imaging modes of the instrument as well as the spacecraft capabilities are described. These capabilities will make cirrus observations with ISOPHOT unprecedented. We discuss the cirrus studies presently proposed for guaranteed observing time with ISO.

CAPABILITIES OF ISOPHOT FOR OBSERVING CIRRUS

ISO will be the first true infrared observatory in space. It carries a cryogenically cooled 60 cm telescope with 4 scientific instruments operating at wavelengths from 2.5 to 240 μm (Kessler, 1992). ISOPHOT will cover this entire wavelength region by means of three different sub-instruments (Lemke, 1992):
PHT-P, a multi-aperture photometer with three detectors. The observer can select from 14 different filters and 12 apertures to measure the infrared spectral distribution in the wavelength range from 3 to 125 μm.
PHT-C, a camera with a 3×3 array operating with 7 filters between 45 and 130 μm, and a 2×2 array with 5 filters between 90 and 240 μm
PHT-S, a 2×64 array grating spectrometer with $\lambda/\Delta\lambda \approx 100$. This sub-instrument with a fixed 25 arcsec entrance aperture operates in the ranges 2.5 to 5 μm and 6 to 12 μm.

The sensitivities for point and extended sources in the broad band filters are presented in Fig. 1. When using proper mapping modes of ISO, it is possible to detect very faint cirrus structures in the order of a few mJy in only a few minutes of integration time.

A number of short wavelength filters of PHT-P are centered on "unidentified" dust features such as the ones at 3.3, 7.7 and 11.3 μm (with $\lambda/\Delta\lambda$ of 17, 10, and 6, respectively), and at 9.7 μm to cover the silicate band. There are 4 filters

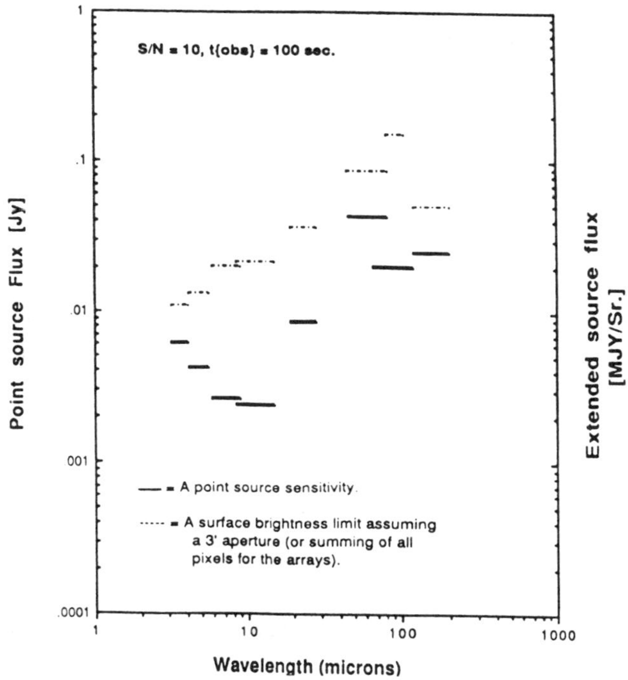

FIGURE I Minimum fluxes that can be detected with S/N=10 in 100 seconds of integration time in the broad band filters.

with spectral responses identical to the 4 IRAS wavelength bands for follow-up observations or cross-calibrations. All filters can be used in the wide beam mode (FOV 180 arcsec) allowing detection of the faintest extended sources. Most of the PHT-C filters are completely new to infrared astronomy and will provide a wealth of new data on infrared cirrus. PHT-S has been designed to perform sensitive measurements on ice and silicate features, as well as on the unidentified dust bands.

ISOPHOT has sets of long and short wavelength polarizers that can be put in the optical path of the PHT-P and PHT-C sub-instruments. Especially in the long wavelength regions, this will open a new class of observations of polarized dust emission. Pioneering work has been done with KAO on the Galactic Center (see e.g. Hildebrand, 1989), it is, however, expected that ISO will yield polarization data on many other sources.

ISO offers the possibility to observe cirrus structures by means of either raster mapping or slew scans requested by the observer. In addition, a 200 μm serendipity mode is foreseen in which the 3×3 array of PHT-C is collecting data while the satellite is routinely slewing between sources (Lemke and Burgdorf, 1992).

To improve the photometric accuracy of the observations, ISOPHOT is equipped with two internal black-body calibration sources and a focal plane chopper. The chopper can deflect the beam onto the calibration sources or can be used to point at off-positions. The observer can even request to use the chopper for scanning purposes.

PROPOSED CIRRUS OBSERVATIONS

The largest proportion of the ISO guaranteed time devoted to infrared cirrus studies will be used to get precise photometry in almost all available ISOPHOT bands for a variety of clouds. Not only positions in relatively faint and quiescent HI clouds have been selected (based on IRAS studies) but also positions in the outskirts of reflection nebulae and star forming regions. High surface brightness regions have been included to allow for PHT-S measurements. The observations will be used to

- obtain a clear determination of the spectral distribution in the near and mid infrared of cirrus clouds; we stress that the small grain features have never been observed in genuine cirrus (they have been inferred from IRAS broad band photometry);

- study the energy budget in the clouds, especially the contribution of the small particle emission with respect to the classical large grains which could lead to new insights in the heating and cooling of clouds;

- observe likely grain abundance variations between different clouds, or between positions inside clouds.

The aim is to improve the ISM-dust model and gain better understanding of the lifecycle of dust and the energy budget of the galaxy.

The cirrus project is closely correlated to a similar study of the interplanetary dust, the emission of which is a considerable brighter foreground component. Moreover, suspected cirrus-free "holes in the sky" will be verified in order to perform extragalactic background measurements there (Fig. 2).

A separate project is the study of diffuse infrared emission from the Galaxy rather than individual clouds. 30 positions along and near the galactic plane have been selected that are representative for the diffuse galactic background emission outside bright discrete sources. Measurements with PHT-S should determine whether there are any dust features, PHT-P photometry at 3.3, 7.7, 25, and 100 μm are necessary for cross calibrations with PHT-S, IRAS, COBE, and possible ISO long wavelength spectrometer (LWS) observations. Slew scans with the 3×3 array are requested to determine FIR properties.

A comprehensive project is foreseen to observe dust absorption features with PHT-S along a variety of sightlines towards continuum sources. A reasonable subset of these observations deals with sightlines through cirrus regions.

More in the realm of star formation and dark clouds is the proposal to determine the polarization of thermal radiation by aligned dust grains in dense star forming clouds and Bok globules. This study enables the investigation of the correlation between field direction within a cloud and the rotational axis or the outflow lobes. The measurements will contribute to our understanding of grain alignment mechanisms. In addition, the alignment of grains belonging to different dust components can be studied, by measuring the alignment at a number of wavelengths.

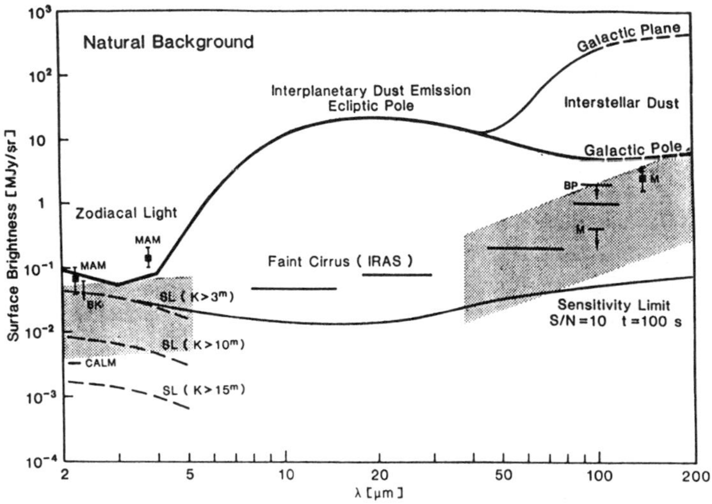

FIGURE II Comparison of the detection limit of ISO with the surface brightness of faint cirrus clouds and the interplanetary dust emission (zodiacal light). The hatched areas indicate the expected brightness of the extragalactic light. Adopted from Lemke et al, 1991.

OPEN TIME PROPOSALS

As a result of the joint collaboration between ESA, ISAS, and NASA for the operations of ISO, astronomers from Europe, Japan, and the USA can compete for open time which comprises 2/3 of the total available observing time during the 18 month mission. The launch of ISO is foreseen in 1995 and the Call for Observing Proposals will be issued in the first half of 1994.

ACKNOWLEDGMENTS

Valuable advice was given by A. Leger, Paris, in planning the PAH studies of cirrus and by K. Mattila, Helsinki in the overall planning of the dust studies. Both are Co-investigators in the project mentioned above.

REFERENCES

Hildebrand, R.H. 1988, *QJRAS*. **93**, 70

Kessler, M. F., 1992 in *Infrared Astronomy with ISO*, Eds. Th. Encrenaz, M. F. Kessler, Nova Science, page 3

Lemke, D. 1992 in *Infrared Astronomy with ISO*, Eds. Th. Encrenaz, M. F. Kessler, Nova Science, page 53

Lemke, D., Burgdorf, M. 1992 in *Infrared Astronomy with ISO*, Eds. Th. Encrenaz, M. F. Kessler, Nova Science, page 69

Lemke, D., Leinert, Ch., Mattila, K. 1991, *Adv. Space Res.* **II**, page 223

The First Symposium on the Infrared Cirrus and Diffuse Interstellar Clouds
ASP Conference Series, Vol. 58, 1994
Roc M. Cutri and William B. Latter (eds.)

COBE[1] OBSERVATIONS OF THE SPECTRUM OF INFRARED CIRRUS

W. T. REACH, J. C. MATHER, M. G. HAUSER, E. DWEK
NASA/GSFC Lab. for Astronomy and Solar Physics

D. J. FIXSEN, T. J. SODROSKI
Applied Research Corporation

The Far-Infrared Absolute Spectrometer (FIRAS) and the Diffuse Infrared Background Experiment (DIRBE) on the *Cosmic Background Explorer (COBE)* provide unique, large-scale measurements of the spectrum of interstellar dust emission from microwave through near-infrared wavelengths. Here we present and discuss the FIRAS spectrum of high-latitude Galactic dust, and we discuss the DIRBE 140-240 μm color temperatures.

[1] The NASA Goddard Space Flight Center (GSFC) is responsible for the design, development, and operation of the *Cosmic Background Explorer (COBE)*. Scientific guidance is provided by the *COBE* Science Working Group. GSFC is also responsible for the development of analysis software and for the production of the mission data sets.

The First Symposium on the Infrared Cirrus and Diffuse Interstellar Clouds
ASP Conference Series, Vol. 58, 1994
Roc M. Cutri and William B. Latter (eds.)

SPATIAL POWER SPECTRAL MEASUREMENTS OF IR CIRRUS

T. N. GAUTIER
JPL/IPAC, Caltech

W. STEWART
University of Manchester and Caltech

We have used the recently released IRAS Sky Survey Atlas (ISSA) image data to investigate the surface brightness distribution of selected regions of infrared cirrus using power spectral analysis techniques. Useful signal to noise ratios have been obtained in all 4 IRAS bands (12 μm, 25 μm, 60 μm and 100 μm) for spatial frequencies between 0.16 deg^{-1} and 6 deg^{-1}. The 2 dimensional power spectra generally show a radially symmetric power law form with spectral index ≥ -3.

This work was carried out in part by the Jet Propulsion Laboratory, California Institute of Technology, under a contract with the National Aeronautics and Space Administration. W. Stewart was supported by a Summer Undergraduate Research Fellowship at The California Institute of Technology.

The First Symposium on the Infrared Cirrus and Diffuse Interstellar Clouds
ASP Conference Series, Vol. 58, 1994
Roc M. Cutri and William B. Latter (eds.)

1 AND 2 MILLIMETER OBSERVATIONS OF INTERSTELLAR DUST FROM ANTARCTICA

G.DALL'OGLIO, L.PIZZO
Dipartimento di Fisica, Università di Roma, Italy

P.ANDREANI
Dipartimento di Astronomia, Università di Padova, Italy

P.MERLUZZI
Istituto di Ingegneria Aereospaziale, Università di Napoli and Osservatorio Astronomico di Capodimonte, NAPOLI, ITALY

L.ROSSI
Area di Ricerca, CNR, FRASCATI, ITALY

Several measurements of the millimetric (1 and 2 *mm*) continuum emission of interstellar clouds in the Galaxy and in the Magellanic Clouds have been gathered at the Italian Base in Antarctica during the antarctic summers 1987-88 and 1990-91. These clouds have been mapped at low ($\sim 1^\circ$) and high ($5'$) resolution by means of a ^{3}He-bolometer fed with a 1m flux collector or a 2.6m telescope.

FIR/mm spectra of the detected regions, constructed with the aid of the IRAS sky-flux maps, show that the millimetric data do not belong to the same thermal component defined by the 60 and 100 μm observations and can be better ascribed to a cold component at temperature of 15-18 K. Surely, free-free emission contributes partially to the millimetric fluxes, but the high resolution data show that this contribution does not exceed 30 %. Moreover, the good spatial correlation between the 100 μm and the 1 and 2 *mm* scans of the detected clouds rather suggests that thermal emission from dust grains is better fitted by a superposition of thermal curves at different temperatures.

A tentative interpretation of these observations has been made: laboratory data on the optical properties of cosmic analogs made of hydrogenate amorphous carbon containing randomly oriented graphitic islands have been used to build a simple dust model. The predicted spectrum matches well with the observed long wavelength spectrum. Some hints on the physical properties of dust grains responsible for this emission are inferred.

The First Symposium on the Infrared Cirrus and Diffuse Interstellar Clouds
ASP Conference Series, Vol. 58, 1994
Roc M. Cutri and William B. Latter (eds.)

THE SUBMILLIMETER TELESCOPE (SMT) - A STATUS REPORT

J. W. M. BAARS, R. N. MARTIN, W. L. PETERS, J. G. MANGUM
Submillimeter Telescope Observatory, Steward Observatory, University of Arizona, Tucson, AZ 85721

The Submillimeter Telescope (SMT) is a collaborative effort of the Max-Planck-Institut für Radioastronomie, Bonn and Steward Observatory, Tucson. It is a 10 m diameter telescope, dedicated to the submillimeter wavelength range and placed in an enclosure which will be opened during observations. The telescope is under construction on Mt. Graham at 3200 m altitude in southeastern Arizona. Construction of the SMT is progressing rapidly and the telescope will go into operation in 1993. This poster will summarize the current construction status. Several technological innovations have been used in the design of the SMT which will assure good performance of the telescope at wavelengths as short as 0.3 mm. These will be illustrated by results of the extensive design optimization and test measurements of the antenna subsystems. In addition, the initial instrumentation capabilities of the SMT facility will be summarized.

CHAPTER III

MOLECULAR LINE AND HI STUDIES OF INTERSTELLAR CLOUDS

The First Symposium on the Infrared Cirrus and Diffuse Interstellar Clouds
ASP Conference Series, Vol. 58, 1994
Roc M. Cutri and William B. Latter (eds.)

MOLECULAR LINE SURVEYS AT HIGH GALACTIC LATITUDES

LORIS MAGNANI
Department of Physics and Astronomy, University of Georgia, Athens, GA 30602

ABSTRACT The IRAS cirrus clouds are the infrared manifestation of diffuse atomic and molecular clouds in the interstellar medium. A fraction of the infrared cirrus is composed primarily of molecular hydrogen gas. These objects are often referred to as high-latitude molecular clouds (since they are identified most easily at high Galactic latitudes) or translucent clouds. Current estimates of the surface filling fraction of molecular gas at $|b| \geq 25°$ based on the CO(J=1−0) transition range from 0.5 - 13%. The molecular fraction of the cirrus is an important quantity for determining the molecular mass surface density of these small clouds. If some of the higher estimates of the mass surface density are correct, then the molecular cirrus contain a significant fraction of the overall molecular gas inventory of the local interstellar medium. In this review we examine the problem in the context of the surveys for molecular gas at high Galactic latitudes which have been carried out in the last decade.

INTRODUCTION

The disk component of spiral galaxies is dominated by stars. The stellar component provides the gravitational potential to which the interstellar medium (ISM) responds. Spitzer (1942) showed that a plane-parallel layer of stars with a scale height substantially greater than that of a second, distributed component forces the second component to assume a Gaussian distribution centered about the midplane of the stellar disk. Since the scale height of the atomic and molecular hydrogen gas which comprise the ISM of the disk (as opposed to the more buoyant gas in the halo) is less than that of the stars, it is natural to assume that the bulk of the cold disk gas should be concentrated along a very narrow layer in the Galactic midplane. CO(J=1−0) surveys appeared to confirm that the majority of the molecular gas in the ISM is concentrated along the midplane. This perception persisted until the early 1980's; the astronomical conventional wisdom postulated that the high-latitude sky (*i.e.*, $|b| \geq 25°$) was rather devoid of molecular gas. In 1983-84 the success of the IRAS mission and the Blitz *et al.* (1984) high-latitude survey forever changed our view of the high-latitude sky.

In this review, we discuss the quantity of molecular gas at high Galactic latitudes. We describe the work prior to the IRAS survey and the systematic searches for high-latitude molecular clouds. Then, we examine the various arguments which lead to the surface filling factor and mass surface density of molecular gas at $|b| \geq 25°$.

SEARCHES PRIOR TO 1983

The association of molecular gas with regions of high-extinction is well-established. In this section we review the extinction surveys and subsequent molecular searches at high latitudes prior to the IRAS mission and the Blitz *et al.* high-latitude survey.

The earliest systematic search for obscured regions was conducted by Barnard (1927) and includes only 3 of 50 photographic plates at Galactic latitudes (old galactic coordinates) $|b^I| > 20°$, and no plates at $|b^I| > 25°$. This attests to the difficulty of detecting obscured areas at high Galactic latitudes where the stellar density drops rapidly.

The survey conducted by P.J. Melotte & K. Lundmark and reported by Lundmark (1926) does contain many dark nebulae at high Galactic latitudes. Melotte and Lundmark used photographic plates taken during the Franklin-Adams 1911 sky survey which have an approximate limiting magnitude of 15. Any fivefold or greater decrease in star density over an area of more than a few square degrees was attributed to the presence of a "dark" nebula. In this manner, Melotte & Lundmark found 1550 dark nebulae in the regions covered by the Franklin-Adams plates, with many of the regions occuring at high Galactic latitudes.

In more recent times, Khavtassi (1955), Roshkovsky (1955), and Schoenberg (1964) have compiled lists of dark nebulae, mostly confined to the galactic plane with few objects at $|b| > 20°$. It was not until the work of Lynds (1962) that a modern, complete survey of all the dark nebulae visible from the northern hemisphere became available. In her work, all of the 879 fields photographed on red and blue sensitive plates with the Palomar Observatory 48-inch Schmidt were examined for the presence of obscuration. The resulting 1801 objects (many of which are related spatially to each other) represent all the easily discernible nebulae at $\delta > -33°$.

Of the 1801 dark clouds detected by Lynds (1962) only 19 are located at $|b| \geq 25°$. Some of the 19 objects are related spatially so that only 7 separate groups or complexes of dark nebulae were identified at high Galactic latitude. L169 is the nebula at highest latitude with $b = 37\overset{\circ}{.}0$. The dozens of obscured regions observed by Melotte & Lundmark at high latitudes were not noted by Lynds in her catalog. The reason for this omission was attributed by Lynds to the greater limiting magnitude of the Palomar plates. The areas which show a decrease in the stellar density on the Franklin-Adams plates do not show a noticeable decrease on the Palomar plates because the greater number of stars visible on the latter produce smaller random fluctuations in the stellar density. Recently, the southern hemisphere has been surveyed in a manner similar to that of Lynds (1962) by Feitzinger and Stüwe (1984) and Hartley *et al.* (1986).

In addition to the Lynds catalog, two other direct indicators of extinction (or lack thereof) at high latitudes existed prior to 1983: the Lick galaxy counts catalog (Shane & Wirtanen 1967), and the reddening studies using Strömgren photometry (*e.g.*, Hillditch, Hill, & Barnes 1976; Knude 1979; Kilkenny 1980; and Perry & Johnson 1982). Neither method revealed the presence of substantial extinction (> 0.5 mag A_V) at high latitudes. The main reasons for this are that the galaxy counts offer poor statistics over 2×2 square degrees of sky and the photometric studies suffer from lack of spatial coverage. Nevertheless, Knude

(1979) identified some 200 low mass dust clouds in the solar vicinity with $E(b-y) \sim 0.03$ mag corresponding to $A_V \sim 0.1$ mag.

Most of the dark clouds in the Lynds catalog (including the high-latitude component) had been searched for several molecular species including CO (Dickman 1975), H_2CO (Dieter 1973), and OH (Heiles 1968; Cudaback & Heiles 1969; Crutcher 1973). The first detections of molecular gas at high Galactic latitudes thus originate from this period. Although there appeared to be little extinction and consequently little molecular gas at high Galactic latitudes aside from the Lynds (1962) objects at $|b| > 25°$, there were five other indications that molecular gas was present at high Galactic latitudes: 1) The bright nebulae noted by Lynds (1965); 2) the absorption observations of OH and HI in the direction of extragalactic background continuum sources (Dickey, Crovisier, & Kazès 1981; Kazès & Crovisier 1981); 3) The CO observations made by Blitz, Fich, & Stark (1982) towards Sharpless HII regions; 4) the CO(J=1−0) emission detected by Knapp & Jura (1976) in interstellar clouds which had been studied previously via optical absorption lines; 5) the two populations of quasars present in a plot of the B−V residuals vs. N(HI) (Teerikorpi 1981).

In 1965, Lynds compiled a catalog of reflection or emission areas as determined from the blue and red POSS plates. There are many more of these Lynds bright nebulae (LBN) at high Galactic latitude than there are Lynds (1962) dark nebulae. Both van den Bergh (1966) and Sandage (1976) note that some of the reflection nebulae at high Galactic latitude could be dust clouds reflecting the integrated starlight of the Galactic plane below them. Assuming normal gas-to-dust ratios, these clouds should contain HI and, perhaps, H_2. In fact, the area studied by Sandage in Ursa Major comprises the North Celestial Pole Loop, a complex of high-latitude clouds (*e.g.*, Magnani, Blitz, & Mundy 1985; de Vries, Heithausen, & Thaddeus 1987).

Dickey *et al.* (1981) searched for OH absorption at 18 cm towards 58 extragalactic radio sources. They detected OH absorption and/or emission in 16 directions including 2 at $|b| > 25°$. Similarly, Kazès & Crovisier (1981) searched for CO(J=1−0) emission from 76 lines of sight showing strong HI 21 cm absorption. CO emission was detected from 21 directions including 4 at $|b| > 25°$ (two of the four sources are included in the sample studied by Dickey *et al.* 1981). The emission from these high-latitude lines of sight was deemed to arise from diffuse interstellar clouds which were not mapped. In fact, two of the sources are probably associated with the MBM53-55 clouds, a complex which extends over 10 degrees in the constellation Pegasus.

A catalog of CO radial velocities toward Galactic HII regions was compiled by Blitz *et al.* (1982). Of the more than 300 Sharpless (1959) HII regions observed, a total of 6 objects at $|b| > 25°$ showed evidence of CO(J=1−0) emission. Despite the fact that these objects are associated with reddish nebulosity on the Palomar plates somewhat similar to that produced by HII regions, we stress that the nebulosity is not attributable to Hα emission but, rather, some other process (see review by R. Guhathakurta, this volume).

Knapp & Jura (1976) observed CO(J=1−0) emission from interstellar clouds along lines of sight to stars which showed CO absorption in the UV. Although none of the detections are at $|b| > 24°$, the study showed that it was possible to detect molecular radio lines from clouds which were not classified as "dark" clouds. Although many directions at high latitudes showed atomic absorption lines from diffuse clouds (*e.g.*, Munch & Zirin 1961; Albert 1983) a systematic search for molecular gas in these directions was not undertaken.

Teerikorpi (1981) determined that a substantial number of molecular clouds had to be present at high-latitudes by plotting the residual B$-$V color of a sample of quasars [defined to be (B–V) – $(\mathrm{B{-}V})_{zo}$; where the latter quantity is the unreddened quasar color] vs. N(HI). The resulting two-branch structure in the quasar population showed greater residuals for the quasars in the direction of spiral arms. Teerikorpi concluded that the more reddened branch of the quasar population arose from greater absorption in the local spiral arms. For a constant gas-to-dust ratio the absorption had to be produced by dust associated with molecular gas. The parameters determined by Teerikorpi for these clouds ($A_V \geq 0.5$ mag, linear sizes less than 2 pc if $n > 50$ cm^{-3}, and 40% of the lines of sight toward the local spiral arm intersect one cloud) are somewhat similar to the parameters for the denser component of the infrared cirrus (*i.e.*, that portion of the cirrus with 100 μm emissivity greater than a few MJy ster^{-1}).

Although the above works indicated that molecular gas was probably present at high Galactic latitudes in addition to the Lynds (1962) dark nebulae, the only published survey for CO(J=1$-$0) emission from high-latitude regions proved negative. Dewey *et al.* (1983) searched for CO from 68 lines of sight at $b \sim 59°$ and $b \sim 84°$. These areas had been chosen on the basis of a high E(B$-$V) value over several square degrees as determined from the Shane-Wirtanen catalog (1967) of galaxy counts.

Ironically, a succesful search for high-latitude CO emission was conducted by Michel Fich and Leo Blitz in January of 1982, with several detections in the region near the North Galactic Pole (M. Fich, private communication). Although the results were never published, this work was the forerunner of the Blitz *et al.* (1984) survey for high-latitude gas.

As of the summer of 1983, the (published) molecular gas inventory at high Galactic latitudes included 7 separate Lynds (1962) dark cloud complexes with visual extinctions in their core regions in all cases greater than one magnitude visual, 6 lines of sight in the vicinity of presumed HII regions, and 4 lines of sight to diffuse, unmapped interstellar clouds.

THE IRAS CIRRUS AND HIGH-LATITUDE MOLECULAR CLOUDS

The year 1983 brought two startling developments for high-latitude studies. The IRAS science team announced the discovery of the infrared cirrus (Low *et al.* 1984) and a team led by Leo Blitz detected CO from several dozen new lines of sight at $|b| > 25°$. The high-latitude cloud catalog presented by Magnani *et al.* (1985) lists 57 objects at high latitudes. All the Lynds clouds at $|b| \geq 25°$ are included and not all the objects are spatially distinct. A subset of the clouds was mapped and several objects with individual MBM designations are really part of spatially-connected complexes (*i.e.*, clouds 27-32 of the MBM catalog correspond to the North Celestial Pole Loop complex mapped in detail by de Vries *et al.* 1987). A survey similar to the Magnani *et al.* (1985) work was conducted in the Southern Hemisphere by Keto & Myers (1986). They detected

15 clouds at $|b| > 20°$ and 3 small clouds in the Chamaeleon complex. The clouds were rather similar to those found in the Magnani *et al.* (1985) survey, though somewhat smaller [masses $\sim$ 2 $M_{\odot}$ compared with 40 $M_{\odot}$ for Magnani *et al.* (1985)]. In addition to the new high-latitude clouds, Keto & Myers (1986) were the first researchers to suggest that the external pressure of the ISM was perhaps able to confine the clouds.

The large spatial extent of some of the high-latitude clouds (many cover several square degrees) naturally led to comparisons with the infrared cirrus. Early work showed that although the majority of the cirrus was atomic in nature (Boulanger, Baud, & van Albada 1985; Burton 1986), the high-latitude molecular clouds had infrared cirrus counterparts (Weiland *et al.* 1986; de Vries 1986; de Vries *et al.* 1987). Moreover, the majority of the high-latitude molecular cirrus represents a type of cloud where molecules are readily detectable by radio means (*i.e.*, their CO abundances are higher than the canonical diffuse clouds studied by optical and UV absorption techniques) but with extinctions low enough that their detection on the Palomar plates is very difficult. This intermediate cloud population is known as translucent molecular clouds (van Dishoeck & Black 1988) and most of the high-latitude molecular clouds are in this category.

With some of the cirrus being primarily atomic and some primarily molecular, questions which naturally arise are: what is the filling factor of each component, and what controls the transition from atomic to molecular components? A discussion of the latter point is presented in the reviews by Ewine van Dishoeck and Bruce Elmegreen (this volume) while the first point is the focus of the remainder of this review.

THE SURFACE FILLING FACTORS

The Low-Latitude Scenario

The fraction of the sky at high or low latitudes covered by cirrus is difficult to determine because at low enough levels, the entire sky shows some 100 μm emission (F. Boulanger, this volume). A quick look at a whole-sky 100 μm map shows that, at a level of 7 MJy ster^{-1} or higher, perhaps 10-20% of the high-latitude sky is covered with cirrus emission (see, for example, the map presented in Heithausen & Thaddeus 1990). Van Buren (1989) studied the cirrus emission from local material heated by stars earlier than B5 and brighter than $m_V \leq 8.0$. The gas corresponding to the cirrus has a density 0.1 cm$^{-3} < n <$ 10^2 cm^{-3} and so it encompasses high-latitude-type clouds and much more diffuse entities. A direct ratio of the number of stars with associated cirrus emission to those which in principle could be detected indicates that the cirrus *volume* filling factor is $\sim$ 0.2. This corresponds to a surface filling factor of about 0.1. A more complete version of this study was reported on by D. Van Buren and J. Gaustad at this meeting with results that the volume filling fraction produced by cirrus clouds with a density greater than 0.5 cm^{-3} is 14.6 $\pm$ 2.4% - corresponding to a surface filling fraction of about 6%.

If the surface filling fraction of the "denser" infrared cirrus is somewhere in the 0.1–0.2 range, it stands to reason that the filling factor of molecular gas at high latitudes should be less than this value.

Perhaps surprisingly, the few systematic surveys for molecular gas at high Galactic latitudes yield different results. Before examining the studies in chronological order below, we must address quantitatively, how much atomic and molecular gas exists locally in the galactic plane. Heiles (1976) estimates that the local HI surface mass density is 8 $M_\odot$ pc^{-2}. This value includes contributions from both the Warm Neutral Medium (WNM) and the Cold Neutral Medium (CNM), and also includes a correction for He which increases the mass surface density by a factor of 1.36 (this correction is usually not included in mass estimates). Liszt (1983) obtains a value of 5 $M_\odot$ pc^{-2} for all the local HI and Dickey & Garwood (1989) obtain a value of 3 $M_\odot$ pc^{-2} for the CNM component.

Local values for the molecular mass surface density [$\equiv \Sigma_m$] are lower and range between 0.5–2.0 $M_\odot$ pc^{-2} [Blitz *et al.* 1984; Dame & Thaddeus 1985; Bloemen *et al.* 1986; Dame *et al.* 1987; Bronfman *et al.* 1988]. All of the molecular mass surface density estimates are based on studies of Giant Molecular Clouds (GMCs) and thus describe the surface mass density only at low latitudes. Feitzinger & Stüwe (1984) extend the Lynds (1962) catalog of dark clouds to the southern hemisphere and thus obtain a whole-sky catalog of dark clouds. Within 250 pc of the Sun, they estimate that the number density of dark clouds is 8×10^{-5} pc^{-3}. If we assume that the scale height of local dark clouds is 100 pc (Blitz *et al.* 1984), we obtain a mass surface density of 0.016M_{ave} , where M_{ave} is the average mass of a dark cloud. For average cloud masses of 50 $M_\odot$, the contribution to Σ_m from dark clouds is $\sim$ 0.8 $M_\odot$ pc^{-2}.

Kazès & Crovisier (1981) refer to their objects as diffuse clouds, but it is likely that they are, for the most part, translucent molecular clouds. We can use their survey as another estimate of the small molecular cloud contribution to Σ_m. Assuming a scale height for these objects of 100 pc and an $N(H_2)/W(CO)$ conversion factor of 2.5×10^{20} $[K\ km\ s^{-1}]^{-1}\ cm^{-2}$, we obtain $\Sigma_{DC} = 0.6$ $M_\odot$ pc^{-2}. We thus conclude that at low latitudes Σ_m ranges from 1-3 $M_\odot$ pc^{-2}. The high-latitude contribution to Σ_m (which we define to be Σ_{HL}) is discussed below.

The High-Latitude Scenario

The first attempt at an unbiased determination of the molecular mass content at high-latitudes was by Magnani, Lada, & Blitz (1986 - hereafter MLB). They conducted a search for CO(J=1−0) emission at high latitudes based on randomly located 100-point grids. Each region was sampled every degree in a 10 × 10 grid with each point observed until a 1σ rms noise level of 0.3–0.4 K was reached. The survey was not completely unbiased, however, because an 8^h gap in sidereal coverage centered on the Taurus clouds existed throughout the observing run. Out of a total of 2500 sampled lines of sight, only 9 sources were detected, of which, only 2 were new objects. Accounting for the uneven sky coverage, the surface filling fraction according to MLB is $\sim$ 0.005; one line of sight out of 200 will, on average, show CO(J=1−0) emission at the $T_A^* \geq 1$ K level. If we assume that the solar neighborhood extends out to 100 pc, the number of clouds contained in a cylindrical region centered about the Sun with radius 100 pc is 120 ± 40. Assuming an average mass for each cloud of 50 $M_\odot$ (Magnani *et al.* 1985), there is $\sim$ 5000 $M_\odot$ of molecular gas in the solar neighborhood attributable to high-latitude clouds. This produces, in turn, a Σ_{HL} of 0.2 $M_\odot$ pc^{-2}, a factor of 5-15 less than the value produced locally by GMCs. In this scenario, the high-latitude molecular clouds are not the dominant constituent of molecular gas locally, although they are a non-negligible component.

In 1988, Désert, Bazell, & Boulanger (hereafter DBB) used the correlation between $N(H_{total})$ [$\equiv N(HI) + 2N(H_2)$] and the infrared brightness of the sky at 100 μm to determine the filling factor of molecular gas at $|b| \geq 5°$. The infrared contribution attributed to N(HI) was subtracted and the left-over infrared emission (dubbed Infrared Excess Clouds or IRECs) was attributed to molecular cirrus. When the DBB survey for regions with $|b| \geq 25°$ is compared to the MBM survey, nearly 47% of the IRECS are associated with high-latitude molecular clouds. In general, the missing MBM clouds were small in angular extent ($<$ 0.5 degrees) which is the resolution of the DBB survey. The only large MBM complex which is missing from the DBB survey is the MBM53-55 complex. This object is embedded in a prominent HI shell and it is possible that virtually all of the infrared emission is attributable to the HI in the region. The results of the DBB survey also support the existence of a 30%–70% asymmetry between the north and south Galactic hemispheres already noted in the high-latitude cloud distribution by Magnani *et al.* (1985). The fraction of the sky covered by IRECs decreases slightly from 0.059 at $|b| > 10°$ to 0.029 at $|b| > 30°$. DBB estimated that the IRECs had a surface filling fraction over the entire sky of 0.04–0.05. The discrepancy between these filling factors and the MLB result could be attributed to molecular clouds not detectable in the CO(J=1−0) transition or, alternatively, to dust-rich HI clouds.

As a follow-up to the DBB survey, Blitz, Bazell, & Désert (1990) surveyed a portion of the DBB IRECs in the CO(J=1−0) transition. They detected CO in 27 out of 201 positions (13.4%) at a 1σ rms noise level of 0.2 K. The 13% detection rate when coupled with the overall DBB results implies that 0.005-0.007 of the sky at $|b| > 10°$ shows CO emission at antenna temperature levels greater than 0.6 K. This result is consistent with the MLB survey and underscores the relative paucity of high-latitude molecular gas detectable via the CO(J=1−0) transition. Blitz *et al.* (1990) speculate that the remainder of the IRECs not detected by their survey are diffuse molecular clouds with CO/H_2 ratios $< 10^{-6}$, and thus not detectable with current millimeter radio technology. In this instance, the DBB catalog would be a complete listing of local, diffuse (as opposed to translucent) molecular clouds.

A more recent CO survey at high latitudes by Heithausen *et al.* (1993 - hereafter HSVMP) contradicts the findings of the two earlier surveys. HSVMP surveyed 620 square degrees in the region $117° \leq l \leq 160°$ and $16° \leq b \leq 44°$ in the CO(J=1−0) transition. The resolution varied from 16′ to 30′ with a beam size of 8′ and a 1σ rms noise level of 0.2 K. A full 13% of the region shows CO emission and for clouds above $b = 25°$, the surface filling factor is 0.092, nearly a factor of 20 higher than the MLB value. However, even in the $b \geq 25°$ region, half of the emission comes from a large cloud called "The Polaris Flare" (Heithausen & Thaddeus 1990). The Polaris Flare is not a typical high-latitude cloud in the sense that its mass is estimated to be 5500 $M_\odot$, a factor of 100 greater than the typical clouds described by Blitz *et al.* (1984). More importantly, the object appears to be connected to the Cepheus Flare, a giant molecular cloud (Grenier *et al.* 1989).

Even without including the Polaris Flare, the surface filling fraction of the CO-emitting clouds is 0.041, nearly an order of magnitude greater than the MLB value. HSVMP attribute some of this discrepancy to the higher sensitivity of their survey which shows that the CO emission is not concentrated in fairly small, isolated clouds, as most previous observations of molecular cirrus have indicated, but is extended over most of the region observed. Moreover, about 2/3 of the area detected in CO during the survey is not included in the DBB catalog. HSVMP interpret that the IRECS are not necessarily "clouds" but can be explained by physical effects such as variations in the gas-to-dust ratio or changes in the dust temperature.

If the survey is complete to a distance of 240 pc, then Σ_{HL} is 0.86 $M_\odot$ pc^{-2}. This value assumes that even at very high latitudes the surface filling fraction does not decrease. If no clouds exist at $b \geq 44°$, then the filling fraction is $\sim 7\%$ and Σ_{HL} is 0.47 $M_\odot$ pc^{-2}.

We note that the above values of Σ_{HL} are obtained by HSVMP using an $N(H_2)/W(CO)$ conversion factor of 0.5×10^{20} [K km s^{-1}]$^{-1}$ cm^{-2}. Magnani *et al.* (1988) have argued that a larger value of $N(H_2)/W(CO)$ ($2\text{–}3 \times 10^{20}$) is appropriate for high-latitude clouds. Use of the larger conversion factor would increase Σ_{HL} by at least a factor of 5. If the results presented by HSVMP are applicable to the whole sky and the higher conversion factor is valid, then the surface density of high-latitude molecular clouds is at least as great as that due to GMCs and dark clouds in the Galactic plane. Even if the low conversion factor is used in deriving Σ_{HL}, the mass surface density of high latitude clouds may be a very important contribution to the Galactic molecular gas inventory.

CONCLUSIONS

It is not at all clear that the HSVMP results are applicable to the whole sky. With a filling factor of at least 0.07, the MLB survey which covered more diverse regions of the sky would have achieved nearly 200 detections if the sensitivity of that survey had been as great as that of HSVMP. Can the factor of 20 difference in detections between the two surveys be attributed solely to a 0.1 K vs. 0.3 K rms noise level? Clearly, this discrepancy must be resolved by a new survey combining the more random search pattern of the MLB effort (which would avoid focusing on perhaps an anomalously CO-rich region) with the higher sensitivity of the HSVMP survey. The very different values of Σ_{HL} which result from the rather few attempts at determining the molecular gas contribution at high latitudes lead to two very different conclusions as to the importance and role played by the high latitude clouds: Are these objects (for the most part translucent molecular clouds) as important a component of the molecular gas inventory as GMCs?

A tantalizing clue may come from a recent study of the Scutum arm region of the Milky Way (Chiar *et al.* 1993). This paper utilizes CO(J=2-1) and CO(J=1−0) data as probes of the temperature and density of molecular cloud ensemble in that region of the Galaxy. Typical lines of sight in their survey intersect regions which are optically thick and cold, implying that much of the CO emission from the region comes from clouds not associated with star formation. The data are analyzed to obtain the relative fraction of the Scutum arm emission which arises from translucent clouds vs. GMCs. The solution yields the result that 58% of observed CO emission comes from a translucent cloud component. Similar conclusions were reached by large-scale studies of the Galactic plane by Polk *et al.* (1988) and Wright *et al.* (1991).

Regardless of which survey is preferred, the molecular content of the high-latitude sky is not negligible. Whether it rivals the GMC component at lower latitudes is still an open question, the resolution of which should be clear in the next several years.

ACKNOWLEDGMENTS

We thank Andreas Heithausen for providing preprints prior to publication of some of the work discussed in this review and Frances Verter for comments which greatly improved the presentation.

REFERENCES

Albert, C.E. 1983, *ApJ*, **272**, 509

Barnard, E.E. 1927, *Photographic Atlas of Selected Regions of the Milky Way*, eds. E.B. Frost & M.R. Calvert, (Washington: Carnegie Institution of Washington)

Blitz, L., Fich, M., & Stark, A.A. 1982, *ApJ*, **49**, 183

Blitz, L., Magnani, L., & Mundy, L. 1984 *ApJ*, **282**, L9

Blitz, L., Bazell, D., & Désert, F.X. 1990, *ApJ*, **352**, L13

Bloemen, J.B.G.M. *et al.*, *A&A*, **154**, 25

Bronfman, L. *et al.* 1988, *ApJ*, **324**, 248

Boulanger, F., Baud, B., & van Albada, G.D. 1985, *A&A*, **144**, L9

Burton, W.B. *et al.* 1986, in *Light on Dark Matter*, ed. F.P. Israel, (Dordrecht: Reidel), p.357

Chiar, J.E., Kutner, M.L., Verter, F., & Leous, J. 1993, *ApJ*, in press

Crutcher, R.M. 1973, *ApJ*, **185**, 857

Cudaback, D.D. & Heiles, C. 1969, *ApJ*, **155**, L21

Dame, T.M. *et al.* 1987, *ApJ*, **322**, 706

Dame, T.M. & Thaddeus, P. 1985, *ApJ*, **297**, 751

Désert, F.X., Bazell, D., & Boulanger, F. 1988, *ApJ*, **334**, 815

de Vries, C.P. 1986, Ph.D. Thesis, University of Leiden

de Vries, H.W., Heithausen, A. & Thaddeus, P. 1987, *ApJ*, **319**, 723

Dewey, R.J. *et al.* 1983, *AJ*, **88**, 1832

Dickey, J.M., Crovisier, J, & Kazès, I. 1981, *A&A*, **98**, 271

Dickey, J.M. & Garwood, R.W. 1989, *ApJ*, **341**, 201

Dickman, R.L. 1975, *ApJ*, **202**, 50

Dieter, N.H. 1973, *ApJ*, **183**, 449

Feitzinger, J.V. & Stüwe, J.A., 1984 *A&AS*, **58**, 365

Grenier, R.P. *et al.* 1989, *ApJ*, **347**, 231

Hartley, M. *et al.* 1986, *A&AS*, **63**, 27

Heithausen, A. & Thaddeus, P. 1990, *ApJ*, **353**, L49

Heithausen, A., Stacy, J.G., de Vries, H.W., Mebold, U., Thaddeus, P. 1993, *A&A*, in press.

Heiles, C. 1968, *ApJ*, **151**, 919

Heiles, C. 1976, *ApJ*, **204**, 379

Hillditch, R.W., Hill, G., & Barnes, J.V. 1976, *MNRAS*, **176**, 175

Kazès, I. & Crovisier, J. 1981, *A&A*, **101**, 401

Keto, E.R. & Myers, P.C. 1986, *ApJ*, **304**, 466

Khavtassi, J.Sh. 1955, *Bull. Abastumani Obs.*, No.18

Kilkenny, D. 1980, *MNRAS*, **191**, 651

Knapp, G.R. & Jura, M. 1976, *ApJ*, **209**, 782

Knude, J. 1979, *A&AS*, **38**, 407

Liszt, H.S. 1983, **275**, 163

Low, F.J. *et al.*, 1984, *ApJ*, **278**, L19

Lundmark, K. 1926, *Uppsala Medd.*, No.12

Lynds, B.T. 1962, *ApJS*, **7**, 1

Lynds, B.T. 1965, *ApJS*, **12**, 163

Magnani, L., Blitz, L., & Mundy, L. 1985, *ApJ*, **295**. 402

Magnani, L., Lada, E.A., & Blitz, L. 1986, *ApJ*, **301**, 395

Magnani, L., Blitz, L., & Wouterloot, J.G.A. 1988, *ApJ*, **326**, 909

Munch, G. & Zirin, H. 1961, *ApJ*, **133**, 11

Perry, C.L. & Johnston, L. 1982, *ApJS*, bf 50, 451

Polk, K.S. *et al.* 1988, *ApJ*, **332**, 432

Roshkovsky, D.A. 1955, *Contr. Alma-Ata 1*, No. 1-2, 136

Sandage, A. 1976, *AJ*, **81**, 954

Schoenberg, E. 1964, *Veroffentl. Stern. Muchen*, Vol. 5, No.21

Shane, W.W. & Wirtanen, C.A. 1967, *Pub. Lick Obs.*, **22**, 1

Sharpless, S. 1959, *ApJS*, **4**, 257

Spitzer, L. 1942, *ApJ*, **95**, 329

Teerikorpi, P. 1981, *A&A*, **98**, 300

van Buren, D. 1989, *ApJ*, **338**, 147

van den Bergh, S. 1966, *AJ*, **71**, 990

van Dishoeck, E.F. & Black, J.H. 1988, *ApJ*, 334, 771

Weiland, J.L. *et al.* 1986, *ApJ*, **306**, L101

Wright, E.L. *et al.* 1991, *ApJ*, **381**, 200

The First Symposium on the Infrared Cirrus and Diffuse Interstellar Clouds
ASP Conference Series, Vol. 58, 1994
Roc M. Cutri and William B. Latter (eds.)

HIGH ALTITUDE MOLECULAR CLOUDS

S. MALHOTRA, G. R. KNAPP
Princeton University Observatory Peyton Hall, Princeton, NJ 08544

ABSTRACT Molecular gas in the Galaxy is mostly confined to a thin layer in the disk with a roughly gaussian distribution perpendicular to the plane. Analysis of the vertical structure of the molecular gas in the first quadrant shows that there is a population of clouds which are smaller than the Giant Molecular Clouds and have a significantly larger scale height. These clouds are presumably the analogue of the local high-latitude molecular clouds.

INTRODUCTION

Molecular gas is confined to a layer much thinner than the stellar disk and is therefore expected to have an equilibrium distribution which is gaussian in z. That the smaller molecular clouds are distributed with larger scale heights has been suggested by Stark (1983). We have found a population of small clouds lying at high altitudes, suggesting a population with a significantly larger scale height. This population is intermediate between GMCs and the local high latitude, high velocity molecular clouds associated with infrared cirrus (Magnani et al. 1985; Desért et al. 1988, 1990; Heiles et al. 1989).

THE VERTICAL DISTRIBUTION OF MOLECULAR GAS

Knapp et al. (1985) observed strip maps of the $^{12}CO(1 \rightarrow 0)$ emission versus Galactic latitude in the first quadrant of the Galaxy using the 7 meter antenna at Bell Laboratories. The survey consists of latitude strips at 38 Galactic longitudes equally spaced in R/R_0. The latitude coverage is $\simeq \pm 2°$, sampled at $\Delta b = 2'$ with a beam of half-power width $100''$. The velocity resolution is 0.65 km/second/channel. The typical rms noise level is 0.3 K. In addition the $^{13}CO(1 \rightarrow 0)$ line was mapped with noise levels of 0.1 K for a few longitudes.

Assuming an axially symmetric model of the inner Galaxy we can obtain distances to the tangent points, and hence find the z-distribution of the molecular gas. We model the distribution as a gaussian in z, fitting both the scale height and the center. Examples of best fit models are shown superposed on the data in Figure I. The models are shown as contours of equal temperature in latitude l and velocity v. While the z-profile is a gaussian at each radius, the model contours are more complex because the velocity structure near the terminal velocity must take into account the velocity dispersion of the gas and the intrusion of emission from gas near the tangent points. The distance to the Galactic center, R_0 is taken to be 8.5 kpc.

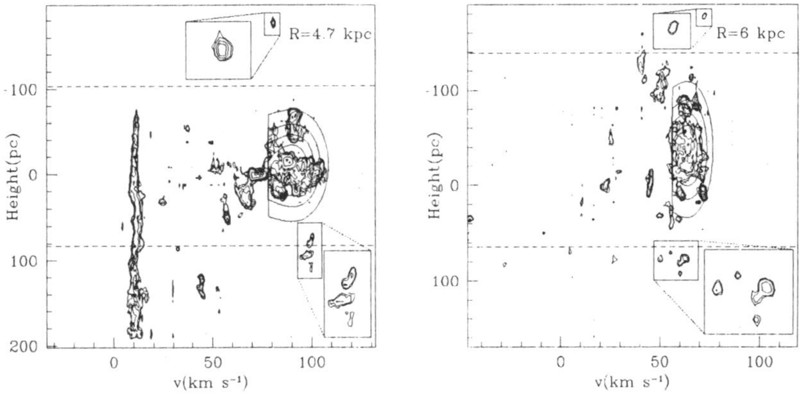

FIGURE I Latitude-velocity maps of ^{12}CO emission at two longitudes (contour levels: 1.6, 2, 3, 4, 6, 8, 10, 12 K). The axes show velocity v and height z at the tangent point. The terminal velocity feature at the extreme right is fitted with a gaussian in height. The best fit model is superposed on the data. The broken lines are $\pm 2.57\sigma$ deviations in height from the center of the distribution.

HIGH ALTITUDE CLOUDS

Selection

After fitting the gaussian distribution in z, we identify clouds that are more than $2.57\sigma_z$ (corresponding to a two-sided significance level of 1%) away from the center of the distribution as high altitude clouds (HACs). We consider only clouds detected in more than one velocity channel and at more than one adjacent latitude.

Properties

We find 23 HACs, most (17) between three and four scale heights from the center of the distribution. The highest altitude cloud found is at z = 200 pc. Most of the HACs are of fairly low mass (average $\simeq 200\ M_\odot$). The mass was found from the integrated CO intensity assuming that $N(H_2) = 3 \times 10^{20} I_{CO}\ cm^{-2}$; the largest mass found is $\sim 10^4 M_\odot$. These clouds are typically smaller than GMCs. The smallest cloud we can detect varies with the longitude. At l=58.21, for example, a cloud must be $\geq$ 5pc to be detected in more than one latitude scan. The largest and the smallest clouds found are 7 and 36 pc in extent (in z). The internal velocity dispersion ranges from 0.49 kms^{-1} to 4.3 kms^{-1}. The HACs have a higher $^{12}CO/^{13}CO$ intensity ratio than the GMCs, as shown in Figure II (cf. Polk et al. 1988).

KINEMATIC DISTANCES

In the above discussion the distances have been determined kinematically, i.e. the gas at the extreme positive velocity has been assumed to be at the tangent

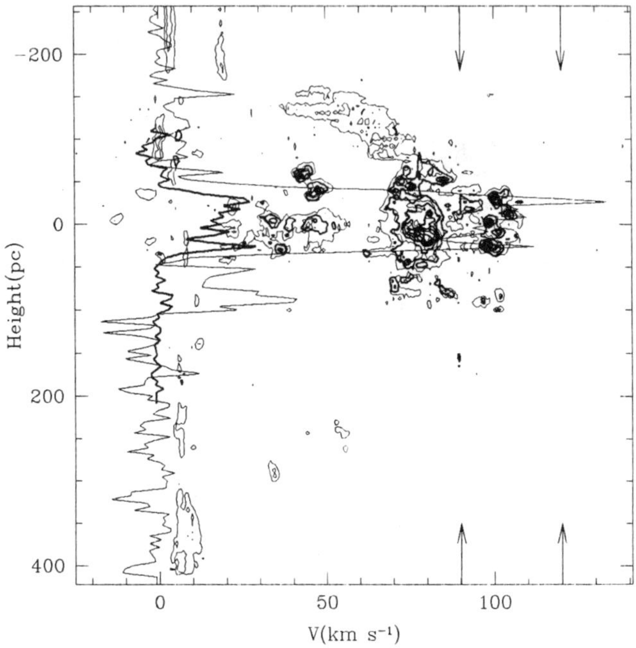

FIGURE II Latitude-velocity maps of ^{12}CO (thin lines) and ^{13}CO (heavy lines) emission at longitude l=28.36. The vertical curves show the ^{12}CO and ^{13}CO emission integrated over the tangent point velocity range (shown by the arrows) versus z.

point distance. We identify high altitude clouds by the ration of their distance z_0 from the center of the molecular cloud distribution to the molecular cloud scale hight; these clouds are outliers in z if they have the same kinematics as the GMCs. Suppose instead the z-distributions are the same but the kinematics are different. Figure III shows the peculiar velocity needed by each HAC to be closer than the tangent point distance (and hence at smaller z) versus this reduced height in units of σ_z. The cloud-cloud velocity dispersion is $5-10$ kms^{-1} (Clemens 1985, Stark 1984, Stark & Brand 1989, Magnani, Blitz & Mundy 1985, Malhotra & Knapp 1993). Figure III thus shows that the peculiar velocities needed to bring the HACs close enough to belong to the GMC cloud population are improbably high.

CONCLUSIONS

There exists a population of intermediate mass clouds with scale height larger than the Galactic plane GMCs. Within the observational uncertainties their properties are found to lie between those of the local high latitude clouds and the GMCs.

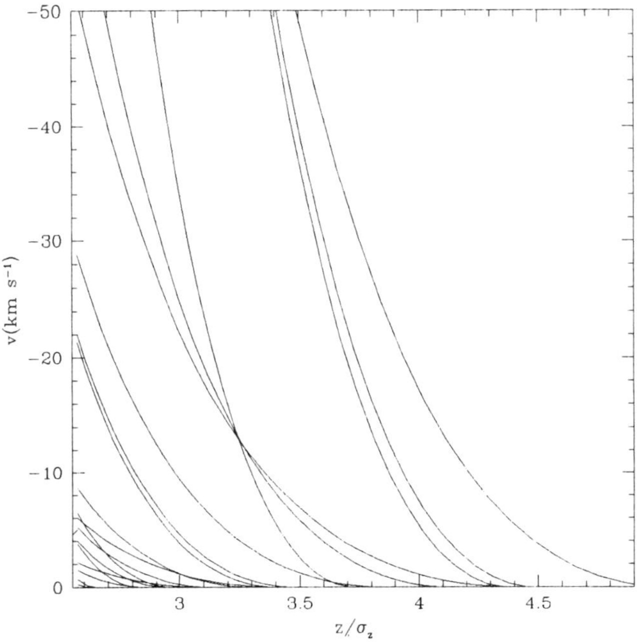

FIGURE III For each cloud in the sample the peculiar velocity needed to reduce the height estimate is plotted against the reduced height in units of σ_z.

ACKNOWLEDGMENTS

This work was supported by NSF grant AST89-21700 to Princeton University. We thank the organizers, Bill Latter and Roc Cutri, for the invitation to this conference and for financial support.

REFERENCES

Clemens, D. P. 1985 *ApJ*, **295**, 422

Desért F. X., Bazell, D., Boulanger, F. 1988 *ApJ*, **334**, 815

Desért F. X., Bazell, D., Blitz, L. 1990 *ApJ*, **352**, 113

Heiles, C., Reach, W. T., Koo, B. C., 1989 *ApJ*, **322**, 313

Knapp, G. R., Stark, A. A., Wilson, R. W. 1985, *AJ*, **90**, 254

Polk, K. S., Knapp, G. R., Stark, A. A., Wilson, R. W. 1988, *ApJ*, **332**, 432

Magnani, L., Blitz, L., Mundy, L. 1985, *ApJ*, **295**, 402

Malhotra, S., Knapp, G. R. 1993, preprint

Stark, A. A. 1983, in *Kinematics, Dynamics and structure of the milky way*, W. H. L. Shuter, Dordrecht: Reidel, 127

Stark, A. A. 1984, *ApJ*, **281**, 624

Stark, A. A., Brand, J. 1989, *ApJ*, **339**, 763

The First Symposium on the Infrared Cirrus and Diffuse Interstellar Clouds
ASP Conference Series, Vol. 58, 1994
Roc M. Cutri and William B. Latter (eds.)

WHERE IS THE MOLECULAR HYDROGEN IN THE DRACO NEBULA?

UWE HERBSTMEIER AND PETER MORITZ
Radioastronomisches Institut der Universität Bonn, 53121 Bonn, FRG

ANDREAS HEITHAUSEN
I. Physikalisches Institut der Universität zu Köln, 50937 Köln, FRG

ABSTRACT CO multitransition studies and independent ROSAT soft X-ray observations confirm that the amount of molecular hydrogen in the Draco nebula is only up to 10 % of the total mass. This result indicates that indirect tracers of molecular hydrogen have to be translated into the molecular hydrogen column densities using non-standard conversion factors.

INTRODUCTION

One of the most important problems in astronomy of the interstellar medium is the determination of the total masses of gas clouds. Since the most abundant constituent of molecular clouds, molecular hydrogen, cannot be directly observed from the ground, molecular hydrogen column densities, $N(H_2)$, are indirectly inferred from the ^{12}CO J= 1 → 0 line integral, $W(^{12}CO)$, by applying a constant conversion factor $x_{wco} = N(H_2)/W(^{12}CO)$. However in the recent years it has been found that clouds showing no CO line emission do contain molecular hydrogen (Martin et al. 1990). Also Blitz et al. (1990) conclude that most of the far-infrared (FIR) excess clouds defined by Désert et al. (1988) are molecular clouds without associated CO emission. Blitz et al. (1990) derive a lower IRAS 100 μm intensity (I_{100}) limit of 4 MJy ster^{-1} for the detectability of CO.

On the other hand Herbstmeier et al. (1993) have found that the Draco nebula, a diffuse molecular cloud at $(l,b) = (91°,38°)$, has a very low molecular content although CO line emission is rather widespread and also found in regions with $I_{100} < 4$ MJy ster^{-1}. The authors have used I_{100}, the column density of the neutral atomic hydrogen, and $W(^{12}CO)$ to show that the FIR-emissivity, $a_{100} = I_{100}/N(H)$, is varying and larger than $1\,10^{-20}$ MJy ster^{-1} cm^{2}. Furthermore they found that x_{wco} is very low ($0.02 - 0.26\,10^{20}$ cm^{-2} (K km s^{-1})$^{-1}$). Now the question is whether these results are biased and whether there is a large amount of molecular hydrogen not associated with significant FIR-emission. To answer this question we checked for abnormal CO excitation conditions by observing several emission lines of higher transitions. Independently, we derived the total hydrogen column density from the comparison of the ROSAT all-sky soft X-ray survey with the IRAS 100 μm image.

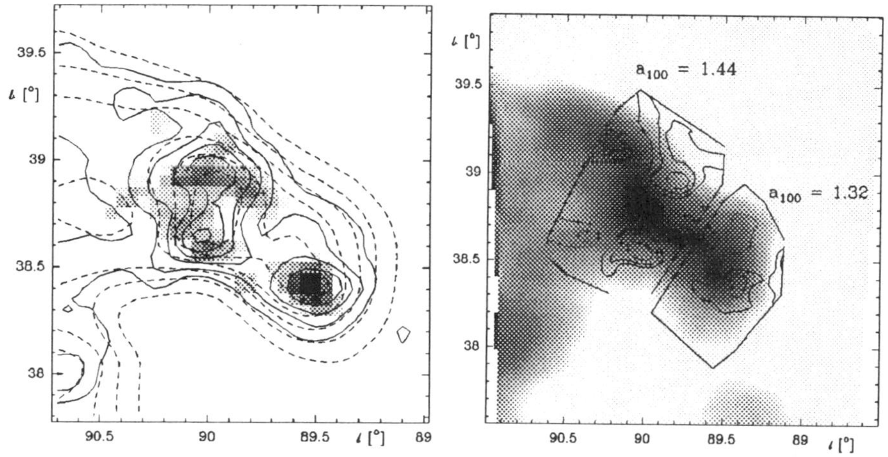

FIGURE I Parts of the Draco nebula: **left:** $W(^{12}CO)$, grey scales from 1 – 7 K km s^{-1}; I_{100} from ISSA, first full contour and step 1 MJy ster^{-1}; Effelsberg 21 cm HI column density distribution, first dashed contour 6.8 10^{19} cm^{-2}, contour step 2.3 10^{19} cm^{-2}. **right:** ROSAT X-ray image, count rates 5.7 – 12.0 10^{-4} s^{-1} arcmin^{-2}, with deviation of the model fit (see text) overlayed (dashed contours: -0.25 – -0.1, full drawn: 0.1 – 0.2); a_{100} in 10^{-20} MJy ster^{-1} cm^2.

THE CO OBSERVATIONS AND RESULTS

We have observed the J= 1 $\rightarrow$ 0, and 2 $\rightarrow$ 1 transition of ^{12}CO and ^{13}CO and the J= 3 $\rightarrow$ 2 transition of ^{12}CO towards the dense clump in the Draco nebula at (l, b) = $(89.5°, 38.4°)$ with the KOSMA 3 m telescope on Gornergrat, Switzerland. Figure I, left, shows I_{100}, the HI column density distribution, and $W(^{12}CO)$ of the observed cloud core and its surroundings. The resolution in (l, b, v) is approximately $(4', 4', 0.08\,\mathrm{km\,s^{-1}})$, $(2', 2', 0.21\,\mathrm{km\,s^{-1}})$, and $(1', 1', 0.14\,\mathrm{km\,s^{-1}})$ for the 1 $\rightarrow$ 0, 2 $\rightarrow$ 1, and 3 $\rightarrow$ 2 transition respectively. All spectra have been observed in total power mode. The intensity has been calibrated in T_R^* using the effectivities given in Kramer (1992). Figure II shows the distributions of the line integrals for the ^{12}CO observations. The line temperatures derived for the centre of the clump are listed in Table I.
A dense core is found at the position of the maximum emission of the J=1 $\rightarrow$ 0 line. Therefore in the centre of the cloud on a scale of $\approx 1' - 3'$ the molecular hydrogen appears to be dominant. This explains the existence of formaldehyd and ammonia in this region (Mebold et al. 1985, 1987). The line ratios found are similar to values observed in other cirrus cloud cores (see Heithausen et al. this volume). This shows that the excitation conditions of the CO molecule are not very different from those in other cloud structures. The very low values for x_{wco} therefore indicate that the CO abundance is larger than for clouds in the galactic plane and/or that the CO molecule is effectively excited by atomic hydrogen.

TABLE I The CO line temperatures for the cloud core

Isotope	1 → 0 [K]	2 → 1 [K]	3 → 2 [K]
^{12}CO	3.78	1.87	0.73
^{13}CO	0.60	0.22	-

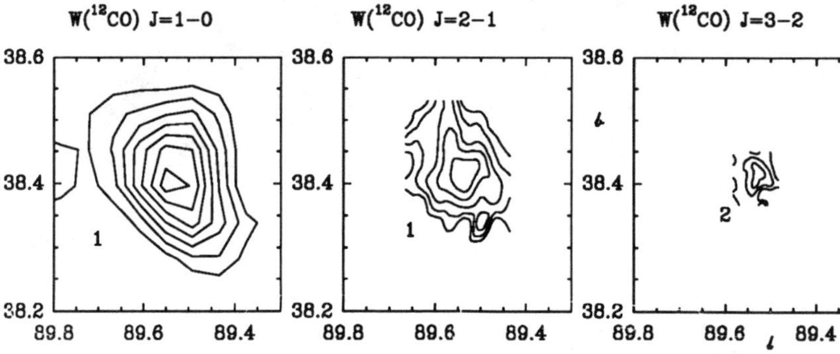

FIGURE II The core of the Draco nebula: The ^{12}CO line integrals for the three indicated transitions. The J= 1 → 0 image is taken from Mebold et al. (1985). The value of the lowest contour is indicated. The contour distance is 1 K km s^{-1}.

THE X-RAY OBSERVATIONS AND RESULTS

The following analysis is based on a part of the ROSAT all-sky X-ray survey in the area of the Draco nebula provided to us by Dr. S.L. Snowden, Max-Planck-Institut für Extraterrestrische Physik. The energy range of the data set is from 100 to 284 eV. Figure I, right, shows the image of the same part of the Draco nebula as the image in Figure I, left. The similarity between the soft X-ray count rate distribution with the I_{100} image is remarkable (see also Mebold et al. this volume). The depression of the countrates where the maxima of the FIR-emission are found can only be interpreted as a shadow in the soft X-ray background cast by the Draco nebula (Snowden et al. 1991).
The correlation analysis of X-ray shadows and HI or FIR-images are a new technique for the determination of the total column density of hydrogen nuclei. From the depth of the shadow it is possible to derive the optical depth of the cloud $\tau = \sigma N(\mathrm{H})$. From τ it is possible to derive the total hydrogen column density, $N(\mathrm{H})$, taking into account the theoretical absorption cross section of the ISM for soft X-ray photons, $\sigma = \sigma(N(\mathrm{H}))$. The very good correspondence of the ROSAT count rate depressions and the IRAS FIR-emission maxima excludes the existence of a significant amount of molecular hydrogen outside the boundaries of the structure seen in the FIR. We compare the X-ray image and the I_{100} intensity distribution quantitatively to derive a_{100} from:

$$C_{\mathrm{obs}} = C_0 + C_1 \mathrm{e}^{\sigma(N(\mathrm{H}))(a_{100}^{-1} I_{100})}. \qquad (1)$$

Here $C_{\rm obs}, C_0$, and C_1 are the observed, the non-attenuated foreground and the original background soft X-ray count rates. The influence of the diffuse, ubiquitous HI layer, as well as the value of C_0 respecitvely are derived in regions outside the Draco nebula and are taken into account. Then, the values of C_1 and a_{100} are derived by a fit of equation (1) to the data. We find that $a_{100} \approx 1.38\,10^{-20}$ MJy ster^{-1} cm^2 is of the same order as the values derived by Herbstmeier et al. (1993), confirming a rather low amount of molecular hydrogen within the Draco nebula. Variations of the FIR-emissivity as found by Herbstmeier et al. (1993) are also confirmed by the analysis of the ROSAT data: The relative deviation of the fitted values to the observed count rates, $(C_{\rm obs} - C_{\rm fit})/C_{\rm obs}$, are shown in Figure I, right, overlayed on the ROSAT soft X-ray image.

CONCLUSIONS

With the help of multitransition observations of the ^{12}CO and ^{13}CO molecules and independently from a comparison of ROSAT soft X-ray data with the IRAS FIR-survey we find that the amount of molecular hydrogen in the Draco nebula is less than expected from standard conversion relations between CO, 100 μm intensity, and molecular hydrogen column density. These results indicate that conclusion based on the standard conversion factors, like the results presented by Blitz et al. (1990) mentioned in the Introduction, have to be reconsidered carefully. We have demonstrated that ROSAT shadowing experiments are a suitable, independent check of the molecular content of such clouds.

ACKNOWLEDGMENTS

We thank Steve Snowden, MPE, for the help with the ROSAT survey data.

REFERENCES

Blitz, L, Bazell, D., Désert, F.X. 1990, *ApJ*, **352**, L13

Désert, F.X., Bazell, D., Boulanger, F. 1988, *ApJ*, **334**, 815

Herbstmeier, U., Heithausen, A., Mebold, U. 1993, *A&A*, **272**, 514

Kramer, C. 1992, PhD Thesis, Universität zu Köln

Martin, C., Hurwitz, M., Bowyer, S. 1990, *ApJ*, **354**, 220

Mebold, U., Cernicharo, J., Velden, L., et al. 1985. *A&A*, **151**, 427

Mebold, U., Heithausen, A., Reif, K. 1987, *A&A*, **180**, 213

Snowden, S.L., Mebold, U., Hirth, W., et al. 1991, *Sci*, 252, 1529

The First Symposium on the Infrared Cirrus and Diffuse Interstellar Clouds
ASP Conference Series, Vol. 58, 1994
Roc M. Cutri and William B. Latter (eds.)

INFRARED CIRRUS AND DISCRETE SOURCES IN THE OUTER GALAXY

MICHEL FICH
Physics Department, University of Waterloo, Waterloo, Ontario, CANADA N2L 3G1

SUSAN TEREBEY
IPAC, California Institute of Technology and Jet Propulsion Laboratory, M/S 100-22, Pasadena, CA 91125

INTRODUCTION

Studies of Galactic infrared cirrus have been based on (1) observations of individual high latitude clouds (e.g. Boulanger, Baud, and van Albada 1985), (2) analysis of large scale emission at high latitudes in the Solar Neighborhood (e.g. Boulanger and Perault 1988), and (3) complete linear decomposition of the global emission from IRAS and other observations over the entire Galaxy (e.g. Sodrowski et al 1989, Bloemen, Deul, and Thaddeus 1990).

Here we present preliminary results using a different approach: a detailed analysis of all of the sources of infrared emission, as detected by IRAS, in two large fields in the outer Galaxy. We study both the extended diffuse component, which is primarily associated with HI gas, and the discrete, 60 micron bright infrared sources, which are generally star forming complexes at distances of several kpc. We identify and carefully measure on the IRAS images a complete flux-limited sample of discrete sources, and make detailed comparisons with the measured values in other IRAS data products. The discrete sources are compact but extended at the IRAS resolution with the result that significantly more flux is measured on the images than is quoted in the Point Source Catalog.

We study the diffuse (infrared cirrus) component by masking out regions containing discrete sources or molecular clouds and derive the infrared colors of the cirrus emission. In practice the masking procedure has little effect on the results because the emission from discrete sources is found to be several orders of magnitude less than the emission from the cirrus component at all IRAS wavelengths.

DATA ANALYSIS

Because confusion problems are much lower in the outer Galaxy we focused on two regions of the galactic plane bounded by (1) $l = 120°$ to $l = 130°$ and (2) $l = 213°$ to $l = 223°$ and by $b = -10°$ to $b = +10°$. These regions are devoid of large optically known HII regions but are otherwise not special. To find all the star forming regions we examined the 60 micron images from the IRAS Sky Flux Plates (now replaced by the IRAS Sky Survey Atlas (ISSA)) and coadded

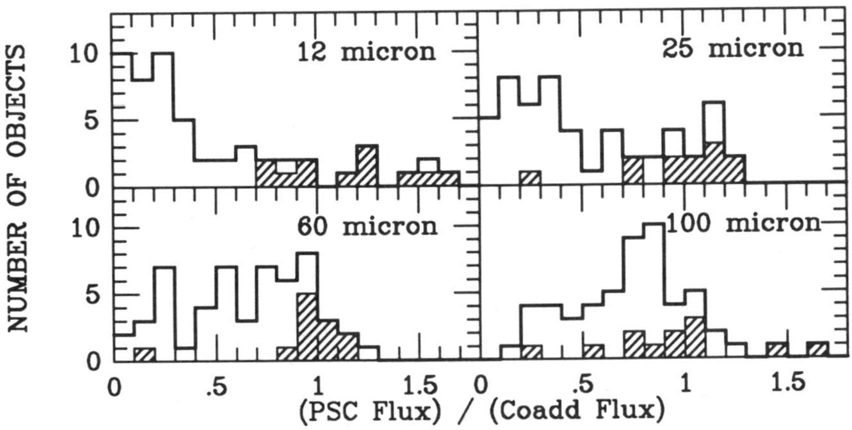

FIGURE I Comparison of PSC and image fluxes.

images resulting in a source list that is complete for objects with 60 micron flux densities greater than 25 Jy and radii less than 10 arcminutes. The structure of the star formation regions was complex and so for detailed analysis we used the 2 degree coadded images produced at IPAC that provided the full resolution of the IRAS survey (FRESCO).

The flux above the local background was determined, where in most cases the background value used was the median flux density found in the pixels along the boundary of each object. The background determination dominates the the uncertainties in the flux densities, which are estimated to be 44%, 20%, 21%, and 32% for 12, 25, 60, and 100 microns.

Comparing our discrete source list with the PSC we find that nearly all sources have a corresponding entry in the PSC. The PSC positions are generally more accurate than we can easily determine from the images but the PSC fluxes are often significantly smaller. This is not surprising since the data on which the PSC is based has been processed with a point source filter which effectively suppresses extended structure.

All of the complex sources and most of the simple sources have very similar Flux Density Distributions (FDD). We expect the complex sources to be mixtures of interstellar clouds and star forming regions and we have defined a distribution of FDD, called Class A that contains all of these sources. Other sources are collectively called Class B and probably are primarily stars.

In Figure I the flux density ratios for all four wavebands are shown as histograms with the Class B sources marked with diagonal lines. It is clear that the PSC is underestimating the total flux for the Class A sources and this effect is most dramatic at 12 and 25 microns where the IRAS spatial resolution is greatest (approximately 1 arcmin) and that there is the least effect at 100 microns (approximately 4 arcmin) resolution. The median flux ratios for the Class B sources are all approximately 1, which would be expected if the sources in this class are unresolved. For the Small Scale Structure Catalog (SSSC) we find matches with fewer of our sources, suggesting that the SSSC is incomplete. The SSSC has poorer positions than the PSC but better correspondence of the fluxes.

After removing all of the discrete sources from our two fields (by masking out these sources on mosaics of ISSA images of the entire (two) fields) we compared the fluxes in each band for each individual field. We first compared the total flux in this diffuse emission to the total in discrete components. For each field and each band Table I lists the total flux for all PSC entries in the field, the total flux from the PSC for all of our selected discrete sources, the total flux as measured in the coadd (FRESCO) images for all our selected sources, and the total diffuse emission in the mosaic (ISSA) images within the fields. A zero level correction has been applied to the ISSA values. However the ISSA images in the g=218 field, which have a relatively low ecliptic latitude of 30 degrees, are clearly contaminated by residual zodiacal emission at the shorter wavelengths. The affected values in Table I are shown in parenthesis to indicate they are not reliable measures of the diffuse galactic emission.

Table I shows: (1) the diffuse emission completely dominates the emission in all bands; (2) the flux measured from the images for discrete sources is a factor of 2 to 4 greater than the PSC flux for all of our objects; and (3) at 12 microns our sample only represents a small fraction of all of the flux detected in the PSC.

TABLE I Flux Density Totals (in Jy).

$l = 125°$ field	12 microns	25 microns	60 microns	100 microns
All PSC entries	4268.6	3190.3	7289.1	14735.2
discrete sources (PSC)	1269.0	1870.5	6181.0	10066.5
discrete sources (Coadd)	2256.5	3679.7	21199.7	38920.1
diffuse emission(ISSA)	8.8×10^4	8×10^4	4×10^5	1.9×10^6

$l = 218°$ field	12 microns	25 microns	60 microns	100 microns
All PSC entries	2549.4	2156.6	4873.0	8952.5
discrete sources (PSC)	229.0	1063.6	4002.8	5596.4
discrete sources (Coadd)	785.2	1700.7	9264.5	17881.9
diffuse emission (ISSA)	(1.1×10^5)	(2.6×10^5)	2.2×10^5	1.6×10^6

For each field we plot the diffuse emission in one band versus another in order to measure the infrared colors (flux ratios) of the diffuse component. To minimize contamination by the star forming regions we have masked out all the discrete sources. For example, Figure II for the $l = 125°$ field shows a linear correlation of the 60 micron versus the 100 micron flux. The derived 60/100 micron ratio confirms our earlier result using the IRAS All Sky Images (Terebey and Fich 1986). However the new ISSA images, which are considerably more sensitive, allow us to extend the analysis to 12 and 25 microns as well.

The derived colors for the $l = 125°$ field are listed in Table II. The values

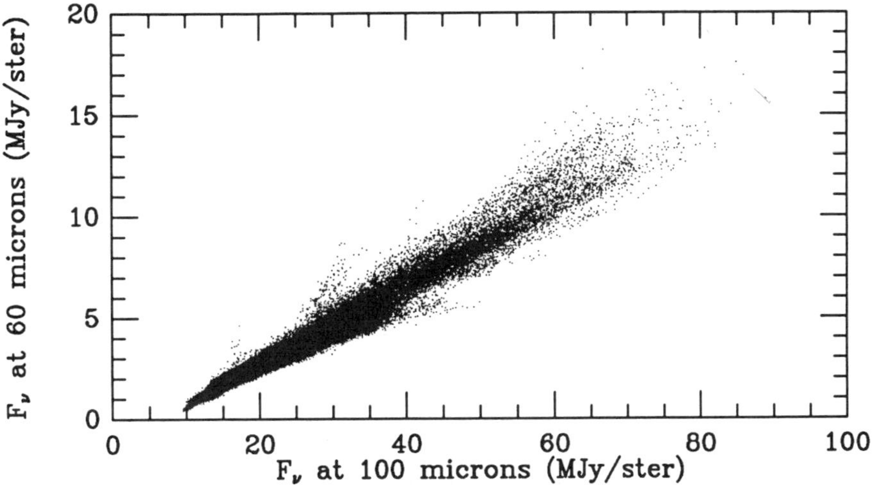

FIGURE II The diffuse emission in the $l = 125°$ field.

shown are the mean of the minimum and maximum slopes of data in the plots and the uncertainties given reflect the maximum range in the fits. The colors are consistent with those derived by Boulanger and Perault (1988) for high latitude cirrus. This suggests that the diffuse galactic emission in the outer Galaxy and the local high latitude cirrus are produced under similar physical conditions.

TABLE II Mean Flux Density Ratios (Colors) of Diffuse Galactic Emission.

$60\mu m/100\mu m$	$25\mu m/100\mu m$	$12\mu m/100\mu m$	$25\mu m/60\mu m$	$12\mu m/25\mu m$
0.203 ± 0.004	0.043 ± 0.012	0.046 ± 0.004	0.216 ± 0.071	1.10 ± 0.10

A much more detailed description of the IRAS analysis is in the final stages of manuscript preparation at this time. A comparison with observations at other wavelengths is currently underway.

REFERENCES

Bloemen, J. B. G. M., Deul, E. R., and Thaddeus, P. 1990, *A&A*, **233**, 437

Boulanger, F., Baud, B., and van Albada, G. D. 1985, *A&A*, **144**, L9

Boulanger, F. and Perault, M. 1988, *ApJ*, **330**, 964

Sodrowski, T. J., Dwek, E., Hauser, M. G., and Kerr, F. J. 1989, *ApJ*, **1989**, 762

Terebey, S. and Fich, M. 1986, *ApJ*, **309**, L73

The First Symposium on the Infrared Cirrus and Diffuse Interstellar Clouds
ASP Conference Series, Vol. 58, 1994
Roc M. Cutri and William B. Latter (eds.)

HIGH-RESOLUTION HI IMAGES OF INTERSTELLAR CIRRUS "CLOUDS"

GERRIT L. VERSCHUUR
Rhodes College, 2000 N. Parkway, Memphis, TN 38112-1690

ABSTRACT A high-resolution study of "normal" high-latitude HI features shows that "clouds" found in a lower resolution survey break up into twisted filaments with angular widths of order arcminutes. These can be related to counterparts seen in IRAS 100 maps. The typical HI profile consists of a very broad component, probably due to stray radiation, and a pervasive feature about 12 km/s wide. These two components, which contain > 50% of the area of the typical profile, must be identified and removed before quantitative comparison between cool HI and cirrus structure can proceed. Maps of the remaining "foreground" cool components show that some of them bear a close relationship to the cirrus while others do not. Until the origin of the multiple component structure of HI profiles is modelled, it may be premature to make quantitative statements about the way dust and gas are related.

INTRODUCTION

Interstellar structure observed in high-resolution HI area surveys may be produced by filament geometry (Verschuur 1991 a/b). Enhanced emission features (EEFs) may often if not always result from column depth increases in directions where filaments twist into the line-of-sight. Such enhancements have traditionally been interpreted to be the signature of "clouds." However, they may be no more than indicators of viewing geometry. This suggests that any attempt to derive "cloud" parameters from the data could be seriously in error, depending on the morphology of the filaments being studied.

THE OBSERVATIONS

To test the twisted filament model as the cause of EEFs ("clouds"), high-resolution observations of three features were made with the Arecibo radio telescope (resolution 4′). 470 directions were observed toward H0827+10 and 1982 gaussians fitted, while 312 directions toward H0928+2 netted 1200 gaussians. Each profile showed a broad underlying component typically 25 – 50 km s^{-1} wide, which is probably due to stray radiation in far sidelobes as described by Verschuur & Schmelz 1989, and Verschuur & Magnani 1993. A second component (linewidth $\sim$ 12 km/s) was found to be pervasive over large areas of sky and appears common to all the profiles (Verschuur & Magnani 1993). After re-

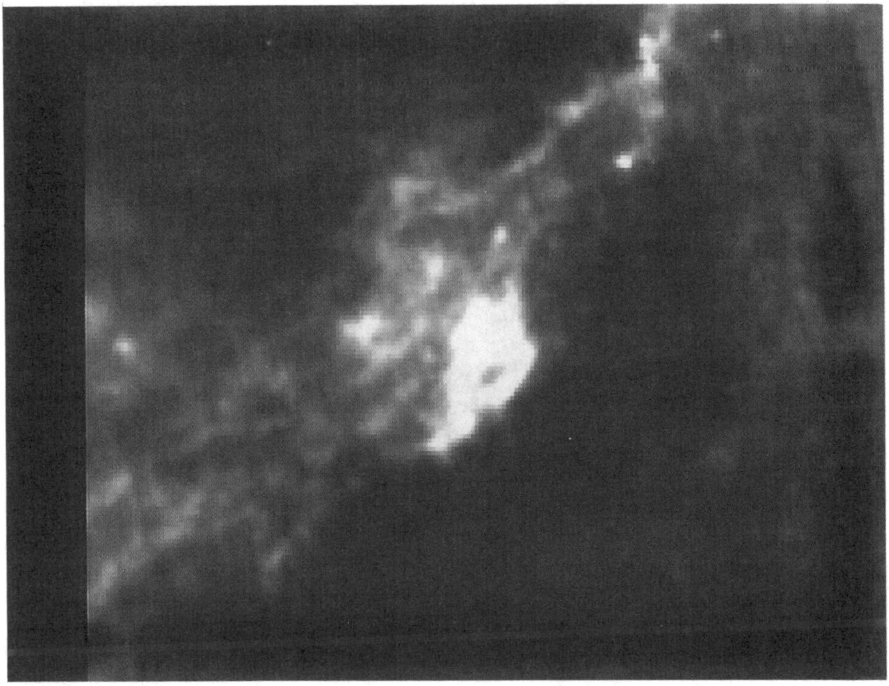

FIGURE 1 A gray-scale image of the 100 emission (from the IRAS ISSA plates) centered on H0827+10. Filamentary structure dominates and it is argued that the brightest emission is due to enhanced emission produced where path lengths are increased in directions where filaments twist into the line-of-sight. The scale of the image is approximately 4 × 5.

moval of Components 1 and 2, each of the profiles was decomposed into narrow gaussians, typically 3 – 4 km/s wide.

Toward H0827+10 discussed below seven individual cool components were identified and separately mapped. Three are major features which together contain 85% of the cool gas. Comparison of their morphology with that of the dust emission shows them to be associated with the dust. Three of the four weaker components appeared to be morphologically unrelated. A fourth very weak one is possibly associated with the major ones.

The IRAS Super Sky Flux (destriped) 100 image for the area around H0827+10 is shown in Fig. 1. An ensemble of twisted filaments crosses the field along an axis consistent with the prediction based on interpretation of HI data given in Verschuur 1991b. Fig. 2 gives the total HI column density for the three major HI components toward H0827+10. Comparison of Figs. 1 and 2 show that the HI structure is comparable to the 100 emission and the HI map appears on the verge of revealing as much information as is present in the

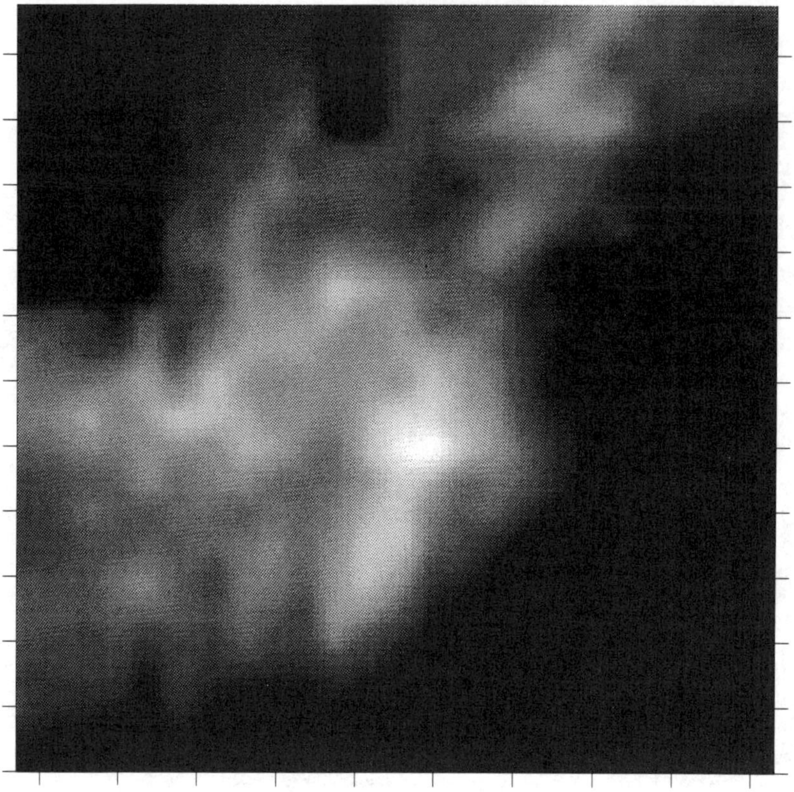

FIGURE 2 Gray-scale image of the total HI column density in the direction of H0827+10 as discussed in the text. Comparison with Plate 1 shows the high degree of morphological similarity between the 100 emission and HI, in particular as regards the existence of filaments in the HI. Tick marks in declination are in 0.2 intervals from 8.8 to 11.0. Ticks marks in right ascension are in 1 m intervals from 8h 23m to 8h 32m.

100 data. The contrast in the HI map is reduced when contributions from the unrelated gaussian components are included.

DISCUSSION

In a model which accounts for EEFs as regions where filaments twist into the line-of-sight, we expect to observe the lowest column densities normal to a filament. In those directions the filaments would be faintest and most difficult to observe. We suspect that the fainter, elongated features in Fig. 2 satisfy the condition that the line-of-sight is nearly normal to the structure.

Filament properties such as gas density can best be derived in directions where a filament is viewed normal to the line-of-sight. An attempt to judge the density in two faint filaments gave values of $> 110(d100)^{-1}$ cm^{-3} and $> 55(d100)^{-1}$ cm^{-3}, where $d100$ is distance in units of 100 pc. The lower limit is set by lack of resolution in both the HI and the IRAS data, because the filament angular widths appear unresolved.

If the filaments of HI and dust are related to the presence of flux tubes in constant motion, the line profiles will have multiple component structure depending on the velocity distribution of the gas. For example, if velocity gradients are present, enhancement of profile brightness at certain velocities will be produced due to velocity crowding. Preliminary work on a model to synthesize the observed profiles shows that such velocity crowding will seriously affect not only how the HI appears to be distributed on the sky, but how we then interpret the data. Work on untangling the observed structures continues.

In general, before interpretation of HI observations of diffuse interstellar structure can be carried out, several steps need to be taken. (a) Remove stray radiation (Component 1) from the profiles. (b) Identify and remove Component 2, the "background" emission. This component was found to contain from 30% to 100% of the profile area toward the "clouds" mapped in the present study. (c) Perform Gaussian analysis of the remaining HI profiles and separately map each cool component before comparing with 100 cirrus structure. This allows determination of which HI components are, in fact, associated with the dust. (d) Make a model to account for the multiple Gaussian structure for the region in question. Such structure may not represent separate "clouds", but rather the presence of complex velocity patterns in HI filaments.

In conclusion, it is suggested that the 21-cm sky cannot be approached like an encyclopedia where one reads a total HI column density from a profile. Reference must be made to a model to account for the velocity structure in the HI, not only in the context of the area being studied, but in terms of its relationship to the larger distribution of HI and dust over the sky.

ACKNOWLEDGMENTS

This work was supported in part by NSF Grant #AST 9022002. I am particularly grateful to L. Magnani who helped guide the author through the Arecibo observations, and Ann Wherle at IPAC who guided the necessary exploration of the IRAS data. Arecibo Observatory is operated by NAIC under contract with the NSF.

REFERENCES

Verschuur, G. L. 1991a, ApSpS, 185, 137

Verschuur, G. L. 1991b, ApSpS, 185, 305

Verschuur, G. L., & Magnani, L., 1993, in preparation

Verschuur, G. L., & Schmelz, J. T. 1989, AJ, 98, 267

The First Symposium on the Infrared Cirrus and Diffuse Interstellar Clouds
ASP Conference Series, Vol. 58, 1994
Roc M. Cutri and William B. Latter (eds.)

ARCMINUTE SCALE HI AND IRAS OBSERVATIONS TOWARD HIGH LATITUDE CLOUD G86.5+59.6

P. G. MARTIN[1], C. ROGERS[2], W. T. REACH[3], P. E. DEWDNEY[2], AND C. E. HEILES[4]

ABSTRACT G86.5+59.6 is a degree-sized high latitude cloud originally selected for investigation by Heiles, Reach, & Koo on the basis of its appearance on the IRAS Skyflux images at 60 and 100 μm. Because of the interesting possibility that this is an intermediate velocity cloud colliding with HI in the Galactic plane, we have examined this region further, both at low resolution over an extended field to provide some context and at higher (arcminute) resolution within the cloud.

INTRODUCTION

IRAS images of infrared cirrus present an all-sky picture of an interstellar environment made up of extended regions each comprising smaller scale substructure. HI 21-cm emission and infrared emission by interstellar dust are correlated over large angular scales (Boulanger & Pérault 1988), and on the finer scale afforded by IRAS images too (Joncas, Dewdney, & Boulanger 1992). There is some cosmic scatter, perhaps the result of variations in grain emissivity, illumination, the dust-to-gas ratio, and unaccounted-for molecular, ionized, or atomic gas. Thus HI and infrared emission provide complementary information to address many interesting questions like: How does the degree of HI – infrared correlation change with angular resolution? How well mixed is the interstellar environment, particularly dust and gas? Do interstellar environments approach homogeneity at some large angular scale?

G86.5+59.6 is a degree-sized high latitude cloud originally selected for investigation by Heiles, Reach, & Koo (1988; hereafter HRK) on the basis of its appearance on the IRAS Skyflux images at 60 and 100 μm. Such regions of bright infrared cirrus might be the result of superposition of unrelated structures separated along the line of sight. However, in the position-velocity plots of the Heiles & Habing (1974) HI survey, HRK found G86.5+59.6 to be a rel-

[1]Canadian Institute for Theoretical Astrophysics, University of Toronto, Toronto, ON, Canada M5S 1A7; Visiting Associate, Theoretical Astrophysics and IPAC, California Institute of Technology, Pasadena, CA 91106

[2]Dominion Radio Astrophysical Observatory, Herzberg Institute of Astrophysics, National Research Council, P.O. Box 248, Penticton, BC, Canada V2A 6K3

[3]NASA/GSFC, Lab for Astronomy and Solar Physics, Code 685, Greenbelt, MD 20771

[4]Astronomy Department, University of California at Berkeley, Berkeley, CA 94720

atively isolated feature in intermediate velocity gas (IVG: ~ -40 km s^{-1}) along with significant "local" low velocity gas (LVG: ~ 0 km s^{-1}) and a connection by a fainter "bridge" (BVG), suggesting that the IVG and LVG components are interacting.

The velocity information available for the gas can be used to "deproject" the effects of superposition implicit in the infrared cirrus; i.e., morphological comparison of observations of dust and gas at sufficient angular resolution might establish the association of dust with particular velocity components of the gas. Interpretation of infrared or HI features seemingly isolated in space and/or velocity can also benefit from the wider context of extensive angular and velocity coverage. Our study illustrates and addresses some of these issues.

STEPPING BACK: THE BIG PICTURE

To provide some context for the G86.5+59.6 degree-sized feature, we have examined a 25° surrounding region using cirrus mosaics of infrared surface brightness (I_{60} and I_{100}) based on the IRAS Sky Survey Atlas (ISSA; resolution 5′). G86.5+59.6 is a relatively bright central feature (still, $I_{100} < 5$ MJy sr^{-1}), superimposed on more extensive IRAS cirrus of even lower surface brightness. We also made a data cube from the Heiles & Habing HI survey (beam sampled at 36′) for the same field and projection, and more recently resurveyed (fully sampled) a 16° field with the DRAO 26-m telescope. The integrated column density, $N_{\rm HI}$, is also not very large, typically a few times 10^{20} cm^{-2}.

Integrated Column Density

The cirrus mosaics were degraded to the same resolution as the HI surveys. A plot comparing I and $N_{\rm HI}$ pixel by pixel shows an obvious linearity (e.g., $I_{100} = 0.85 N_{\rm HI} - 1.4$) whose slope is typical of this relationship at high latitudes (Boulanger & Pérault 1988). This correlation was used to create consistent "colour tables" with which to display the cirrus and $N_{\rm HI}$ images, making the morphological correspondence between infrared brightness and $N_{\rm HI}$ all the more striking. In addition to the central G86.5+59.6 feature, many "arm" and "ring" features appear in each with identical "colours". Below we shall investigate whether there is as good (or better) correspondence at higher resolution.

Radial Velocity Slabs from the Survey Data

To seek an association of the infrared cirrus with particular velocity systems (deprojection), we made images from the HI survey data cubes integrated over three velocity slabs (IVG, -70 to -29 km s^{-1}; BVG, -29 to -13 km s^{-1}; and LVG, -13 to 35 km s^{-1}). The colour representation was kept identical to show correctly the relative importance of different features.

G86.5+59.6 is clearly the brightest (largest column density) feature in the IVG, but it is not totally isolated; there are other IVG features with clear counterparts in the infrared cirrus, for example the NW part of the ring to the E and several arms to the W.

The LVG is much more widespread, as might be expected for the "local" material, and is actually brighter at positions other than G86.5+59.6. It too has structure seen in the cirrus, for example, defining the S part of the cirrus ring with extensions to the SE. There is a slight brightening at G86.5+59.6, but perhaps a more interesting feature is a ridge of emission running to the NE. This ridge lies parallel to bright IVG which lies just to its NW.

The BVG is the least spectacular component, but very interesting. Because the division into three slabs is rather arbitrary, it naturally blends in some regions with features of the IVG and in others with the LVG. In particular, it contributes to the N and S of the cirrus ring. At G86.5+59.6, however, it is less bright than in other parts of the ring, and therefore seems less likely to be the result of an interaction.

The BVG also helps illustrate the effects of superposition on the perception of features in the infrared cirrus (or total $N_{\rm HI}$). Although the features in the BVG are generally weak they can be identified in the cirrus; this is because they add to any smooth pedestal present from other velocity components; in particular, a bright patch of cirrus to the SE of the ring is created by BVG on a much smoother background of LVG. In this context, it is interesting that G86.5+59.6 is situated near the conjunction of three arms in the IVG. Thus in addition to the LVG contribution (pedestal), the G86.5+59.6 bright feature that stands out might be the result of superposition of different systems within the IVG. This can be investigated at higher (arcminute) resolution.

60/100 μm Brightness Ratio

HRK measured the overall 60/100 μm brightness ratio to be about 0.3, higher than (0.2) for most of the other clouds in their study; this diagnostic suggested that grains in this line of sight have been modified by shocks. From the mosaics we found that although there is some range in the ratio in the larger field (0.25 – 0.34) quite similar ratios occur for dust associated with IVG and LVG. The ratio at G86.5+59.6, near 0.32 overall, is not unusual in comparison to its surroundings. This value 0.32 is like the Galactic average for atomic gas (Deul 1988) and that found on a large scale in the northern Galactic cap (Boulanger & Pérault 1988; $b > 50°$, which encompasses G86.5+59.6). There are regions in the Galaxy where the brightness ratio is lower (closer to 0.20), including much of the HRK sample, but this ratio is more typical of molecular clouds (Deul). While there is no persuasive evidence from this broad perspective that indicates that G86.5+59.6 is a special region of interaction, there are suggestions of local colour changes near G86.5+59.6 related to different velocity components. The sense of these appears to be a somewhat lower ratio for LVG and higher ratio for IVG; whether this distinction is generally true is still being investigated.

ZOOMING IN ON G86.5+59.6 WITH DRAO ST AND HIRES

The issue of interaction can also be pursued by looking at higher angular resolution. To this end a 2° field has been observed with the DRAO Synthesis Telescope, with velocity resolution 0.8 km s^{-1}and with angular resolution 1′. This angular resolution is about 35 times better than in the survey data and about 5 times that in the ISSA plates. We have reprocessed the IRAS data in this field with the High Resolution (HiRes) software package at IPAC, which uses the Maximum Correlation method to enhance the resolution to about 1′.

Integrated Column Density

As mentioned above, the column density and infrared surface brightness at G86.5+59.6 are quite low, and so this is an interesting test of observational and processing capabilities in faint diffuse fields. By simple inspection, it can be seen that the morphological similarity of the cirrus and integrated N_{HI} images down to about $1'$ is quite remarkable. *At this finest scale yet achieved, this agreement is quite quantitative.* There is a good pixel by pixel correlation between I and N_{HI} and the slope of this linear relationship is similar to that found in the above and other studies at larger angular scales. This suggests that there is dust associated with each of the two major velocity components (even without velocity slicing of the cube) and that there is a general absence of H_2. Emission in CO has not been detected at positions within G86.5+59.6 (HRK; W. Reach, private communication).

Radial Velocity Slabs from the DRAO ST Data

To investigate the association of velocity structure in HI with cirrus morphology on this finer scale, the DRAO ST data have been integrated over the same velocity ranges as defined for the survey data radial velocity slabs. The HI images were displayed for comparison with the HiRes images, again using a consistent colour table.

A striking discovery is that the IVG provides most of the detailed morphological structure seen in this infrared cirrus. *This deprojection analysis therefore shows definitively that there is dust in the IVG component*, of note because the dust has either survived the acceleration to -40 km s^{-1} or more or has reformed since (dust destruction is predicted in shocks of sufficiently high velocity).

As in the survey data, the LVG is more diffuse; but it has a hook-like feature (with NE spur) which can be located in the cirrus too. This hook and spur do not align exactly with a similar pattern in the IVG. Nor does the LVG appear at the ends of the SE–NW bar in the IVG.

The BVG has weak features that perhaps correlate (with modest lateral displacement) most with the IVG. Again, the division between velocity slabs is somewhat arbitrary. There are no obvious bridges between common features in both the IVG and LVG as would provide circumstantial evidence for interaction.

EXAMINATION OF THE HI DATA CUBES

The detailed relationships between HI and infared cirrus are arguably best studied by examining the images of individual velocity channels. This work is still in progress on both the survey data and ST data and no final conclusions will be presented. An unresolved question is whether the three velocity components are interacting now or whether they interacted in the past. Because the relative velocity vector would not necessarily be lined up along our line of sight, then potential connections between the morphologies in the different velocity components need not be exactly superimposed. Instead, the patterns in each slab might be translated versions of one another (translating the LVG pattern to the W does seem to line up the structures better with the IVG) or one pattern could be embedded in another. Investigation of velocity gradients and fine velocity structure in the cube is important.

Volume Rendering

One analysis tool with which we are working is volume rendering of the HI data cube. In the ST cube the tapering by the 9-m primary beam pattern is not removed so that noise near the cube edges is not amplified. While this permits a clear view of the interior of the cube, it does tend to heighten the perception of central concentration in all velocity ranges. The 26-m cube emphasizes how the LVG is more widely distributed while the IVG is concentrated. Near G86.5+59.6 the LVG and IVG components of $N_{\rm HI}$ appear as clearly separated components (whether there is spatial separation is another question). These renderings also reveal that there are different morphological features in the different velocity components, as was described above using the three coarse velocity slabs. Here one has the potential of observing the "three-dimensional" structure, including any systematic velocity gradients. This approach can in principle be used to deproject, or at least interpret, the infrared cirrus image. For example, the relatively sharp NE boundary in the infrared cirrus image of G86.5+59.6 reflects that of the IVG; to the SW the IVG boundary is itself less sharp and in the integrated cirrus it is even less apparent because of the underlying diffuse LVG.

Animation

We have also found it instructive to make an animation showing successive velocity slices of the data cubes (0.8 km s^{-1} per slice). The directional orientation of the velocity gradients near G86.5+59.6 within the LVG and IVG components can be noted: toward negative radial velocities, LVG runs from the NE to the SW, while IVG appears (in broad brush) to have subcomponents running from NE and SE to the W. We also have experimented with a false-colour animation in which the LVG and IVG components appear simultaneously against a background image of I_{100}. This is an effective means of intercomparing the morphologies of the different velocity components and infrared emission at small angular scales and with better velocity resolution than in the broad velocity slabs. *The animation shows several striking correspondences of features in the infrared image with features in HI which appear over only a few velocity channels* (thus these features have small line widths). Also the LVG slice with maximum column density (peak in the integrated line profile toward G86.5+59.6) appears to bisect and fill in a spatial gap present in the IVG slice at its peak velocity; the opportunity to investigate such potentially interesting interrelationships is washed out in the radial velocity slabs.

REFERENCES

Boulanger, F., & Pérault, M. 1988, *ApJ*, **330**, 964

Deul, E. 1988, Ph.D. thesis, Leiden

Heiles, C., & Habing, H. 1974, *A&AS*, **14**, 1

Heiles, C., Reach, W.T., & Koo, B-C. 1988, *ApJ*, **332**, 313 (HRK)

Joncas, G., Dewdney, P.E., & Boulanger, F. 1992, *ApJ*, **397**, 165

The First Symposium on the Infrared Cirrus and Diffuse Interstellar Clouds
ASP Conference Series, Vol. 58, 1994
Roc M. Cutri and William B. Latter (eds.)

PROBING THE ATOMIC ISM: IRAS, HI, [CII], AND OPTICAL OBSERVATIONS OF AN ISOLATED CIRRUS FILAMENT

DAVID L. SHUPE
Physics Department, Caltech, Pasadena, CA 91125

TERRY HERTER, DAVID F. CHERNOFF, GORDON J. STACEY
Astronomy Department, Cornell University, Ithaca, NY 14853

ROC M. CUTRI
Steward Observatory, University of Arizona, Tucson, AZ 85721

ABSTRACT We present comprehensive radio, infrared, and optical observations of emission from a diffuse interstellar cloud. The filament is very sharply defined both spatially and in velocity, providing a unique laboratory for study of the atomic interstellar medium. The measured column density of HI indicates that the cloud is relatively tenuous (with estimated $A_V \sim 0.1 - 0.2$), while the lack of detectable CO suggests that little molecular material is present. The cloud's *IRAS* colors require that the proportion of small grains is larger than that in standard grain models. A marginal measurement of the [CII] 158 μm flux is consistent with theoretical expectations when the enhanced numbers of small particles are taken into account.

INTRODUCTION

Study of diffuse clouds and infrared cirrus is often complicated by the presence of confusing foreground and background emission along a particular line of sight. A particularly isolated and well-defined cirrus cloud was serendipitously discovered by Herter, Shupe, and Chernoff (1990) in coadded *IRAS* images and 21 cm data. The object is relatively tenuous, so that it is uniformly illuminated by the incident radiation field, and observations do not average over multiple "layers" with varying properties. This cloud presents an ideal laboratory for study of the atomic interstellar medium.

We have extended our study of this object with the goal of constraining the properties of the gas and dust as much as possible (further details may be found in Shupe et al. 1994). Ultimately, we wish to explore the physical processes at work in the atomic phase of the interstellar medium. The dust plays a pivotal role by heating the gas via photoejection of electrons and enabling the formation of H_2. The dominant coolant of the gas is collisional excitation and emission in the [CII] 158 μm line.

OBSERVATIONS

HI and CO lines

The filament was initially mapped in HI with 4′ resolution at Arecibo. Subsequent radio observations include higher-resolution mapping in the 21 cm line at the VLA, and a search for CO (J=1-0) emission using the NRAO 12m. The results of these observations include:

Column density: The peak column density is 4×10^{20} H-atom cm^{-2}. The extinction is estimated from a standard value of A_V/N_H (Bohlin, Savage, and Drake 1978) to be $A_V \sim 0.22$ at the peak, and $A_V \leq 0.1$ over most of the filament.

Dust emission/gas emission: The excellent correlation between the gas and dust emission is seen in Figure 1. At the positions of the Arecibo pointings,

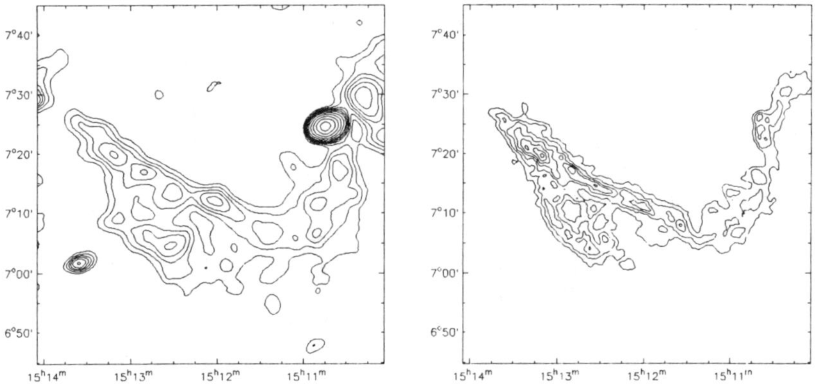

FIGURE 1 *Left:* 100 μm *IRAS* map produced by five iterations of IPAC's HiRes processor. Bright sources are infrared galaxies. *Right:* Integrated intensity in the 21 cm line as mapped at the VLA.

the HI column density was compared to the 100 μm surface brightness, and is constant within the errors with a value of 0.52 ± 0.05 MJy sr$^{-1}/10^{20}$ H-atom cm^{-2}.

Temperature: As no strong continuum sources lie behind the cloud, the gas temperature cannot be found from an absorption measurement. The measured velocity dispersion of the HI line sets an upper limit to the kinetic temperature of 120-275 K.

Molecular content: A search for CO at the HI peak failed to detect any emission, setting an upper limit of 0.036 K km s^{-1}.

[CII] Observations

Observations in the [CII] 158 μm line were made in April 1992 using the MPE/UCB Far-Infrared Imaging Fabry-Perot Interferometer (Poglitsch et al. 1991) aboard the Kuiper Airborne Observatory. This instrument employs a 5×5 array of detectors with a total field of view of $200'' \times 200''$. The beam size is

55″. Two pointing positions, one at the HI peak and the other 200″ to the southwest, were observed for a total of 64 minutes. No clear emission features corresponding to the HI map were detected. Averaging over all 25 detectors yields a marginal positive flux of $12.7 \pm 4.1 \times 10^{-7}$ ergs s^{-1} cm^{-2} sr^{-1}.

The average HI column density over this region is 2×10^{20} H-atoms cm^{-2}, which leads to a cooling rate limit of 8×10^{-26} ergs s^{-1} H-atom^{-1}.

Optical Photometry

Deep CCD optical imaging in the *BVRI* bands was carried out in April 1991 at the Kitt Peak 0.9 m and the Steward Observatory 90-inch telescopes. Comparison with the 21 cm map reveals faint emission/scattering from the cloud at the level of 27 – 28 magnitudes per square arcsecond. The $V - R$ color ranges from 0.5 to 0.7 mag, and the measured $R - I$ values are 0.3-0.6 mag for different cuts across the cloud. The colors measured by Guhathakurta and Tyson (1989; see also Guhathakurta & Cutri, this volume) for brighter cirrus fields are significantly redder.

DISCUSSION

Dust Properties

Comparisons with galactic cosecant fits (Boulanger and Pérault 1988) and samples of isolated clouds (Laureijs 1989) show that $I_\nu(60)/I_\nu(100) \approx 0.32$ measured for the filament is significantly higher than average values, as is $I_\nu(12)/I_\nu(100) \approx 0.088$ somewhat higher. To find the implications of these differences for the dust properties in the filament, we have calculated emission from a graphite-silicate model using optical constants from Draine and Lee (1984) and the computational method of Guhathakurta and Draine (1989). We find that the radiation field intensity must be within a factor of roughly 2 of the standard ISRF of Mathis, Mezger and Panagia (1983). However, the colors cannot be fit by a power-law size distribution, even when it is extended to radii of 5 Å. Rather, large amounts of the smallest particles must be added.

A more realistic dust model is that of Désert, Boulanger, and Pérault (1990; DBP), which includes PAHs, very small grains (VSGs), and big grains. By scaling the amounts of each of these components, the filament colors can be matched for radiation field strengths of 0.5 to 1 only if the proportions of PAHs and VSGs relative to the big grains are increased by about 50% - 100% over that of DBP.

Although grains larger than 2000 Å are too cold to contribute to any *IRAS* band, they significantly affect the optical scattering. Calculations of isotropic scattering from a grain model similar to that described above show that if the maximum grain size is 2000 Å, the cloud shows a modest "red excess"; but most of the excess may be accounted for by extending the distribution to 10000 Å. The "red excess" seen in other cirrus clouds, as well as reflection nebulae and planetary nebulae, is commonly attributed to either PAHs (d'Hendecourt et al. 1986) or HAC (Witt and Schild 1988). If these particles are present in this filament, they must be severely dehydrogenated so they do not produce red emission.

Heating and Cooling

The cooling rate limit derived above is largely consistent with other estimates of heating and cooling in the diffuse ISM. UV observations of C^+ produce cooling estimates of 10×10^{-26} ergs s^{-1} H-atom^{-1} (Pottasch et al. 1979) or lower (Gry, Lequeux, and Boulanger 1992). A recent rocket experiment measuring the [CII] line directly finds 3×10^{-26} ergs s^{-1} H-atom^{-1} (Bock, this volume). Considering heating from PAHs, Verstraete et al. (1990) find a heating rate of 8×10^{-26} ergs s^{-1} H-atom^{-1}. Bakes (this volume) has computed heating from a size distribution of grains and small particles, deriving a rate of 3×10^{-26} ergs s^{-1} H-atom^{-1}; however, since this filament contains an excess of small particles, its heating rate should be proportionally larger.

ACKNOWLEDGMENTS

We thank the staff of the Infrared Processing and Analysis Center (IPAC), for assistance in obtaining and interpreting the *IRAS* images. This work was supported in part by NSF Award AST-8657467 to D.F.C. and an NSF Fellowship to D.L.S.

REFERENCES

Bohlin, R.C., Savage, B.D., and Drake, J.F. 1978, *ApJ*, **224**, 132.

Boulanger, F., and Pérault, M. 1988, *ApJ*, **330**, 964.

Désert, F.X., Boulanger, F., and Puget, J.L. 1990, *A&A*, **237**, 215.

Draine, B.T., and Lee, H.M. 1984, *ApJ*, **285**, 89.

Gry, C., Lequeux, J., and Boulanger, F. 1992, *A&A*, **266**, 457.

Guhathakurta, P., and Draine, B.T. 1989, *ApJ*, **345**, 230.

Guhathakurta, P., and Tyson, J.A. 1989, *ApJ*, **346**, 773.

d'Hendecourt, L.B., Leger, A., Olofsson, G., and Schmidt, W. 1986, *A&A*, **170**, 91.

Herter, T., Shupe, D.L., and Chernoff, D.F. 1990, *ApJ*, **352**, 149.

Laureijs, R.J. 1989, Ph.D. Dissertation, University of Groningen, The Netherlands.

Mathis, J.S., Mezger, P. G., and Panagia, N. 1983, *A&A*, **128**, 212.

Poglitsch, A., Beeman, J.W., Geis, N., Genzel, R., Haggerty, M., Haller, E.E., Jackson, J., Rumitz, M., Stacey, G.J., and Townes, C.H. 1991, *Internat. J. Infrared Millimeter Waves*, **12**, 859.

Pottasch, S.R., Wesselius, P.R., and van Duinen, R.J. 1979, *A&A*, **74**, L15.

Shupe et al. 1994, *ApJ*, in preparation

Verstraete, L., Léger, A., d'Hendecourt, L., Dutuit, O., and Défourneau, D. 1990, *A&A*, **237**, 436.

Witt, A.N., and Schild, R.E. 1988, *ApJ*, **325**, 837.

The First Symposium on the Infrared Cirrus and Diffuse Interstellar Clouds
ASP Conference Series, Vol. 58, 1994
Roc M. Cutri and William B. Latter (eds.)

PHASE TRANSITIONS IN INTERSTELLAR CIRRUS

WILLIAM T. REACH
NASA/GSFC Code 685, Greenbelt, MD 20771

ABSTRACT The infrared surface brightness, H I column density, and CO line brightness of a sample of isolated, degree-sized infrared clouds are compared. Two phase transitions are observed in the clouds, separating three distinct regions: the H I halo, the H_2 cloud, and the molecule-rich cores. Assuming the infrared emission traces the total column density and the H I traces the atomic column density, we obtain a map of H_2 that is much more extended than the observed CO, due to variations in the CO abundance.

INTRODUCTION

The hydrogen in an interstellar cloud exposed to the interstellar radiation field is primarily atomic down to a certain depth, where the extinction (by dust and self-shielding) of the radiation field is large enough that the H_2 formation rate exceeds its photodissociation rate. Molecules other than H_2 form deeper into the cloud, and some, such as CO, are capable of self-shielding. We use the term "phase transition" here to refer to a distinct chemical abundance transition. The goal of this project is to determine where the phase transitions occur for local clouds.

We had previously identified a sample of isolated interstellar clouds for detailed study (Heiles, Reach, & Koo 1988; hereafter HRK). The selection criteria were that each cloud be an isolated, centrally condensed, $\sim$ degree-sized feature on the *IRAS* 100μm map that is furthermore isolated from extensive H I structures.

OBSERVATIONS

CO survey of selected cirrus clouds

We surveyed all of our clouds for CO using the NRAO 12-m and CfA 1.2-m telescopes. The ratio of cloud size to beam size is so large (40–120 for the 12-m) that it was impossible to map the clouds fully. Instead, we observed the peaks in the infrared maps, and made one-dimensional scans through the cloud cores. Some clouds that were not detected with the 12-m were confirmed as nondetections using the 1.2-m (whose larger beam would certainly contain the CO-emitting regions). Half (15/30) of the clouds were detected in CO, with characteristic CO-emitting regions of 3-10$'$ in size.

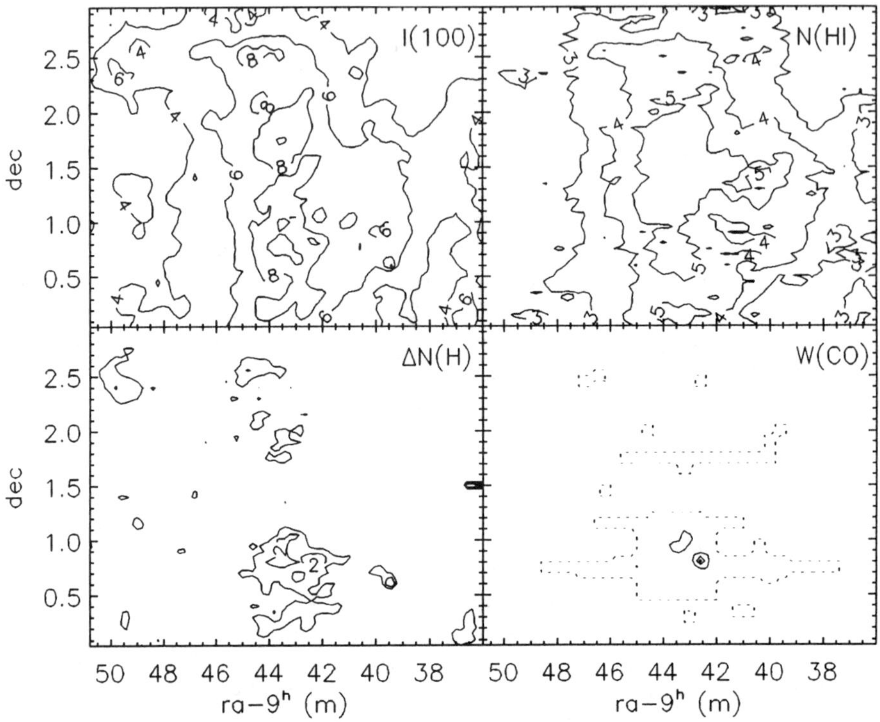

FIGURE I Maps of G236+39 in 100μm surface brightness [MJy sr^{-1}] (*top left*), H I column density [10^{20}cm^{-2}] (*top right*), excess column density [10^{20}cm^{-2}] (*bottom left*), and integrated CO line [K km s^{-s}] (*bottom right*). The infrared map is more centrally peaked than the H I map, but not so clumped as the CO map. If the infrared excess is due to dust associated with H_2, then the southern part of the cloud contains H_2 but very little CO.

Complete CO and H I maps of G236+39

In order to investigate the detailed relationship between atomic and molecular gas and dust emission, we desire sensitive observations of all three components at the best common resolution.

Only one of our clouds with infrared excess is accessible to the Arecibo 305-m, the only telescope with sensitivity and resolution matched to the *IRAS* survey. This cloud, G236+39, was completely mapped using drift scans sampled every 12 sec (half-beam), for a total of over 9000 spectra.

The CO(1–0) 115.2 GHz emission from G236+39 was mapped using the NRAO 12-m. All 100μm surface brightness peaks were observed; and the southern cloud core was mapped at 2′ spacing. Additional observations of CO(2–1), ^{13}CO(1–0), and ^{13}CO(2–1) were made toward the CO peaks the infrared excess peak.

DISCUSSION

Infrared excess as a tracer of molecular gas

The infrared, H I, and CO maps of G236+39 are shown in Figure I. The infrared map is morphologically identical to the H I map in the peripheries of the cloud, and the locations of the CO peaks agree with the strongest local maxima in the infrared map. However, a detailed comparison reveals that the infrared emission cannot be accounted for using linear combinations of the H I and CO maps.

Let us define the infrared excess as the difference between the 100μm map and a scaled H I map. The infrared excess map is shown in Figure I. The infrared excess is mostly within the southern cloud core, with some in a bright rim around the northern cloud core. We consider three hypotheses for the infrared excess:
(1) Enhancement of the grain temperature does not seem to apply because the $60/100\mu$m color temperature *decreases* in regions with higher infrared excess. Further there are no bright visible stars nearby, nor are there any 12μm point sources in the cloud.
(2) Enhancement of the dust/gas ratio is unlikely because there are no evident dust production sites (12μm sources) within the cloud.
(3) Existence of molecular gas is the most likely explanation, which we pursue below.

The H_2/H I phase transition

If H_2 is formed on grains, destroyed by photodissociation, and protected by self-shielding in the wings of its absorption bands, then the total column density is predicted to be a quadratic function of the H I column density. From the quadratic fit of the 100μm and H I maps, we find the column density $N_c = (3.6 \pm 0.5) \times 10^{20}\text{cm}^{-2}$, where the H I and H_2 columns are equal. The critical column may be expressed in terms of the collision rate, γ, the relative strength of the radiation field, G_o, and the photoabsorption rate in the Lyman and Werner bands, β (Jura 1975):

$$N_c = 2 \times 10^{20}(\frac{\beta G_o}{n\gamma})^2 = 4.5 \times 10^{19}(\frac{100\text{cm}^{-3}}{n})^2(\frac{100\text{K}}{T})\ \text{cm}^{-2}.$$

This value is roughly within the range predicted for plausible densities and temperatures. For example, with a gas pressure of $p/k = 4000\text{cm}^{-3}\text{K}$, we find $n = 30\text{cm}^{-3}$ and $T = 240\text{K}$.

Regions of enhanced CO abundance

The CO morphology is clearly different from the infrared morphology, in the sense that the CO emission is more clumped. The sensitivity of the observations was high enough to detect CO over much larger areas, *if* the CO were distributed as the infrared emission. There is no simple relation between CO brightness and infrared brightness. There are cloud-to-cloud variations in the infrared brightness threshold above which CO emission is detected, as well as large scatter in the ratio of CO to infrared brightness within a given cloud.

Because they are optically thin, the ^{13}CO lines are a good tracer of the ^{13}CO column density; they are not, however, a tracer of H_2. The most extreme case is in the middle of the cloud, where the infrared excess is highest but neither CO nor ^{13}CO were detected. Here the abundance $[^{13}CO/H_2] < 1 \times 10^{-7}$, an order of magnitude lower than elsewhere in the cloud.

We identify the regions where the CO abundance is high as a separate phase. A calculation of CO photodissociation and chemistry yield abundances $[CO/H_2] \sim 10^{-4}$ for clouds with $A_V > 1$ and $\sim 10^{-6}$ for $A_V < 0.3$ (van Dishoeck and Black 1988). At positions with detected CO, our abundance is similar to that of an opaque cloud, while at nondetections the abundances are even lower than diffuse clouds. The strong variations could be due to shadowing effects on the photodissociation, time-dependent chemistry, or thermochemical instability (Glassgold and Langer 1976).

CONCLUSIONS

The infrared brightness is proportional to the H I column density in the peripheries of cirrus clouds. At column densities higher than $\sim 4 \times 10^{20}$cm^{-2}, there is a substantial infrared excess that we interpret as emission from dust associated with H_2. Our 21-cm observations can be fitted using a model of H_2 formation including self-shielding. Over the $\sim 30'$ cloud core of G236+39, some 60% of H nuclei are in H_2 molecules. The regions of infrared excess are larger than the regions with detectable CO lines. The abundance of CO must increase by at least an order of magnitude in the regions of detectable CO lines. This explains the observations by Blitz *et al.* (1990) that there are infrared excess clouds (Désert *et al.* 1989) without detectable CO: substantial fractions of cirrus clouds have most H nuclei in H_2 molecules, but the CO abundance is more than an order of magnitude lower than its fully self-shielded value.

ACKNOWLEDGMENTS

This work was done partially as a National Reach Council-NASA/GSFC Research Associate, and partially as a research assistant at the Berkeley Astronomy Department, Univ. of California supported in part by NSF Award No. 91-23362 to Dr. Carl Heiles.

REFERENCES

Blitz, L., Bazell, D., & Désert, F.-X. 1990, *ApJ*, **352**, L13

Désert, F.-X., Bazell, D., & Boulanger, F. 1988, *ApJ*, **334**, 815

Glassgold, A. E., & Langer, W. D. 1976, *ApJ*, **204**, 403

Heiles, C., Reach, W. T., & Koo, B.-C. 1988, *ApJ*, **332**, 313 (HRK)

Jura, M. 1975, *ApJ*, **197**, 575

van Dishoeck, E. F., & Black, J. H. 1988, *ApJ*, **334**, 771

The First Symposium on the Infrared Cirrus and Diffuse Interstellar Clouds
ASP Conference Series, Vol. 58, 1994
Roc M. Cutri and William B. Latter (eds.)

THE HI GAS KINETIC TEMPERATURE OF HIGH GALACTIC LATITUDE CIRRUS CLOUDS

RONALD STARK
Sterrewacht Leiden

JOHN M. DICKEY
University of Minnesota

W.B. BURTON
Sterrewacht Leiden

LEX WENNMACHER
Radioastronomisches Institut der Universität Bonn

We present the results of a combined absorption and emission study of the HI 21 cm line towards several cirrus clouds at high galactic latitude. The absorption measurements were obtained with the Very Large Array (VLA) synthesis telescope, the emission data with the Effelsberg telescope. The three cirrus complexes studied were selected from the Deul and Burton (1990) study. Deul and Burton had shown that these cirrus clouds have associated HI emission with radial velocities near 0 km s^{-1} as well as with more extreme velocities of $v < -20$ km s^{-1}. Absorption against background continuum sources was detected in 40% of the lines of sight investigated. The HI line was always found to be optically thin, $\tau < 0.6$. The 0 km s^{-1} component was found in each direction where absorption was detected; in addition, two directions showed absorption at intermediate negative velocities. The derived apparent spin temperatures vary between 20 K and > 350 K. The emission profiles reveal much more widely distributed gas than is visible in absorption. Only the cool clumps in the cirrus show up in absorption. The observations suggest a collision of a high velocity cloud with the galactic disk gas, by which different parts of the cirrus are at present seen in different phases of shock completion. This model may explain the large observed differences in the HI emission and absorption profiles over small distances on the sky, as well as the variation of the far-infrared brightness ratio.

The First Symposium on the Infrared Cirrus and Diffuse Interstellar Clouds
ASP Conference Series, Vol. 58, 1994
Roc M. Cutri and William B. Latter (eds.)

DIGITAL SELECTION OF STRUCTURAL FEATURES IN NEUTRAL HYDROGEN EMISSION MAPS

F. R. CHROMEY
Vassar Col. and IBM/Watson

BRUCE G. ELMEGREEN
IBM Watson Research Center

The 140-foot telescope at NRAO was used to map the intensity of neutral hydrogen emission in a 20 × 14 degree area in Cepheus and Cassiopeia. The velocity resolution is 0.25 km s^{-1}, and the angular resolution is 20 arc minutes. Three digital filtering algorithms are described. These quantify the presence of cloud-like, shell-like, and filamentary structures of a particular scale size in parsecs. This sort of objective and uniform definition of what constitutes a feature permits meaningful comparisons between the structure of neutral hydrogen and the structure of other objects in the region, such as molecular clouds, infrared and Hα emission, and young stars, and it also quantifies the proportion of the total emission that is in various geometric forms and sizes.

CHAPTER IV

CIRRUS IN EXTERNAL GALAXIES

The First Symposium on the Infrared Cirrus and Diffuse Interstellar Clouds
ASP Conference Series, Vol. 58, 1994
Roc M. Cutri and William B. Latter (eds.)

THE ORIGIN OF THE FAR INFRARED LUMINOSITY WITHIN SPIRAL GALAXIES

NICK DEVEREUX
New Mexico State University, Astronomy Department, Box 30001/Dept. 4500, Las Cruces, NM 88003

ABSTRACT

The IRAS $S_{100\mu m}/S_{60\mu m}$ dust color temperature profiles are presented for two nearby spiral galaxies M 101 and M 81. The radial dust temperature profiles provide an important constraint on the origin of the far infrared luminosity. The observed dust temperature is compared with that expected for diffuse interstellar dust heated by the general interstellar radiation field within each galaxy. The implications for the contribution of cirrus to the far infrared luminosity of M 101 and M 81 are discussed.

INTRODUCTION

The origin of the far infrared (40 - 120 μm) luminosity in spiral galaxies is a subject of continuing speculation. At issue, of course, is whether or not the far infrared fluxes that were measured for literally thousands of galaxies by the Infrared Astronomical Satellite (IRAS) can be utilized to measure high mass star formation rates. Although use of the Hα emission line luminosity is a less controversial way to measure high mass star formation rates, the number of galaxies for which the Hα emission line luminosity has been measured is limited to about 200 galaxies at present (Kennicutt and Kent 1983; Bushouse 1987; Romanishin 1990). In contrast, IRAS measured far infrared fluxes for literally thousands of galaxies. Consequently, the IRAS database is regarded as a potentially valuable resource for quantifying high mass star formation rates (Young et al., 1986; Helou et al., 1985; Gavazzi et al., 1986; Leech et al., 1988; Devereux and Young 1990; 1991, 1992).

Evidence supporting high mass stars as the origin of the far infrared luminosity includes the fact that the O and B stars, which are required to generate the Hα emission line radiation in galaxies, can also supply the luminosities measured in the far infrared, independent of the $S_{100\mu m}/S_{60\mu m}$ dust color temperature (Devereux and Young 1990) and spiral type (Devereux and Young 1991). Nevertheless, the ambiguity surrounding the origin of the far infrared luminosity in spiral galaxies remains due to the perception that 70 - 90 % of the far infrared luminosity may be produced by cirrus, or more specifically, diffuse interstellar dust heated by the radiation field of non-ionizing stars (Walterbos and Schwering 1987; Lonsdale-Persson and Helou 1987; Bothun, Lonsdale and Rice 1989; Helou., this conference).

We know that the general interstellar radiation field can, in principle, supply the luminosities measured in the far infrared because it requires only a moderate amount of extinction of the optical light, corresponding to A_v = 0.5 - 1.0 magnitudes to account for the L_{fir}/L_{blue} luminosity ratios observed for spiral galaxies. What may be more difficult to explain are the warm dust temperatures indicated by the $S_{100\mu m}/S_{60\mu m}$ ratio. The purpose of this contribution is to show how the radial dependence of the $S_{100\mu m}/S_{60\mu m}$ dust color temperature can provide an important constraint on the origin of the far infrared luminosity within spiral galaxies.

The IRAS High Resolution images (HiRes) provide the opportunity to quantify the radial dependence of the $S_{100\mu m}/S_{60\mu m}$ dust color temperature for several nearby spiral galaxies of large angular size. Since the angular resolution of the HiRes $100\mu m$ images is about 100 arc sec and lower than the $\sim$ 60 arc second resolution of the HiRes 60 μm images, it is necessary, for the purposes of calculating dust color temperatures, to convolve the 60 and 100 μm HiRes images to the same resolution.

The azimuthally averaged $S_{100\mu m}/S_{60\mu m}$ radial dust color temperature profiles are illustrated in Figure 1 for the nearby spiral galaxies M 101 and M 81. The dashed lines in Figure 1 represent the temperatures expected for dust grains of a variety of sizes and compositions when heated by the general interstellar radiation field. The intensity of the general interstellar radiation field was determined from the extinction corrected visual surface brightness profile for each galaxy. The dashed line labelled cirrus indicates the temperature expected for a dust grain size distribution that is required to explain IRAS measurements of Galactic cirrus (Desert, Boulanger and Puget 1990). The dust temperatures are calculated from the $S_{100\mu m}/S_{60\mu m}$ flux ratio assuming a λ^{-1} emissivity law. The dust temperatures expected for large, $\geq 0.1\mu m$, Si and C grains are based on the models of Jura (1982).

RESULTS

M 101

M 101 is a late type (SABcd) spiral galaxy. The radial $S_{100\mu m}/S_{60\mu m}$ dust color temperature profile is remarkably uniform. The mean dust temperature is 31 $\pm$ 2 K for a λ^{-1} emissivity law (see Figure 1a). The observed dust temperature is well in excess of the temperature expected for large, $\geq 0.1\mu m$, dust grains heated by the general interstellar radiation field. The observed dust temperature is also higher than expected for cirrus. Evidently the general interstellar radiation field is unable to heat cirrus up to the high temperature that is observed for the dust in M 101. The surface brightness of M 101 would, in fact, have to be at least 1.5 magnitudes brighter in the V band in order for the general interstellar radiation field to be able to heat the cirrus up to the observed temperature.

M 81

M 81 is an early type (SAab) spiral galaxy. The radial $S_{100\mu m}/S_{60\mu m}$ dust color temperature profile varies with radius from a peak value of about 35 K (λ^{-1}) in the nucleus to about 28 K (λ^{-1}) in the disk (see Figure 1b). The observed dust temperature is well in excess of the temperature expected for large, $\geq 0.1\mu m$,

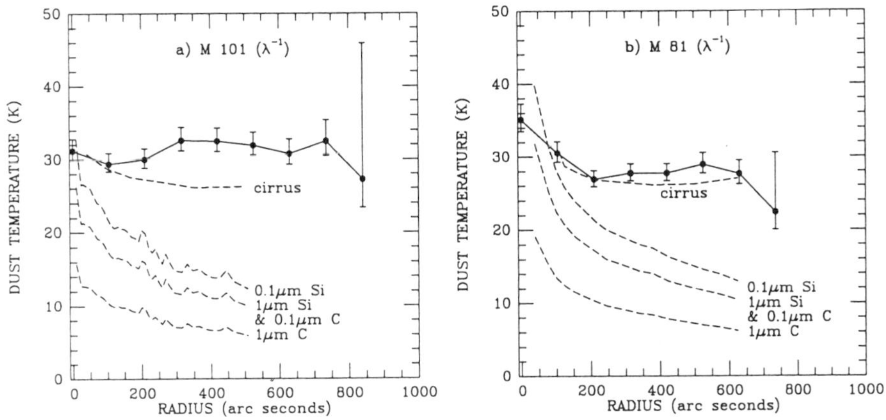

FIGURE I The solid lines indicate the azimuthally averaged $S_{100\mu m}/S_{60\mu m}$ dust color temperature profiles observed for the nearby spiral galaxies M 101 (left panel) and M 81 (right panel). The dashed lines indicate the temperatures expected for dust grains of a variety of sizes and compositions when heated by the general interstellar radiation field within each galaxy.

dust grains heated by the general interstellar radiation field. On the other hand, the observed dust temperature appears to be consistent with that expected for diffuse interstellar dust, or equivalently cirrus, that is heated by the general interstellar radiation field within M 81.

DISCUSSION

The results presented for M 101 and M 81 highlight the importance of the radial dust temperature profiles in constraining the origin of the far infrared luminosity within spiral galaxies. In the case of the late type spiral galaxy M 101, an explanation of the origin of the far infrared luminosity in terms of cirrus, or more specifically, diffuse interstellar dust heated by the radiation field of non-ionizing stars, is untenable because the intensity of the general interstellar radiation field is too low, by at least 1.5 magnitudes, to heat the dust up to the observed temperatures. The limit is necessarily a lower limit because of the inevitable contribution of ionizing stars to the V band surface brightness profile that was used to estimate the intensity of the non-ionizing interstellar radiation field within M 101. On the other hand, O and B stars do provide a viable alternative explanation for the origin of the far infrared luminosity because the dust temperature is comparable to that observed for Galactic and extragalactic HII regions (Scoville and Good 1989; Sodroski et al., 1989; Rice et al., 1990) at all locations within M 101.

In the case of the early-type spiral galaxy M 81, the far infrared luminosity may be attributed to diffuse dust heated by the general interstellar radiation field, because the expected temperature is very similar to the temperature that is actually observed. The distinction between ionizing and non-ionizing stars as the origin of the far infrared luminosity would, however, require a further determination of the relative contribution of ionizing and non-ionizing stars to the general interstellar radiation field within M 81.

REFERENCES

Bothun, G.D., Lonsdale, C.J., and Rice, W., 1989, *ApJ*, **341**, 129

Bushouse, H., 1987, *ApJ*, **320**, 49

Desert, F.X, Boulanger, F., and Puget, J.L., 1990, *A&A*, **237**, 215

Devereux, N.A., and Young, J.S., 1990, *ApJ*, **350**, L25.

Devereux, N.A., and Young, J.S., 1991, *ApJ*, **371**, 515

Devereux, N.A., and Young, J.S., 1992, *AJ*, **103**, 1536

Gavazzi, G., Cocito, A., and Vettolani, G., 1986, *ApJ*, **305**, L15

Helou, G., Soifer, B.T., and Rowan-Robinson, M., 1985, *ApJ*, **298** L7.

Jura, M., 1982, *ApJ*, **254**, 70

Kennicutt, R.C., and Kent, S.M., 1983, *AJ*, **88**, 1094

Leech, K.J., Lawrence, A., Rowan-Robinson, M., Walker, D., and Penston, M.V., 1988, *MNRAS*, **231**, 977

Lonsdale-Persson, C.J., and Helou, G., 1987, *ApJ*, **314**, 513

Rice, W., Boulanger, F., Viallefond, F., Soifer, B.T., and Freedman, W.L., 1990, *ApJ*, **358**, 418

Romanishin, W., 1990, *AJ*, **100**, 373

Scoville, N.Z., and Good, J.S., 1989, *ApJ*, **339**, 149

Sodroski, T.J., Dwek, E., Hauser, M.G., and Kerr, F.J., 1989, *ApJ*, **336**, 762

Walterbos, R.A.M., and Schwering, P.B.W., 1987, *A&A*, **180**, 27

Young, J.S., Schloerb, F.P., Kenney, J.D., and Lord, S.D., 1986, *ApJ*, **304**, 443

The First Symposium on the Infrared Cirrus and Diffuse Interstellar Clouds
ASP Conference Series, Vol. 58, 1994
Roc M. Cutri and William B. Latter (eds.)

COOL DUST IN THE SPIRAL GALAXY NGC 3079

ROB PRICE, NEBOJSA DURIC
Institute for Astrophysics, University of New Mexico, Albuquerque, NM 87131

W. D. DUNCAN
Royal Observatory, Blackford Hill, Edinburgh, EH9 3HJ, Scotland

INTRODUCTION

Dust in galaxies plays an important role in controlling the temperature of the ISM by reradiating high energy optical and UV photons at FIR/submm wavelengths, where the broad emission feature constitutes ~30% of the total luminosity from galaxies. The dust absorption of the high energy photons prevents the dissociation of molecular gas from which new stars are formed. Since the ISM of galaxies is composed of different gas phases it is necessary to study the dust emission associated with each of these gas phases. Our current understanding of dust emission from galaxies is largely based on IRAS observations which may be dominated by warm dust emission (Devereux and Young 1990).

We perform a comprehensive study of the FIR/submm emission from NGC 3079, an edge-on, spiral galaxy. We estimate the temperature, the total mass, the fraction of cool dust, and the emissivity of the dust by combining JCMT 800 and 1100 μm observations with IRAS, mm, radio, and CO observations.

SUBMM OBSERVATIONS

Continuum observations of the nuclear region of NGC 3079 were obtained at 800 and 1100 μm using the UKT14 bolometer on the 15 meter James Clerk Maxwell Telescope (Duncan et al., 1990). The integration time of 400 seconds per pointing provided an rms sensitivity of ~45 mJy for the 12 flux measurements at 800 μm and ~20 mJy for the 6 pointings at 1100 μm. The calibration uncertainty was ~20%. The beam width was ~17" and the pointings were separated by 8", with a ~3" pointing error.

FLUX DISTRIBUTIONS

Flux measurements of NGC 3079 at 800 and 1100 μm are shown in Fig I. These are among the first off nucleus submm observations of dust emission in a spiral galaxy.

A 2-D Gaussian fit to the 800 μm distribution has a peak flux of ~500 mJy, a deconvolved FWHM of 23×16" (2.3×1.6 kpc), and is oriented at 10° east of the vertical, which is consistent with the plane of the galaxy. From this fit we estimate the integrated 800 μm flux within a 72" beam to be 1.1±0.4 Jy, which

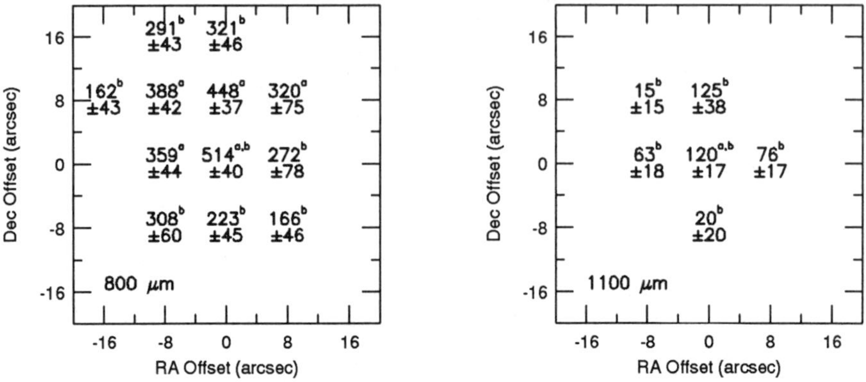

FIGURE I The JCMT flux measurements of NGC 3079 at 800 and 1100 μm in mJy. The errors correspond to the rms scatter in 40 individual 10 sec integrations per pointing and do not include the ~20% calibration errors. Superscripts a and b refer to the observing dates Jan. 14 and 17, 1992, respectively.

is in agreement with earlier UKIRT observations (0.8± 0.3; Eales et al 1989). The 1100 μm distribution is consistent with a point source of 140 mJy.

The ratio of the 800 and 1100 μm peak fluxes yields an emissivity index, β (where emissivity scales as ν^{β}), of 2.5±1.0 consistent with the value of 2 that is commonly used in the submm regime (Hildebrand 1983).

CO:SUBMM COMPARISON

To constrain the mass and temperature of the dust in the nuclear region of NGC 3079 we use the CO map from the Nobeyama 45m telescope (Fig IIa, Sofue and Nakai, 1993 private communication). Direct comparison of our 800 μm fluxes with the CO map shows that the mean CO:submm flux ratio is 1.3±0.3 without substantial spatial variation (Fig IIb).

Since the uncertainty of the ratio is approximately equal to the calibration uncertainty of the 800 μm fluxes, we suggest that the CO:submm flux ratio is constant within the errors. In the usual interpretation this implies that the molecular gas to dust mass ratio is constant across the nuclear region of the galaxy on a 1.6 kpc scale.

Assuming β=2 and a molecular gas-to-dust mass ratio of 100, we find the temperature of the dust dominating the emission at 800 μm is ~20K.

THE CONTINUUM SPECTRUM OF NGC 3079

Combining the integrated 800 and 1100 μm fluxes with our NRAO (12 meter) 3 mm, IRAS, and radio observations and the UKIRT submm observations of Eales et al. (1989) we are able to form the continuum spectrum of the nuclear region of NGC 3079 over 4 orders of magnitude, from 12 μm to 20 cm (Fig III).

We fit the FIR/submm emission assuming a single dust temperature and

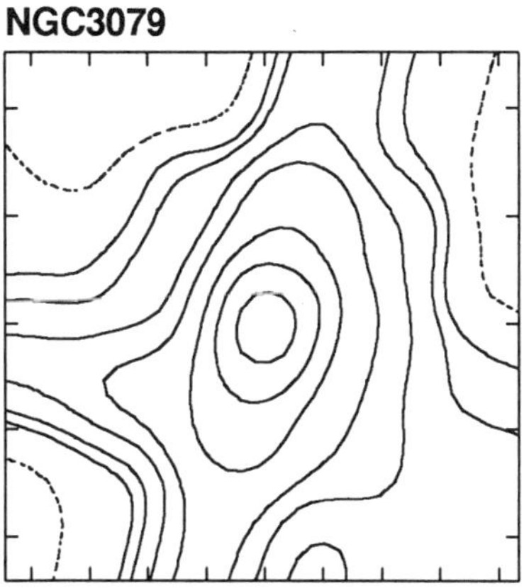

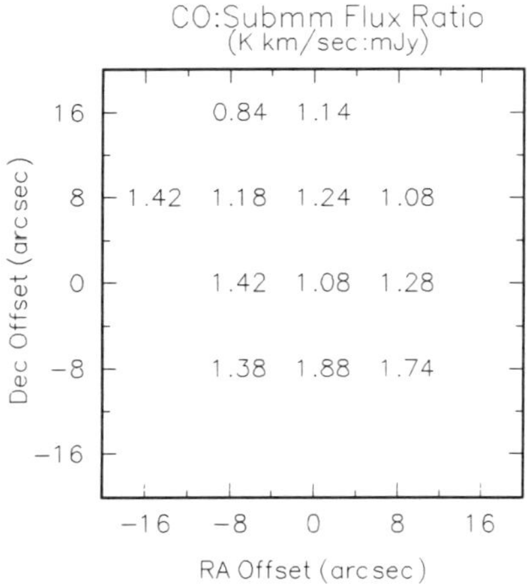

FIGURE II a) This CO map from the Nobeyama 45 m telescope (Sofue and Nakai, 1993 private communication) is very well matched to our JCMT observations since the map has 15" resolution on 7.5" grid. The right ascension is centered is at 09 58 35 with 1 second intervals between tick marks and the declination is centered at 55 55 15 with 15" intervals. The contours intervals are -5, 5, 10, 20, 30, 50, 70, and 90% of the peak flux of 132.2 K km s^{-1}. b) The CO:submm flux ratios are presented at each 800 μm pointing.

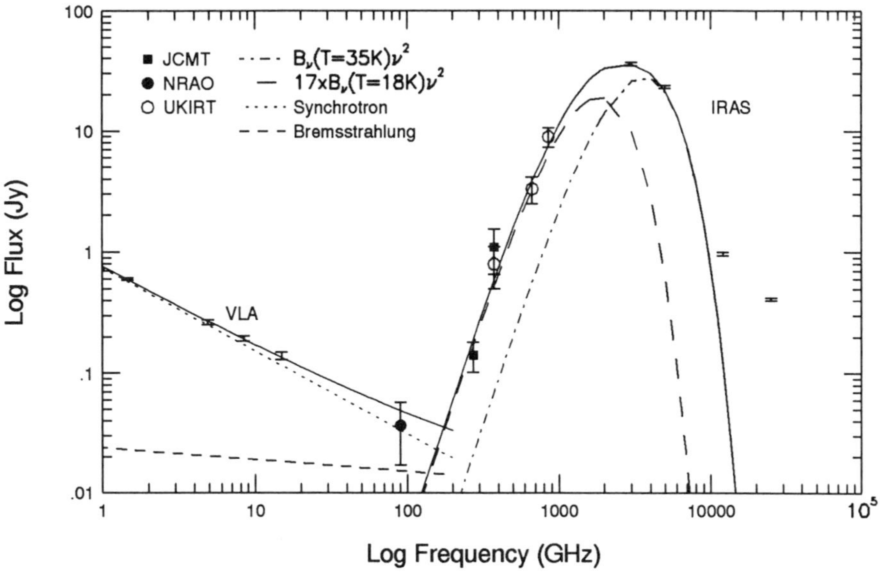

FIGURE III The continuum spectrum of the nuclear region of NGC 3079 is presented from 12 μm to 20 cm. The FIR/submm regime is fit with two dust temperature components. The fit is improved relative to the single temperature fit and provides a mass-to-gas dust ratio that is consistent with Galactic molecular clouds.

find that T=37 K and the emissivity is β=1.2. But the existence of a second (cool) dust component is evident since the fit lies below the points in the submm regime and the total dust mass calculated from the fit is only a small fraction (1/2000) of the mass of the molecular hydrogen as calculated from the CO measurements (Fig II).

We refit the submm/FIR emission with two temperature components, while constraining the cool component to be 95% of the dust mass and β=2. The two component fit (Fig III) results in dust temperatures of 35 K and 18 K and a gas-to-dust mass ratio (~170) that is similar to those found in Galactic molecular clouds. The luminosity of the warm dust component makes up ~70% of the total luminosity between 40 and 1000 μm, despite the fact that it constitutes only 5% of the mass. Deriving SFRs from either the warm luminosity or the total luminosity are therefore very similar and both are consistent with the SFR calculated using the thermal bremsstrahlung component in Fig III (1-2 $M_{\odot}$ yr^{-1}).

SUMMARY

We have detected extended emission at 800 μm in the nuclear region of the galaxy NGC 3079. Comparison of the 800 μm flux distribution with a CO map implies that the molecular gas-to-dust mass ratio is constant across the nuclear region of the galaxy on ~1.6 kpc scale. Assuming a molecular gas-to- dust mass ratio of 100 and β=2, the temperature of the dust is quite cool (~20 K). We present the continuum spectrum of the nuclear region of NGC 3079 from 12 μm to 20 cm. The submm/FIR portion is best fit with two dust components (T=35 K and 18 K), which provides a gas-to-dust mass ratio (~170) as found in Galactic molecular clouds. The warm component is ~5% of the dust mass and ~70% of submm/FIR luminosity and dominates the IRAS bands, while the massive cool dust component dominates the submm emission. The SFR determined from the FIR emission is consistent with the SFR calculated from the thermal radio component (a few $M_{\odot}$ yr^{-1}).

ACKNOWLEDGMENTS

We are grateful to Drs. Sofue and Nakai for the use of their CO image prior to publication and to Shawn Gordon for his help with the 3 mm observations.

REFERENCES

Devereux, N. A. and Young, J. S. 1990, *ApJ*, **350**, L25

Duncan, W. D. et al. 1990, *MNRAS*, **243**, 126

Eales, S. A., Wynn-Williams, C. G., and Duncan, W. D. 1989, *ApJ*, **339**, 859

Hildebrand, R. H. 1983, *QJRAS*, **24**, 267

The First Symposium on the Infrared Cirrus and Diffuse Interstellar Clouds
ASP Conference Series, Vol. 58, 1994
Roc M. Cutri and William B. Latter (eds.)

A NEW FIR STUDY OF M31 WITH HIRES MAPS: EVIDENCE FOR DEFICIENCY OF VERY SMALL GRAINS?

CONG XU
Max-Planck-Institut für Kernphysik, W6900 Heidelberg, Germany

GEORGE HELOU
IPAC 100-22, California Institute of Technology, Pasadena, CA 91125

ABSTRACT This is a report on first results of a study on the FIR emission of M31 using new high resolution (HiRes) IRAS maps ($\sim 1'$). Significant differences are found in the IRAS color-color diagrams of small cells ($2' \times 2'$) within the disk of M31 compared to Galactic cirrus, in the sense of a deficiency in the 25 and 60 μm emission. We interpret the results as evidence for deficiency of very small grains (but not Polycyclic Aromatic Hydrocarbons) in M31.

INTRODUCTION

M31, being the nearest spiral galaxy outside the Milky Way, is one of a few galaxies resolved by IRAS. The original IRAS maps of M31, taken both in the Additional Observations program (AO) and in the all-sky Sky Survey (Sky Flux), were throughly studied by Walterbos and Schwering (1987). Earlier studies of IRAS maps of M31 can be found in Habing et al. (1984) for the disk, and Soifer et al. (1986) for the central bulge. These pioneering works have provided valuable knowledge on the general FIR morphology of spiral galaxies. We are carrying out a new study on the FIR emission of M31 using the high resolution (HiRes) IRAS maps ($\sim 1'$) recently produced at IPAC. Our major motivation is to study the sources (HII regions) and the diffuse emission in M31 separately. This is not possible in the previous studies because of the coarse resolutions of the original IRAS maps. Here we report the first results from this project, namely the IRAS color-color diagrams of a complete sample of small areas ('cells' of $2' \times 2'$) within the disk of M31.

IRAS COLOR-COLOR DIAGRAMS

The resolution achieved in the new HiRes maps of M31 is $\sim 0.5' \times 0.9'$ (in-scan and cross-scan half-power diameters respectively) for the 12 and 25 μm maps, $\sim 0.8' \times 1'$ for the 60 μm map, and $\sim 1.5' \times 1.5'$ for the 100 μm map. In order to simplify the comparison between maps, we smooth all of them to a $1.7'$ circular beam on a grid with $0.5'$ pixels. We then select a sample of small areas (cells), each of size $2' \times 2'$, within the M31 field. The surface brightness at wavelength λ (λ =12, 25, 60, and 100 μm), $B_\nu(\lambda)$ in Jy sr^{-1}, of each cell is calculated from the

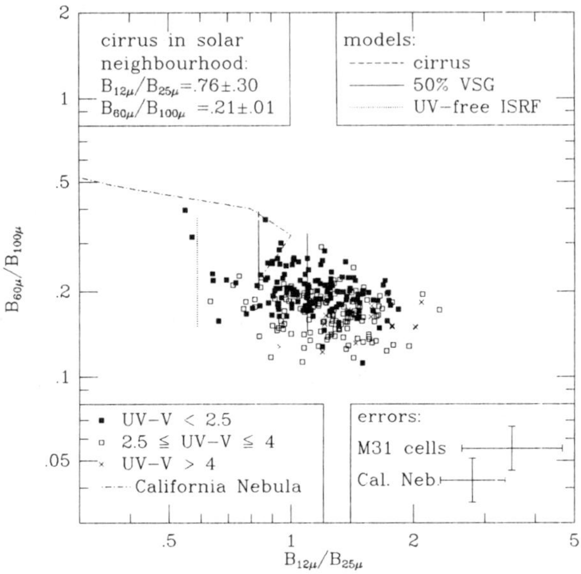

FIGURE 1 $R(60,100)$ vs $R(12,25)$ diagram for cells $(2' \times 2')$ in M31.

corresponding smoothed map by averaging the surface brightness of a 4×4 array of adjacent pixels. We have included only those pixels which are within $80'$ from the center in the plane of M31 ($r \leq 16$ kpc for the assumed distance of 690 kpc). Furthermore we only consider cells with surface brightness sufficiently high that each of the IRAS color ratios plotted in the following figures is significant at least at the 3σ level.

Figure 1 shows the color-color diagram $R(60,100)$ vs $R(12,25)$ for the M31 cells, with various symbols indicating different UV–V colors as explained in the legend. The dashed vertical line shows the colors calculated by Désert et al. (1990, hereafter DBP90) for Galactic cirrus using the three–population dust model with Polycyclic Aromatic Hydrocarbons (PAHs), very small 3–dimensional grains (VSG), and large dust grains. The color variation along the line reflects a range in the intensity of the heating radiation from 0.05 to 10 times the ISRF in the Solar Neighbourhood. The line is vertical because in the DBP90 model the spectral shapes of PAH and VSG emission and their relative intensity are almost independent of the intensity of heating radiation. There is a clear trend in this diagram that cells with bluer UV – V colors (the solid squares) show warmer $R(60,100)$ *and* cooler $R(12,25)$ ratios. This is consistent with the two-component model first suggested by Helou (1986), who interpreted the anti-correlation between $R(60,100)$ and $R(12,25)$ of *galaxies* as resulting from the superposition of two components of FIR emission from interstellar dust: a warm component with high $R(60,100)$ and low $R(12,25)$, and a cool component with low $R(60,100)$ and high $R(12,25)$. The warm component is in general associated with massive star formation regions, while the cool component is associated

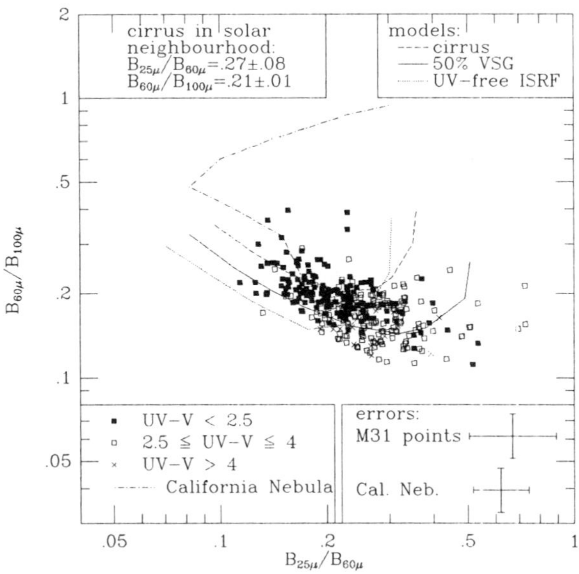

FIGURE 2 $R(60,100)$ vs $R(25,60)$ diagram for cells in M31 disk.

with "cirrus" or with quiescent molecular clouds heated by the interstellar radiation field (ISRF). In the case of M31 cells, bluer UV − V colors indicate a younger stellar population, i.e., a more important contribution from the warm component.

On the other hand, most of the emission in M31 disk has cooler $R(60,100)$ and warmer $R(12,25)$ ratios than the cirrus in the Solar Neighbourhood ($R(60,100) = 0.21 \pm 0.01$ and $R(12,25) = 0.76 \pm 0.30$, Boulanger and Pérault 1988). This cannot be due to the difference in the intensity of ISRF because the entire trajectory of the cirrus model (the dashed line) which spans a very wide range of ISRF intensity (0.05 — 10 times of the local ISFR), lies near the upper-left border of the data domain.

The problem is presented even better in the $R(60,100)$ vs $R(25,60)$ diagram which is plotted in Figure 2. The points are coded by the UV − V color as in Figure 1. Clearly, this color-color diagram allows better discrimination at lower heating intensities than Figure 1, with the points separating out in UV − Vcolor, and spreading out in a weak anticorrelation $R(60,100)$ *vs* $R(25,60)$. The data are clearly inconsistent with the model prediction for Galactic cirrus.

INTERPRETATION

The diffuse emission from the disk of M31 displays IRAS colors which are different from Galactic cirrus. Scaled to 100 μm, the dust emission is depressed at 25 μm and at 60 μm. The ratio $R(12,100) = B_\nu(12\ \mu m)/B_\nu(100\ \mu m)$ in M31 is indistinguishable from its value in cirrus.

One possible reason for this difference between Galactic cirrus and M31 dust is that the UV radiation field is weaker in M31 than in the Galaxy. Weaker UV could result in a smaller contribution from small grains at 60 μm, therefore reducing $R(60,100)$. However, weak UV heating will also reduce $R(12,25)$ and $R(25,60)$ because VSG and PAH will be 'cooler' (smaller amplitude of the temperature fluctuations), in conflict with the data. In Figures 1 and 2, this interpretation fails primarily because its prediction misses most of the data points in Figure 1 [too low $R(12,25)$], and it does not generate sufficiently elevated values of $R(25,60)$ in Figure 2.

Another possibility is that dust in M31 is deficient in "very small grains." Since 25 μm is most closely associated with the VSG population, we test in Figures 1 and 2 (solid lines) the prediction by the DBP90 model modified by reducing the VSG abundance to half its value in Galactic cirrus. This variation on the model provides the *best fit* to the data. $R(60,100)$ is reduced because of a smaller contribution from VSG to $B_\nu(60\ \mu\text{m})$, whereas $R(12,25)$ is enhanced because of the increased abundance of PAHs relative to VSG, in agreement with the data in Figure 1. As the intensity of the heating radiation drops below the Solar Neighbourhood value, large grains cool down, so their reduced contribution at 60 μm causes the increased $R(25,60)$ values, thus aligning model predictions and data in Figure 2. At the lowest heating levels the large grains are so cold that their emissivity at 100 μm drops enough to cause the up-turn in the solid line on Figure 2.

The VSG deficiency in M31 would remain a model-dependent result if based solely on the IRAS data, since no direct VSG signature is involved, and DBP90 may have oversimplified PAH and VSG properties, especially at low heating levels. However, an independent argument in favor of the VSG deficiency would be provided by a weaker 2175 Å bump, in line with the suggestion (Mathis 1990) that VSG are responsible for that bump. Recent HST data do indeed reveal narrower and weaker 2175 Å bumps in the extinction spectra derived for two stars at opposite sides of M31 located about 8 and 11 kpc from the nucleus (Hutchings *et al.* 1992). We therefore conclude that there is significant evidence for a deficiency of VSG in M31 compared to the Galaxy.

REFERENCES

Boulanger, F., and Pérault, M., 1988, *ApJ*, **330**, 964

Désert, F.A., Boulanger, F., and Puget, J.L., 1990, *A&A*, **273**, 215 (DBP90)

Habing, H.J., *et al.*, 1984, *ApJ*, **278**, L59

Helou, G., 1986, *ApJ*, **311**, L33

Hutchings, J.B. *et al.*, 1992, *ApJ*, **400**, L35

Mathis, J.S., 1990, *ARA&A*, **28**, 37

Soifer, B.T., *et al.*, 1986, *ApJ*, **304**, 651

Walterbos, R.A.M., and Schwering, P.B.W. 1987, *A&A*, **180**, 27

The First Symposium on the Infrared Cirrus and Diffuse Interstellar Clouds
ASP Conference Series, Vol. 58, 1994
Roc M. Cutri and William B. Latter (eds.)

^{13}CO EMISSION FROM MOLECULAR CLOUDS IN M33

C. E. WALKER[1]
McDonald Observatory, University of Texas, Austin, TX 78712, USA

C. D. WILSON[1]
Dept of Phys/Astron, McMaster U, Hamilton, Ontario L8S 4M1, Canada

ABSTRACT Due to the proximity of the nearly face-on spiral galaxy, M33, individual giant molecular clouds (GMC's) can be resolved. With single-dish data from the NRAO 12 m telescope and interferometric data from OVRO, we have obtained CO data to determine: a) the variation, if any, in the ^{12}CO/^{13}CO intensity ratio (R) toward 7 GMC's in M33; b) whether R is a sensitive function of metallicity; c) what the lower limit is to both R in diffuse gas and the fraction of CO emission from diffuse gas. Comparisons are made to individual GMC's in the Milky Way.

INTRODUCTION

There has recently been a renewed interest in using observations of the isotopomers of CO to study the physical conditions in the molecular interstellar medium in galaxies. The ^{12}CO/^{13}CO line ratios observed in normal galaxies are about 10 (Encrenaz et al. 1979; Rickard & Blitz 1985; Young & Sanders 1986; Sage & Isbell 1991), a factor of 2-3 larger than what is observed in molecular clouds in the Milky Way (Solomon, Scoville, & Sanders 1979). Observations of starburst galaxies (cf., Casoli, Dupraz, & Combes 1992) show even larger line ratios, in the range of 20-40. Since observations of other galaxies average the emission over large areas of the disk, explanations for the observed intensity ratio include different beam filling factors in the two lines (Encrenaz et al. 1979), different intrinsic ratios for CO emission from diffuse gas and from giant molecular clouds (Polk et al. 1988; Young & Sanders 1986), or a change in the ^{12}CO/^{13}CO abundance ratio (Casoli, Dupraz, & Combes 1992). Since the ^{13}CO line is more likely to be optically thin than the ^{12}CO line, variations in the line ratios among galaxies may have important implications for the use of CO emission as a tracer of molecular hydrogen gas.

M33 is close enough (0.84 Mpc) that millimeter-wave interferometers can resolve individual molecular clouds. We can combine inteferometric and single dish observations of the ^{12}CO and ^{13}CO emission to distinguish between the different explanations offered for the high line ratios seen in more distant galaxies. We can also place limits on the relative importance of emission from diffuse molecular gas in M33, as well as constraining the ^{12}CO to ^{13}CO ratio in this diffuse gas.

---

[1]Visiting Astronomer, National Radio Astronomy Observatory 12 m telescope which is operated by Associated Universities Inc., under cooperative agreement with the National Science Foundation

OBSERVATIONS

Seven positions in the disk of the Local Group spiral galaxy M33 have been observed with the NRAO 12 m telescope in the ^{12}CO and ^{13}CO J=1–0 lines. The telescope was centered on previously identified giant molecular clouds (Wilson & Scoville 1990, 1992). One of these positions has also been mapped with the OVRO interferometer in the ^{13}CO J=1–0 line.

- ^{13}CO emission is detected at all seven positions at a level of $> 3\sigma$. The average integrated intensity ratio $R = S(^{12}CO)/\, S(^{13}CO)$ is 9.9±0.7. Integrated intensity ratios for the individual positions are given in Table 1.

- The giant molecular cloud MC20 is detected in ^{13}CO at the 6σ level in the interferometer map. By combining these data with earlier ^{12}CO data (Wilson & Scoville 1990), the integrated intensity ratio for this individual cloud is 7.0 ± 2.0, where the uncertainty comes from the calibration uncertainty in the two lines. The ratio of the peak brightness temperatures in the two lines is 5.8 ± 1.6 (Table 2).

- MC20 appears very similar in both the ^{12}CO and ^{13}CO lines (Table 2; Figures 3-5). In particular, the diameter and velocity width agree very well.

- The lower limit to the amount of ^{12}CO emission from diffuse gas in the field around MC20 is $F = S(diffuse)/S(GMCs) > 0.3$. Adopting the upper limit $F = 1.0 \pm .2$ from the NRAO and OVRO ^{12}CO fluxes, we obtain a lower limit for the integrated intensity ratio in the diffuse gas of $R > 17 \pm 13$.

- The value of R obtained for M33 from the NRAO data is in good agreement with values measured previously in other normal galaxies (Encrenaz et al. 1979; Young & Sanders 1986; Rickard & Blitz 1985; Sage & Isbell 1991), and is roughly a factor of 2–3 larger than the value obtained for Galactic giant molecular clouds (Gordon & Burton 1976; Solomon, Scoville, & Sanders 1979; Polk et al. 1988; Lee, Snell, & Dickman 1990). Although the single dish and interferometer values of the ratio agree within the errors, the interferometer value is somewhat smaller and more consistent with the value for Galactic GMCs. The slightly lower ratio obtained for the peak brightness temperatures is also consistent with Galactic observations, where lower ratios are obtained for the peak than for the average emission of GMCs.

- The much larger value of the ratio obtained for the diffuse gas, although again consistent with the previous two measurements within the errors, is similar to the value of 30±13 obtained for diffuse clouds in the Galaxy (Knapp & Bowers 1988). Interferometric observations in ^{13}CO of a few additional molecular clouds in M33 would reduce the uncertainty in the ratios to the point where any differences in the ratio should be statistically significant.

DISCUSSION

Estimating the Importance of Diffuse Gas in M33

Polk et al. (1988) used ^{12}CO and ^{13}CO observations of the Galactic plane to argue that diffuse molecular clouds contribute a significant fraction of the ^{12}CO emission of the Galaxy and, furthermore, that these clouds have a significantly larger value of the ratio, $R = S(^{12}CO)/S(^{13}CO)$. Following their arguments, we can use our ^{12}CO and ^{13}CO observations to set a lower limit on both the

TABLE 1 ^{12}CO and ^{13}CO NRAO 12m Observations of M33 Centered on Molecular Clouds

Clouds ($> 10^5 M_\odot$)	α(1950) ($^{h\ m\ s}$)	δ(1950) ($^{\circ\ \prime\ \prime\prime}$)	$S(^{12}CO)$ (K km s^{-1})	$S(^{13}CO)$ (K km s^{-1})	$\frac{S(^{12}CO)}{S(^{13}CO)}$[a]	# of Chan.[b]
MC1-4	01 31 03.0	30 23 54	1.72±0.19	0.22±0.05	7.9±3.0	3
MC6,7,10	01 31 03.0	30 21 54	3.20±0.16	0.25±0.05	13.4±3.0	5
MC13	01 31 10.3	30 20 40	2.22±0.18	0.15±0.06	10.1±1.0	3
MC19	01 31 13.4	30 23 52	2.32±0.26	0.16±0.05	11.3±1.9	5
MC20	01 31 10.9	30 25 24	3.10±0.19	0.35±0.06	9.0±0.8	5
MC32	01 30 51.8	30 23 52	2.89±0.26	0.25±0.05	8.9±2.3	6
NGC 604[c]	01 31 44.5	30 31 15	4.43±0.23	0.35±0.06	8.5±2.6	8
-220 km s^{-1}	01 31 44.5	30 31 15	1.94±0.16	0.21±0.04	7.4±1.5	6
-240 km s^{-1}	01 31 44.5	30 31 15	2.49±0.16	0.14±0.04	11.9±2.8	2

[a]Calculated using only channels with signal-to-noise in ratio > 2.

[b]Number of 2.6 km s^{-1} channels used to calculate the ratio.

[c]Contains 3 molecular clouds; data for the two separate velocity components also shown.

TABLE 2 **Properties of the Molecular Cloud MC20**

Property	^{12}CO	^{13}CO
Deconvolved Diameter[a] (FWHM) (pc)	31x36	31x41
Peak Temperature (K)	3.68	0.64
Integrated Intensity[b] (Jy km s^{-1})	36.4	5.2
Velocity Width (FWHM) (km s^{-1})	9.2	8.7
Virial Mass[c] ($M_\odot$)	3.9×10^5	3.8×10^5
Molecular Mass[d] ($M_\odot$)	4.1×10^5	1.7×10^5

A distance to M33 of 0.84 Mpc is assumed throughout.

[a]Beam size is 32x25 pc.

[b]Integrated over 15 km s^{-1}; the ^{12}CO emission has broader faint wings.

[c]$M_{vir} = 99 \Delta V^2_{FWHM} \overline{D}$, where $\overline{D} = 0.7 \times \overline{D_{FWHM}}$; assumes a $1/r$ density profile.

[d]For ^{12}CO, assumes $\alpha = 3 \times 10^{20}$ cm^{-2} (K km s^{-1})$^{-1}$; for ^{13}CO, assumes optically thin emission, $[^{13}CO]/[H_2] = 10^{-6}$, and $T_{ex} = 10$ K.

amount of CO emission from diffuse gas in M33 and the value of R in the diffuse clouds in M33.

Let $R_{GMC} = 7.0 \pm 2.0$ be the intrinsic value of R in giant molecular clouds (adopting the value determined for MC20), $R_{obs} = 9.9 \pm 0.7$ be the average value of R observed with the NRAO 12 m telescope (where the uncertainty is the uncertainty in the mean), and F be the ratio of the ^{12}CO flux from the diffuse gas to the ^{12}CO flux from giant molecular clouds. The intrinsic value of R in the diffuse gas, R_{diff}, is then given by

$$R_{diff} = \frac{F R_{obs} R_{GMC}}{(1+F)R_{GMC} - R_{obs}}$$

The constraint that R_{diff} be positive yields the lower limit $F > R_{obs}/R_{GMC} - 1 = 0.4 \pm 0.1$. On the other hand, assuming that all the flux missed by the interferometer is contained in diffuse gas (rather than small giant molecular clouds) yields an upper limit to F, which for the MC20 field is 1.0 ± 0.2. Using this value of F gives a lower limit $R_{diff} > 17 \pm 13$.

Thus the data suggest that at least 30% and at most 50% of the ^{12}CO emission from M33 is due to a diffuse gas component. However, the uncertainties in R_{GMC} and R_{obs} are sufficiently large that the two measurements still agree within the error bars. If the two values were the same, that result would imply either (1) there is not a significant amount of diffuse molecular gas in this region of M33, or (2) the value of R is the same in both diffuse and giant molecular clouds. In this case, using the CO-to-H_2 conversion factor determined in individual GMCs to determine the total gas mass from single dish data would definitely be valid. On the other hand, if there IS a significant difference between the ^{12}CO/^{13}CO ratios in GMCs and diffuse gas, this difference could lead to systematic errors in the estimation of global gas masses of galaxies. However, it would be important to understand physically why the ratios are so different before attempting to determine a correction to galactic global gas masses.

ACKNOWLEDGMENTS

C.E.W. wishes to acknowledge support from the W. M. Keck Foundation.

REFERENCES

Casoli, Dupraz, & Combes 1992, *A&A*, **264**, 55
Encrenaz *et al.* 1979, *A&A*, **78**, L5
Gordon & Burton 1976, *ApJ*, **208**, 346
Knapp & Bowers 1988, *ApJ*, **331**, 974
Lee, Snell, & Dickman 1990, *ApJ*, **335**, 536
Polk *et al.* 1988, *ApJ*, **332**, 432
Rickard & Blitz 1985, *ApJ*, **292**, L57
Sage & Isbell 1991, *A&A*, **247**, 320
Solomon, Scoville, & Sanders 1979, *ApJ*, **232**, L89
Wilson & Scoville 1990, *ApJ*, **363**, 435
Wilson & Scoville 1992, *ApJ*, **385**, 512
Young & Sanders 1986, *ApJ*, **302**, 680

The First Symposium on the Infrared Cirrus and Diffuse Interstellar Clouds
ASP Conference Series, Vol. 58, 1994
Roc M. Cutri and William B. Latter (eds.)

CIRRUS IN OTHER GALAXIES

GEORGE HELOU
Infrared Processing and Analysis Center, Caltech

In the absence of detailed spatial information, it is useful to define infrared cirrus in other galaxies as the emission from dust heated by non-ionizing stars. The advantages and shortcomings of this definition are discussed.

Several empirical tests – more or less directly related to that definition – could be used to estimate the contribution of cirrus to the total emission from a galaxy, or to identify galaxies dominated by cirrus emission. Some of the more convenient ones are reviewed:

1. the total infrared excess, or any estimate of the ratio of the heating flux to the ionizing flux;
2. the IRAS colors;
3. the infrared surface brightness, or a reasonable estimate thereof;
4. the ratio of infrared to blue luminosities;
5. the infrared emissivity per mass in the ISM;
6. the ratio of infrared to radio luminosities.

For galaxies in the Local Group it is possible to identify and study cirrus directly, by virtue of the spatial resolution achievable; examples of such recent studies are used to compare the properties of cirrus in M31 and M33 to its properties in the Milky Way.

A proper accounting of the contributions of cirrus is critical to a correct interpretation of the integrated emission from galaxies, and therefore critical to understanding the history of star formation over the lifetime of galaxies.

6

38
Scattering by Diffuse Clouds in the G
Searching for the Cosmic Backgrou
37
CO Emission from
Molecular Clouds in M33

CHAPTER V

DUST AND ORGANIC MATTER IN DIFFUSE INTERSTELLAR CLOUDS

The First Symposium on the Infrared Cirrus and Diffuse Interstellar Clouds
ASP Conference Series, Vol. 58, 1994
Roc M. Cutri and William B. Latter (eds.)

DUST IN DIFFUSE INTERSTELLAR CLOUDS

B. T. DRAINE
Princeton University Observatory, Princeton NJ 08544-1001, U.S.A.

ABSTRACT The properties of the dust in diffuse interstellar clouds are reviewed, based on evidence from interstellar depletions, extinction, polarization, scattering, and emission. Various contemporary grain models are summarized. Some difficult problems confronting grain models are discussed.

1. INTRODUCTION

Interstellar dust makes its presence known in many rather direct ways: by attenuating radiation reaching us from distant objects (interstellar extinction); by altering the spectra of distant objects (interstellar reddening), by linear dichroism of the interstellar medium (interstellar polarization of starlight); by scattering of light (reflection nebulae, diffuse galactic light, X-ray haloes); and by emission of radiation, mainly in the infrared and far-infrared. In addition there are a number of indirect clues to the presence of dust: the "depletion" of various elements from the gas phase, and the role which dust is inferred to play in the formation of H_2. The above interstellar evidence is complemented by direct meteoritic evidence, which provides tantalizing clues, as well as the direct observation of interstellar grains entering the solar system. These separate strands of observational evidence are discussed in §§2-4.

A number of different grain models are currently under consideration. These models are briefly described in §5; from their similarities one can infer what seems to be relatively secure, while from their differences the uncertainties are emphasized. A number of outstanding problems confronting our understanding of interstellar dust are outlined in §6, and §7 describes avenues for future progress.

2. OPTICAL EVIDENCE

2.1 Extinction

Most of what we know concerning interstellar dust comes from its interactions with photons. Observations of interstellar reddening give us information about the wavelength-dependent extinction (scattering + absorption) cross sections for dust grains.

Observations of the wavelength-dependent extinction, yielding "extinction curves", have been reviewed by many authors (e.g., Mathis 1990; Whittet 1992; Cardelli 1994). Cardelli, Clayton & Mathis (1989) recently discovered that observed extinction curves can be fitted remarkably well by a one-parameter family

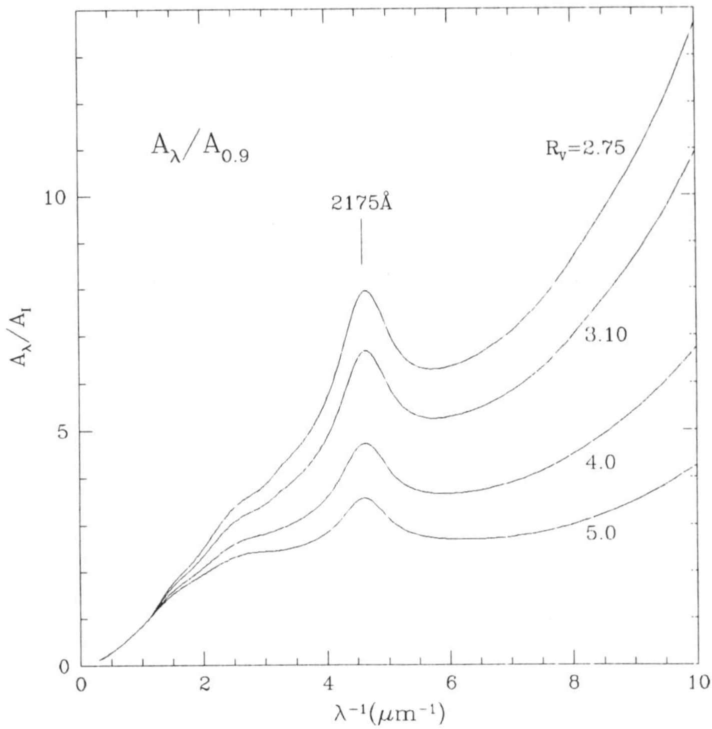

FIGURE 1 Representative extinction curves, obtained from the mean R_V-dependent extinction law found by Cardelli, Clayton & Mathis (1989). Extinction is shown relative to the extinction at $I = 0.9\mu$m. Curves are labelled by $R_V \equiv A_V/E(B-V)$

of functions. The parameter may be taken to be the so-called ratio of total to selective extinction: $R_V \equiv A_V/(A_B - A_V)$, where A_B and A_V are the extinction in the B and V bands. Observed values of R_V extend from as small as 2.75 to greater than 5; $R_V \approx 3.1$ is typical in diffuse clouds. Some representative extinction curves, normalized to the I band extinction, are shown in Figure 1. While the one-parameter family does a good job of fitting observed extinction curves, systematic deviations do exist (Mathis & Cardelli 1992). The rapid rise of the extinction at the shortest wavelengths requires that a substantial part of the extinction be contributed by particles with radii $a \lesssim 0.02$Å.

It is remarkable (see §6.5) that the extinction longward of I appears to have a universal wavelength-dependence!

2.2 Spectral Features in Absorption

Table 1 contains a list of the stronger spectral features associated with grains in diffuse clouds. The 2175Å feature – conspicuous in Fig. 1 – is by far the strongest; identification of the carrier remains uncertain (see §6.4) although graphite is a plausible candidate.

There are a number of additional spectral features, known for decades, which still resist identification: the so-called "diffuse interstellar bands" (DIBs) [see Krelowski (1989, 1994) for more detail]. There are now approximately 200 DIBs known (Jenniksens 1992; Jenniksens & Désert 1993). The strongest DIB (at 4430Å) is listed in Table 1. The FWHM of the 4430Å band is 20Å, but some of the weaker bands are considerably narrower (e.g., $\lambda\lambda$6270Å, with FWHM = 1.4Å). The DIB with the longest wavelength (1.32μm) is also listed in Table 1.

Moving further into the infrared, we encounter a strong absorption feature at 3.4μm which occurs at the frequency expected for C-H stretching modes in aliphatic hydrocarbons. Unfortunately, it is not possible to identify which hydrocarbon; it is therefore difficult to estimate how much material is required to account for this feature, but it probably incorporates from 2 – 50% of the cosmic carbon abundance.

Infrared studies of diffuse cloud extinction show another very strong feature peaking near 9.7μm; this is generally attributed to a Si–O stretching mode in some form of silicate material, perhaps amorphous silicate with the composition of olivine ($Mg_xFe_{2-x}SiO_4$); there is another broad feature at 18μm which is attributed to the O–Si–O bending mode. The strength of the feature implies that silicate material incorporates a large fraction ($\gtrsim$ 50%) of the interstellar silicon (Draine & Lee 1984).

Silicon carbide (SiC) grains would be expected to produce an absorption feature near 11.2μm (an emission feature at this wavelength is observed in the spectra of some carbon stars). Upper limits on the strength of an interstellar absorption feature at 11.2μm imply that $\lesssim$ 5% of the interstellar silicon is in SiC (Whittet, Duley & Martin 1991).

It is somewhat embarassing that we have been provided with so much spectral information and yet have thus far failed to definitively identify the strongest feature (the 2175Å bump) or any of the hundreds of diffuse bands.

2.3 Polarization

In addition to being reddened, starlight which has passed through dusty interstellar clouds is generally found to be linearly polarized, with the degree of polarization peaking at a wavelength $\lambda_p \approx 5500$Å. The direction of polarization is believed to coincide with the component of the interstellar magnetic field perpendicular to the line of sight. This linear dichroism of the interstellar medium requires the presence of nonspherical dust grains, aligned with the interstellar magnetic field. Attempts to model the linear dichroism produced by nonspherical grains of effective radius $a_{\rm eff}$ conclude that it peaks at $\lambda_p \approx 2\pi a_{\rm eff}$; therefore the aligned nonspherical grains must typically have $a_{\rm eff} \approx 0.1\mu$m (if smaller grains were aligned λ_p would move toward shorter wavelengths). Therefore extinction and polarization observations together require a range of particle sizes from $a \lesssim 0.02\mu$m to $a \gtrsim 0.1\mu$m. Polarization studies in the 10μm region show strong polarization in the 9.7μm feature, so that at least part of the silicate material is in aligned grains. Attempts to measure polarization in the 2175Å feature, however, find at most weak polarization associated with the feature (Clayton *et al.* 1992), so the grains responsible for the 2175Å bump are either nearly spherical or at most weakly aligned.

TABLE 1 **Solid-State Absorption Features in Diffuse Clouds**[1]

λ (μm)	$\frac{\Delta\tau_\lambda}{N_H}$ (10^{-23}cm^2)	$\frac{\int \Delta\tau d\tilde{\nu}}{N_H}$ (10^{-20}cm)	X	σ_λ (10^{-19}cm^2)	$\int \sigma d\tilde{\nu}$ (10^{-16}cm)	$\frac{N(X)}{N_H}$ (10^{-5})	Notes
0.22	53.	820.	graphite?	100.	1600.	5.3	2,3
0.443	1.6	1.8	?	?	$8850 f_{.443}$	$2\times10^{-4} f_{.443}^{-1}$	4
1.32	0.18	0.004	?	?	$8850 f_{1.32}$	$5\times10^{-6} f_{1.32}^{-1}$	5
3.4	0.23	0.023	C in HAC?	0.11?	.009?	26?	6,7,8
"	"	"	C in hydroc.?	?	0.1?	2?	9
9.7	2.5	0.62	silicates	6.	1.6	4	10,11,12
19.	1.0	0.2	silicates	2.	0.5	4	12,13

---

[1] For mean extinction law with $N_H/E(B-V) = 5.8 \times 10^{21}$ cm^{-2} (Bohlin, Savage & Drake 1978), $E(B-V)/E(J-K) = 1.91$ (Rieke & Lebofsky 1985), $A_{9.7} = 0.35E(J-K)$ (Roche & Aitken 1984; Draine 1989*b*), $\Delta A_{9.7} \approx 0.30E(J-K)$ (cf. Draine 1989*b*). Here $\tilde{\nu} \equiv \lambda^{-1}$, and σ is the absorption cross section per molecule X.

[2] $\Delta\tau/N_H$ is based on mean Drude fit parameters of Fitzpatrick & Massa (1986).

[3] σ_λ is for 2:1 oblate spheroids with $a_{eff} < 0.01$ μm (Draine 1989*a*), and assumed density $\rho = 2$ g cm^{-3}.

[4] From $W_{.443}/E(B-V) = 2.04$ Å for 25 stars, and $\Delta\tau/W = .051$/Å toward HD183143 (Herbig 1975). Here $f_{.443}$ is the (unknown) oscillator strength for the 0.443 μm band.

[5] From $W_{1.32}/E(B-V) = 0.42$ Å and FWHM=4.0 Å (Joblin *et al.* 1990). Here $f_{1.32}$ is the (unknown) oscillator strength for the 1.32 μm band.

[6] $\Delta\tau/N_H$ is estimated from $\Delta\tau_{3.4}$ toward Galactic Center (McFadzean *et al.* 1989) and VI Cyg 12 (Adamson, Whittet & Duley 1990), with N_H estimated from $\Delta\tau_{9.7} \approx 3.6$ toward GC (Roche & Aitken 1985) and 0.6 toward VI Cyg 12 (Roche & Aitken 1984).

[7] $\int \Delta\tau d\tilde{\nu}/\Delta\tau_{3.4} = 100$ cm^{-1} from GC-IRS7 (Butchart *et al.* 1986; Adamson, Whittet & Duley 1990).

[8] σ_λ and $\int \sigma d\tilde{\nu}$ for hydrogenated amorphous carbon with H/C≈0.5 (Dischler, Bubenzer & Koidl 1983).

[9] $\int \sigma d\tilde{\nu} = A\, 3.82 \times 10^{-17}$ cm, where $A \approx 0.1 - 0.5$ per C in saturated hydrocarbons (Wexler 1967); we take $A \approx 0.25$.

[10] $\Delta A_{9.7}/E(J-K) \approx 0.30$ (see note 1). Note, however, that Cohen, Tielens & Bregman (1989) argue for a significantly smaller value.

[11] $\int \Delta\tau d\tilde{\nu}/\Delta\tau_{9.7} \approx 250$ cm^{-1} based on μ Cep emissivity of Roche & Aitken (1984).

[12] σ_λ and $\int \sigma d\tilde{\nu}$ are estimates based on assumption that features are produced by silicates with approximately cosmic silicon abundance $\sim 4 \times 10^{-5}$.

[13] $A_{19}/A_{9.7} \approx 0.4$ (cf. Draine 1989*b*). The 19 μm feature is broad and blended with long-wavelength tail of 8.7 μm feature; choice of baseline is therefore difficult. We take $\Delta A_{19} \approx A_{19}$. $\int \tau d\tilde{\nu}$ is integrated from 15-23 μm.

2.4 Scattering

Scattering of light by interstellar dust is evident from the existence of reflection nebulae. The diffuse galactic light shows that the entire galaxy is a reflection nebula. Attempts have been made to infer the albedo and scattering phase function of interstellar dust from reflection nebulae; unfortunately, these studies are difficult since they generally require assumptions concerning the geometry (see the review by Witt 1989). Recent work has been directed toward trying to determine the scattering properties of dust in the extreme ultraviolet (e.g., Witt *et al.* 1993).

An entirely different type of reflection phenomenon is found in the X-ray region: here the scattering by the grains is predominantly in the extreme-forward-scattering direction. This results in "halos" around X-ray point sources, observed with *Einstein* by Catura (1983) and Mauche & Gorenstein (1986), and with *ROSAT* (Predehl *et al.* 1991). The observed scattering is consistent with the Mathis, Rumpl, & Nordsieck (1977, hereafter MRN) model of interstellar dust, but is not a sensitive test of different grain models (Mathis & Lee 1991).

2.5 Emission

Interstellar dust is a powerful source of far-infrared emission, as was dramatically revealed by *IRAS* (Low *et al.* 1984) and now *COBE* (Wright *et al.* 1991). The *COBE FIRAS* results show that the interstellar dust emission L_ν peaks at $\sim$ 140μm. If the dust is approximated as being at a single temperature T_d then $L_\nu \propto Q_\nu B_\nu(T_d)$, where $Q_\nu \pi a_{\rm eff}^2$ is the absorption cross section at frequency ν of a dust grain of volume $(4\pi/3)a_{\rm eff}^3$. Wright *et al.* showed that the $1\,{\rm cm} < \lambda <$ 110μm spectrum could be approximated by $Q_\nu \propto \nu^{1.65}$ and $T_d = 23.3\,{\rm K}$; since there is undoubtedly a spread in grain temperatures, this gives a lower bound on the emissivity slope, $d\ln Q_\nu/d\ln\nu > 1.65$. The observed spectrum includes emission from star-forming regions and mass-losing stars – dust which is hotter than diffuse cloud dust. It is probably correct to suppose that the emission from diffuse interstellar clouds has L_ν peaking at $\sim 160 \pm 20\mu$m. If $Q_\nu \propto \nu^{1.9\pm0.1}$, then one can show that $Q_\nu B_\nu$ peaks at $h\nu/kT_d = 4.9 \pm 0.1$, so the characteristic dust temperature $T_d = 18 \pm 2$K. This is a reasonable estimate for the "typical" temperatures of dust grains in diffuse clouds in the solar neighborhood.

Although the emission from dust at $300 > \lambda > 110\mu$m can be satisfactorily represented as $\propto \nu^{1.9} B_\nu(18{\rm K})$, this functional form significantly underestimates the emission at wavelengths $\lambda \lesssim 60\mu$m. Figure 2 shows the emission measured by *IRAS* (Boulanger & Perault 1988) and *FIRAS* (Wright *et al.* 1991).

It is clear that grains at $T \lesssim 20$ K fall short of being able to reproduce the emission observed by *IRAS* in the 60μm band, and completely fail to account for the emission at 25μm and 12μm.

Interstellar grains are heated mainly by absorption of photons. The thermal energy content of a very small grain at $T \approx 18$K may be small compared to the heat deposited when a starlight photon is absorbed, in which case the grain temperature will be a strong function of time. If absorption of a single photon raises the grain temperature to substantially above 20K, then the resulting infrared emission at wavelengths $\lambda < 100\mu$m will be substantially stronger (relative to the emission at longer wavelengths); this appears to be the explanation for the emission observed in the *IRAS* 60, 25, and 12μm bands (Draine & Anderson

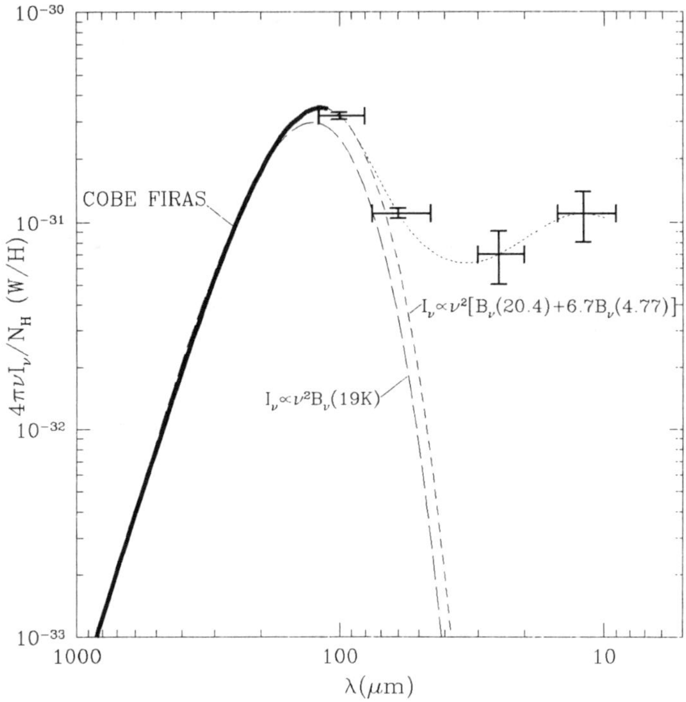

FIGURE 2 Emission spectrum of interstellar dust. Data for *IRAS* 100, 60, 25, and μm emission per H nucleus is from Boulanger & Perrault (1988). *COBE FIRAS* spectrum for $\lambda > 110\mu$m is from Wright *et al.* (1991), normalized to join to the 100μm *IRAS* flux. The dashed lines show two extrapolations to shorter wavelength: one based on a ν^2 emissivity and two temperatures (Wright *et al.* 1991), the other showing the emission which would result if dust with a ν^2 emissivity and $T_d = 19\,$K accounted for 100% of the emission at 200μm. The dotted line is a smooth curve representing the combined COBE and *IRAS* data; the total power at $\lambda > 10\mu$m is 5.2×10^{-31}W/H $= 1.6 \times 4\pi[\lambda I_\lambda/N_{\rm H}]_{100\mu m}$

1985).

The total emission at $\lambda > 10\mu$m is estimated from Figure 1 to be $\sim 5.2 \times 10^{-31}$W/H. There is additional emission at $\lambda < 10\mu$m which is estimated to be $\sim 0.6 \times 10^{-31}$W/H (see Désert *et al.* 1990), so that the total power per H is $\sim 5.8 \times 10^{-31}$W/H.

A very small fraction ($\sim 1\%$) of the emission by grains is in the near-IR and visible; this "extended red emission" (ERE) was observed in reflection nebulae (Witt & Schild 1988) and possibly in diffuse clouds (Guhathakurta & Tyson 1989).

2.6 Spectral Features in Emission

The 9.7μm silicate feature has been observed in emission (Gillett *et al.* 1975), but only where dust grains are heated to $T \gtrsim 200\,$K in the neighborhood of a luminous star. A number of other spectral features appear in emission in reflection nebulae, HII regions, and planetary nebulae (see the review by Sellgren 1994); the strongest occur at 3.3, 6.2, 7.7, 8.6 and 11.3μm. Léger & Puget (1984) tentatively identified these features as C-H and C-C stretching and bending modes in polycyclic aromatic hydrocarbons (PAHs), although a number of other identifications have been proposed (cf. Sellgren 1994). The 3.3μm emission has been observed by Giard *et al.* (1989); at $|b| \approx 4°$ they find that $I(3.3\mu\mathrm{m}) \approx .006(\lambda I_\lambda)_{100\mu\mathrm{m}}$, or approximately 0.4% of the total $\lambda > 10\mu$m emission.

Désert, Boulanger & Puget (1990) have estimated that $\sim$ 20% of the total power emitted by interstellar dust is radiated by PAH particles. Since the radiated energy must be supplied by absorbed photons, this implies that PAHs must account for $\sim$ 20% of the absorption by interstellar grains. In order to contribute this much opacity, the PAHs must contain $\sim$ 9% of the interstellar carbon.

3. INDIRECT EVIDENCE

3.1 Depletions

A strong clue to the composition of interstellar dust is provided by observations of underabundances (depletions) of various elements in the gas phase. The depletions are most pronounced for elements like Ca, Fe, and Si, which may have gas-phase abundances which are low by factors of 10^2 or 10^3 (Jenkins 1989; Meyer 1994). It is gratifying that such depletion is observed for elements like Ca, Fe, and Si which are expected to form into refractory silicates. The depletion of carbon has been uncertain, but now there are several measurements indicating that perhaps as much as 2/3 of the carbon may be missing from the gas (Hobbs, York, & Oegerle 1982; Cardelli *et al.* 1991; Cardelli *et al.* 1993).

3.2 Formation of H_2

An indirect, but important, constraint on interstellar dust is that the dust be able to catalyze the formation of H_2 with an effective rate coefficient $\sim 3 \times 10^{-17}\,\mathrm{cm^3\,s^{-1}}$ (Jura 1975). This implies that the dust grain collision cross section per H atom must be $\sim 5 \times 10^{-22}\alpha^{-1}\,\mathrm{cm^2/H}$, where $\alpha < 1$ is the fraction of colliding H atoms which are converted to H_2.

4. METEORITIC EVIDENCE

There is now unequivocal evidence that some interstellar grains have been preserved in meteorites [see recent reviews by Anders & Zinner (1993) and Ott (1993)]: the evidence is based on isotopic compositions which deviate so greatly from solar abundances as to preclude formation in the solar nebula. Table 2 lists the three principal meteoritic grain types which have been identified as of interstellar origin. All 3 are carbon-based, and exhibit quite extraordinary values of $^{12}C/^{13}C$! In addition, TiC has been found within some graphite (Bernatowicz *et al.* 1991) and SiC (Bernatowicz *et al.* 1992) grains.

A small number of Al_2O_3 grains with extraordinary inferred ^{26}Al content have been found (Huss *et al.* 1992; Nittler *et al.* 1993), but no interstellar silicate

TABLE 2 **Interstellar Grains found in Meteorites**[1]

composition	size (μm)	abundance in C2 chondrites (ppm)	$^{12}C/^{13}C$ rel. to solar[2]
Diamond	~ 0.001	400	1.04
SiC	0.03–10	7	0.02–28
Graphite	0.8–7	2	0.02–80

[1]From Anders & Zinner (1993)
[2]Solar $^{12}C/^{13}C$=89

grains have been identified in meteorites. Why not? The answer is simple: the presolar grains are rare, and the procedures used to isolate them are designed to chemically destroy the bulk of the meteoritic matrix, leaving behind only grains which are resistant to this chemical attack. Since the matrix is largely silicate, the isolation procedures will dissolve any presolar silicate grains which might be present.

The isolation procedures therefore leave us with a very interesting – but very nonrepresentative – sample of interstellar dust. The graphite and SiC grains – with their extraordinary $^{12}C/^{13}C$ ratios – probably formed in outflows from AGB carbon stars. The origin of the microdiamonds is less clear (Lewis, Anders & Draine 1989).

The local interstellar gas, with $n_H \approx 0.3\,\mathrm{cm}^{-3}$, is moving at $\sim 20\,\mathrm{km\,s}^{-1}$ relative to the Sun (Lallement *et al.* 1991). While interstellar ions undergo a collisionless shock at the heliopause, interstellar grains with sufficiently low charge-to-mass ratio enter the interplanetary region relatively undeflected. With heliocentric velocities $\sim 20\,\mathrm{km\,s}^{-1}$, techniques for nondestructive capture of such grains have not been devised, but a detector on the Ulysses spacecraft (Grün *et al.* 1993) has measured a mass flux $6 \times 10^{-21}\,\mathrm{g\,cm}^{-2}\,\mathrm{s}^{-1}$ in particles with $M > 4 \times 10^{-15}\,\mathrm{g}$ $[a > 0.07(3\,\mathrm{g\,cm}^{-3}/\rho)^{1/3}\mu\mathrm{m}]$. For a dust-to-hydrogen mass ratio of 0.01, the total grain flux would be expected to be $1 \times 10^{-20}\,\mathrm{g\,cm}^{-2}\,\mathrm{s}^{-1}$, and grain models (e.g., the MRN model) typically have $\gtrsim 50\%$ of the grain mass in grains with $M > 4 \times 10^{-15}\,\mathrm{g}$. The Ulysses result is therefore in excellent agreement with expectations.

5. GRAIN MODELS

In this section I summarize the features of a number of different grain models which are currently under consideration. Every grain model includes silicate material to explain the 9.7μm absorption/emission feature, but the models differ in other respects.

5.1 Bare Graphite + Bare Silicate (+ PAH)

In the course of trying to reproduce the observed interstellar extinction curve, MRN discovered that the observed optical-UV extinction could be quite well reproduced by a simple mixture of bare graphite and bare silicate grains, each with a power-law size distribution $dn/da \propto a^{-3.5}$, extending from $0.005-0.25\mu m$. In this model the 2175Å feature is attributed to graphite.

Draine & Lee (1984) estimated the dielectric functions of graphite and silicate from the far-infrared to vacuum ultraviolet, and found that the graphite-silicate model was consistent with observations in the infrared. They calculated that $0.1\mu m$ silicate and graphite grains would be heated by interstellar starlight to temperatures of 15 and 19K, respectively. The predicted far-infrared emission is in excellent agreement with the *COBE FIRAS* results.

Emission by interstellar dust at 60, 25, $12\mu m$, and in the IR emission features requires a grain component in addition to the $.005 - 0.25\mu m$ grains of the MRN model. The emission into the 60, 25, and $12\mu m$ *IRAS* bands could be due to graphite or silicate grains which are small enough to undergo "temperature fluctuations" in response to stochastic heating by starlight photons (Draine & Anderson 1985), although failure to observe the silicate feature in emission requires that the very smallest particles, undergoing temperature excursions up to $\gtrsim 250$ K, cannot be silicates (Désert, Boulanger, & Shore 1986). The emission in the various IR bands seems to require that the presence of PAH-like particles with as few as ~ 24 carbon atoms (e.g., coronene, $C_{24}H_{12}$).

Draine & Anderson (1985) have shown that the mass in ultrasmall grains must be considerably larger than what one would estimate by simply extending the MRN size distribution to smaller sizes. It is important to recognize that when the requisite numbers of ultrasmall grains and PAHs are added, the size distribution of the larger grains must now be changed from the MRN power law if the overall grain mixture is to be consistent with observations of interstellar extinction.

5.2 Coated silicate + bare graphite + PAH

Greenberg and co-workers (e.g., Greenberg 1989) have advocated grain models consisting of coated silicate grains plus small bare graphite grains (to account for the 2175Å feature). The coating on the silicate grains is imagined to be a carbonaceous residue produced by UV photolysis of ices which are deposited on the grains in molecular clouds.

This model has been extended by Désert, Boulanger, and Puget (1990) to include a PAH component, containing 9% of the cosmic carbon abundance. The graphite particles have radii $0.0012 - 0.015\mu m$. The coated silicate "big grains" have radii in the range $0.015 - 0.11\mu m$. The coating on the silicate grains is presumed to account for the $3.4\mu m$ absorption feature and the ERE.

5.3 Coated Silicate

Duley, Jones, & Williams (1989) have advocated a grain model with a single component: silicate grains with a coating of hydrogenated amorphous carbon (HAC). The particles are imagined to have a bimodal size distribution: a substantial mass in very small particles $a \leq 0.01\mu m$, plus big grains with an MRN power law distribution extending from $0.05 - 0.25\mu m$. Duley *et al.* do *not* resort to graphite – they attribute the 2175Å feature to OH^- on the surfaces of small silicate grains.

The ERE is attributed to the HAC coatings, which is supported by the fact that the observed spectrum resembles laboratory observations of fluorescence from HACs (Witt & Schild 1988).

There seem to be two possible problems with this model. The first is the attribution of the 2175Å bump to OH^-:silicate. As discussed by Draine (1989a), the strength of the 2175Å bump requires that a relatively large fraction of the O atoms must be in low-coordination sites *and* be hydrogenated *and* the 2175Å transition must have a relatively large oscillator strength. While not impossible, these requirements seem somewhat extreme. The claim that OH^-:silicate produces such an absorption feature is based not on direct measurement but rather on indirect arguments based on observed photoluminescence (Steel & Duley 1987); direct measurement of the absorption profile and its strength would be of great value.

The second problem has to do with the IR emission features. Duley *et al.* challenge the notion that some sort of PAH-like particles are required to explain the NIR emission features. Duley & Williams (1988) argue that HAC material actually consists of small domains ($\sim 10^2$ atoms) which can be individually heated by absorption of a UV photon, with the heat localized long enough ($\gtrsim 1$ s) for the domain to cool by IR emission rather than by heat diffusion. This proposal seems implausible: it requires HAC to have a thermal diffusivity *far* below that measured for laboratory materials. The hypothesis is easily tested in the laboratory: if Duley *et al.* are correct, then bulk samples of HAC material should efficiently convert UV illumination into $3 - 11\mu$m emission. I am not aware of any experiments showing efficient conversion: on the contrary, an experiment attempting to measure this for Escarpelle coal (Papoular, Reynaud & Nenner 1990) found an efficiency $\sim 10^{-7}$ for conversion of incident $3000 - 6000$Å radiation into emission in the 3.3μm feature. Such efficiencies, unless they increase by many orders of magnitude for radiation at shorter wavelengths, or are many orders of magnitude greater for HACs than for coal, are incapable of accounting for the observed interstellar 3.3μm emission, which accounts for $\sim 0.3\%$ of the overall emission. This appears to be a fatal objection to this model, or any other model which does not attribute the IR emission bands to ultrasmall particles.

5.4 Rowan-Robinson's Grain Model

Rowan-Robinson (1992) has put forward a grain model with 5 distinct components: (1) 0.0005μm graphite (or PAH); (2) $0.002 - 0.03\mu$m graphite; (3) 0.1μm amorphous carbon; (4) $0.01 - 0.1\mu$m silicate; and (5) 30μm amorphous carbon (containing 12% of the cosmic carbon abundance).

This model attributes the 2175Å bump to small graphite grains. The first four components are similar to other grain models; the really distinctive feature is the inclusion of 30μm amorphous carbon grains. These grains would be much cooler than the $\sim 0.1\mu$m grains. Since the giant grains would absorb only $\sim 3 \times 10^{-4}$ of the total absorbed starlight, they would contribute only this fraction of the infrared emission, but in the form of an "excess" at long wavelengths which would not be inconsistent the the *COBE* observations (Wright *et al.* 1991). Nevertheless, since they would play such a minor role in the energetics, the giant grains would, for most purposes, be relatively unimportant.

5.5 Composite Grains

Mathis & Whiffen (1989) have proposed that the typical interstellar grain consists of a coagulated structure built up from small ($\sim 0.005\mu$m) particles of silicate, graphite, and amorphous carbon. In addition, to explain the 2175Å feature, they include a component of isolated small graphite grains. Mathis (1986) has shown that conglomerate grains can account for the observed wavelength-dependence of interstellar polarization if (a) a small fraction of the subunits are superparamagnetic, and (b) a conglomerate grain is aligned by the magnetic field if and only if it contains at least one superparamagnetic subunit.

The fluffy grains in this model produce a far-infrared emissivity varying as $d\ln Q_\nu/d\ln\nu = 1.60$ in the $100 - 1000\mu$m region; as discussed in §2.5, this is not steep enough, since the *COBE* spectrum requires $d\ln Q_\nu/d\ln\nu > 1.65$, and probably closer to ~ 2. Furthermore, the grain temperatures calculated in this model are probably somewhat too low (because the FIR opacities are too large). The model does not explicitly deal with the IR emission features; they are attributed either to an additional PAH-like component or possibly to emission from amorphous carbon/HAC along the lines suggested by Duley & Williams (1988).

6. DIFFICULT ISSUES

6.1 Grain Survival and Interstellar Depletions

Interstellar grains (even the microdiamonds!) don't last forever: they are subject to destruction (return of the solid to the gas phase) by a number of processes, most importantly sputtering and grain-grain collisions in interstellar shock waves. McKee (1989) has recently reviewed grain destruction, and estimates a grain lifetime (averaged over all interstellar phases) $\sim 4 \times 10^8$ yr.

The observed large depletions of elements like Fe and Si require that the bulk of the solid material in grains actually condensed in the interstellar medium rather than in stellar outflows (Draine & Salpeter 1979; McKee 1989; Draine 1990). It is not easy to understand how such effective recondensation can take place; Draine (1990) argued that it required quite rapid exchange of matter between the dense molecular phases (where condensation is rapid) and the less dense phases (where the grain destruction primarily occurs). This inferred interchange rates between phases is more rapid than one might have estimated based on existing models of the interstellar medium. Draine (1990) argued that less dense phases are being incorporated into molecular clouds at a rate $\gtrsim 140 M_\odot\,\mathrm{yr}^{-1}$.

6.2 Grain "Purity"

As discussed above, observations of the 9.7μm feature suggest that it is due to silicate material, while many grain models attribute the 2175Å feature to relatively pure graphite. If most of the interstellar grain mass condensed in the cold interstellar medium, rather than in warm stellar outflows, then it is a challenge to understand how the condensing matter can be segregated into silicate and non-silicate material (e.g., graphite): why didn't the impinging metal atoms simply stick where they landed, leaving a mixture of C, Si, Fe, etc?

One can imagine ways in which segregation of condensates could occur. It might be that, once a part of the growing surface develops either "silicate" or "graphite" character, impinging "foreign" atoms (e.g., Si impinging on graphite) are removed (after a short residence) by either reaction with impinging O and H (e.g., graphite : Si+O $\rightarrow$ graphite+SiO) or photodesorption (e.g., graphite : Si+ $h\nu \rightarrow$ graphite + Si. Alternatively, perhaps prolonged UV irradiation leads to gradual separation into separate chemical phases.

6.3 PAH Formation

The PAH hypothesis requires that PAHs contain $\gtrsim 5\%$ of the interstellar carbon. It is not easy to understand how PAHs can be so ubiquitous and abundant, given their apparent vulnerability to destruction by sputtering in interstellar shock waves, heating in intense radiation fields, and chemical attack by atomic oxygen. Furthermore, coagulation of PAHs into larger entities is expected to greatly impair the efficiency of conversion of stellar photons into 3.3μm emission. Given these removal mechanisms, the required ubiquity of PAHs is difficult to understand. Injection of PAHs in carbon star outflows may take place, but cannot account for the average PAH abundance (Latter 1991). Therefore one wonders whether PAHs might be formed *in situ* by ion-molecule reactions (Herbst 1991). Alternatively, perhaps PAHs are injected into the gas from organic grain mantles as the result of grain-grain collisions or even grain mantle explosions as envisioned by Greenberg & Yencha (1973).

6.4 The 2175Å Feature

It is embarassing that we have been unable to convincingly identify the carrier of the very strong 2175Å feature. This identification is made even more difficult by the fact that observations (Fitzpatrick & Massa 1986) show that (1) the feature profile is very well reproduced by a Drude absorption profile; (2) the FWHM of the interstellar feature is observed to vary by as much as a factor 1.6 from one line-of-sight to another, although the wavelength λ_0 of the peak shows very little ($< 1\%$) variation; and (3) variations in λ_0 seem to be uncorrelated with variations in FWHM. Any identification must be able to account for this. None of the hypotheses seem able to do so in a natural way.

Graphite is often considered, but Draine & Malhotra (1993) have shown that changes in the size, shape, coating, or coagulation of the grains would be expected to produce correlated changes in λ_0 and FWHM. They conclude that if the feature is due to graphite, the variations in profile parameters must be due to changes in the dielectric function (due to changes in composition, or surface effects).

The possibility that the 2175Å feature is due to some mixture of PAHs has also been raised (Léger *et al.* 1989), but this seems to face serious problems: (1) It would be extremely fortuitous for a mixture of PAHs to produce a composite absorption profile so closely matching a Drude profile. (2) Furthermore, it would be even more fortuitous if variations in the mixture from one line of sight to another somehow cause changes in the profile width with hardly any effect on λ_0.

6.5 Extinction Curve Variations

Cardelli, Clayton & Mathis (1989) discovered that the interstellar extinction appears to be described remarkably well by a one-parameter family of curves. It is now incumbent on grain modellers to account for this variability within their models.

Cardelli *et al.* also found that A_λ/A_I appears to be a universal function for $3\mu m > \lambda > 0.9\mu m$, even as R_V varies. This result, confirmed by Martin & Whittet (1990), is remarkable: it suggests that variations in grain size do not affect the near-infrared extinction! Does this mean that scattering does not contribute significantly to the extinction at $\lambda > 0.9\mu m$? This would be somewhat surprising, since at least some grain models predict an albedo ~ 0.5 at $0.9\mu m$ (Draine & Lee 1984), so that changes in the size distribution should affect the wavelength-dependence of the extinction. Martin & Whittet argue that variations in extinction shortward of I are due to changes in the population of small grains, which do not contribute to $\lambda > 0.9\mu m$ extinction or polarization. It will be of great interest to see if the apparent universality of the infrared extinction is confirmed by future studies; if so, then it will be a challenge to understand, and may prove a crucial test for grain models.

7. SUMMARY AND PROSPECTUS

Recent observational studies have led to new constraints on dust models, based on: (1) *IRAS* and *COBE* observations, which demonstrate that the grains which account for most of the absorption of starlight are heated to temperatures $T \approx 18$ K. (2) The IR emission features at 3.3, 6.2, 7.7, 8.6 and $11.3\mu m$, which point to the presence of polycyclic hydrocarbons in ultrasmall grains or molecules; these may well be PAHs. They must account for a substantial part of the UV extinction. (3) The ERE, which strongly suggests the presence of HAC as a grain constituent.

Progress on understanding interstellar dust requires sustained efforts on many different fronts. On the observational side, it will be important to continue studies of extinction curves; this is perhaps particularly important in the infrared, to test whether there really is a tendency for dust to have a "universal" extinction law between 0.9 and $3.3\mu m$. Thus far spectroscopy of the IR emission features has been limited to regions of high surface brightness – planetary nebulae, HII regions, and reflection nebulae. When operational, *ISO* and *SIRTF* will enable detailed spectroscopy to be carried out on normal diffuse clouds. This should help clarify the origin of these features.

Laboratory studies are of great importance. For example, vacuum UV measurement of the dielectric properties of thin films of candidate materials, such as graphite (with possible impurities) or OH^-:silicate could help identify the still-controversial carrier of the 2175Å feature. Spectroscopic studies on candidate materials (e.g., PAH molecules or ions) might lead to identification of some of the DIBs. Studies of the fluorescence and luminescence of HAC materials can test our understanding of the ERE; as discussed in §5.3, it should be relatively easy to either confirm or reject the proposal of Duley *et al.* (1989) that HAC in large grains produces the IR emission features.

Theoretical work remains important. Improved calculational tools, such as the discrete dipole approximation, now permit calculations of absorption and scattering by grains which may be inhomogeneous and anisotropic, and with arbitrary geometries (cf. Draine & Flatau 1993). This enables detailed studies of polarization by aligned grains, the optical properties of composite structures produced by coagulation, etc. More detailed comparisons will be possible between grain models and measured extinction and polarization curves, scattering phase functions, etc.

ACKNOWLEDGMENTS

I thank the organizing committee for a very successful conference, and for the opportunity to give this talk. This work was supported in part by NSF grant AST90-17082.

REFERENCES

Adamson, A.J., Whittet, D.C.B., & Duley, W.W. 1990, *MNRAS*, **243**, 400

Anders, E. & Zinner, E. 1993, *Meteoritics*, **28**, 490

Bernatowicz, T.J., Amari, S., Zinner, E.K., & Lewis, R.S. 1991, *ApJ*, **373**, L73

Bernatowicz, T.J., Amari, S., & Lewis, R.S. 1992, *Lunar Planet. Sci.*, **XXII**, 5

Bohlin, R.C., Savage, B.D., & Drake, J.F. 1978, *ApJ*, **224**, 132

Boulanger, F., & Perault, M. 1988, *ApJ*, **330**, 964.

Butchart, I., McFadzean, A.D., Whittet, D.C.B., Geballe, T.R., & Greenberg, J.M. 1986, *A&A*, **154**, L5

Cardelli, J.A. 1994, this volume

Cardelli, J.A., Clayton, G.C., & Mathis, J.S. 1989, *ApJ*, **345**, 245

Cardelli, J.A., Mathis, J.S., Ebbets, D.C., & Savage, B.D. 1993, *ApJ*, **402**, L17

Cardelli, J.A., Savage, B.D., Bruhweiler, F.C. *et al.* 1991, *ApJ*, **377**, L57

Catura, R.C. 1983, *ApJ*, **275**, 645

Clayton, G.C., Anderson, C.M., Magalhães, *et al.* 1992, *ApJ*, **385**, L53

Cohen, M., Tielens, A.G.G.M., & Bregman, J.D. 1989, *ApJ*, **344**, L13

Désert, F.-X., Boulanger, F., & Puget, J.L. 1990, *A&A*, **237**, 215

Désert, F.-X., Boulanger, F., & Shore, S. 1986, *A&A*, **160**, 295

Dischler, B., Bubenzer, A., & Koidl, P. 1983, *Appl. Phys. Lett.*, **42**, 636

Draine, B.T. 1989a, in *Interstellar Dust*, ed. L.J. Allamandola & A.G.G.M. Tielens, Dordrecht: Kluwer, 313

Draine, B.T. 1989b, in *Infrared Spectroscopy in Astronomy*, ed. B.H. Kaldeich (Noordwijk: ESA), 93

Draine, B.T. 1990, in *The Evolution of the Interstellar Medium*, ed. L. Blitz, Astr. Soc. Pac., 193

Draine, B.T., & Anderson, N. 1985, *ApJ*, **292**, 494

Draine, B.T., & Flatau, P.J. 1993, *J. Opt. Soc. Am. A*, submitted

Draine, B.T., & Lee, H.M. 1984, *ApJ*, **285**, 89

Draine, B.T., & Malhotra, S. 1993, *ApJ*, **414**, 632

Draine, B.T., & Salpeter, E.E. 1979, *ApJ*, 231, 438

Duley, W.W., Jones, A.P., & Williams, D.A. 1989, *MNRAS*, **236**, 709

Duley, W.W., & Williams, D.A. 1988, *MNRAS*, **231**, 969

Fitzpatrick, E.L., & Massa, D. 1986, *ApJ*, **307**, 286

Giard, M., Pajot, F., Lamarre, J.M., Serra, G., Caux, E. 1989, *A&A*, 215, 92

Gillett, F.C., Forrest, W.J., Merrill, K.M., Capps, R.W., & Soifer, B.T. 1975, *ApJ*, **200**, 609

Greenberg, J.M. 1989, in "Evolution of Interstellar Dust and Related Topics", ed. A. Bonnetti, J.M. Greenberg, S. Aiello (Amsterdam: North-Holland), 7

Greenberg, J.M., & Yencha, A.J. 1973, in *Interstellar Dust and Related Topics*, ed. J.M. Greenberg & H.C. Van de Hulst, (Dordrecht: Reidel), 369.

Grün, E., Zook, H.A., Baguhl, M., Balogh, A., Bame, S.J., *et al.* 1993, *Nature*, **362**, 428

Guhathakura, P., & Tyson, J.A. 1989, *ApJ*, **346**, 773

Herbig, G.H. 1975, *ApJ*, **196**, 129

Herbst, E. 1991, *ApJ*, **366**, 133

Hobbs, L.M., York, D.G., & Oegerle, W. 1982, *ApJ*, **252**, L21

Huss, G.R., Hutcheon, I.D., Wasserburg, G.J. & Stone, J. 1992, *Lunar Plan. Sci.*, **XXIII**, 563

Jenkins, E.B. 1989, in *Interstellar Dust*, ed. L.J. Allamandola & A.G.G.M. Tielens, Dordrecht: Kluwer, 23

Jenniksens, P. 1992, Ph.D. thesis, Univ. of Leiden

Jenniksens, P., & Désert, F.-X. 1993, in preparation

Joblin, C., Maillard, J.P., d'Hendecourt, L., & Léger, A. 1990, *Nature*, **346**, 729

Jura, M. 1975, *ApJ*, **197**, 575

Krelowski, J. 1989, in *Interstellar Dust*, ed. L.J. Allamandola & A.G.G.M. Tielens, Dordrecht: Kluwer, 67

Krelowski, J. 1994, this volume

Lallement, R., Bertaux, J.L., Chassefière, E., and Sandel, B. 1991 *A&A*, **252**, 385.

Latter, W.B. 1991, *ApJ*, **377**, 187

Léger, A., & Puget, J.L. 1984, *A&A*, **137**, L5

Léger, A., Verstraete, L., d'Hendecourt, L., Défourneua, D., Dutuit, O, Schidt, W., & Lauer, J.C. 1989, in *Interstellar Dust*, ed. L.J. Allamandola & A.G.G.M. Tielens, Dordrecht: Kluwer, 173

Lewis, R.S., Anders, E., & Draine, B.T. *Nature*, **339**, 117

Low, F.J., Beintema, D.A., Gautier, T.N. *et al.* 1984, *ApJ*, **278**, L19

Martin, P.G., & Whittet, D.C.B. 1990, *ApJ*, **357**, 113

Mathis, J.S. 1986, *ApJ*, **308**, 281

Mathis, J.S. 1990, *ARA&A*, **28**, 37

Mathis, J.S., & Cardelli, J.A. 1992, *ApJ*, **398**, 610

Mathis, J.S., & Lee, C.-W. 1991, *ApJ*, **376**, 490

Mathis, J.S., Rumpl, W, & Nordsieck, K.H. 1977, *ApJ*, **215**, 425 (MRN)

Mathis, J.S., & Whiffen, G. 1989, *ApJ*, **341**, 808

Mauche, C.W., & Gorenstein, P. 1986, *ApJ*, **302**, 371

McFadzean, A.D., Whittet, D.C.B., Longmore, A.J., Bode, M.F., & Adamson, A.J. 1989, *MNRAS*, **241**, 873

McKee, C.F. 1989, in *Interstellar Dust*, ed. L.J. Allamandola & A.G.G.M. Tielens, Dordrecht: Kluwer, 431

Meyer, D. 1994, this conference

Nittler, L.R., Walker, R.M. & Zinner, E. 1993, *Lunar Planet. Sci.*, **XXIV**, 1087

Ott, U. 1993, *Nature*, **364**, 25

Papoular, R., Reynaud, C. & Nenner, I. 1990, *Phys. Rev. Lett.*, **64**, 1781

Predehl, P., Bräuninger, H., Burkert, W., & J.H.M.M. Schmitt 1991, *A&A*, **246**, L40

Rieke, G.H, & Lebofsky, M.J. 1985, *ApJ*, **288**, 618

Roche, P.F., & Aitken, D.K. 1984, *MNRAS*, **208**, 481

Roche, P.F., & Aitken, D.K. 1985, *MNRAS*, **215**, 425

Rowan-Robinson, M. 1992, *MNRAS*, **258**, 787

Sellgren, K. 1994, this volume

Steel, T.M., & Duley, W.W. 1987, *ApJ*, **315**, 337

Wexler, A.S. 1967, *Applied Spectroscopy Reviews*, **1**, 29

Whittet, D.C.B. 1992, *Dust in the Galactic Environment*, Bristol: Institute of Physics.

Whittet, D.C.B., Duley, W.W., & Martin, P.G. 1990, *MNRAS*, **244**, 427

Witt, A.N. 1989, in *Interstellar Dust*, ed. L.J. Allamandola & A.G.G.M. Tielens, Dordrecht: Kluwer, 87

Witt, A.N., & Schild, R.E. 1988, *ApJ*, **325**, 837

Witt, A.N., Petersohn, J.K., Holberg, J. Murthy, J., Dring, A., & Henry, R.C. 1993, *ApJ*, **410**, 714

Wright, E.L., Mather, J.C., Bennett, C.L. *et al.* 1991, *ApJ*, **381**, 200

The First Symposium on the Infrared Cirrus and Diffuse Interstellar Clouds
ASP Conference Series, Vol. 58, 1994
Roc M. Cutri and William B. Latter (eds.)

TINY GRAINS, LARGE MOLECULES, AND THE INFRARED CIRRUS

K. SELLGREN
Astronomy Department, Ohio State University, 174 W. 18th Av., Columbus, OH 43210, USA

ABSTRACT The infrared cirrus clouds discovered by IRAS shows mid-infrared emission at 12 and 25 μm which is far in excess of thermal emission due to dust in equilibrium with the interstellar radiation field. Two general classes of models have arisen to explain this. In the first model, the mid-infrared emission is due to thermal emission from tiny grains, 10–20 Å in size, stochastically heated by single ultraviolet (UV) photons to high temperatures for short periods of time; these tiny grains are probably composed largely of aromatic hydrocarbons. In the second model, the mid-infrared emission is due to UV-excited fluorescence from large polycyclic aromatic hydrocarbon molecules in the gas phase. Studies of the composition and emission mechanism for the material responsible for the mid-infrared cirrus emission have been limited by the low surface brightness of the cirrus, which has made detailed spectroscopic studies difficult. Regions of higher column density and higher UV radiation density, such as reflection nebulae and H II regions, may be similar to cirrus in the characteristics of their mid-infrared emission. Even these bright, well-observed sources, however, have so far been unable to unambiguously distinguish between models for the emission mechanism and composition, partly because interstellar conditions are so difficult to reproduce in the laboratory. It may turn out that both tiny grains and large molecules are important for the mid-infrared emission of the cirrus.

DISCOVERY OF 12 MICRON CIRRUS EMISSION

Infrared cirrus clouds, discovered in IRAS images by Low et al. (1984), are thought to be diffuse or translucent interstellar clouds whose dust is heated by the interstellar radiation field. Observations of cirrus and other sources, along with theoretical and laboratory studies, have led to a view of dust as containing not only large classical grains but also a component of tiny grains and/or large molecules which are critical to understanding the infrared emission from cirrus.

Traditional grain models, such as the Mathis, Rumpl & Nordsieck (1977, hereafter MRN) model in which the grains range in radius from 100 Å to 2500 Å, assume time-independent grain temperatures determined by the equilibrium with the radiation field. These types of models predicted strong emission at 60 and 100 μm from the cirrus but negligible emission at 12 and 25 μm. After Low et al. (1984) detected the infrared cirrus for the first time at 60 and 100 μm, however, two groups immediately suggested that if the composition of the cirrus was not too different from that of denser regions, then the cirrus emission should be observable at 12 and 25 μm due to emission from tiny grains (Draine & Anderson 1985) and/or polycyclic aromatic hydrocarbon (PAH) molecules (Puget, Léger & Boulanger 1985).

The predicted existence of mid-infrared cirrus emission (Draine & Anderson 1985; Puget, Léger, & Boulanger 1985) was quickly confirmed by observations (Boulanger, Baud & van Albada 1985; Gautier & Beichman 1984) detecting 12 and 25 μm emission from cirrus. Since then, a number of excellent studies of the properties of 12 μm emission from cirrus clouds have been published, including Weiland et al. (1986), Leene (1986), Odenwald & Rickard (1987), Heiles, Reach, & Koo (1988), Laureijs, Chlewicki, & Clark (1988), Laureijs et al. (1989), Herter, Shupe, & Chernoff (1990), Paley et al. (1991), and Welty & Fowler (1992). The diffuse galactic emission has also been studied, particularly in a definitive IRAS study by Boulanger & Pérault (1988) who found the 12 μm emission of the diffuse interstellar medium to be similar in nature to that of the cirrus clouds. Some studies of cirrus and the diffuse galactic emission have concluded that some of the 60 μm emission is also due to tiny grains (Draine & Anderson 1985; Boulanger & Pérault 1988; Boulanger et al. 1988; Heiles, Reach, & Koo 1988; Désert, Boulanger, & Puget 1990; Herter, Shupe, & Chernoff 1990; Paley et al. 1991). Up to 40% of the total emission of the infrared cirrus, as well as of the diffuse interstellar medium, is due to the 12 and 25 μm emission. Thus the tiny grains and/or large molecules are an important component of the interstellar medium (ISM), both in absorption in the visible and ultraviolet (UV) and in emission in the infrared.

The tiny grains and PAHs envisioned by Draine & Anderson (1985) and Puget, Léger, & Boulanger (1985) were required as a component of the ISM in order to explain the near-infrared (1–5 μm) emission from reflection nebulae. Extended continuum emission from reflection nebulae was observed at 1–13 μm, along with emission features at 3.3, 6.2, 7.7, 8.6, and 11.3 μm (Sellgren, Werner, & Dinerstein 1983; Sellgren 1984; Sellgren et al. 1985). This emission could not be explained by reflected starlight or by thermal emission from dust in radiative equilibrium with the central star. Sellgren and her co-workers proposed that the continuum was due to thermal emission from tiny grains, $\sim$ 10 Å in radius, which undergo large excursions in temperature due to stochastic heating by single stellar photons.

The emission features at 3.3 μm, 6.2, 7.7, 8.6, and 11.3 μm in reflection nebulae had been previously detected in a wide range of sources, including planetary nebulae, protoplanetary nebulae, H II regions, and galaxies dominated by star formation rather than an active nucleus (see Aitken 1981, Willner 1984, Bregman 1989, Puget & Léger 1989, and Allamandola, Tielens, & Barker 1989 for reviews). These five features are always seen together; they are observed in emission but not in absorption. At high spectral resolution the features are observed to be broad ($\lambda/\Delta\lambda \sim$ 10–100) and smooth. The association of near-infrared continuum emission, attributed to tiny grains, with the infrared emission features at 3.3, 6.2, 7.7, 8.6, and 11.3 μm inspired two groups to propose that the infrared emission features were due to PAH molecules (Léger & Puget 1984; Allamandola, Tielens, & Barker 1985). Tiny grains and PAH molecules both briefly emit infrared radiation after excitation by single UV photons, and are similar in size. What is special about PAHs is that they are observed in the laboratory to have absorption features that more or less coincide in wavelength with the infrared emission features seen in the ISM.

The original PAH models were aimed particularly at the infrared emission features, attributing them to UV-excited vibrational fluorescence from PAH molecules in the gas phase. It is less clear how PAHs might produce a near-infrared continuum, perhaps by a vibrational quasi-continuum due to many weak combination and overtone emission features (Léger & Puget 1984; Allamandola, Tielens, & Barker 1985; Puget & Léger 1989), or perhaps by an electronic fluorescence or phosphorescence (Allamandola, Tielens, & Barker 1985, 1989). The tiny grain model (Sellgren 1984) was proposed to explain the near-infrared continuum emission as thermal emission from stochastically heated grains. The tiny grain model does not specifically address the origin of the associated emission features, except as solid state features in the material of which the tiny grains are composed. At present it seems possible that either tiny grains or PAHs (or both) may account for either the infrared emission features or their associated near-infrared continuum emission (or both). What both models have in common is a small size for the emitting particles, and excitation of the tiny particles by single stellar photons.

MODELS FOR THE 12 MICRON CIRRUS EMISSION

The 12 and 25 μm emission from cirrus clouds has been modeled with a variety of grain and PAH components. The Draine & Anderson (1985) model, while successful in predicting the presence of detectable 12 μm emission from cirrus, actually predicts less 12 μm emission than is observed in cirrus clouds. They used two models, one in which the MRN distribution was extended to smaller grain radii of 3 Å with no change in the power law index of the grain size distribution, and another in which there was an enhanced population of tiny grains above the extrapolation of the MRN distribution to small grain sizes. Both of these models predict less 12 μm emission than observed in cirrus, when models and observations are both normalized to the 100 μm emission. A similar result was found by Désert, Boulanger, & Puget (1990), who found that they could only match the observed 12 μm to 100 μm intensity ratio in cirrus by including emission from PAHs as well as from tiny grains and MRN type grains. Siebenmorgen & Krügel (1992) have fit five different grain components to the diffuse galactic emission; their model attributes the 12 μm emission to PAH clusters and PAH molecules. Chlewicki & Laureijs (1988) argue that only large grains and PAHs are needed to fit IRAS observations, and do not include tiny grains in their model.

The only spectroscopic observation of either the cirrus or the diffuse ISM to date is the ground-breaking work of Giard et al. (1988, 1989), who used a balloon-borne telescope and narrow-band filters on and off the 3.3 μm emission feature to detect this emission feature in the diffuse galactic emission. All mid-infrared (5–30 μm) measurements to date of the infrared cirrus, however, have been made with broad-band filters, such as the IRAS 12 and 25 μm filters. Observations in H II regions and planetary nebulae show that there are a large number of spectral features due to large molecules and/or tiny grains within the

IRAS 12 μm filter bandpass. Bregman et al. (1989), in a spatial and spectral study of the Orion Nebula, find that the relative strengths of the 7.7, 8.6, and 11.3 μm features, the broad emission bump at 6–9 μm, and the plateau at 11–13 μm all vary spatially within Orion. Roche, Aitken, & Smith (1991) also find that the relative strengths and widths of the 7.7, 8.6, 11.3, and 12.7 μm features and 11-13 μm plateau vary from source to source within our Galaxy. Bregman et al. (1989) attribute the narrow emission features at 7.7, 8.6, and 11.3 μm to PAHs, the broad emission bump at 6–9 μm and the plateau at 11–13 μm to PAH clusters, and the underlying mid-infrared continuum to tiny hydrogenated amorphous carbon grains; they explain the varying strengths of different features by changes in the excitation and relative abundances of these three components. Each of these spectral components may contribute, in varying amounts, to the total emission observed within the IRAS 12 μm filter.

The low surface brightness of the infrared cirrus has so far prevented spectroscopic observations in the mid-infrared, which could critically test models which make assumptions about the relative contributions of PAH features and tiny grain continuum to the IRAS 12 μm emission. In higher surface brightness regions, however, such as reflection nebulae, H II regions, and planetary nebulae, the spectroscopic observations are available to determine the contributions of different spectral features. Thus in these regions, much more detailed models can be constructed, and tested vigorously through spectroscopic and spatial studies. The limitation of comparing these results to the infrared cirrus is that these regions have significantly higher column densities of gas and dust than in a typical cirrus cloud, and furthermore are heated (and in some cases ionized) locally by an embedded star rather than being heated solely by the diffuse interstellar radiation field.

Reflection nebulae provide a better analog to cirrus clouds than do H II regions or planetary nebulae, due to the lack of ionizing radiation. The IRAS energy distributions of reflection nebulae (Sellgren, Luan, & Werner 1990), cirrus clouds (Weiland et al. 1986; Heiles, Reach, & Koo 1988; Laureijs et al. 1989; Paley et al. 1991), and the diffuse galactic emission (Boulanger & Pérault 1988) show that the 60 μm to 100 μm intensity ratio is significantly higher in reflection nebulae, due to the presence of localized heating by the central star. However, the 12 μm to 100 μm intensity ratio and the 25 μm to 100 μm intensity ratio are very similar in reflection nebulae, cirrus clouds, and the diffuse galactic emission. This suggests that the mix of large grains with tiny grains and/or large molecules is similar in reflection nebulae and cirrus clouds.

Some of the models which have been compared to cirrus observations have also made predictions for reflection nebulae. Désert, Boulanger & Puget (1990) predict not only the IRAS energy distribution of a reflection nebula illuminated by a B3 star, but also the emission at 2.2, 3.4, 200, 400, and 800 μm. Comparison of their model to observations of the reflection nebula NGC 7023 (Whitcomb et al. 1981; Sellgren et al. 1983, 1985) gives an excellent fit to data longward of 60 μm, but worse agreement at the shortest wavelengths; including only large grains and tiny grains predicts far too little 2.2 and 3.4 μm emission but adding PAHs to their model predicts too much emission at 2.2 and 3.4 μm. Siebenmorgen & Krügel (1992) have done a more complicated fit to NGC 7023, using five different grain components: large silicate grains, large amorphous carbon grains, small graphite grains, PAH clusters, and PAH molecules. They fit their model to the emission spectrum of the nebula and find a very good fit to the observations from 3 to 300 μm, although they do not show the fit to the observations at 1–2 μm or at 400 μm, nor do they compare their model calculations to the observed extinction curve for the central star.

LABORATORY ANALOGS

How well do laboratory materials, which have been proposed to explain the infrared interstellar emission features and associated continuum emission, fit the astronomical observations? The best constraints on the laboratory analogs come not from cirrus clouds, but from higher surface brightness regions such as reflection nebulae and H II regions, where spectroscopy aids in identifying the composition of the tiny grains and/or large molecules. Spectral studies of these regions show that the strongest infrared emission features are at 3.3, 6.2, 7.7, 8.6, and 11.3 μm; these features are always seen together in the ISM. Their relative emission feature strengths vary by a factor of a few from source to source, but all five main features are strongly correlated. A number of weaker emission features are also seen in many, but not all, sources showing the main five features, including wavelengths of 3.40, 3.46, 3.52, 3.57, 5.2, 5.7, 6.9, 11.05, 11.5, 11.9, and 12.7 μm (Bregman et al. 1983; de Muizon et al. 1986; Allamandola et al. 1989; Roche, Aitken, & Smith 1991). There are also some broad emission features: a plateau of emission at 3.2–3.6 μm, a broad bump of emission at 6–9 μm, and a plateau of emission at 11–13 μm. Some of the weaker features and very broad features are correlated with the main features; others are more variable.

Any successful laboratory analog *must* predict all five main features, with central wavelengths, emission feature profiles, and relative feature strengths that match the ISM data to within the observational variations. A successful laboratory analog also *must not* predict features which are not observed in the ISM. Differences in the spatial distributions of the main features from those of some of the weaker features and broad features indicate that there may be several distinct ISM components required to explain all the near-infrared and mid-infrared emission from tiny grains/large molecules. This complicates the job of the laboratory astrophysicist. It seems likely, however, that the five main features arise from a single ISM component, although it may be a component with a range of grain or molecular sizes.

Table 1 summarizes most of the current proposed identifications for the infrared ISM emission features. For each of the five main features, the proposed spectral identification for the feature is given, along with the laboratory analogs which show the same feature. Some of the papers on laboratory analogs have only discussed a subset of the five main features (Saperstein, Metin, & Kaufman 1989; Balm & Kroto 1990), but most present laboratory spectra or theoretical calculations covering all five features. Several points are immediately apparent from Table 1. First, the majority of the laboratory analogs agree on the identification of the 3.3, 6.2, and 11.3 μm features as aromatic hydrocarbon features. This is extremely encouraging and indicates that major progress has been made toward an identification of the features. Second, it is equally clear from Table 1 that none of the laboratory analogs agree on a consistent interpretation of the 7.7 or 8.6 μm features. This is particularly troubling for the 7.7 μm feature,

TABLE 1 Proposed Identifications for the Infrared Emission Features

λ (μm)	Identifications	Material
3.3	aromatic C–H stretch	1, 3, 7, 8
6.2	Aromatic C=C in-plane stretch	1, 3, 4, 6, 7, 8
	Aromatic amine $-NH_2$ deformation	1
	$7F_{1u}$ mode of $C_{60}H_{60}$	2
	Conjugated carbonyl C=O	3
	Nitrogen-substituted graphitic carbon	5
	Cross-conjugated ketone C=O stretch	8
7.7	Aromatic amine CN stretch	1
	$5F_{1u}$ mode of $C_{60}H_{60}$	2
	Aromatic and aliphatic C–O stretch	3
	Nitrogen-substituted graphitic carbon	5
	Blend, aromatic C=C stretch	7
	Cross-conjugated ketone	8
8.6	$4F_{1u}$ mode of $C_{60}H_{60}$	2
	Terminal ether or alcohol groups	4
	Aromatic C–H in-plane bend	7
	Cross-conjugated ketone	8
11.3	Aromatic C–H out-of-plane bend (solo H)	1, 3, 4, 6, 7, 8
	$3F_{1u}$ mode of $C_{60}H_{60}$	2
	Aromatic C=C vibration	4
	Internal hydrogen C–H	9

Laboratory Materials for Proposed Identifications — *(1)* Amorphous carbon: Duley & Williams 1981; Borghesi, Bussoletti, & Colangeli 1987. *(2)* $C_{60}H_{60}$: Webster 1991. *(3)* Coal (Vitrinite): Papoular *et al.* 1989. *(4)* Hydrogenated amorphous carbon: Blanco, Bussoletti, & Colangeli 1988; Ogmen & Duley 1989. *(5)* Nitrogenated amorphous carbon: Saperstein, Metin, & Kaufman 1989. *(6)* Carbonaceous residue from Orgueil meteorite: Wdowiak, Flickinger, & Cronin 1988. *(7)* Polycyclic aromatic hydrocarbons: Allamandola, Tielens, & Barker 1989; Léger, d'Hendecourt, & Défourneau 1989. *(8)* Quenched carbonaceous composite: Sakata *et al.* 1987. *(9)* Spiral carbonaceous microparticle with internal hydrogen: Balm & Kroto 1990.

because it is by far the strongest of the ISM emission features, yet it is not at all dominant or even easily identified in the published spectra of most laboratory analogs. The lack of consensus on the origin of the strong 7.7 μm feature shows that while progress has been made on an identification for the features, the issue is far from settled. Third, a careful look at Table 1 shows that only two of the nine laboratory analogs listed (PAHs and quenched carbonaceous composite) predict all five of the ISM emission features. Even these two materials have other difficulties explaining the ISM observations, however, as discussed below. Thus while a number of laboratory analogs provide approximate matches to the ISM features, none so far is completely satisfactory as an identification of the feature material.

Another requirement for a successful identification of the ISM features is that the laboratory analog *not* predict any features which are not observed in the ISM. Several of the substances in Table 1 have this problem. The most persistent laboratory feature with no corresponding ISM feature is at 7.3 μm, which is seen in amorphous carbon, aromatic char at 733 K, hydrogenated amorphous carbon, the Orgueil meteorite, quenched carbonaceous composite and vitrinite (Borghesi, Bussoletti, & Colangeli 1987; Sakata et al. 1987; Blanco, Bussoletti, & Colangeli 1988; Ogmen & Duley 1988; Wdowiak, Flickinger, & Cronin 1988; Papoular et al. 1989).

A stringent test of the success of a laboratory analog as an identification for the ISM features is that the laboratory wavelength match the ISM wavelength within the observational variation of the ISM feature. Table 2 compares the wavelengths of laboratory analogs to the wavelengths of the ISM features. All of the laboratory analogs listed, with the exception of quenched carbonaceous composite, fail to match the ISM wavelengths within the observational variation of the ISM feature.

Difficulties with the wavelength agreement have been discussed extensively for PAHs. Sakata et al. (1990), Balm & Kroto (1990), and Colangeli et al. (1992) all conclude from detailed comparisons of the observed central wavelengths of PAHs to the ISM features that individual PAHs or mixtures of PAHs are unable to match the observed ISM wavelengths. However, this discussion is complicated by the difference between laboratory and ISM conditions. Interstellar PAHs, if present in the ISM, would be in the gas phase at densities where collisions are negligible, and may be ionized and/or partially dehydrogenated; they emit after excitation by ultraviolet photons. Many published PAH laboratory spectra are absorption spectra of neutral, fully hydrogenated, solid PAHs at room temperature. Laboratory groups have worked to improve the simulation of ISM conditions, by obtaining laboratory data on PAHs isolated in rare gas matrices in absorption (Léger, d'Hendecourt, & Défourneau 1989; Joblin et al. 1993; Hudgins, Sandford, & Allamandola 1993*a*), heated solid PAHs in absorption (Bernard, d'Hendecourt, & Léger 1989; Blanco, Fonti, & Orofino 1990; Flickinger & Wdowiak 1990; Flickinger, Wdowiak, & Gómez 1991; Colangeli, Mennella, & Bussoletti 1992; Mennella, Colangeli, & Bussoletti 1992), heated gas-phase PAHs in absorption (Flickinger, Wdowiak, & Gómez 1991; Kurtz 1992; Mennella, Colangeli, & Bussoletti 1992; Joblin et al. 1993), heated gas-phase PAHs in emission, (Kurtz 1992), gas-phase PAHs in emission after excitation by UV radiation (Cherchneff & Barker 1989; Shan, Suto, & Lee 1991; Brenner & Barker 1992), and positively ionized PAHs (d'Hendecourt &

TABLE 2 Wavelengths Observed in Lab and ISM

Source	Wavelengths (μm)					Ref
ISM	3.293 $\pm$0.004	6.20 $\pm$0.06	7.8 $\pm$0.2	8.62 $\pm$0.03	11.22 $\pm$0.10	1
Amorphous carbon	...	6.29	...	...	11.3	2
$C_{60}H_{60}$	...	6.50	7.86	8.83	10.4	3
Aromatic char	3.29	6.29	...	...	11.4	4
Coal	3.30	6.21	...	...	11.6	5
Coal	3.29	6.22	...	...	11.5	6
HAC	...	6.21	...	8.05	11.55	7
HAC	...	6.29	...	...	11.3	8
Orgueil meteorite	...	6.17	...	...	11.3	9
PAH	3.275+3.311	6.25	7.64	8.81	11.77	10
QCC	3.29	6.25	7.7	8.6	11.4	11

Notes to Table: Error bars on wavelengths of features in the interstellar medium are the observational variation from source to source. Laboratory materials with no wavelength given for a particular feature did not show that feature in laboratory spectra.

References — *(1)* Interstellar medium (ISM): Tokunaga et al. 1991; Russell, Soifer, & Willner 1978; Cohen et al. 1989; Roche, Aitken, & Smith 1991; Witteborn et al. 1989. *(2)* Amorphous carbon: Borghesi, Bussoletti, & Colangeli 1987. *(3)* $C_{60}H_{60}$: Webster 1991. *(4)* Aromatic char (heated to 733 K): Blanco, Bussoletti, & Colangeli 1988. *(5)* Coal (vitrinite): Papoular et al. 1989. *(6)* Coal (semi-anthracite): Papoular, Reynaud, & Nenner 1991. *(7)* Hydrogenated amorphous carbon (HAC): Ogmen & Duley 1988. *(8)* Hydrogenated amorphous carbon (HAC): Blanco, Bussoletti, & Colangeli 1988. *(9)* Orgueil meteorite: Wdowiak, Flickinger, & Cronin 1988. *(10)* Polycyclic aromatic hydrocarbon (PAH; gas-phase coronene at 720 K): Kurtz 1992. *(11)* Quenched carbonaceous composite (QCC): Sakata et al. 1987.

Léger 1993; Hudgins, Sandford, & Allamandola 1993*b*).

So far, laboratory work is inconclusive as to whether closer approximations to ISM conditions will resolve discrepancies between observations of laboratory analogs and the ISM; sometimes the fit is better, sometimes worse. One result of heating PAHs is to substantially improve the fit to the ISM 3.3 μm feature by removing much of the substructure in the 3.3 μm feature of laboratory PAHs (Flickinger & Wdowiak 1990; Shan, Suto, & Lee 1991; Joblin & et al. 1993). Another effect of heating PAHs, or exciting them by UV radiation, is to shift the central wavelength to longer wavelengths (Colangeli, Mennella, & Bussoletti 1992; Brenner & Barker 1992; Joblin et al. 1993). This improves the wavelength agreement between laboratory and ISM for the 3.3 μm feature, but presumably worsens it for the 11.3 μm feature. A study by d'Hendecourt & Léger (1993) of positively ionized PAHs finds that their spectra are a substantially worse fit to the ISM than those of neutral PAHs; features appear in the laboratory spectra of ionized coronene which are not seen in the ISM, and features observed in the ISM disappear from the laboratory spectra. This is disturbing, because regions of high UV energy density, such as reflection nebulae, H II regions, and planetary nebulae, where all spectra to date have been obtained, are predicted to contain positively charged PAHs (Omont 1986; d'Hendecourt & Léger 1987; Allamandola, Tielens, & Barker 1989), or a mixture of neutral and positively charged PAHs (Lepp & Dalgarno 1988; Bakes & Tielens 1993). It also becomes unclear whether the infrared spectra of the cirrus will resemble that of regions of high UV energy density, since regions illuminated by the diffuse interstellar radiation field, such as cirrus clouds, have been variously predicted to contain neutral PAHs (d'Hendecourt & Léger 1987; Verstraete et al. 1990), a mixture of neutral and positively charged PAHs (van der Zwet & Allamandola 1985; Allamandola, Tielens, & Barker 1989), or a mixture of neutral and negatively charged PAHs (Lepp et al. 1988; Lepp & Dalgarno 1988; Bakes & Tielens 1993). More laboratory and theoretical work is needed.

So far no one substance, or mixture of substances, successfully (1) predicts the existence of all five main ISM features, (2) avoids predicting additional strong emission features not seen in the ISM, (3) fits the central wavelengths of all five main ISM features, (4) fits the wavelength widths of all five main ISM features, and (5) fits the relative strengths of all five main ISM features. The common thread that runs through most proposed laboratory analogs, however, is that the material is some sort of aromatic hydrocarbon. This convergence of opinion is encouraging for hopes that a definitive identification of the material responsible for the ISM emission features will someday be demonstrated.

THE FUTURE

Detailed spectroscopy of the infrared cirrus is currently difficult due to the low surface brightness of the cirrus. The ultimate identification of the composition of the tiny grains and/or large molecules that produce the 12 and 25 μm emission from the infrared cirrus, therefore, must await the successful launch of infrared satellites such as ISO, SIRTF, and IRTS. These satellites will have the needed sensitivity to low surface brightness emission to obtain spectra of the infrared cirrus, and thus to allow studies of the spatial and spectral variations in the infrared cirrus emission in the kind of detail which has so far been only possible in higher surface brightness regions of the ISM.

ACKNOWLEDGMENTS

The author is grateful to an Alfred P. Sloan Research Fellowship and the organizing committee for financial support.

REFERENCES

Aitken, D. K. 1981, in *IAU Symposium 96, Infrared Astronomy*, ed. C. G. Wynn-Williams & D. P. Cruikshank (Dordrecht: Reidel), p. 207.

Allamandola, L. J., Tielens, A. G. G. M., & Barker, J. R. 1985, *ApJ*, **290**, L25.

Allamandola, L. J., Tielens, A. G. G. M., & Barker, J. R. 1989, *ApJS*, **71**, 733.

Allamandola, L. J., Bregman, J. D., Sandford, S. A., Tielens, A. G. G. M., Witteborn, F. C., & Wooden, D. H. 1989, *ApJ*, **345**, L59.

Bakes, E. L. O. & Tielens, A. G. G. M. 1993, submitted to *ApJ*.

Balm, S. P., & Kroto, H. W. 1990, *MNRAS*, **245**, 193.

Bernard, J. P., d'Hendecourt, L. B., & Léger, A. 1989, *A&A*, **220**, 245.

Blanco, A., Bussoletti, E., & Colangeli, L. 1988, *ApJ*, **334**, 875.

Blanco, A., Fonti, S., & Orofino, V. 1990, *ApJ*, **364**, 152.

Borghesi, A., Bussoletti, E., & Colangeli, L. 1987, *ApJ*, **314**, 422.

Boulanger, F., Baud, B., & van Albada, G. D. 1985, *A&A*, **144**, L9.

Boulanger, F., & Pérault, M. 1988, *ApJ*, **330**, 964.

Boulanger, F., Beichman, C., Désert, F. X., Helou, G., Pérault, M., & Ryter, C. 1988, *ApJ*, **332**, 328.

Bregman, J. D. 1989, in *IAU Symposium 135, Interstellar Dust*, ed. L. J. Allamandola & A. G. G. M. Tielens (Dordrecht: Kluwer), p. 109.

Bregman, J. D., Dinerstein, H. L., Goebel, J. H., Lester, D. F., Witteborn, F. C., & Rank, D. M. 1983, *ApJ*, **274**, 666.

Bregman, J. D., Allamandola, L. J., Tielens, A. G. G. M., Geballe, T. R., & Witteborn, F. C. 1989, *ApJ*, **344**, 791.

Brenner, J. D., & Barker, J. R. 1992, *ApJ*, **388**, L39.

Cherchneff, I., & Barker, J. R. 1989, *ApJ*, **341**, L21.

Chlewicki, G., & Laureijs, R. J. 1988, *A&A*, **207**, L11.

Cohen, M. et al. 1989, *ApJ*, **341**, 246.

Colangeli, L., Mennella, V., & Bussoletti, E. 1992, *ApJ*, **385**, 577.

Colangeli, L., Mennella, V., Baratta, G. A., Bussoletti, E., & Strazzulla, G. 1992, *ApJ*, **396**, 369.

de Muizon, M., Geballe, T. R., d'Hendecourt, L. B. & Baas, F. 1986, *ApJ*, **306**, L105.

Désert, F. X., Boulanger, F., & Puget, J. L. 1990, *A&A*, **237**, 215.

d'Hendecourt, L. B. & Léger, A. 1987, *A&A*, **180**, L9.

d'Hendecourt, L. & Léger, A. 1993, paper presented at The First Symposium on the Infrared Cirrus & Diffuse Interstellar Clouds, April, Tucson.

Draine, B. T., & Anderson, N. 1985, *ApJ*, **292**, 494.

Duley, W. W., & Williams, D. A. 1981, *MNRAS*, **196**, 269.

Flickinger, G. C., & Wdowiak, T. J. 1990, *ApJ*, **362**, L71.

Flickinger, G. C., Wdowiak, T. J., & Gómez, P. L. 1991, *ApJ*, **380**, L43.

Gautier, T. N., & Beichman, C. A. 1984, *BAAS*, **16**, 968.

Giard, M. et al. 1988, *A&A*, **201**, L1.

Giard, M., Pajot, F., Lamarre, J. M., Serra, G., & Caux, E. 1989, *A&A*, **215**, 92.

Heiles, C., Reach, W. T., & Koo, B. C. 1988, *ApJ*, **332**, 313.

Herter, T., Shupe, D. L., & Chernoff, D. F. 1990, *ApJ*, **352**, 149.

Hudgins, D. M., Sandford, S. A., & Allamandola, L. J. 1993*a*, paper presented at The First Symposium on the Infrared Cirrus & Diffuse Interstellar Clouds, April, Tucson.

Hudgins, D. M., Sandford, S. A., & Allamandola, L. J. 1993*b*, paper presented at The First Symposium on the Infrared Cirrus & Diffuse Interstellar Clouds, April, Tucson.

Joblin, C., Léger, A., d'Hendecourt, L., & Défourneau, D. 1993, paper presented at The First Symposium on the Infrared Cirrus & Diffuse Interstellar Clouds, April, Tucson.

Kurtz, J. 1992, *A&A*, **255**, L1.

Laureijs, R. J., Chlewicki, G., & Clark, F. O. 1988, *A&A*, **192**, L13.

Laureijs, R. J., Chlewicki, G., Clark, F. O., & Wesselius, P. R. 1989, *A&A*, **220**, 226.

Leene, A. 1986, *A&A*, **154**, 295.

Léger, A. & Puget, J. L. 1984, *A&A*, **137**, L5.

Léger, A., d'Hendecourt, L., & Défourneau, D. 1989, *A&A*, **216**, 148.

Lepp, S., Dalgarno, A., van Dishoeck, E. F., & Black, J. H. 1988, *ApJ*, **329**, 418.

Lepp, S., & Dalgarno, A. 1988, *ApJ*, **335**, 769.

Low, F. J., et al. 1984, *ApJ*, **278**, L19.

Mathis, J. S., Rumpl, W., & Nordsieck, K. H. 1977, *ApJ*, **217**, 425 (MRN).

Mennella, V., Colangeli, L., & Bussoletti, E. 1992, *ApJ*, **391**, L45.

Odenwald, S. F., & Rickard, L. J 1987, *ApJ*, **318**, 702.

Ogmen, M. & Duley, W. W. 1988, *ApJ*, **334**, L117.

Omont, A. 1986, *A&A*, **164**, 159.

Paley, E. S., Low, F. J., McGraw, J. T., Cutri, R. M., & Rix, H. W. 1991, *ApJ*, **376**, 335.

Papoular, R., Conard, J., Guiliano, M., Kister, J., & Mille, G. 1989, *A&A*, **217**, 204.

Papoular, R., Reynaud, C., & Nenner, I. 1991, *A&A*, **247**, 215.

Puget, J. L., Léger, A., & Boulanger, F. 1985, *A&A*, **142**, L19.

Puget, J. L. & Léger, A. 1989, *ARA&A*, **27**, 161.

Roche, P. F., Aitken, D. K., & Smith, C. H 1991, *MNRAS*, **252**, 282.

Russell, R. W., Soifer, B. T., & Willner, S. P. 1978, *ApJ*, **220**, 568.

Sakata, A., Wada, S., Onaka, T., & Tokunaga, A. T. 1987, *ApJ*, **320**, L63.

Sakata, A., Wada, S., Onaka, T., & Tokunaga, A. T. 1990, *ApJ*, **353**, 543.

Saperstein, D. D., Metin, S. S., & Kaufman, J. H. 1989 *ApJ*, **342**, L47.

Sellgren, K. 1984, *ApJ*, **277**, 623.

Sellgren, K., Werner, M. W., & Dinerstein, H. L. 1983, *ApJ*, **271**, L13.

Sellgren, K., Allamandola, L. J., Bregman, J. D., Werner, M. W., & Wooden, D. H. 1985, *ApJ*, **299**, 416.

Sellgren, K., Luan, L., & Werner, M. W. 1990, *ApJ*, **359**, 384.

Shan, J., Suto, M., & Lee, L. C 1991, *ApJ*, **383**, 459.

Siebenmorgen, R. & Krügel, E. 1992, *A&A*, **259**, 614.

Tokunaga, A. T., Sellgren, K., Smith, R. G., Nagata, T., Sakata, A., & Nakada, Y. 1991, *ApJ*, **380**, 452.

van der Zwet, G. P., & Allamandola, L. J. 1985, *A&A*, **146**, 76.

Verstraete, L., Léger, A., d'Hendecourt, L., Dutuit, O., & Défourneau, D. 1990, *A&A*, **237**, 436.

Wdowiak, T. J., Flickinger, G. C., & Cronin, J. R. 1988, *ApJ*, **328**, L75.

Webster, A. 1991, *Nature*, **352**, 412.

Weiland, J. L., Blitz, L., Dwek, E., Hauser, M. G., Magnani, L., & Rickard, L. J 1986, *ApJ*, **306**, L101.

Welty, D. E. & Fowler, J. R. 1992, *ApJ*, **393**, 193.

Whitcomb, S. E., Gatley, I., Hildebrand, R. H., Keene, J., Sellgren, K., & Werner, M. W. 1981, *ApJ*, **246**, 416.

Willner, S. P. 1984, in *Galactic and Extragalactic Infrared Spectroscopy*, ed. M. F. Kessler & J. P. Phillips (Dordrecht: Reidel), p. 37.

Witteborn, F. C., Sandford, S. A., Bregman, J. D., Allamandola, L. J., Cohen, M., Wooden, D. H., & Graps, A. L. 1989, *ApJ*, **341**, 270.

The First Symposium on the Infrared Cirrus and Diffuse Interstellar Clouds
ASP Conference Series, Vol. 58, 1994
Roc M. Cutri and William B. Latter (eds.)

INFRARED SPECTROSCOPY OF ORGANIC MATERIAL IN DIFFUSE INTERSTELLAR CLOUDS

YVONNE J. PENDLETON[1]
Theoretical Studies Branch, NASA Ames Research Center, Moffett Field, CA 94035

ABSTRACT Spectra of objects which lie along several lines of sight through the diffuse interstellar medium (ISM) all contain an absorption feature near 3.4 μm which has been attributed to saturated aliphatic hydrocarbons on interstellar grains. The similarity of the absorption bands near 3.4 μm (2950 cm^{-1}) along different lines of sight reveal that the carrier of this band lies in the diffuse dust. Several materials have been proposed as "fits" to the 3.4 μm feature over the years. A comparison of these identifications is presented. A remarkable similarity between the spectrum of the diffuse dust and an organic extract from the Murchison meteorite suggests that some of the interstellar organic material may be preserved in primitive solar system bodies. The optical depth/extinction τ/A_V ratio for the 3.4 μm (2950 cm^{-1}) band is higher toward the Galactic center than toward sources which sample the interstellar medium in the local neighborhood. A similar trend has been observed previously for silicates, indicating that the two materials may be simultaneously enhanced in the Galactic center.

INTRODUCTION

Infrared spectral studies provide insight into the chemical identity of the organic component of the interstellar medium because the fundamental vibrational frequencies of the common chemical bonds between the most chemically abundant elements occur in the mid-infrared (5000 – 400 cm^{-1}; 2 – 25 μm). Infrared spectroscopic techniques have been used to detect organic material in the interstellar medium (Wickramasinghe & Allen 1980, 1983; Adamson, Whittet, & Duley 1990; Butchart et al. 1986; Sandford et al. 1991; Pendleton 1993) as well as in primitive solar system bodies such as comets, meteorites, and asteroids (Tokunaga & Brooke 1990; Cronin and Pizzarello 1990; Brooke, Tokunaga, & Knacke 1991; Cruikshank & Brown 1987; Ehrenfreund et al. 1991; Mueller et al. 1992).

Unequivocal identification of the specific molecules responsible for absorption and/or emission features observed in the ISM has met with limited success, in part due to technical constraints on the quality and completeness of astronom-

[1]Visiting Astronomer at the Infrared Telescope Facility, which is operated by the University of Hawaii under contract from the National Aeronautics and Space Administration.

ical spectra. Until fairly recently, adequate wavelength coverage at sufficiently high resolution and good signal-to-noise ratios were not available to distinguish between possible "fits" to the 3.4 μm feature. Consequently, the existing data through the CH region were "fit" with a variety of organic material, including E-Coli bacteria (Hoyle & Wickramasinghe 1982), organic grain mantles (Greenberg 1982; Schutte & Greenberg 1988), and Hydrogenated Amorphous Carbon (HAC) (Jones, Duley, & Williams 1987). Through comparisons of these materials to the higher resolution, high signal-to-noise data available today, it has become apparent that some of the proposed "fits" are not as similar to the diffuse dust spectra as had been assumed.

Over the years the instrumentation has improved dramatically, so that we are now able to see structure previously hidden in the 3.4 μm region. The 3.4 μm absorption band is actually composed of several bands which correspond to the stretching modes of C-H in saturated aliphatic hydrocarbons (Sandford et al. 1991; Pendleton 1993). Through a combined effort of infrared observational and laboratory analyses, the profile of the interstellar feature, as seen in the higher S/N data of Sandford et al. 1991, has been shown to be better fit by the spectra of residues produced through laboratory experiments which simulate chemical processing in the ISM than by other suggested materials, including HAC (see figures 1a and b). The organic extract of the Murchison meteorite (deVries et al. 1992) and the diffuse ISM (Figure 1a) reveals a striking similarity between the two. This was pointed out by Ehrenfreund et al. (1991) and Pendleton (1993). The similarity of the spectra suggest that some of the interstellar organic material may have been preserved in the carbonaceous meteorites. Comparisons to the Orgueil meteorite (Ehrenfreund et al. 1991) also show a strong similarity to the Galactic center data in the 3.4 μm region.

EVIDENCE FOR ORGANICS IN THE DIFFUSE DUST

The Galactic center

Until fairly recently, the only clear detection of the 3.4 μm absorption feature was along a line of sight toward the Galactic center (Butchart et al. 1986 and others). Two absorption bands are seen in the spectra of Galactic center sources (IRS 3, 6, & 7), the broad 3.0 μm feature is attributed to OH molecules and the 3.4 μm feature is attributed to the stretching vibration of the CH molecule. The CH feature does not vary from source to source, suggesting that the CH feature arises from dust in the diffuse ISM (MacFadzean et al. 1990; Sandford et al. 1991; Pendleton 1993). The OH feature is attributed to the presence of a molecular cloud which is fairly local to the Galactic center. The Galactic center sources are seen through varying amounts of the edge of the molecular cloud.

Figure 2a shows the 3.4 μm band seen toward the Galactic center source IRS6. The higher resolution spectra of the Galactic center sources [IRS7 (Sandford et al. 1991) and IRS6 (Pendleton et al. 1993)] show subfeatures at 3.38, 3.42, 3.48, and possibly 3.51 μm, which match the symmetric and asymmetric C-H stretching frequencies of CH_3 (methyl) and -CH_2 (methylene) groups in saturated aliphatic hydrocarbons. The relative peak heights imply that the average CH_2/CH_3 ratio of interstellar hydrocarbon is 2.5 +/- 0.4. The diffuse medium probably contains short chains of hydrocarbons such as -CH_2-CH_2-CH_3.

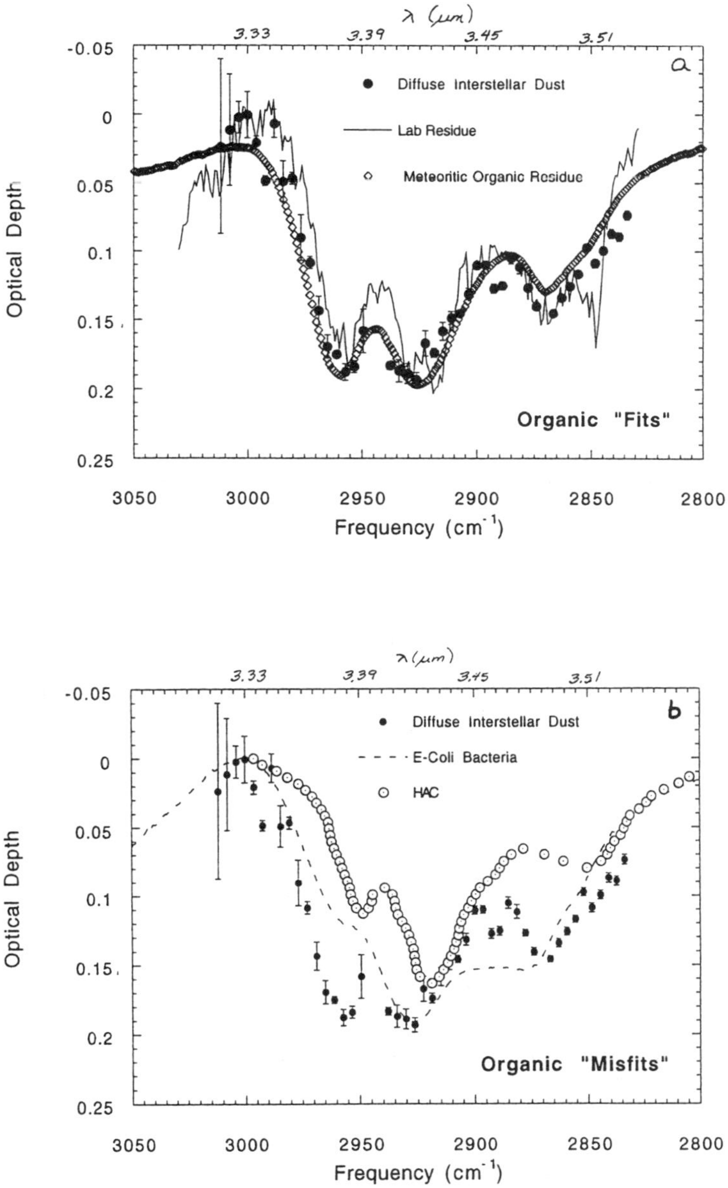

FIGURE 1 Infrared spectra of various materials that have been proposed as "fits" to the diffuse dust spectrum as seen toward the Galactic center source, IRS 7, (solid points) compared to a) lab residue produced through UV photolysis of an interstellar ice mixture that was then warmed to 200K (solid line), the near infrared spectrum of the acid insoluble organic extract from the Murchison meteorite (diamond symbol; deVries et al. 1992), b) Hydrogenated Amorphous Carbon (circles) and an infrared spectrum of E-Coli Bacteria (dashed line) taken at NASA Ames Research Center.

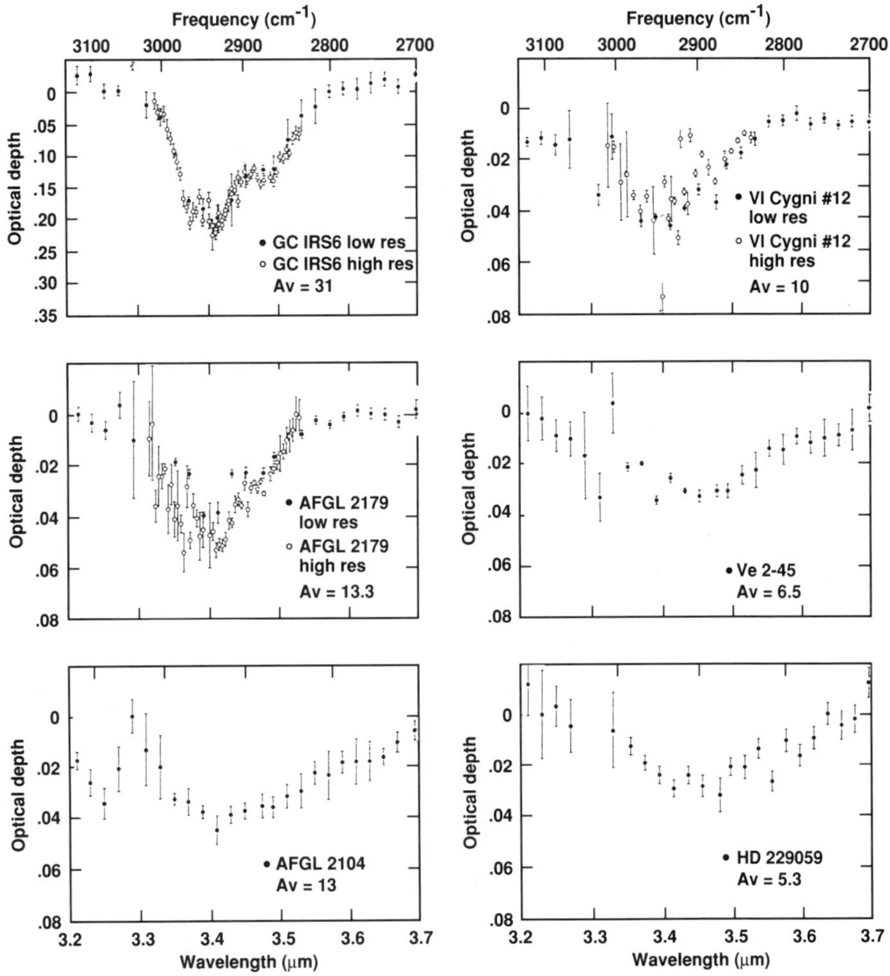

FIGURE 2 A comparison of optical depth plots for several different sightlines through our galaxy. The flux spectra that these plots were made from were obtained at the NASA IRTF using the 32 channel CGAS. The high resolution (R~700) data are superposed on the low resolution data (R~250). The figures (a-f) are ordered in terms of decreasing amounts of intervening extinction. The objects shown are: a) Galactic center source IRS 6 (A_V = 31), b) AFGL 2179 (A_V=13.3), c) AFGL 2104 (A_V=13), d) VI Cygni #12 (A_V=10, e) Ve 2-45 (A_V=6.5), and f) HD 229059 (A_V=5.3).

Other Sightlines Through the Galaxy

Adamson, Whittet, & Duley (1990) and Sandford et al. (1991) detected the 3.4 μm feature a line of sight away from the Galactic center toward the luminous supergiant, VI Cygni #12. These observations supported the assertion that the carrier resides in the diffuse medium. The observations shown in Figure 2 and others shown in Sandford et al. 1991, show that the feature appears along several different lines of sight toward objects which suffer varying amounts of interstellar extinction.

Figure 2 indicates the presence of the feature even in objects which suffer only small amounts of visual extinction. In order to be a good candidate for this type of project, the background "candle" object must not have interfering bands originating from the stellar photosphere. The objects used in this study (Pendleton et al. 1993) were primarily WC stars which have little H in their photospheres and therefore a CH band seen towards these objects likely arises from intervening material. The objects were also carefully selected so that they were not obscured by dense molecular cloud material, although some of the objects are still in the vicinity of the remnant parental cloud material. The absence of a strong OH feature at 3.0 μm provides a good indication that the line of sight does not intercept dense cloud material.

DUST IN THE DENSE AND DIFFUSE MEDIA

Ice mantles are observed in clouds with densities greater than $n_H \sim 10^4$ cm^{-3} (Whittet et al. 1988). As a production mechanism for solid carbon, UV photolysis of interstellar ice mantles in dense clouds is an important source of solid organic material in the diffuse ISM, since this process is efficient enough to surpass the output from AGB stars (Jenniskens et al. 1993). The evolution of organic refractory mantles has been discussed by several authors (Greenberg 1982; Moore & Donn 1982; Strazzulla et al. 1983; Tielens & Allamandola 1987; Schutte 1988). The schematic representation shown in Figure 3 (Pendleton & Cruikshank 1993) illustrates a possible scenario for the cycling of dust between the dense and the diffuse media. Figure 3 shows that with time, the simple ice mantle on a grain undergoes ultraviolet and/or thermal processing from nearby protostars. Upon cloud fragmentation and dissipation, the complex mantles are further heated which causes the volatile material to evaporate. The organic refractory material remains, and becomes incorporated into the diffuse medium. Further processing of the organic refractory grains by UV photons increases the refractory nature of the material as functional groups are removed (Jenniskens et al. 1993). Material from evolved stars mixes with the left over dense cloud material, and a new generation of dense clouds form out of the processed material. Note that the dense cloud scenario is completely analogous to the laboratory astrophysics experiments which produce the organic residue shown in Figure 1a. Infrared spectra of mixtures of molecular interstellar ice analogs reproduce many of the major spectral features attributed to ice in dense clouds (Tielens & Allamandola 1987; Allamandola & Sandford 1988; Allamandola, Sandford & Valero 1988). UV photolysis of such simple molecules produces new species, including radicals, within the ice. Upon heating, these chemically active species become free to move and react to form more complex compounds. Warming to

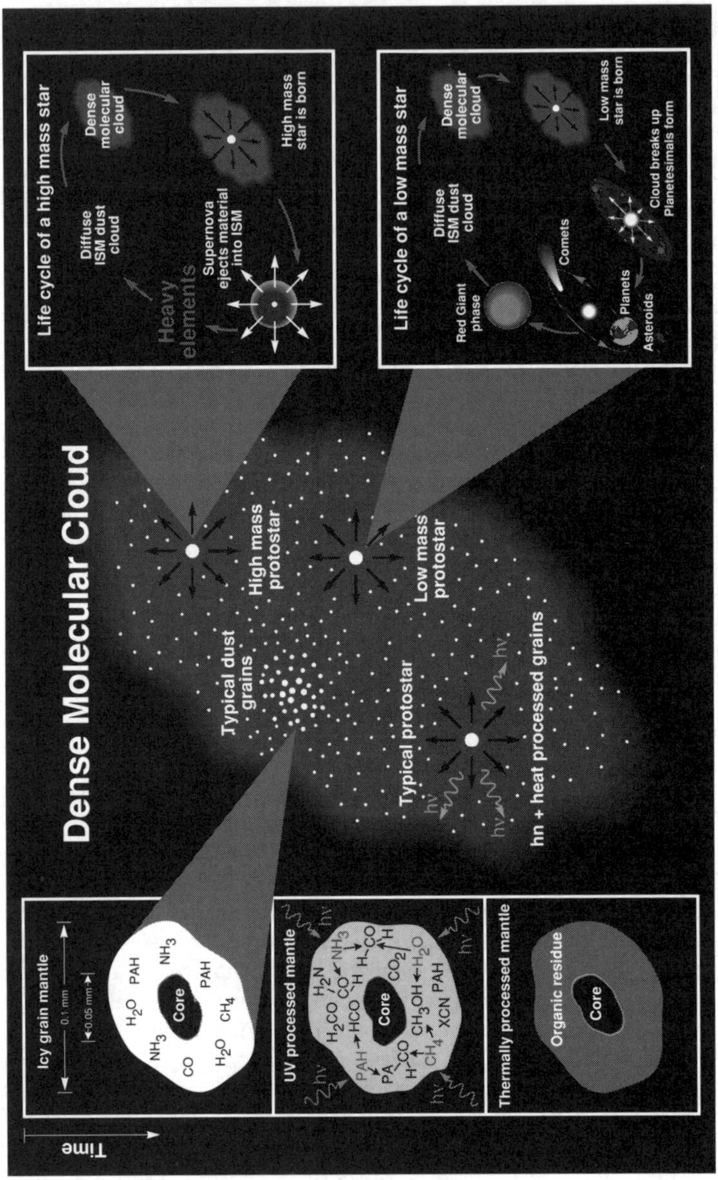

FIGURE 3 A schematic representation of the evolution of dust grains within a dense molecular cloud, and the subsequent enrichment of the diffuse interstellar medium by grains from evolved stars and dissipated molecular cloud material (Pendleton & Cruikshank 1993).

200K leads to the evaporation of the original volatile ices. The remaining residue has a 3 μm infrared spectrum which closely matches the Galactic center source spectra. Therefore, the match shown in Figure 1a supports a model in which the diffuse medium hydrocarbon component contains a mixture of interlinked, aliphatic hydrocarbons which were produced by UV photolysis of interstellar ices.

A puzzling aspect to this scheme, however, is that the organic refractory component does not appear in the spectra of dense cloud sources. Given the relatively short cycling times between the dense and the diffuse clouds (McKee 1989), and the strength of the CH feature, the absence of the aliphatic component in the dense cloud spectra is surprising. Allamandola et al. (1993) reported the detection of a 3.47 μm (2880 cm^{-1}) absorption band in several protostellar sources which they attribute to interstellar diamond-like material. The diffuse medium aliphatic hydrocarbons are rich in methyl (-CH_3) and methylene (-CH_2-) groups which fall shortward and longward of this band, respectively, and they are not seen in the spectra of the embedded protostellar sources. Allamandola et al. (1993) point out that the absence of the -CH_3 and -CH_2- bands in the dense cloud sources implies that C-rich materials in the diffuse medium do not become incorporated into, or do not survive incorporation into, dense molecular clouds. Smith, Sellgren, & Brooke (1993) have obtained spectra of dense cloud sources (and background field stars) for the Taurus cloud. Some of their low resolution observations show a 3.4 μm wing on the strong 3.1 μm water ice band. As expected from the higher resolution Allamandola et al. (1993) results for similar objects, the 3.4 μm wing does not show the substructure seen in the organic component of the diffuse medium. The Smith, Sellgren, & Brooke (1993) study also demonstrates that the 3.4 μm wing has the same A_V threshold as the 3.1 μm ice band, which is surprising if the wing were due to the hardier organic refractory material since the organic material should remain longer than the more volatile water ice component. Higher resolution data and a complete survey of both dense clouds and the diffuse medium are needed to better understand the evolution of dust from one regime to the other.

BAND DEPTH TO EXTINCTION RATIO

The τ/A_V relationship for the hydrocarbon feature is higher toward the Galactic center than toward sources in the local interstellar neighborhood. Figure 4 demonstrates the similarity of this trend with that of the τ/A_V relationship for the 9.7 μm silicate absorption band Roche & Aitken (1984; 1985). A paucity of carbon stars at the Galactic center has been suggested as an explanation for the enhanced silicate abundance (Roche 1988). However, the increasing $\tau_{3.4\mu m}/A_V$ towards the Galactic center may not be consistent with such an explanation. The fact that both materials track each other so well in abundance and the apparent increase of both materials in the Galactic center suggests that grains composed of both silicate cores and organic refractory mantles may be enhanced in the Galactic center. A very recent detection of the 3.4 μm absorption feature toward the nucleus of the Seyfert galaxy NGC 1068 (Bridger, Wright, & Geballe 1993) is the first extragalactic detection of the hydrocarbon feature. Preliminary results suggest that the ratio of $\tau_{3.4\mu m}/\tau_{9.7\mu m}$ for our galaxy (0.06; Adamson et

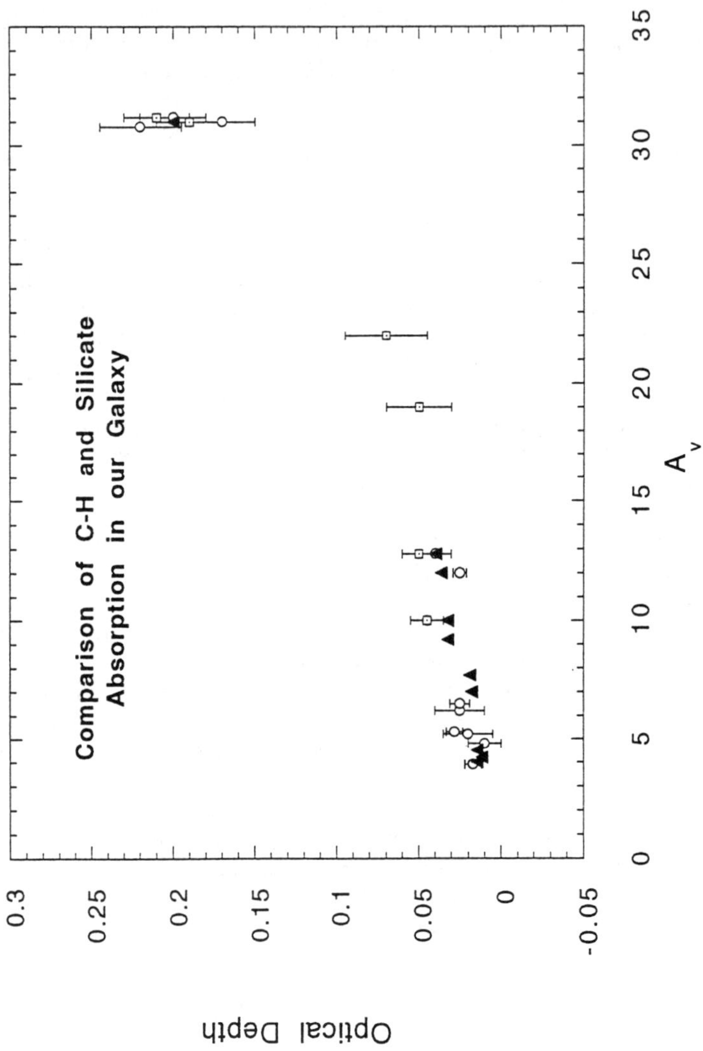

FIGURE 4 Comparison of the optical depth versus extinction data for the hydrocarbon and silicate absorption features. The high A_V sources are all at the Galactic center. The silicate feature optical depth data shown by the filled triangles (Roche & Aitken 1984), was divided by 18 for comparison with the hydrocarbon feature. Low resolution (R=250) and high resolution (R=700) hydrocarbon feature optical depth data (Pendleton et al. 1993) are indicated by open circles and squares, respectively.

al. 1990) is significantly different from that found for NGC 1068 (about 0.16; Bridger, Wright, & Geballe, 1993) which might indicate a higher fraction of organic to silicate dust in the dense torus around the nucleus of NGC 1068. A survey of the relative abundances of these materials in other spiral galaxies may aid in the understanding of the composition of our own galaxy.

An increase in grain size in the Galactic center might account for the enhanced C-H and silicate abundances. Lequeux & Jourdain de Muizon (1990) have detected the C-H (3.4 μm) and SiC (12 μm) bands in the circumstellar (carbonaceous) envelope of a proto- planetary nebula, CRL 618. The ratio of band depth to extinction is smaller in CRL 618 than toward the Galactic center, which they attribute to a lesser degree of hydrogenation of the material. Observations of objects with intermediate A_V values (15-20) would be extremely useful as a probe to the conditions which give rise to this interesting trend.

CONCLUSIONS

The carrier of the 3.4 μm band resides in the diffuse interstellar medium, as evidenced by the presence of the absorption feature along many lines of sight toward a variety of background candles. Higher resolution, higher signal-to-noise data has provided a new level of detail which shows that the 3.4 μm band is actually a combination of several C-H stretching bands. Due to the higher quality data, materials that were previously suggested as fits to the 3.4 μm feature have not proven to be well matched to the diffuse medium dust. Laboratory residues produced through the ultraviolet photolysis of simple interstellar ices and the organic insoluble extract from the Murchison meteorite provide a better fit to the diffuse dust than hydrogenated amorphous carbon material. A remarkable similarity exists between the high resolution diffuse dust spectra and the organic extract from the Murchison meteorite. The correlation between optical depth and A_V for both the hydrocarbon (3.4 μm) and silicate (9.7 μm) absorption features indicates that the two materials may be enhanced in the Galactic center.

ACKNOWLEDGMENTS

This work benefitted from collaborations with the following researchers: Louis Allamandola, Sherwood Chang, Dale Cruikshank, Scott Sandford, Kris Sellgren, and A.G.G.M. Tielens as well as the excellent services of the support staff at the NASA IRTF.

REFERENCES

Adamson, A. J., Whittet, D. C. B., & Duley, W. W. 1990, *MNRAS*, **243**, 400

Allamandola, L. J., Sandford, S. A., & Valero, G. J. 1988, *Icarus*, **76**, 225

Allamandola, L. J., Sandford, S. A., Tielens, A. G. G. M., & Herbst, T. M. 1993, *Science*, **260**, 64

Bridger, A., Wright, G., & Geballe, T. 1993, in *Infrared Arrays*, conference held at UCLA, July 1993.

Brooke, T. Y., Tokunaga, A. T., & Knacke, R. J. 1991, *AJ*, **101**, 268

Butchart, I., McFadzean, A. D., Whittet, D. C. B., Geballe, T. R., & Greenberg, J. M. 1986, *A&A*, **154**, L5

Cronin, J. R., & Pizzarello, S. 1990, *Geochem. et Cosmochim. Acta*, **54**, 2859

Cruikshank, D. P. & Brown, R. H. 1987, *Science*, **238**, 183

deVries, M. S., Reihs, K., Wendt, H. R., Golden, W. G., Hunziker, H., Fleming, R., Peterson, E., & Chang, S. 1993, *Geochim. et Cosmochim. Acta*, **57**, 100

Ehrenfreund, P., Robert, F., d'Hendecourt, L., & Behar, F. 1991, *A&A*, **252**, 712

Greenberg, J. M. 1982, in Submillimetre Wave Astronomy, eds. J. E. Beckman & J. P. Phillips, 261

Hoyle, F. & Wickramasinghe, N. C., 1962, *MNRAS*, **124**, 417

Jenniskens, P., Baratta, G. A., Kouchi, A., deGroot, M. S., Greenberg, J. M., and Strazulla, G. 1993, *A&A*, **273**, 583

Jones, A. P., Duley, W. W., & Williams, D. A. 1987, *MNRAS*, **229**, 213

Lequeux, J. & Jourdain de Muizon, M. 1990, *A&A*, **240**, L19

McFadzean, A. D., Whittet, D. C. B., Longmore, A. J., Bode, M. F., & Adamson, A. J. 1989, *MNRAS*, **241**, 873

McKee, C. F. 1989, in *Interstellar Dust*, eds. L. J. Allamandola and A.G.G.M. Tielens (Kluwer: Dordrecht), 431

Moore, M. H. and Donn, B. 1982, *ApJ*, **257**, L47

Mueller, B., Tholen, D., Hartmann, W., & Cruikshank, D. 1992, *Icarus*, **97**, 150

Pendleton, Y. 1993, in *Infrared Astronomical Observations: Future Observational Directions*, ed. S. Kwok (San Francisco: P.A.S.P. Conference Series), 171

Pendleton, Y. & Cruikshank, D. 1993, *Sky and Telescope*, in press

Pendleton, Y., Sandford, S. A., Allamandola, L. A., Tielens, A.G.G.M., & Sellgren, K. 1993, *ApJ*, submitted

Roche, P. F.,& Aitken, D. K. 1984, *MNRAS*, **208**, 481

Roche, P. F.,& Aitken, D.K. 1985, *MNRAS*, **215**, 42

Roche, P. F. 1988, in *Dust in the Universe*, eds. M. E. Bailey & D. A. Williams (Cambridge University Press), 42

Sandford, S. A., Allamandola, L. A., Tielens, A. G. G. M., Sellgren, K., Tapia, M., & Pendleton, Y. 1991, *ApJ*, **331**, 607

Schutte, W. A. 1988, *The Evolution of Interstellar Organic Grain Mantles*, Thesis, Leiden University

Schutte, W. A.& Greenberg, J. M. 1988, in *Dust in the Universe*, eds. M. E. Bailey & D. A. Williams (Cambridge University Press), 403

Smith, R. G., Sellgren, K., & Brooke, T. Y. 1993, *MNRAS*, **263**, 749

Strazzulla, G., Calcagno, L., Foti, G. 1983, *MNRAS*, **204**, 59

Tielens, A. G. G. M. & Allamandola, L. J. 1987, in *Physical Processes*, eds. G. E. Morfill & M. Scholer (Dordrecht; Reidel), 333

Tokunaga, A.T., & Brooke, T. Y. 1990, *Icarus*, **86**, 208

Whittet, D. C. B., Bode, M. F., Longmore, A. J., Adamson, A. J., McFadzean, A. D., Aitken, D. K., and Roche, P. F. 1988, *MNRAS*, **233**, 321

Wickramasinghe, D. T., & Allen, D. A. 1980, *Nature*, **287**, 518

Wickramasinghe, D. T. & Allen, D. A. 1983, *Ap & SS*, **97**, 369

The First Symposium on the Infrared Cirrus and Diffuse Interstellar Clouds
ASP Conference Series, Vol. 58, 1994
Roc M. Cutri and William B. Latter (eds.)

MODELING COMPOSITE AND FLUFFY GRAINS: THE EFFECTS OF POROSITY

MICHAEL J. WOLFF
Space Astronomy Laboratory, University of Wisconsin, 1150 University Ave., Madison, WI 53706

GEOFFREY C. CLAYTON
Center for Astrophysics and Space Astronomy, Campus Box 389, University of Colorado, Boulder, CO 80309

PETER G. MARTIN
Canadian Institute for Theoretical Astrophysics, University of Toronto, Toronto, ON M5S 1A1, Canada

REGINA E. SCHULTE-LADBECK
Department of Physics and Astronomy, University of Pittsburgh, 3941 O'Hara Street, Pittsburgh, PA 15260

ABSTRACT Recent studies of interplanetary and interstellar dust provide evidence that cosmic dust grains are fluffy, composite objects, highlighting the need for models of the electromagnetic scattering by these grains. Effective Medium Theory (EMT) with Mie-type series solutions has been used to explore the effects of porosity which would be important in composite dust particles. While this indirect approach is both flexible and computationally efficient, it is not necessarily a good approximation. The need for EMT and its rather restrictive assumptions may be circumvented through a direct computation of the scattering properties via finite element methods, such as the Discrete-Dipole Approximation (DDA).

Extensive calculations with the DDA method are used to examine more directly the effects of porosity. DDA cross sections for grains with a range of porosity are compared to those computed by the EMT/series expansion technique to examine the applicability of several mixing rules, including two extensions of the Bruggeman rule. We show that for particles with Rayleigh vacuum inclusions the extension proposed by Rouleau & Martin is quite successful. We also investigate the effects of larger, non-Rayleigh vacuum inclusions for various levels of porosity and find that they can be significant.

An article containing a full presentation of the results and a detailed discussion of the computational methodologies has been submitted to the Astrophysical Journal.

INTRODUCTION

An often used model of interstellar grains consists of a power-law size distribution of separate populations of bare spherical silicate and graphite grains (Mathis, Rumpl, & Nordsieck 1977). While this kind of simple model has been valuable, it is likely that real interstellar grains are more complicated. Mathis & Whiffen (1989) model the electromagnetic scattering by loose, porous aggregates containing using Effective Medium Theory (EMT) combined with Mie-type series solutions. Although the EMT method is attractive, the literature contains conflicting reports of its failures and successes (cf., Bohren & Huffman 1983). With the recently improved feasibility of finite element methods, comparisons of direct and indirect approaches have become possible, as described above. However, the lack of sufficient coverage of parameter space due to the computational intensity of finite element methods and, in some publications, the lack of numerical details leave several basic questions unanswered. Below we address whether there is an EMT mixing rule which is valid over a large range of size parameter and porosity, and the effects of having vacuum inclusions that are not in the Rayleigh limit.

RESULTS AND DISCUSSION

In Figure 1 we compare the DDA results for extinction to those of the EMT/series solution method for 60% porosity. We treat the cases of individual vacuum inclusions in both the Rayleigh (single dipole) and non-Rayleigh (cluster removal) regimes. The overall agreement between DDA and the mixing rule results is quite good. With either approach there are a number of systematic changes that can be followed through the sequence of increasing porosity. At small x, Q_e decreases with increasing porosity. The first peak in Q_e shifts to higher x with decreasing porosity. There is a systematic departure of the results for the DDA non-Rayleigh case from those for Rayleigh inclusions and the various EMT solutions. The magnitude of the departure increases with porosity, with the fractional diameter of the non-Rayleigh inclusions, and, for small x, with x (entering the Mie regime). The sense of the departure is that the non-Rayleigh values of Q_e rise more quickly with x toward their peak. The systematic effect of increasing the diameter of the porous inclusions appears to be opposite that of increasing the porosity: with increasing inclusion diameter the peak shifts to smaller x.

SUMMARY

The effects of grain porosity on the scattering properties of cosmic dust grains are of great interest. Despite the number of previous investigations, basic questions regarding the nature of porous grains and the applicability of EMT to this problem remain unanswered. In an effort to resolve these issues, we present extensive DDA and EMT/series solution calculations in the Mie regime for several porosity levels (10, 20, 40, 60, and 80%) and DDA-grain inclusion sizes (Rayleigh and non-Rayleigh). We are able to make the following conclusions:

1. It is possible to construct an EMT which agrees well quantitatively with finite

FIGURE 1 Extinction for 60% porosity for two orientations of the incident electric vector, E and H, which are parallel and perpendicular to the major axis of a 2:1 prolate spheroid, respectively. The efficiency factor Q_e is defined with respect to the projected area; grain alignment is "picket fence". The error bars are twice the average deviation as computed from a sequence of grains with a unique random number seed (same porosity type and number of dipoles). Any error bars which would be within the boundary of a symbol are omitted. The size parameter is with respect to the semi-minor axis, b. The base grains have 8320, 20606, and 34336 dipoles for the ranges $x \leq 3$, $4 \leq x \leq 7$, and $x \geq 8$, respectively.

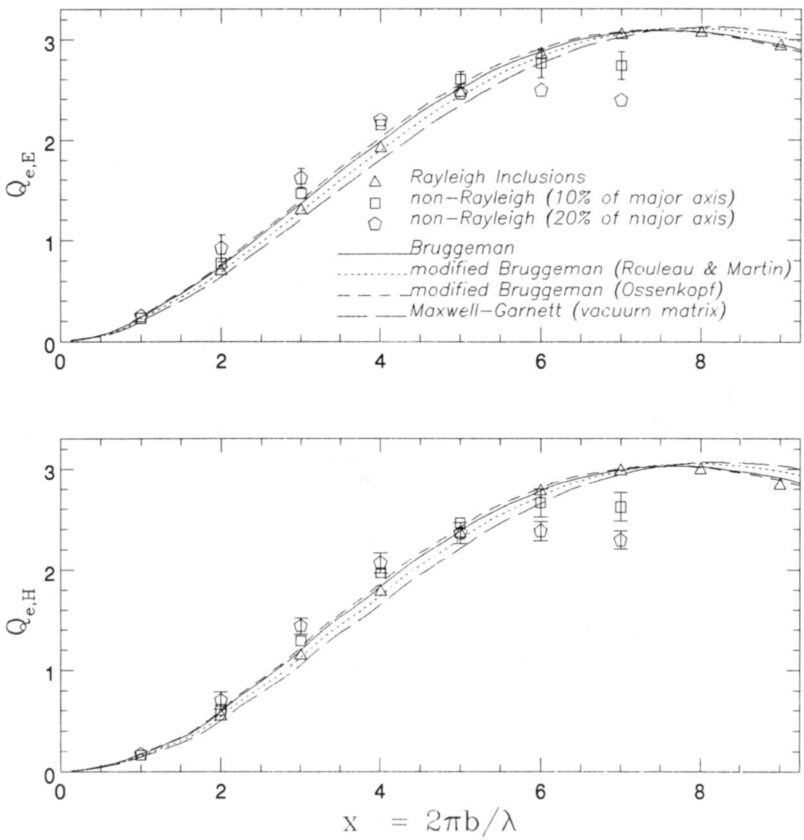

element methods over a large range of porosity and size parameter. The characteristics of the finite element vacuum inclusions are restricted to those which preserve the EMT assumptions of statistical uniformity and small size compared to the wavelength. This can be accomplished most easily by grains with small concentrations of Rayleigh vacuum inclusions (single dipole removal in DDA). Larger concentrations of Rayleigh inclusions require empirically-calibrated modifications to the traditional EMT mixing rules. Of those tested, the most successful mixing rule is the extension to the Bruggeman rule proposed by Rouleau & Martin (1993). It reproduces the DDA results for Rayleigh inclusions through a size parameter value of 9 (the maximum tested) and a porosity level of at least 60%.

2. The effects of non-Rayleigh vacuum inclusions whose diameters are appreciable fractions (10 and 20%) of the grain size can be significant. They produce a shift in the position of maximum extinction toward smaller x, opposite the trend caused by simply increasing the level of porosity. Perhaps not surprisingly, no EMT studied was able to reproduce these results in an entirely satisfactor manner. Thus, finite element methods play a crucial role for grain models embodying these types of porosity.

ACKNOWLEDGMENTS

We acknowledge productive conversations with Bruce Draine, John Mathis, Ken Nordsieck, and Francois Rouleau. Bruce Draine kindly provided DDSCAT4.b The majority of our DDA computations were made possible by a grant from the Pittsburgh Supercomputing Center and the excellent support of its technical staff. MJW was supported by a fellowship from the NASA Graduate Student Researchers' Program and NASA grant NAS 5-2677. NASA grant NAGW-2338 provided support for GCC. Research support to PGM from the Natural Science and Engineering Research Council of Canada is acknowledged.

REFERENCES

Bohren, C. F., & Huffman, D. R. 1983, *Aborption and Scattering of Light by Small Particles*, New York: Wiley

Mathis, J. S., Rumpl, W., & Nordsieck, K.H. 1977, *ApJ*, **217**, 425

Mathis, J. S., & Whiffen, G. 1989, *ApJ*, **341**, 808

Ossenkopf, V. 1991, *A&A*, **251**, 210

Rouleau, F., & Martin, P. G. 1993, *ApJ*, submitted

The First Symposium on the Infrared Cirrus and Diffuse Interstellar Clouds
ASP Conference Series, Vol. 58, 1994
Roc M. Cutri and William B. Latter (eds.)

STABILITY OF SMALL GRAINS IN HII REGIONS AND REFLECTION NEBULAE

M. W. WERNER AND T. N. GAUTIER III
Jet Propulsion Laboratory, California Institute of Technology, Pasadena, CA 91109

T. CAWLFIELD
Principia College, Elsah, IL 62028

ABSTRACT We have analyzed IRAS data to assess the relative amounts of small and large grains in HII regions and reflection nebulae. Our most important finding is that no evidence for small grain destruction is seen in reflection nebulae, even for [high] values of the radiation energy density at which significant grain destruction apparently occurs in HII regions. This suggests that it is not only the total radiant energy density but also the energy per photon which determines the stability of small grains in astrophysical environments.

INTRODUCTION

The mid-infrared emission from interstellar dust is in many cases considerably enhanced by a large contribution due to non-equilibrium emission from very small grains. Sellgren (1989; this volume) discusses the evidence that the 3 to 25 μm radiation from reflection nebulae must be mainly due to non-equilibrium emission from small grains, and many authors have shown that non-equilibrium emission must be responsible for much of the 12, 25 and even 60 μm radiation seen by IRAS from the diffuse cirrus (Weiland *et al.*, 1986), from interstellar dust in other regions of the Milky Way, and from external galaxies as well. For the present discussion we need not distinguish between the "macromolecule" and "micrograin" picture of these small particles, and refer to them simply as "small grains". Ultimately, however, we hope that improved understanding of the stability of these small grains under varying conditions will provide useful constraints on their nature.

Although the prevalence of the enhanced short wavelength emission suggests that the small grains are a hardy and ubiquitous species, there is evidence that this emission is suppressed under certain circumstances. In particular, Helou, Ryter and Soifer (1991) summarize data on three extended interstellar sources showing that the 12um to (60um + 100um) brightness ratio, gamma, defined here as

$$\Gamma = \frac{\nu I_\nu(12\mu\text{m})}{\Delta\nu I_\nu(60\mu\text{m}) + \Delta\nu I_\nu(100\mu\text{m})} \quad (1)$$

decreases as the 60/100um brightness ratio, theta

$$\Theta = \frac{I_\nu(60\mu m)}{I_\nu(100\mu m)} \tag{2}$$

increases. Because Γ is related to the abundance ratio of small to large grains, and Θ to the temperature of the large grains which are in equilibrium with the heating starlight, this correlation suggests that the small grains are preferentally destroyed in regions of higher radiation energy density.

Helou *et al.* find that this effect is much more pronounced for two HII regions (σ Sco and the California Nebula) than for the reflection nebula associated with HD147889, which is also included in their sample. The current study was, in fact, motivated by our observation that a group of reflection nebulae studied by Sellgren, Luan and Werner (1990) showed no variation of Γ with Θ. We therefore inferred that there might be differences in the behavior of small grain populations in reflection nebulae and in HII regions. We have explored this conjecture by using the IRAS data to study the spatial distribution of the infrared radiation from HII regions and reflection nebulae and to investigate the relative small grain population as a function of both the stellar temperature and energy density of the illuminating radiation. A more complete discussion of these data and their implications is presented by Gautier *et al.*, (1993).

RESULTS

We used the recently released Infrared Sky Survey Atlas (ISSA) data to obtain surface brightness measurements of reflection nebulae and HII regions at wavelengths of 12, 60 and 100 microns. Our surface brightness measurements were confined to the visible boundary of the nebulae, as determined from the Palomar prints. Background emission was removed based on measurements in the surrounding regions and linear interpolation across the program targets; the errors in the brightness measurements were estimated from the deviations of the background emission from this linear fit. For each nebula, a series of measurements spaced by about 6 arcmin were made using a 4.5×4.5 arcmin square aperture.

Figure 1 shows measurements of Γ vs. Θ for two HII regions [the California nebula and the Rosette nebula], while Figure 2 shows similar measurements for the Pleiades and NGC7023 reflection nebulae. Typical errors are shown for a few points in NGC7023; for the other sources, the errors are considerably smaller, as can be seen as well from the systematic clustering of the measurements. Note that the reflection nebula and HII region data extend over the same range of log Θ, from < -0.7 to > 0.0, corresponding empirically to starlight energy densities in the range < 1 to > 80 ev/cm^3 (Boulanger *et al.*, 1988).

Over this range of Θ, the relative amount of emission from small grains, as measured by the parameter Γ, decreases by about an order of magnitude in each of the HII regions but is essentially constant in the reflection nebulae. The points shown in Figures 1 and 2 are scattered throughout the nebulae under study, but the variation [or lack thereof] of Γ with Θ is seen systemtatically on radial cuts from the edge to the center of the individual nebulae as well.

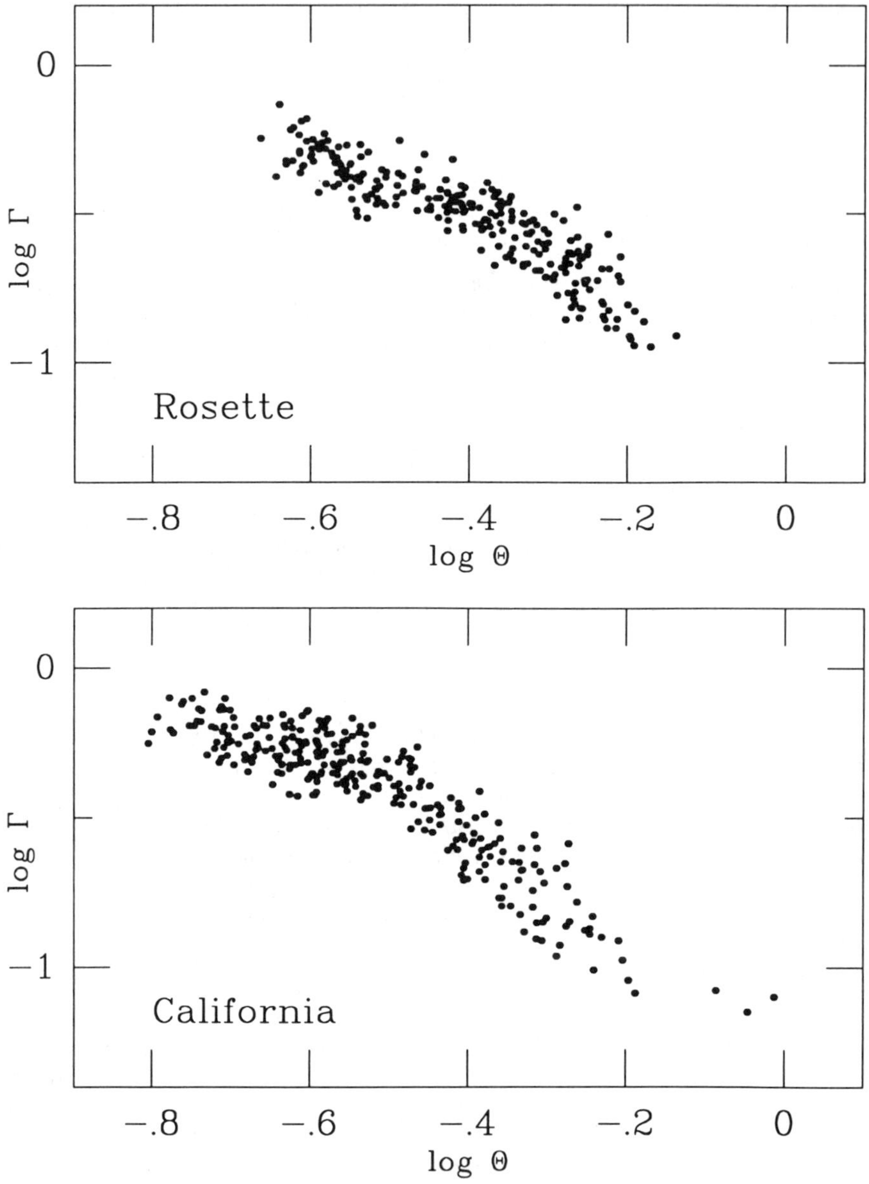

Figure 1. Measurements of gamma vs. theta for two HII regions, the Rosette and California nebulae. $\Gamma = \nu I_\nu(12\mu m)/\Delta\nu I_\nu(60\mu m) + \Delta\nu I_\nu(100\mu m)$, is a measure of the relative abundance of small grains, while $\Theta = I_\nu(60\mu m)/I_\nu(100\mu m)$ is a measure of the overall radiant energy density. The measurements were made on the IRAS Sky Survey Atlas data with a 4.5 arcmin square aperture at points space 6 arcmin apart across the entire nebulae.

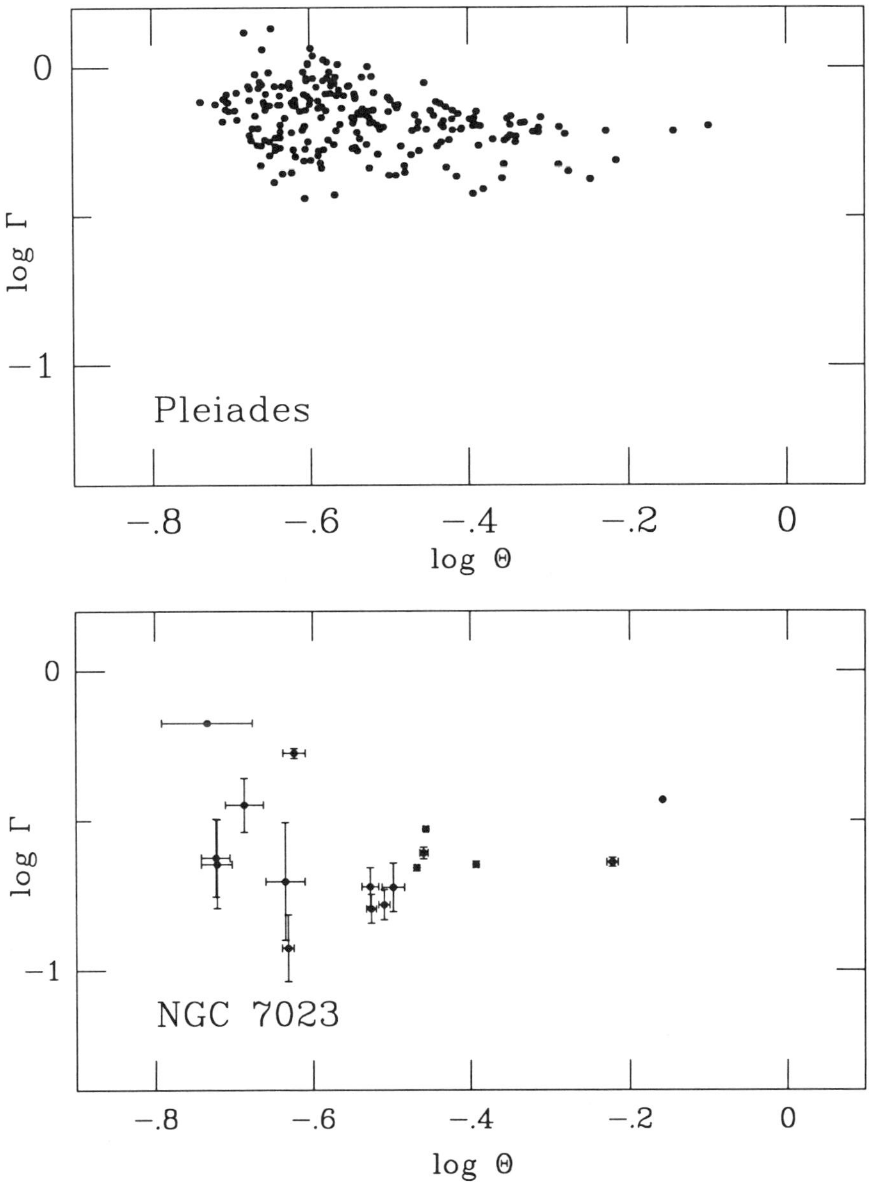

Figure 2. Same as Figure 1 for two reflection nebulae, NGC7023 and the Pleiades. Typical errors are shown for several points in NGC7023. For the Pleiades and the HII regions in Figure 1, the errors are smaller than this.

The most natural explanation of this result is that the small grains are preferentially destroyed in regions of increased radiation energy density within HII regions but are able to survive in similar radiation fields in the reflection nebulae. For this purpose, the most significant difference between the HII regions and the reflection nebulae is probably the presence of higher energy photons in the former. The effective temperatures of the O6 stars which excite the HII regions in our study are about twice those of the B3-B7 stars which illuminate the selected reflection nebulae. This suggests that it is not only the total energy density, but also the energy per photon, which determines whether the small grains will be stable, and thus that one-or-two-photon processes may be responsible for destroying the small particles. Two processes which should be considered in this context are double ionization of large molecules followed by dissociation (Leach, 1989), or enhanced evaporation resulting from thermal pulsing induced by absorption of two or more photons in rapid succession (Leger and Puget, 1984).

ACKNOWLEDGMENTS

This work was carried out at the Jet Propulsion Laboratory, California Institute of Technology, under contract with the National Aeronautics and Space Administration. T. Cawlfield was supported by a Caltech Summer Undergrduate Research Fellowship.

REFERENCES

Boulanger, F., Beichman, C., Desert, F.X., Helou, G., Perault, M., and Ryter, C. 1988, *ApJ*, **332**, 328

Gautier, T.N., Werner, M.W., and Cawlfield, T. 1993, in preparation.

Helou, G., Ryter, C. and Soifer, B.T. 1991, *ApJ*, **376**, 505

Leach, S. 1989, in *Interstellar Dust, IAU Symposium 135*, eds L.J. Allamandola and A.G.G.M. Tielens, Dordrecht: Kluwer, 155

Leger, A., and Puget, J.L. 1984, *A&A*, **137**, L5

Sellgren, K. 1989, in *Interstellar Dust, IAU Symposium 135*, eds L.J. Allamandola and A.G.G.M. Tielens, Dordrecht: Kluwer, 103

Sellgren, K., Luan, L, and Werner, M.W. 1990, *ApJ*, **359**, 384

Weiland, J., Blitz, L., Dwek, E., Hauser, M.G., Magnani, L., and Rickard, L.J. 1986, *ApJ*, **306**, L101

The First Symposium on the Infrared Cirrus and Diffuse Interstellar Clouds
ASP Conference Series, Vol. 58, 1994
Roc M. Cutri and William B. Latter (eds.)

ON THE NATURE OF ORGANIC MATTER IN THE DIFFUSE MEDIUM

PETER JENNISKENS
NASA/Ames Research Center, Mail-Stop 239-4, Moffett Field, CA 94035

ABSTRACT Accretion of H, C, O, and N on silicate cores in dense medium environments, and subsequent photolysis of the dirty ice mantles formed, is an efficient process for creating reduced organic matter. However, the initial product consists of small 2C–3C compounds only and is rich in oxygen (O/C$\sim$2.0). This organic refractory residue is processed in the diffuse medium into a more complex and more refractory matter by UV photons. Laboratory simulations of the processing show the residue to evolve into a hydrogenated amorphous carbon with O/C<0.2, H/C $\sim$0.6 and an optical gap of 0.6–1.1 eV on a typical diffuse medium timescale of 3×10^7 yrs.

INTRODUCTION

Organic matter in the diffuse medium shows up as a highly absorbing compound in visual extinction, as an emitter of mid-infrared emission bands in reflection nebulae and as an absorber at 3.4 micron a.o. towards IRS-7. Organic matter may also be responsible for the bump in the extinction curve at 220 nm and the Diffuse Interstellar Bands (DIB).

There are two possible sources of reduced organic matter: nucleation of freshly synthesised carbon in carbon rich stellar outflows and accretion of existing gas-phase carbon (C, CO) in the interstellar medium. Abundance constraints over time favour the latter mechanism, unless there is a continuous infall of carbon poor extragalactic gas. For reviews on the accretion mechanism see Tielens & Allamandola (1987), Tielens et al. (1989), and Greenberg (1991).

Accretion occurs in our galaxy at a rate of 1×10^{24} g/s and mainly in the dense (as opposed to diffuse) medium where accretion timescales are less than 10^6 yrs. Photolysis of the resulting dirty ice mantles results in an organic refractory residue when the ice mantles sublime and the stored radicals HCO, OH and NH_2 are allowed to combine (Greenberg et al. 1972, Hagen et al. 1979). The product consists mainly of 2C–3C compounds like glycolic acid, hydroxyacetamide, glyceramide etc. (Argarwal et al. 1985). The abundant presence of oxygen and nitrogen in carboxylic acid and amine groups cause the material to be non-volatile at low pressure and at room temperature because of hydrogen bonding.

The signature of organic matter in the diffuse medium suggests it to be highly unsaturated and complex. Some of the arguments that link it to complex organic matter found in meteorites and IDP's are summarized in Jenniskens

(1992). To get to this complex a product, processing of the initial organic refractory residue by UV photons or cosmic ray ions is needed. Pyrolysis will evaporate the organic residue before being processed.

THE PROCESSING

The processing of organic refractory residue was studied for cosmic ray analogue 3 keV He–ion bombardment at the Catania Institute of Astronomy (G. Strazzulla, G.A. Baratta) and subsequently for 10 eV UV photon photolysis at the Leiden Laboratory Astrophysics (A. Kouchi, M.S. de Groot, J.M. Greenberg). The experiments are described in Jenniskens et al. (1993). UV sources are currently not strong enough to give the high dose of deposited energy characteristic of diffuse medium conditions. UV photon processing is the dominant source of carbonisation (loss of H, O and N) and graphitization (formation of polyaromatic structures), although the latter by only two orders of magnitiude while depositing five orders of magnitude more energy per unit volume.

Carbonisation is fast initially, and then slows down when the number of alternative energy release channels increases. On a timescale of some 200 yrs, the amount of oxygen over carbon has decreased to 1/e. Obviously, the freshly synthesised organic refractory residue is not found in the diffuse medium. Instead, the matter is processed and oxygen (and nitrogen) poor.

Polymersiation by UV photons is not an efficient process. Although carbonisation leads to unsaturated bonds in the material and possibly to cross linking, the formation of a graphitic lattice proceeds only very slowly. *In situ* Raman spectroscopy was applied to follow the D and G band lattice vibrations. During ion bombardment the bands are first detected at a dose of 1×10^{24} eV/cm^3. UV photolysis does not lead to detectable Raman bands before at least 7×10^{26} eV/cm^3. The formation of unsaturated bonds and polyaromatic structure is associated with the optical darkening of the residue. Optical spectroscopy shows the residue to have amorphous solid properties: the collection of carbon atoms (each contributing 4 energy states) has a parabolic distribution of densities of states in the valence band. This shows up in a Tauc diagram, When the absorbance (α) is plotted in a Tauc diagram, in which $\sqrt{\alpha E}$ is plotted versus E $\sim 1/\lambda$, a series of linear slopes are found. Processing causes a progressive shift of the slope with a decreasing intercept at α=0. The intercept is called the optical gap. This optical gap as a function of amount of processing (total energy deposited per unit volume) can be compared to the amount of processing on a typical diffuse medium timescale of 3×10^7 yrs. Extrapolation of the (still very limited) data for UV processsing show the solid in the diffuse medium to have a relatively large optical gap of 0.6–1.1 eV. This corresponds to polyaromatic structures of at most 30–60 fused rings (Robertson 1986).

DIFFUSE MEDIUM EXTINCTION

Although it may be difficult to define the diffuse medium as an object, interstellar extinction curves in lines of sight with no apparent heating (increase of 60/100 micron flux ratio in IRAS Skyflux Maps) and low total extinction ($A_V < 3$ mag) do stand out among all extinction curves in the sample of Aiello et al. (1988).

Put in the parameterisation scheme of Fitzpatrick & Massa (1988), the diffuse medium lines of sight have a small range in values for both the linear rise (c_2) and bump width (γ), with $c_2 = 0.74 \pm 0.13$ and $\gamma = 0.943 \pm 0.064$.

The average diffuse medium extinction curve has somewhat higher FUV non-linear rise (c_4) than in the average curves of Seaton (1979) and Savage & Mathis (1979). The average curve amounts to:

$$k(x) = -0.19 + 0.74x + \frac{3.46x^2}{(x^2 - 4.583^2)^2 + 0.943^2 x^2}$$

$$5.9 < x < 8.0: \qquad +0.45[0.5392(x-5.9)^2 + 0.0564(x-5.9)^3]$$

where $k(x) = E_{\lambda-V}/E_{B-V}$ and $x = 1/\lambda \mu\text{m}^{-1}$ (Jenniskens & Greenberg 1993). Higher values of c_4 correlate with a higher abundance of molecular hydrogen with respect to all matter in the lines of sight (Jenniskens, Ehrenfreund & Désert, 1992).

DIFFUSE INTERSTELLAR BANDS

The reduced carbon mantles in the diffuse medium may erode into small particles and ultimately into PAHs. It still has to be proven that his can happen in diffuse medium conditions. However, the building blocks are present in the processed material. PAHs are thought to account for the mid-infrared emission bands observed in reflection nebulae, the continuum mid-infrared excess emission, and are possibly responsible for the diffuse interstellar bands (Salama & Allamandola 1992). For a comparisson of diffuse band strength and UV extinction, see Désert et al. (1993).

Laboratory studies of DIBs (i.e. Salama & Allamandola 1992, Ehrenfreund et al. 1992) may have found some support from a systematic survey of DIBs (Jenniskens & Désert), which shows the visual and near infrared spectrum towards obscured early type stars at 0.3 Å resolution. The total number of (unidentified) absorption features is currently some 200. DIBs are found throughout the diffuse medium, even at very low reddening.

There is now evidence that DIBs are weakened in very low reddened lines of sight (Jenniskens, Ehrenfreund & Foing 1993). The data are interpreted in terms of large molecules being the carriers of DIBs and different ranges for the ionization potential of two families of DIBs are inferred. Some remarkable structure in the two strongest DIBs, λ6284 and λ5780 may also point towards an origin in large molecules (Jenniskens & Désert 1993). I refer to these papers for a broader context and references that give more appropriately credit to other workers in the field.

ACKNOWLEDGMENTS

I am indepted to Gianni Strazzulla, Guiseppe Baratta, Akira Kouchi, and Menno de Groot for their support in the experiments described here and to Mayo Greenberg, Harm Habing and Xander Tielens for guidance.

REFERENCES

Aiello S., Barsella B., Chlewicki G., Greenberg J.-M., Patriarchi P., Perinotto M., 1988, *A&AS*, **73**, 195

Agarwal V.K., Schutte W., Greenberg J.M., Ferris J.P., Briggs R., Connor S., van de Bult C.P.E.M., Baas F., 1985, *Origins of Life*, **16**, 21

Désert F.-X., Jenniskens P., Dennefeld M., 1993, *A&A*, in press.

Ehrenfreund P., d'Hendecourt L., Verstraete L., Léger A., Schmidt W., Defourneau D., 1992, *A&A*, **259**, 257

Fitzpatrick E.L., Massa D., 1988, *AJ*, **328**,734

Greenberg J.M., Yencha A.J., Corbett J.W., Frisch H.L., 1972, *Mem. Soc. Roy. des Sci. de Liege 6e ser.*, 3, 425

Greenberg J.M., 1991, in *Chemistry in Space*, Greenberg & Pirronello eds., p. 227

Hagen W., Allamandola L.J., Greenberg J.M., 1979, *Ap&SS*, **65**,215

Jenniskens P., 1992, in *Proceedings of the 30th Liege Intern. Astroph. Coll.*, A. Brahic & J. Surdej eds., p. 335

Jenniskens P., Ehrenfreund & Désert F.X., 1992, *A&A*, **265**, L1

Jenniskens P, Baratta G.A., Kouchi A., de Groot M.S., Greenberg J.M., Strazzulla G., 1993, *A&A*, in press.

Jenniskens P., 1993, *A&A*, in press.

Jenniskens P., Greenberg J.M., 1993, *A&A*, in press.

Robertson J., 1986, *Adv. Phys.*, **35**, 317

Salama F., Allamandola L.J., 1992, *Nature*, **358**,42

Savage B.D., Mathis J.S., 1979, *ARA&A*, **17**, 73.

Seaton M.J., 1979, *MNRAS*, **187**, 73P.

Tielens A.G.G.M., Allamandola L.J., 1987, in *Interstellar Processes*, Hollenbach D.J., Thronson H.A. eds., p. 397

Tielens A.G.G.M., Allamandola L.J., Sandford S.A.,1989, in *Solid State Astrophysics*, Bussoletti E. & Strazzulla G. eds., North-Holland, p. 29

The First Symposium on the Infrared Cirrus and Diffuse Interstellar Clouds
ASP Conference Series, Vol. 58, 1994
Roc M. Cutri and William B. Latter (eds.)

PAHS, DIBS, AND INTERSTELLAR EXTINCTION

FARID SALAMA AND LOUIS ALLAMANDOLA
NASA-ARC, Space Science Division, MS: 245-6, Moffett Field, CA 94035

ABSTRACT Neutral naphthalene ($C_{10}H_8$), phenanthrene ($C_{14}H_{10}$), and pyrene ($C_{16}H_{10}$) absorb strongly in the ultraviolet and may contribute to the extinction curve. The cations of these PAHs absorb in the visible. $C_{10}H_8^+$ has 12 discrete absorption bands which fall between 5000 and 6800 Å . The strongest band at 6741 Å falls close to the weak 6742 Å diffuse interstellar band (DIB). Six other weaker bands also match DIBs. If $C_{10}H_8^+$ is indeed responsible for the 6742 Å feature, it accounts for 0.3% of the cosmic C. The spectrum of $C_{16}H_{10}^+$ is dominated by a strong band at 4435 Å in an Ar matrix and 4395 Å in a Ne matrix, a position which falls very close to the strongest DIB, that at 4430 Å. If $C_{16}H_{10}^+$, or a closely related pyrene-like ion is indeed responsible for the 4430 Å feature, it accounts for 0.2% of the cosmic C. We also report an intense, very broad, UV-to-visible continuum which is associated with both ions and could explain how PAHs convert interstellar UV and visible radiation into IR.

INTRODUCTION

PAHs have recently been recognized as an important component of the carbon containing material in the ISM. PAHs are extremely stable in their neutral and ionized forms and their ubiquitous presence in the ISM make them attractive candidates for the diffuse interstellar absorption bands (DIBs) /1/. The UV-visible-NIR absorption properties of the *free, unperturbed species*, are not known. Matrix isolation spectroscopy provides a perfect tool to study these properties by simulating the low temperature (4 - 15 K) and isolation conditions in space. Considering the need for spectroscopic data on neutral and ionized PAHs under conditions relevant to astrophysical environments to test the existing theories and their implications, we have initiated a systematic program to investigate their UV-Visible-NIR spectroscopic properties. Here, the spectroscopic characteristics of the small PAHs, naphthalene ($C_{10}H_8$) /2/, phenanthrene ($C_{14}H_{10}$) and pyrene ($C_{16}H_{10}$) /3/, in their neutral and positive ion forms are briefly reviewed and the aspects of this study which are relevant to astrophysics are discussed. This work has been described in reference 4.

RESULTS AND DISCUSSION

A. Spectroscopy of Neutral PAHs
The experimental apparatus and technique are described in reference 2 (a). Naphthalene ($C_{10}H_8$). Over the wavelength range covered in these experiments

(1800-9000 Å), neutral naphthalene isolated in Ne exhibits 3 distinct vibronic band systems with different intensities peaking at 2116, 2685, and 3076 Å. The strongest system, that at 2116 Å, has an oscillator strength, f of 1.0 and dominates the neutral spectrum /2/. No absorption is observed in the visible.

Phenanthrene ($C_{14}H_{10}$). The UV-to-NIR absorption spectrum of phenanthrene isolated in Ar consists of several distinct band systems with different intensities. Over the wavelength range covered in these experiments (2400 - 9000 Å), neutral phenanthrene exhibits 3 band systems peaking at 2894, 3436, and 3875 Å . No absorption is observed in the visible range.

Pyrene ($C_{16}H_{10}$). Neutral pyrene isolated in Ne shows only three distinct band systems over the wavelength range covered in our experiments (1800-9000 Å). The strongest absorptions have similar intensities and peak at 2320, 2655, and 3258 Å in a Ne matrix. Neutral pyrene does not absorb in the visible.

Astrophysical Implications of the Neutral PAH Spectra. From the astrophysical perspective, these results indicate that: (i) Small, neutral PAHs (less than 20 C atoms) do not absorb in the visible and therefore cannot contribute to the known DIBs. (ii) The 3 PAHs studied have strong UV absorption bands with oscillator strengths in the range 0.2 - 1.0. If present, they could contribute substructure to the UV portion of the interstellar extinction curve. The cross sections of the strongest bands are 4.0×10^{-16} , 1.1×10^{-16} and 1.5×10^{-16} cm^2 $molecule^{-1}$ in the case of naphthalene, phenanthrene and pyrene /2, 3/ respectively. In order for these species to produce substructure on the extinction curve at the 10% level would require neutral PAH abundances of respectively 1.3×10^{-7}, 2.0×10^{-7}, and 1.5×10^{-7} with respect to hydrogen.

B. Spectroscopy of Ionized PAHs

The Naphthalene Cation ($C_{10}H_8^+$). The new spectral features associated with the $C_{10}H_8^+$ that arise upon VUV irradiation fall into 7 band systems peaking at 6741, 4560, 3769, 3050, 2760, 2443, and 2227 Å respectively in Ne. A continuum is also produced which is apparently associated with the cation /2/. Two features of the cation absorption spectrum are particularly relevant to astrophysical questions: (i) the visible-NIR vibronic system consisting of several discrete bands with the strongest at 6741 Å; and (ii) the broad and strong continuum extending from the UV to the visible (2000 - 3800 Å, with a weak tail to ~ 5100 Å).

3. The Pyrene Cation ($C_{16}H_{10}^+$). The new discrete spectral features associated with $C_{16}H_{10}^+$ peak at 3606, 4293, 4395, 4546, and 4840 Å in a Ne matrix. The entire UV-Vis-NIR spectrum of $C_{16}H_{10}^+$ is dominated by the strong absorption band which falls at 4395 ± 5 Å in Ne and 4435 ± 5 Å in Ar, remarkably close to the peak position of the well-known 4430 Å DIB /3/. The effect of matrix material on band positions is dramatically illustrated here where a red shift of 40 Å is measured when going from Ne to Ar. A continuum is also produced which extends from 2000 to 3600 Å having a weak tail out to ~ 5100 Å. In view of the proximity of the matrix isolated pyrene cation band to the well-known 4430 Å DIB, it is of interest to study the influence of side-groups on the pyrene moiety to determine if pyrene-like species might indeed be responsible for this DIB. In order to investigate the influence of side groups on the pyrene cation spectrum, we have measured the 1-pyrenecarboxaldehyde cation ($C_{17}H_{10}O^+$) spectrum. The aldehyde side group (HC=O) perhaps

represents an extreme (and unlikely ?) case for an interstellar molecular side-group. Its interest, however, lies in the fact that it should provide an *upper* limit to the magnitude of the pyrene cation absorption shift that one might expect when pyrene is substituted with other, perhaps more likely, side groups such as CH_2, CH_3. Indeed, alkylated pyrene derivatives have been recently detected in Murchison meteorite samples /7/. Apart from a 122 Å red-shift, the overall spectral appearance is remarkably unchanged from that of $C_{16}H_{10}^+$.

4. Astrophysical Implications of the PAH Cation Spectra. Contrary to their neutral precursors, *ionized PAHs absorb in the visible and thus could contribute to the DIBs*. The comparison of the PAH cation discrete features with the DIBs indicates that the $C_{10}H_8^+$ and $C_{16}H_{10}^+$ bands show a remarkable agreement with some of the DIBs. In the case of $C_{10}H_8^+$, 6 out of 12 of its discrete features match DIBs. One extreme approach is to require that *all* of the $C_{10}H_8^+$ bands, regardless of their intensity, have known DIB counterparts. Another approach is to only require that the strongest $C_{10}H_8^+$ band(s) be present. The former approach *requires* a high fractional naphthalene abundance among the PAHs, the latter, a substantially lower abundance. To test for $C_{10}H_8^+$ in the ISM, observations have been carried out /5 b, c/ to search for the stronger bands of $C_{10}H_8^+$ -which were predicted at 6520 Å, 6151 Å, and 5965 Å /2/- *along the same lines-of-sight with the 6741 Å DIB*. The results of the search have been positive for some objects /5 c/ and negative for some other objects /5 b/. It should be kept in mind that a specific ion may be present in abundance along one line-of-sight and considerably weaker along another. The second approach would be to soften the requirements for a molecular species to contribute to the DIBs by requiring that only the strongest PAH ion bands be evident and that subsequent searches be made for the less intense bands. If this criterion is adopted, and using the $C_{10}H_8^+$ oscillator strengths measured in our experiments ($f = 4.0 \times 10^{-4}$) /2 a/, we calculate that 0.3 % of the available carbon would be required to be in the form of $C_{10}H_8^+$ if the naphthalene cation is responsible for the 6742 Å DIB along the line-of-sight to HD183143.

Regarding the pyrene cation, $C_{16}H_{10}^+$, its strongest absorption feature falls intriguingly close to the strong 4430 Å DIB when isolated in Ne and matches it when isolated in Ar. As discussed earlier, vibronic band positions in Ne are believed to better represent the free gas-phase values which are presumably the values one should use in comparisons with the DIBs. These results raise the interesting possibility of a connection between $C_{16}H_{10}^+$ and the 4430 Å DIB. i.e.: Is the proximity of the 4395 Å absorption of $C_{16}H_{10}^+$ to the 4430 Å DIB only a coincidence? or is it an indication that the carrier of the 4430 Å DIB may be the pyrene cation ($C_{16}H_{10}^+$) or a *closely related, pyrene-like species*?. Side-groups which do not participate in ring aromaticity (such as CH_2, CH_3) are known to induce a red-shift of PAH cation vibronic transitions on the order of 30-50 Å /3/ while, as shown in the previous section, an aldehyde side group can induce a red-shift of about 120 Å. Thus this result raises the serious possibility that a substituted pyrene cation is responsible for the 4430 Å DIB. The suggestion that a mixture of different PAH ions may well explain some (or most?) of the DIBs seems to be supported by the present results on the pyrene cation. Using the $C_{16}H_{10}^+$ oscillator strength of 0.126 in Ar, implies that 0.2

% of the cosmic carbon must be in the form of $C_{16}H_{10}^+$ if the pyrene cation is responsible for the 4430 Å DIB.

CONCLUSION

Neutral PAHs show strong absorptions in the UV and may produce weak structures on the extinction curve. PAH *cations* absorb in the visible and NIR, making them good candidates for the carriers of the DIBs. The strongest band of $C_{10}H_8^+$ falls at 6741 Å, matching the weak 6742 Å DIB. Six other weaker bands (6520, 6376, 6314, 5848, 5784, and 5503) also fall very close to the positions of known DIBs. The possibility that $C_{10}H_8^+$ is responsible for some of the diffuse interstellar bands has been tested by searching for new DIBS at 6520, 6151, and 5965 Å, other strong naphthalene cation band positions. If $C_{10}H_8^+$ is indeed responsible for the 6742 Å feature, it accounts for 0.3% of the cosmic carbon. In the case of the pyrene cation ($C_{16}H_{10}^+$), the spectrum is dominated by a single intense band which falls remarkably close to the well-known 4430 Å DIB. This suggests that a closely related, pyrene-like molecular ion could be responsible for this band. If $C_{16}H_{10}^+$ carries the 4430 Å feature, it accounts for 0.2% of the cosmic carbon. Both cation spectra show a remarkably intense, broad continuum extending from the UV to the visible. If a continuum is a general property of PAH cations, this characteristic will have a strong impact on our understanding of how PAHs convert interstellar UV and visible radiation into IR radiation. The ultimate solution to the problem of the DIBs can only be obtained through a detailed and systematic comparison of the DIBs with appropriate laboratory data. Before this can be achieved, much more laboratory work coupled with a detailed and systematic DIB observing program is needed. We hope that we have demonstrated the importance of performing laboratory studies under conditions strictly relevant to those reigning in the diffuse ISM. From the observational point of view it is important that a new, high-resolution survey of the DIBs be carried out along as many lines of sight as possible.

REFERENCES

1. (a) G. P. Van der Zwet and L. J. Allamandola, *A. & A.*, **146**, 76 (1985); (b) A. Léger and L. B. d' Hendecourt, *A. & A.*, **146**, 81 (1985).
2. (a) F. Salama and L. J. Allamandola, *J. Chem. Phys.*, **94**, 6964 (1991); (b) F. Salama and L. J. Allamandola, *J. Chem. Phys.*, **95**, 6190 (1991); (c) F. Salama and L. J. Allamandola, *Ap. J.*, **395**,301 (1992).
3. F. Salama and L. J. Allamandola, *Nature*, **358**, 42 (1992).
4. F. Salama and L. J. Allamandola, *Faraday Trans.*, in press (1992).
5. (a) G. H. Herbig, *Ap. J.*, **196**, 129 (1975); (b) J. Krelowski, in this issue (1993); (c) P. Jenniskens and X. Désert, *A. & A.*, in press (1993).
6. (a) J. Krelowski and G. A H. Walker, *Ap. J.*, **312**, 860 (1987); (b) T. P. Snow and C. G. Seab, *Ah. J.*, **382**, 189 (1991).
7. M. S. De Vries, H. R. Wendt, H. E. Hunziker, E. Peterson, and S. Chang, *Lunar Planet. Sci.*, **315** (1991).

The First Symposium on the Infrared Cirrus and Diffuse Interstellar Clouds
ASP Conference Series, Vol. 58, 1994
Roc M. Cutri and William B. Latter (eds.)

INFRARED SPECTROSCOPY OF MATRIX ISOLATED PAHS

D. M. HUDGINS, S. A. SANDFORD, and L. J. ALLAMANDOLA
NASA Ames Research Center, MS 245-6, Moffett Field, CA 94035

ABSTRACT Although free-molecular PAHs are postulated to account for a number of interstellar emission and absorption features, laboratory spectra of PAHs under truly isolated conditions are not available. To remedy this situation we have undertaken a study of the infrared properties of matrix isolated PAHs. Here we present mid-infrared spectra (4000 - 500 cm^{-1}, 2.5 - 20 μm) of the PAHs naphthalene ($C_{10}H_8$), anthracene ($C_{14}H_{10}$), phenanthrene ($C_{14}H_{10}$), pyrene ($C_{16}H_{10}$), tetracene ($C_{18}H_{12}$), and chrysene ($C_{18}H_{12}$) isolated in solid argon matrices. The results are compared to other available PAH spectra and discussed in the framework of the current astronomical data. Some examples are given of how spectra of isolated PAHs resolve problems associated with the interstellar PAH model, including that of the dominance of the 885 cm^{-1} (11.23 μm) band in the C-H out-of-plane bending region.

INTRODUCTION

Since their discovery in 1973 by Gillett, Forest and Merrill, the family of infrared features at 3050, 1610, 1310, 1150, and 885 cm^{-1} emitted from many different astronomical objects has been the focus of much interest. While the specific carriers of these features are unknown, extensive observations have revealed some of their general characteristics. First, the observed emission features are non-thermal in nature suggesting that the carrier is a free molecule rather than an agglomeration (e.g. grain). Second, there is a strong correlation between the observed emission intensity and the object's carbon abundance indicating that the molecule is carbon-rich. Third, the features are observed even in highly energetic regions of the interstellar medium indicating that the carbonaceous molecule is highly stable. Of course, the spectrum of the highly stable carbonaceous molecule responsible for this emission must be consistent with the astronomical spectrum. Among those species which have been suggested as potential carriers, polycyclic aromatic hydrocarbons (PAHs) come closest to satisfying all of the criteria.

One of the difficulties in testing the PAH hypothesis is the lack of appropriate laboratory spectra of free, isolated PAHs. The great majority of the existent PAH infrared spectra have been taken from pure PAH crystals or PAHs suspended in solvents or KBr pellets. Intermolecular interactions between the PAHs and these media are known to significantly alter the positions, profiles, and intensities of many of the infrared spectral bands. However, by isolating PAH molecules in an inert matrix (e.g. Ar) many of these complications are greatly reduced or eliminated. Hence, the true spectroscopic characteristics and general spectral trends of isolated PAHs can be learned. Furthermore, matrix spectra accurately reproduce the rotational temperatures of free molecular PAHs in the interstellar medium.(~10K). In order to provide an empirical basis for modelling the interstellar emission from PAHs, we have undertaken a systematic study of the infrared spectroscopic properties of matrix-isolated PAHs. In this paper we present some of our results for six PAHs, focusing on a few astrophysically relevant issues.

EXPERIMENTAL

We employ the matrix isolation technique to trap and isolate individual PAH molecules in an inert matrix while they are probed spectroscopically. In a typical experiment, the substrate is cooled to 10K and rotated to face a vacuum deposition furnace (oven) and argon deposition port . By warming the pure, solid PAH in the vaporization oven, a slow stream of gaseous PAH molecules is generated. These molecules are frozen simultaneously onto the cold substrate with a great over-abundance of argon gas (PAH/argon ≈ 1/1000). The sample is deposited to a thickness appropriate for the measurement of the isolated PAH's infrared spectrum. After sample deposition is complete, the matrix is returned to the scanning position and its spectrum is recorded. Subsequent controlled warming of the sample allows the spectral effects of PAH clustering to be monitored as the effective sample concentration increases.

DISCUSSION

In Figures 1 and 2 we present the mid-infrared spectra of the PAHs naphthalene ($C_{10}H_8$), anthracene ($C_{14}H_{10}$), phenanthrene ($C_{14}H_{10}$), pyrene ($C_{16}H_{10}$), tetracene ($C_{18}H_{12}$), and chrysene ($C_{18}H_{12}$) isolated in solid Argon matrices. Spectra were measured from 4000 to 500 cm^{-1}. The spectral regions shown are those which contain PAH features. The vertical scale in the plot of the C-H stretching region (3200-2800 cm^{-1}) has been expanded by a factor of ~10 relative to the rest of the spectrum. These particular PAH spectra were chosen to demonstrate some general properties of PAH spectra which are of astronomical interest. Two specific applications of these spectra are presented here. A full detailed, discussion of the isolated PAH spectra and their aforementioned implications can be found elsewhere (Hudgins et al., 1993).

The C-H Stretching Region, Figure 1.

This spectral region between 3150 and 2800 cm^{-1} is dominated by the well-known aromatic C-H stretching bandswhich fall between 3100 and 3025 cm^{-1}. Co-adding these spectra produces a band whose profile is remarkably similar to the interstellar 3050 cm^{-1} feature. Of particular astrophysical interest are the C-C overtone/combination features that fall in the vicinity of and at slightly lower frequencies than the C-H stretch fundamentals (3025 - 2800 cm^{-1}). This is significant since a number of minor interstellar emission features in this spectral window. Two alternative explanations for the interstellar features have been suggested: (1) they are due to chemical side groups on PAHs; and/or (2) they are overtone/combinations of lower frequency modes and "hot bands" of the anharmonic C-H stretching modes. Figure 1 establishes that overtone/combination bands in this region are not merely a convenience for the second model but are, in fact, a necessary consequence of PAH emission. PAHs in general possess a large number of such overtone/combinations in the C-H stretching region. Overlap of these features from a mixture of PAHs in the interstellar medium readily accounts for the observed broad plateau and associated substructure. Hence, while some features in this region may well be attributed to sidegroups, others must be attributed to overtone-combinations if the PAH hypothesis is correct.

The C-H Out-of-Plane Bending Region, Figure 2.

The absorption frequencies of the C-H out-of-plane bending modes in PAHs fall roughly in the spectral region between 900 and 600 cm^{-1} and are highly characteristic for the number of adjacent hydrogen atoms on an aromatic ring. Due to strong coupling between the bending vibrations of adjacent hydrogen atoms, the peak frequencies of these modes shift to lower frequencies when the number of adjacent hydrogen atoms increases. For aromatic rings with nonadjacent hydrogen atoms (no hydrogen atoms on either of the neighboring positions on the ring) the bending vibration falls in the 900 - 860 cm^{-1} frequency range. For doubly adjacent hydrogen atoms (one hydrogen atom in a neighboring position) the vibration shifts to the 860 - 800 cm^{-1} range. Triply adjacent hydrogen atoms appear in the 810 - 750 cm^{-1} range and quadruply adjacent hydrogen atoms fall in the 770 - 735 cm^{-1} range. PAHs may have several absorption bands in this region because different rings can have different numbers of adjacent hydrogen atoms. In Figure 2, the adjacency of the hydrogen atoms is indicated next to their corresponding band. This region, therefore, places important constraints on the molecular structure of PAHs in the interstellar medium. In this region the interstellar spectra are dominated by a feature at 885 cm^{-1} and a plateau extending from 950 - 770 cm^{-1}. The range of the plateau indicates that the interstellar PAHs are predominantly made up of aromatic rings with nonadjacent, and doubly or triply adjacent hydrogen atoms. PAHs with four or five hydrogen atoms are excluded. The dominance of the 885 cm^{-1} feature, due to non-adjacent hydrogen atoms, has been taken to imply that interstellar PAHs are severely dehydrogenated. However, these laboratory spectra demonstrate that the integrated absorption strength per hydrogen atom is at least a factor of two stronger for non-adjacent hydrogen atoms than for multiply adjacent hydrogen atoms. Thus it is not necessary to invoke severe dehydrogenation to account for the dominance of the 885 cm^{-1} feature in the interstellar spectra.

REFERENCES

Gillett, F.C., Forrest, W.J., and Merrill, K.M., *Ap.J.*, **183**, 87 (1973)
Hudgins, D.M., and Allamandola, L.J., *Ap.J.*, (in preparation)

Figure 1. PAH Spectra in the C-H Stretching Region

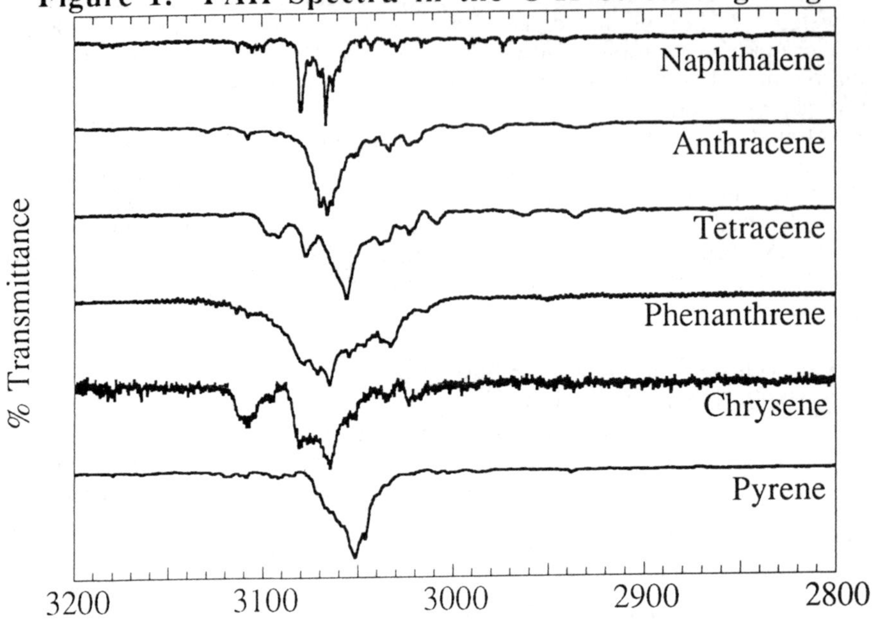

Figure 2. PAH Spectra Through the C-C Stretching and C-H Bending Regions

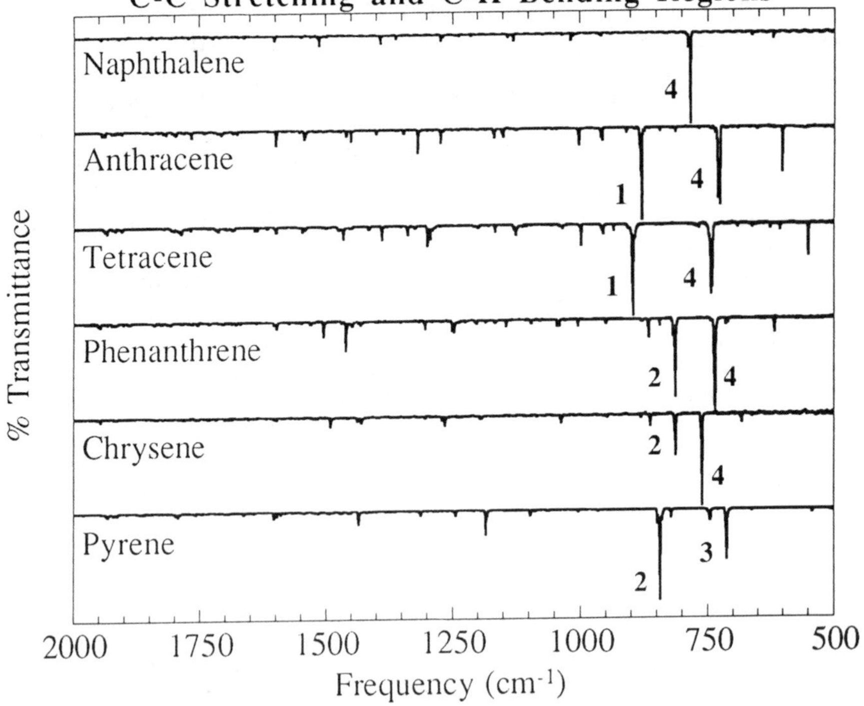

The First Symposium on the Infrared Cirrus and Diffuse Interstellar Clouds
ASP Conference Series, Vol. 58, 1994
Roc M. Cutri and William B. Latter (eds.)

MID-INFRARED SPECTROSCOPY OF PAH CATIONS: THE NAPHTHALENE ION, $C_{10}H_8^+$.

D. M. HUDGINS, S. A. SANDFORD, and L. J. ALLAMANDOLA
NASA Ames Research Center, MS 245-6, Moffett Field, CA 94035

ABSTRACT Although ionized polycyclic aromatic hydrocarbons (PAHs) may contribute to such observed phenomena as the diffuse interstellar bands (DIBs), the Infrared cirrus, and a well-known family of mid-infrared emission features, their IR spectroscopic properties are almost unknown. Here we present the mid-infrared spectrum of the naphthalene cation, $C_{10}H_8^+$, the first result in our ongoing study of the infrared spectroscopic properties of matrix isolated PAH cations. The important specific features of the spectrum between 4000 cm^{-1} and 500 cm^{-1} (2.5 μm and 20 μm) are discussed. More general astrophysical implications of PAH^+ spectra in the mid-infrared are considered. This and other cation spectra resolve one of the major problems associated with the PAH model. Namely, the relative intensity of the "1310" cm^{-1} ("7.7"μm) band with respect to the 885 cm^{-1} (11.23 μm) feature.

INTRODUCTION

Since their discovery in 1973 by Gillett, Forest and Merrill, the family of infrared features at 3050, 1610, 1310, 1150, and 885 cm^{-1} emitted from many different astronomical objects has been the focus of much interest. The ubiquity and intensity of these features indicates that the carriers may comprise as much as 10-20% of the total carbonaceous material in the interstellar medium. While the specific carriers of these features are unknown, extensive observations have revealed some of their general characteristics. First, the observed emission features are non-thermal in nature suggesting that the carrier is a free molecule rather than an agglomeration (e.g. grain). Second, there is a strong correlation between the observed emission intensity and the object's carbon abundance indicating that the molecule is carbon-rich. Third, the features are observed even in highly energetic regions of the interstellar medium indicating that the carbonaceous molecule is highly stable. Of course, the spectrum of the highly stable carbonaceous molecule responsible for this emission must be consistent with the astronomical spectrum. Among those species which have been suggested as potential carriers, polycyclic aromatic hydrocarbons (PAHs) come closest to satisfying all of the criteria. However, while the general similarity between the spectra of neutral PAHs and the interstellar infrared emission spectra does indicate that the carrier is PAH-*related*, important discrepancies in the finer details of these spectra show that neutral PAHs are insufficient to adequately reproduce the interstellar spectrum. This is not surprising since PAHs characteristically have low (~6-7 eV) ionization potentials and, thus, are expected to be largely ionized in the energetic emission regions where they are observed. Consequently, laboratory spectra of free, ionized PAHs are crucial to the verification and exploitation of the PAH model. To achieve this end, we have undertaken a systematic study of the infrared spectroscopic properties of matrix isolated PAH cations. Here we present our results for the naphthalene cation.

EXPERIMENTAL

Our goal is to obtain laboratory spectra of PAHs under conditions comparable to those under which they are found in the interstellar medium. Since the species responsible for the infrared emission features exist as free molecules (e.g. isolated from one another), the PAHs of our laboratory study must satisfy this criterion also.

We employ the matrix isolation technique, whereby a species is isolated by freezing it in great excess of a noble gas on a cold substrate. The experimental apparatus consists of a CsI window suspended from a cryogenically cooled cold finger and housed in a high vacuum chamber. The cold finger is cooled by a closed cycle helium refrigerator and mounted in such a way that the window may be rotated under vacuum to face any of the available ports on the vacuum chamber. The substrate can be maintained at any temperature between 10 and 300K. The entire vacuum chamber is positioned in the sample compartment of a Fourier Transform Infrared Spectrometer

A typical experiment begins with the substrate cooled to 10K and then rotated to face the sample deposition line. A gaseous sample of ≤0.1% naphthalene in argon flows through this port and freezes on the cold CsI window. Samples are commonly deposited to a thickness which saturates the largest naphthalene bands, facilitating detection of the weaker ion features.The cold head is subsequently returned to the scanning position so that its spectrum prior to irradiation can be recorded. The sample is then rotated to face the UV lamp for ionization. This is accomplished with Lyman-α radiation generated by microwave discharge in a 10% H_2/He mixture. The sample is finally returned to the scanning position where the post-photolysis spectrum is recorded. Comparison of this spectrum with the neutral spectrum permits identification of those features which appeared during photolysis. Space does not permit a discussion of how the ion bands are assigned. Such a discussion can be found in Hudgins et al. (1993 - in press).

RESULTS

A typical ion spectrum is shown in Figure 1. To enhance detail only regions of direct relevance to the ion spectrum are shown. Detailed comparison of the unirradiated sample spectrum (top) with the irradiated spectrum (below) between 3500 and 500 cm^{-1} reveals several new features which can be attributed to the naphthalene cation. These fall at 1525.7, 1518.8, 1400.9,

1218.0, 1214.9, 1023.2, and 758.7 cm^{-1} (6.554, 6.584, 7.138, 8.210, 8.231, 9.773, and 13.180 μm). No new features were found in the C-H stretching region (3150 - 2900 cm^{-1}). The new bands detected all fall in the C-C stretching and C-H bending regions of the PAH spectrum.

Bands corresponding to the naphthalene cation are identified by their distinctive behavior as a function of irradiation time. Bands whose intensities peak during the first 10 minutes of photolysis and then remain constant or decrease slightly upon further irradiation are assigned to the naphthalene cation since they follow the same behavior reported by Salama and Allamandola (1991) for the visible bands of this species.

ASTROPHYSICAL IMPLICATIONS

(1) The strongest observed ion bands all fall in the C-C stretching and C-H in-plane bending region of naphthalene. These bands are evident in spite of the fact that only a small percentage of the parent neutral naphthalene has been converted to the cation. This remarkable behavior demonstrates that these modes are greatly enhanced with respect to their counterparts in the neutral molecule. For the strongest N^+ feature, lying at 1218.0 cm^{-1} , this enhancement is greater than a factor of ten.

(2) The lack of any new features in the C-H stretching region, and only a weak new band corresponding to the C-H out-of-plane bend (758.7 cm^{-1}), the most intense mode in the neutral molecule, demonstrates that the IR cross sections of the C-H modes in the ion are, at best, comparable to or, more likely, weaker than the corresponding cross sections of the neutral species.

(3) This behavior is consistent with that predicted by Pauzat et al. (1992) as a general property of the IR spectra PAH ions. Our results are also in general agreement with the recent experimental results reported by Szczepanski et al. (1992). However, Szczepanski et al. did not report the 758.7 cm^{-1} ion band.

(4) The strongest ion feature, the 1218 cm^{-1} (8.210 μm) band, falls in the envelope of the strongest interstellar PAH emission feature, the so-called "7.7" μm feature.

(5) If this great enhancement of the intensities of the C-C stretching modes with respect to the C-H modes is a general property of PAH ions, the 'ad-hoc' assumption of severe dehydrogenation of PAHs made by some to rationalize the dominance of the interstellar "7.7" μm feature over the 11-13 μm C-H out-of-plane bending feature is unnecessary.

These results are presented in detail by Hudgins et al. (1993)

REFERENCES

Hudgins, D.M., Sandford, S.A., and Allamandola, L.J., *J.Phys.Chem.*, (in press, 1993)

Pauzat , F., Talbi, D., Miller, M.D., DeFrees, D.J., and Ellinger, Y., *J.Phys.Chem.*, **96**, 7882 (1992)

Salama, F., and Allamandola, L.J., *J.Chem.Phys.*, **94**(11), 6964-6977 (1991)

Szczepanski, J., Roser, D., Personette, W., Eyring, M., Pellow, R.,and Vala, M., *J.Phys.Chem.*, **96**, 7876 (1992)

Figure 1. Spectrum of the Naphthalene Cation.

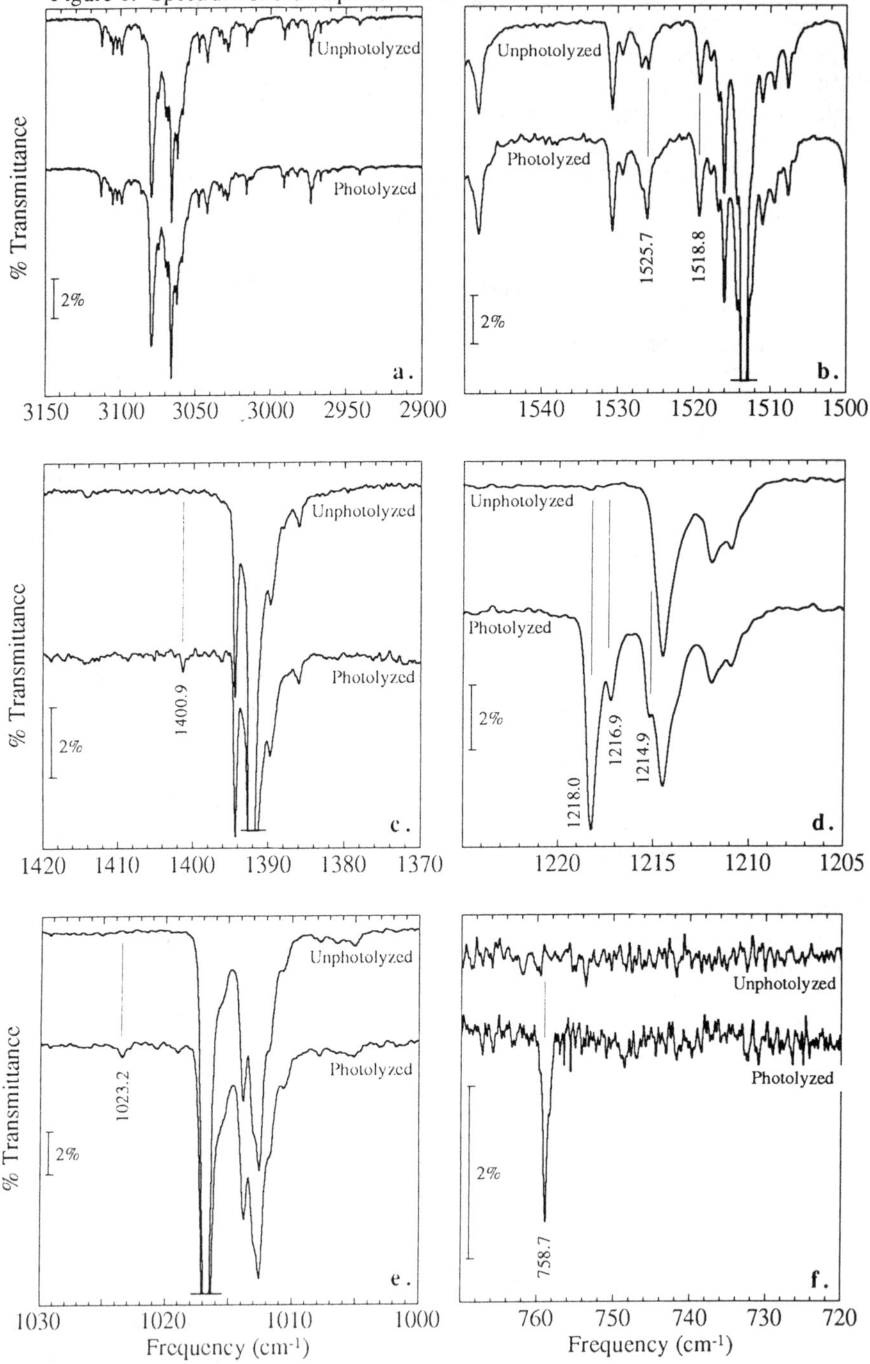

The First Symposium on the Infrared Cirrus and Diffuse Interstellar Clouds
ASP Conference Series, Vol. 58, 1994
Roc M. Cutri and William B. Latter (eds.)

IR SPECTROSCOPY OF LABORATORY-SIMULATED INTERSTELLAR PAHS

CHRISTINE JOBLIN
NASA ARC, MS: 245-6, Moffett Field CA 94035, and GPS Paris

LOUIS D'HENDECOURT, ALAIN LÉGER
IAS, bât 121, Université Paris Sud, 91405 ORSAY, and GPS Paris

DANIEL DÉFOURNEAU
GPS, Université Paris 7, 2, place Jussieu, 75231 PARIS Cédex 05

INTRODUCTION

In the PAH model (Léger and Puget, 1984), polycyclic aromatic hydrocarbons are an ubiquitous component of the interstellar medium and are responsible for the mid-IR flux observed in cirrus. The emission mechanism requires the molecules to be isolated; after the absorption of a UV photon, there is a fast conversion into the electronic ground state with a high vibrational energy, and finally the cooling takes place by slow IR emission ($t_{1/2}$ = 5 s., Léger et al. 1989b). Although the IR spectra of PAH molecules can in first approximation account for the unidentified IR (UIR) features, the spectral fit has to be improved. However, most laboratory spectra have been obtained for condensed molecules at low temperatures, whereas the PAH model states that the interstellar molecules are isolated and at high temperatures ($T \sim 1000$ K). Here, we discuss the IR absorption spectra (3-20 μm) measured for high-temperature gas phase PAHs. In that experiment, molecules are evaporated and thermally excited in an oven. The spectral information obtained with thermally excited molecules should be relevant for comparisons with astronomical spectra. Indeed, for an isolated photoexcited molecule, most of the IR emission is correctly described by a simple thermal model with a vibrational temperature T_{vib} (Léger et al. 1989a and 1989b, Schutte et al. 1993).

EFFECT OF ENVIRONMENT ON THE BAND INTENSITIES

The spectra obtained in gas phase are compared to those obtained in solid and rare gas (neon) matrices (see Joblin et al. 1993). The relative intensities of the CH modes compared to the CC modes are affected by the physical environment (Table 1). This effect mostly concerns the CH in plane modes in the solid (CH stretch at 3.3 μm and CH bending modes around 8.6 μm) and the CH out of plane modes (11–14 μm region) in the rare gas matrices.

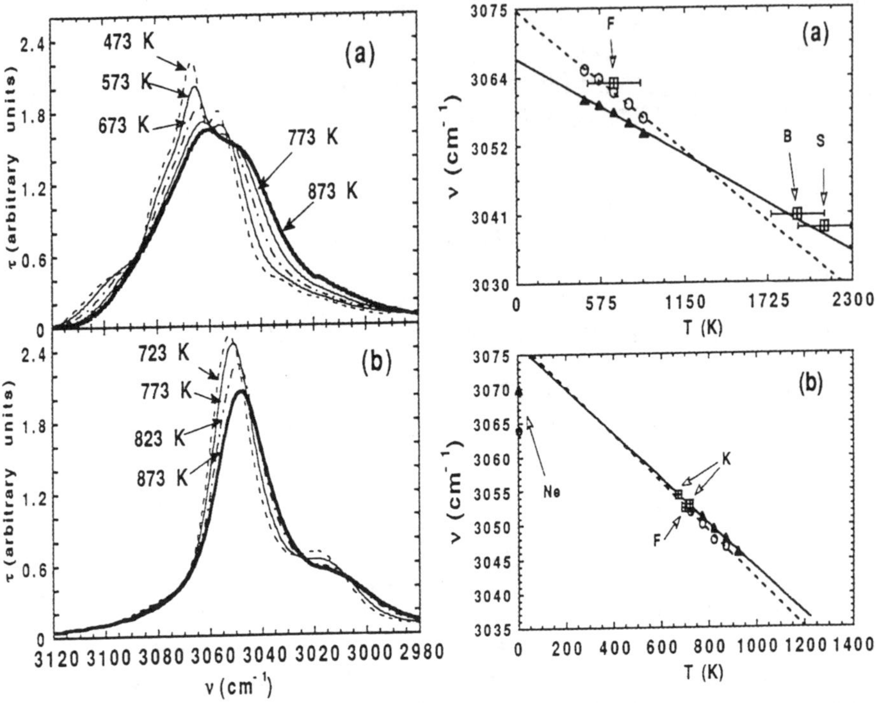

FIGURE I Evolution of the 3.3 μm band of (a) naphtalene ($C_{10}H_8$) and (b) coronene ($C_{24}H_{12}$) with the molecular vibrational temperature.
Left diagrams: band positions at the peak (triangles), and at half maximum (circles) versus temperature. Ne: data obtained in neon matrices (Joblin et al. 1993). Published data are also reported: K (Kurtz 1992), F (Flinckinger et al. 1991), B (Brenner and Barker 1992), S (Shan et al. 1991).

TABLE I Variation of the band intensity ratios ($A = \sigma\Delta\lambda$) in various environments, solid (300 K), gas phase ($T \sim 800$ K), and Ne matrix (4 K).

			$\frac{A_{3.3\,\mu m}}{A_{7.7\,\mu m}}$	$\frac{A_{6.2\,\mu m}}{A_{7.7\,\mu m}}$	$\frac{A_{8.6\,\mu m}}{A_{7.7\,\mu m}}$	$\frac{A_{11.3-13\,\mu m}}{A_{7.7\,\mu m}}$
	Pyrene	$C_{16}H_{10}$	0.2	1.0	0.2	0.8
Solid/Gas	Coronene	$C_{24}H_{12}$	0.3	1.0	0.4	1.2
	Ovalene	$C_{32}H_{14}$	0.35	0.9	0.4	0.9
	Pyrene	$C_{16}H_{10}$	0.6	0.9	0.6	0.2
Matrix/Gas	Coronene	$C_{24}H_{12}$	0.6	< 1.3	0.6	0.2
	Ovalene	$C_{32}H_{14}$	0.6	< 1.1	0.5	0.15

EFFECT OF TEMPERATURE ON THE BAND POSITIONS

A temperature increase involves a band shift towards longer wavelengths and a band broadening (see for example the 3.3 μm band in Fig. 1). Whereas the broadening results from a combination of vibrational and rotational effects, the shift is mostly due to a vibrational (anharmonic) effect. This is confirmed by the fact that the linear extrapolation of our data (Fig. 1) leads to a good agreement with the positions measured for photoexcited molecules with $T_{vib} \sim 2000$ K and $T_{rot} \sim 300$ K (Shan et al. 1991, Brenner and Barker 1992).

The linear dependence of the band frequencies with temperature can be explained by taking into account the first anharmonic terms of the vibrational energy.

$$\frac{E_v^{anharm}}{hc} = \sum_i \omega_i(n_i+\frac{1}{2})+\sum_i x_{ii}(n_i+\frac{1}{2})^2 + \sum_{i;k\ k\neq i} x_{ik}(n_i+\frac{1}{2})(n_k+\frac{1}{2})+... \quad (1)$$

x_{ii} : anharmonicity constant of the ν_i mode

x_{ik} : coupling constant between ν_i and ν_k modes

For a transition involving one quantum of the ν_i mode ($\Delta n_i = 1$), the observed frequency is given by:

$$\nu_i(T) = \nu_i(0) + 2x_{ii} < n_i > + \sum_{i;k\ k\neq i} x_{ik} < n_k > \quad (2)$$

where $< n >$ is the mean population of the ν mode: $< n >= \frac{1}{exp(\frac{hc\nu}{k_B T})-1}$.

By developing (2) in the high-temperature range, a linear dependence of the band positions with temperature is found.

ASTROPHYSICAL IMPLICATIONS

The values of the band intensity ratios, measured for gaseous PAHs, can be used to derive informations on the PAH distributions in objects where the UIR bands are observed (see Joblin et al. 1993). However, at present, spectral information is available only in strongly irradiated regions. PAHs could be ionized in these regions and laboratory experiments indicate that the IR spectra of ions are rather different from the spectra of neutrals (Szczepanski et al. 1992, see also the presentations of d'Hendecourt et al. and Hudgins et al.). Anyway, laboratory data on neutral molecules should be of interest for future observations and studies of the IR cirrus and more generally for the diffuse ISM where PAHs are expected to be neutral.

Why does the 3.3 μm band usually occur at 3.289 μm?

We have seen that the positions of the IR bands depend on the vibrational temperature of the molecule. Using diagrams similar to those of Fig. 1, we can estimate the position of the 3.3 μm band emitted by a PAH molecule, after the absorption of a 10 eV photon (see Fig. 2). In fact, a mean temperature over the cooling of the molecule has been used (see Léger et al. 1989b). Although this temperature strongly depends on the molecular size, the position of the resultant 3.3 μm is in good agreement with the position of the interstellar feature at 3.289 μm (3040 cm^{-1}) (Tokunaga et al. 1991) . *The observed position for the 3.3 μm band is then consistent with the carriers being free PAHs at high temperatures.*

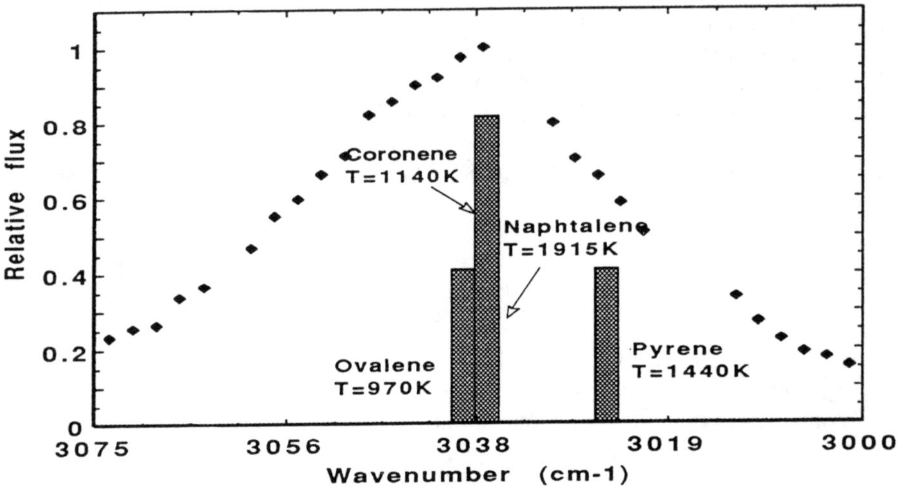

FIGURE II Expected position of the 3.3 μm band emitted by different PAHs after absorption of a 10 eV photon, compared to the mean interstellar feature of Type 1 from Tokunaga et al. (1991). The expected band width is not given because of strong rotational broadening in laboratory spectra which has no counterpart in astrophysical conditions.

REFERENCES

Brenner, J.D., Barker, J.R., 1992, *ApJ*, **388**, L39.

Flinckinger, G.C., and Wdowiak, T.J., and Gomez, P.L., 1991, *ApJ*, **380**, L43.

Joblin, C., d'Hendecourt, Léger, A., and Défourneau, D.,1993 *A&A*, submitted.

Kurtz, J., 1992, *A&A*, **255**, L1.

Léger, A., Boissel, P., Désert, F.X., and d'Hendecourt, L., 1989a, *A&A*, **213**, 351.

Léger, A., d'Hendecourt, L., and Défourneau D., 1989b, *A&A*, **216**, 148.

Léger, A., and Puget, J. L., 1984,*A&A*, **137**, L5.

Shan, J., Suto, M., and Lee, L.C., 1991, *ApJ*, **383**, 459.

Schutte W.A., Tielens A.G.G.M, and Allamandola L.J.,1993, *ApJ*, in press.

Szczepanski, J., Roser, D., Personette, W., Eyring, M., Pellow, R., and Vala, M., 1992, *J.P.C.* **96**, 7876.

Tokunaga, A., Sellgren, K., Smith, R., Nagata, T., Sakata, A., and Nakada, Y., 1991, *ApJ*, **380**, 452.

The First Symposium on the Infrared Cirrus and Diffuse Interstellar Clouds
ASP Conference Series, Vol. 58, 1994
Roc M. Cutri and William B. Latter (eds.)

SPECTRAL PROPERTIES OF POLYCYCLIC AROMATIC HYDROCARBON IONS AND THE INTERSTELLAR EXTINCTION CURVE

WEI LEE AND THOMAS J. WDOWIAK
Department of Physics, University of Alabama at Birmingham, Birmingham, AL 35294-1170

ABSTRACT Ultraviolet-visible absorption measurements are performed on neutral and ionized forms of PAHs that have been proposed as the possible carriers of the UIR bands. Pronounced decrease in the strength of the UV absorption bands upon ionization reveals that the ionized interstellar PAHs might not add structure to the UV extinction curve. Our experiments also indicate that a UV-visible absorption continuum exists for PAH ions providing an explanation of how PAHs can convert interstellar radiation to IR emission even in ultraviolet-poor regions.

INTRODUCTION

The origin of the principal UIR emission bands between 3.2 and 14.0 μm (3100 and 700 cm^{-1}) has been widely speculated upon and a general hypothesis now exists involving emission by vibrationally excited interstellar PAH molecules (Gillett, Forrest, and Merrill 1973; Duley and Williams 1981; Léger and Puget 1984; Allamandola, Tielens, and Barker 1985; Blanco, Bussoletti, and Colangeli 1988). PAHs are attractive as a major constituent in the interstellar medium (ISM) as they are extremely stable against UV induced photodissociation and their structure allows for the conversion of UV radiation into discrete IR emission bands through infrared fluorescence (Allamandola et al. 1989). Although PAHs are thought to be ubiquitous and abundant throughout the ISM, and are widely proposed to be responsible for the UIR by the correlation of their general IR spectral characteristics, one of the criticisms of the PAH model raised by Leach (1987) and Donn et al. (1989) is that the strong absorption peaks of PAH molecules in the near-UV are not evident in the UV extinction curve. This important criticism requires an answer if the PAH hypothesis is to remain plausible.

Since a large fraction of the PAHs in the general ISM are expected to be ionized, it is important for the PAH hypothesis that an investigation be made of the spectral properties of ionized PAHs. This includes electronic transitions yielding the UV-visible spectrum as well as vibrational transitions leading to the IR features. A recent laboratory study by Salama and Allamandola (1992a, b) of the UV-visible spectra of naphthalene and pyrene cations, which they consider to be the possible carriers of the diffuse interstellar bands (DIBs), has provided a significant clue toward finding an answer to the objection of Leach (1987) and Donn et al. (1989). They report a continuum developed in the UV-visible along with discrete visible bands when the PAHs were isolated in inert cryogenic matrices and ionized. The UV absorption bands of the neutral forms of these two smaller PAHs appear to weaken under the vacuum UV irradiation.

We find the loss of the UV absorption bands is a general phenomenon when compact PAH species, namely, coronene, benzo[ghi]perylene, perylene, benzo(e)pyrene, and pyrene, are isolated in boron oxide glass matrices and ionized. This indicates that ionized interstellar PAHs might not add structure to the interstellar extinction curve in the region from the 4428 ÅDIB to the 2175 Åbump. Also the broad continuum apparently associated with of the cation could play an important role in UV pumping of UIR emission. Details of our UV-visible spectral measurements of the isolated PAHs in their neutral and ionized forms and discussion of the astrophysical implications are published elsewhere (Lee and Wdowiak 1993; Wdowiak and Lee 1993).

EXPERIMENTAL

Samples are prepared by dissolving PAH molecules in molten boron oxide glass that is then sandwiched between two quartz plates. Measurements of the spectral range from 190 to 820 nm of the quartz-PAH/glass-quartz samples against a quartz-glass-quartz reference are made at 295 K, before and after ionization by gamma irradiation. A thorough description of the experimental technique is reported elsewhere (Lee and Wdowiak 1993).

RESULTS AND DISCUSSION

All of the neutral PAH molecules included in this study exhibit benzenoid and ethylenic absorption bands in the ultraviolet region arising from the transition of an electron from a π-bonding orbital to a π-antibonding orbital (Silverstein and Bassler 1967). The multiplicity of vibrational sublevels of non-symmetric PAH molecules, such as benzo[ghi]perylene, perylene, and benzo(e)pyrene, cause the discrete peaks to coalesce, and broad absorption bands are observed in the ultraviolet. However, matrix-isolated PAH samples lose the near-UV features upon gamma irradiation, demonstrating that π-bonding to π-antibonding transitions present in the neutral molecules disappear in the ionic state. Figure I shows the UV-visible spectra of neutral and irradiated perylene. It is clear that the bands of the neutral perylene decrease in strength dramatically upon gamma irradiation. Similar results are found for the other four species (Lee and Wdowiak 1993). The disappearance of the UV absorption features of PAHs when ionized indicates why PAH species are not observed in the ISM at UV wavelengths while being evident from the UIR. Although recent IR measurements on ionized naphthalene obtained by Szczepanski et al. (1992) and the coronene cation as reported by d'Hendecourt and Léger (1993) indicate that the IR spectra of the cations differ from those of the neutral species, no general conclusion as yet can be drawn from this limited number of studies regarding the determination of neutral or ionized PAH species as the more possible UIR carrier. This is in terms of the wavelengths and relative intensities of the laboratory spectral features correlating with those observed in IR emission from most stellar sources.

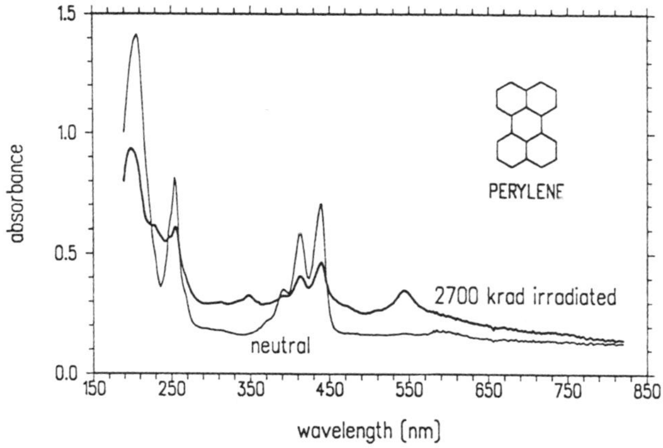

FIGURE I Absorption spectra of the neutral and irradiated perylene isolated in boron oxide glass matrix sandwiched between two quartz plates. It is estimated that 62% of the perylene molecules in this sample are ionized after a 2700 krad dose. PAH concentration = 1.16×10^{-7} mol cm^{-2}; perylene:boron oxide glass = 1:1350.

An IR spectrum of an ion at cryo-temperatures, as reported by these authors, also should not be expected to be representative of one vibrationally excited as suggested by experiments with neutral PAH species at 300 K and elevated temperatures approaching 800 K (Flickinger and Wdowiak 1990). Therefore, it is highly desirable to obtain the spectrum of PAH cations at similarly high temperatures.

Neutral PAHs generally absorb strongly in discrete bands in the mid- and near-UV, however, to account for the luminosities of the observed IR bands, conversion of longer wavelengths to the UIR is likely. In most of the sources which emit the UIR bands, the discrete features carry 30 - 50% of the total IR energy, implying a similar fraction of the UV-visible radiation is absorbed by the carrier. Yet such absorption is not evident in the interstellar extinction curve. Even the 2175 Åbump may not account for such a large fraction. Salama and Allamandola (1992a, b) show the ionized naphthalene and pyrene to possess a broad continuum from the UV to the visible. This continuum absorption, as also shown for the perylene cation in Figure II, may answer the problem. Our study indicates that the continuum absorption is general for PAH ions including larger ones. The structured UV absorption which is characteristic of neutral PAHs is "converted" into the continuum absorption when these PAHs are ionized (Lee and Wdowiak 1993).

We have found that the combination of the infrared emission bands (UIR) and the lack of band structure in the UV portion of the interstellar extinction curve is best explained by the interstellar PAHs being ionized. This strongly suggests that PAH molecules in the interstellar medium are ionized and the ejected photoelectrons are heating the ISM (Bakes and Tielens 1993).

We are grateful to Marla Moore for irradiating some of the samples with the Co^{60} at GSFC, and acknowledge the support of NASA grant NAGW-749.

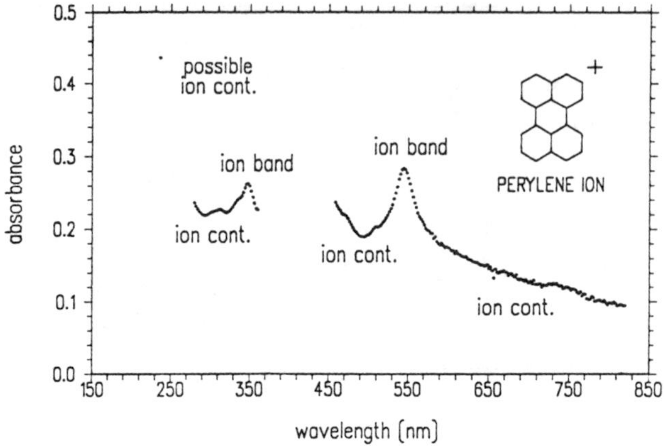

FIGURE II The absorption continuum and bands of the perylene ion obtained by substracting the spectrum of the neutral (× 38%) from that of the irradiated sample (see Fig. I). Spectral regions covered by neutral bands are not shown. Cross-section (cm^2 per ion) can be obtained by multiplying the absorbance by 2.31×10^{-17}.

REFERENCES

Allamandola, L. J., Tielens, A. G. G. M., and Barker, J. R. 1985, *ApJ*, **290**, L25; 1989, *ApJS*, **71**, 733

Bakes, E. L. O., and Tielens, A. G. G. M. 1993, this symposium.

Blanco, A., Bussoletti, E., and Colangeli, L. 1988, *ApJ*, **334**, 875

d'Hendecourt, L., and Léger, A. 1993, this symposium.

Donn, B. D., Allen, J. E., and Khanna, R. K. 1989, in *Interstellar Dust*, L. J. Allamandola and A. G. G. M. Tielens, Dordrecht: Kluwer, 181

Duley, W. W., and Williams, D. A. 1981, *MNRAS*, **196**, 269

Flickinger, G. C., and Wdowiak, T. J. 1990, *ApJ*, **362**, L71

Gillett, F. C., Forrest, W. J., and Merrill, K. M. 1973, *ApJ*, **183**, 87

Leach, S. 1987, in *Polycyclic Aromatic Hydrocarbons and Astrophysics*, L. d'Hendecourt, A. Léger, and H. Boccara, Dordrecht: Reidel, 99

Lee, W., and Wdowiak, T. J. 1993, *ApJ*, **420**, L127

Léger, A., and Puget, J. L. 1984, *A&A*, **137**, L5

Salama, F., and Allamandola, L. J. 1992a, *ApJ*, **394**, 301; 1992b, *Nature*, **358**, 42

Silverstein, R. M., and Bassler, G. C. 1967, *Spectrometric Identification of Organic Compounds*, New York: John Wiley & Sons, 149

Szczepanski, J. et al. 1992, *J. Phys. Chem.*, **96**, 7876

Wdowiak, T. J., and Lee, W. 1993, *J. Chem. Soc. Faraday Trans.*, **89**, in press.

The First Symposium on the Infrared Cirrus and Diffuse Interstellar Clouds
ASP Conference Series, Vol. 58, 1994
Roc M. Cutri and William B. Latter (eds.)

A NEW METHOD TO CALCULATE THE EXTINCTION PROPERTIES OF IRREGULARLY SHAPED PARTICLES

FRANÇOIS ROULEAU
Department of Astronomy, University of Toronto, Toronto, ON Canada M5S 1A7

P. G. MARTIN
Canadian Institute for Theoretical Astrophysics, University of Toronto, Toronto, ON Canada M5S 1A7

A new method to calculate the extinction of light by irregularly shaped particles, the generalized semi-analytical (GSA) method, is described. It allows a better approximation of the shape of a particle by interpolating explicitly the polarization field sampled at various points within the particle. The performance of the GSA method is compared to that of the discrete dipole approximation (D-DA) method and of two other methods which solve for the *surface* polarization field in the zero frequency limit: the Fuchs method, and the semi-analytical (SA) method. In that limit various test cases considered include a single sphere, a single spheroid, a pair of touching spheres, a single cube, and a single short cylinder. The GSA method yields an improved accuracy compared to the other methods for moderate optical constants and shows less variability as a function of grid resolution than the DDA method. It is also found that the DDA method tends to systematically overestimate the extinction, especially when the incident wave polarization vector is perpendicular to the major axis of elongated structures. In addition, for finite size parameters, the GSA method and the DDA method are compared to Mie spheres and spheroids. Again, the DDA method is shown to overestimate the extinction systematically, whereas the GSA method yields better accuracy for small size parameters. The likely sources of this discrepancy are discussed.

The First Symposium on the Infrared Cirrus and Diffuse Interstellar Clouds
ASP Conference Series, Vol. 58, 1994
Roc M. Cutri and William B. Latter (eds.)

PROXIMITY EFFECTS IN CLUSTERS OF PARTICLES

FRANÇOIS ROULEAU
Department of Astronomy, University of Toronto, Toronto, ON Canada M5S 1A7

P. G. MARTIN
Canadian Institute for Theoretical Astrophysics, University of Toronto, Toronto, ON Canada M5S 1A7

When two spheres are brought together, there is an enhancement in the orientation-averaged extinction per unit volume compared to isolated spheres; "proximity effects" such as this are expected to be most important at close separations. Using finite element methods for a pair of nearby Rayleigh spheres, this is both verified and quantified for the first time. The results are also expressed in terms of a normal mode analysis. As more spheres are added to the system there is a further growth in the extinction enhancement, which, however, depends on the topology. Arrangements of touching spheres examined included stacks, close packings, radiating structures, rings, chains, and clusters. Effective medium theory models are often used to describe composite or porous particles. It is shown that with these models the extinction predicted is significantly higher than with finite element treatment of roughly equivalent particles. Furthermore, the effects of topology, not accounted for explicitly in effective medium theory models, are found to be important. The modified prescription of effective medium theory proposed by Ossenkopf does not provide better agreement with finite element methods unless the filling factor is high, higher than typical of even compact clusters. Comparison between finite element methods and effective medium theory models in the case of a laboratory soot sample with high extinction yields a paradox that is discussed briefly.

The First Symposium on the Infrared Cirrus and Diffuse Interstellar Clouds
ASP Conference Series, Vol. 58, 1994
Roc M. Cutri and William B. Latter (eds.)

DISTRIBUTION OF SMALL DUST PARTICLES INSIDE MOLECULAR CLOUDS

J.P. BERNARD
IPAC, Caltech 100-22, Pasadena CA 91125, U.S.A. and Radio Astronomy Laboratory, Nagoya University, Chikusa-ku, Nagoya 464-01, Japan

F. BOULANGER, J.L. PUGET
IAS, bât 120, Campus d'Orsay, 91405 Orsay CEDEX, France.

We have used the numerical model developed by Bernard et al. 1992 to compute the penetration of heating radiation and the resulting IR emission of individual molecular clouds in a self consistent way. In the model, thermal fluctuations of small dust particles consisting of 2D Polycyclic Aromatic Hydrocarbon molecules (PAH) and 3D Very Small Grains (VSG) are treated as proposed by Désert et al. 1990. We have investigated physical conditions of some selected clouds in ρ-ophiuchi and Chamaeleon, heated either by the galactic diffuse Inter Stellar Radiation Field or receiving radiation from nearby young stars (Bernard et al. 1993). The individual clouds studied were selected among those showing short IRAS wavelengths Limb Brightening effect.

The model used allows to quantify the abundance variations of the smallest dust components which appear necessary, inside the clouds, to reproduce the spatial structure of the IR emission. For the clouds where IR brightness profiles can be used reliably down to the cloud center, it is found that the smallest dust particles must be present only in a halo (PAH halo) surrounding the cloud. Inner regions of the clouds detected in ^{13}CO ($A_v > 0.5$ mag) show very low abundances of these particles, compatible with almost no small dust grains in the corresponding dense medium. This conclusion is reached both for isolated clouds and nearby recently formed OB stars.

The inner position of the PAH halo is found at a visible extinction of $\simeq 0.5$ mag independent of the external radiation field, indicating that some other physical parameter(s) may be responsible for the halo. Several hypothesis are examined. The comparison with ^{13}CO emission data also shows that a substantial amount of dust is present in the limb brightening region where no ^{13}CO is observed with the current sensitivity. The use of the IR emission model allow to determine the cloud density distribution in these regions.

References

Bernard J.P., Boulanger F., Désert X., Puget J.L., 1992, A&A 263, 258.
Bernard J.P., Boulanger F., Puget J.L., 1993, A&A accepted.
Désert X., Boulanger F., Pujet J.L., 1990, A&A 237, 215.

The First Symposium on the Infrared Cirrus and Diffuse Interstellar Clouds
ASP Conference Series, Vol. 58, 1994
Roc M. Cutri and William B. Latter (eds.)

THE ORGANIC REFRACTORY COMPONENTS OF DENSE AND DIFFUSE CLOUDS

L.J. ALLAMANDOLA, S.A. SANDFORD
NASA Ames Research Center, Moffett Field, CA 94035-1000

The questions raised by the newly detected infrared absorption band in the CH stretch region at 2880 cm^{-1} (3.4 μm) in the spectra of dense molecular clouds will be discussed. This spectral signature is in sharp contrast with that of the diffuse interstellar medium. The spectrum of carbon dust in the diffuse ISM is characterized by strong bands at about 2965, 2930, and 2865 cm^{-1}, frequencies which straddle the new feature's position and indicate that diffuse ISM dust is dominated by saturated aliphatic hydrocarbons. It therefore appears that there are two very different and independent solid hydrocarbon interstellar dust components: saturated hydrocarbons in the diffuse ISM and diamond-like hydrocarbons in the dense (and possibly diffuse) ISM. The absence of $-CH_3$ and $-CH_2-$ bands in the spectra of embedded protostars implies that diffuse medium, C–rich materials do not get incorporated in dense clouds. This is puzzling in view of the 'short' times presumed for cycling between the dense and diffuse media.

The First Symposium on the Infrared Cirrus and Diffuse Interstellar Clouds
ASP Conference Series, Vol. 58, 1994
Roc M. Cutri and William B. Latter (eds.)

INFRARED SPECTROSCOPY OF PAH CATIONS: CORONENE$^+$

LOUIS D'HENDECOURT, ALAIN LÉGER
Institut d'Astrophysique Spatiale, Université de Paris–Sud

The PAH hypothesis identifies a certain number of strong astronomical infrared emission bands with vibrational transitions from thermally excited large aromatic molecules. This identification, together with the understanding of the emission mechanism, is based on strong similarities between IR spectra of such molecules (usually measured in absorption, but in some cases also in emission) and these so-called unidentified infrared bands.

However, in the environments where they are observed, PAHs are most probably ionised. The IR spectra of these cations are not known and recent theoretical calculations suggest that they may differ markedly from the spectra of neutral species.

Using classical matrix isolation spectroscopy methods, we have produced cations of molecules such as pyrene, coronene and ovalene. For coronene, quantitative results have been obtained. The IR spectrum of the ion is quite different from that of the neutral: some bands do almost totally vanish, in particular the 3.28 μm one, whereas others do appear around 6.4, 7.2, and 8.2 μm where no prominent astronomical lines are observed. Similar results are found for other species. Astrophysical implications will be dicussed in the poster.

CHAPTER VI

CLOUD CHEMISTRY

The First Symposium on the Infrared Cirrus and Diffuse Interstellar Clouds
ASP Conference Series, Vol. 58, 1994
Roc M. Cutri and William B. Latter (eds.)

THE PHYSICS AND CHEMISTRY OF CIRRUS CLOUD CORES AND OTHER TRANSLUCENT CLOUDS

B. E. TURNER
National Radio Astronomy Observatory, Charlottesville, Virginia, U.S.A.

ABSTRACT Detailed observations of the J=1–0 and J=2–1 lines of $C^{18}O$ and ^{13}CO, the 2 and 6 cm lines of H_2CO, and the (1,1) and (2,2) lines of NH_3 have been made in 11 cirrus cloud cores and 27 small galactic plane clouds. The CO lines are all modelled successfully in terms of hydrostatic equilibrium polytropic structures, which establish the run of density and temperature, the electron abundance, and the radiation field, which controls the ^{13}CO to $C^{18}O$ line ratios. These models are applied to NH_3 and H_2CO, whose excitation depends on electrons as well as neutrals. H_2CO is found to have a flat radial abundance distribution and a large abundance that cannot be explained by gas-phase chemistry but requires a photocatalytic formation on grains. Similarly NH_3 is too abundant to be explained by current gas-phase chemistry although its distribution is more centrally peaked. The properties of cirrus cores and small galactic plane clouds are indistinguishable. $n \sim r^{-a}$ structures can also formally fit the observations (without altering the chemical conclusions), but hydrostatic equilibrium models are preferred because they are intrinsically more restrictive, yet are as successful as the power-law models.

1. INTRODUCTION

Until recently, only diffuse clouds have been amenable to detailed studies of astrochemistry, because the physical conditions can be determined in them by a variety of molecular and atomic diagnostics, and by the fact that the clouds are essentially transparent and insignificantly clumped. Only diatomic molecules occur in them, however.

By contrast, more massive, dense molecular clouds are poorly understood because of their complex structure ("clumping") on all scales, uncertain radiation fields within them, and their association often with energetic phenomena (shocks, outflows, ionization fronts). Although complex molecules occur in them, their chemistry is relatively poorly understood.

Translucent clouds (total extinction $\sim 2-5$ mag) avoid the limitations of both diffuse and massive molecular clouds. We have studied two catagories in detail: (i) the "core" regions of the high-latitude cirrus clouds, first identified by Magnani et al (1985); (ii) the small galactic plane clouds identified by Clemens and Barvainis (1988). Observations of these objects in two transitions of $C^{18}O$, ^{13}CO, H_2CO, NH_3, CS, C_3H_2, and HC_3N show that these objects are spatially resolved, are quiescent (linewidths $\lesssim 0.5$ km/s), and are not clumped. Reliable physical conditions can therefore be derived. The variety of complex molecules

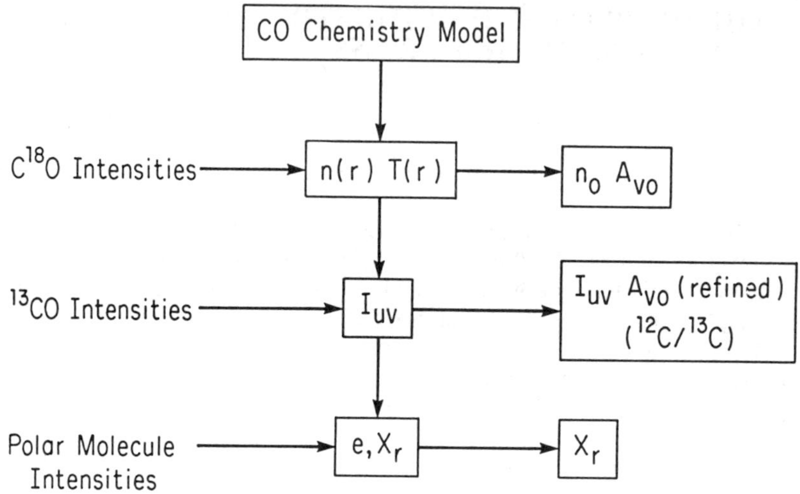

FIGURE 1

observed have allowed many aspects of astrochemistry to be studied.

2. MODELING THE STRUCTURE AND CHEMISTRY OF MOLECULAR CLOUDS

To understand the chemistry, evolution, and physical environment of molecular clouds, their physical structure must be known. The chemistry depends on the physical conditions, and vice versa. Current knowledge does not permit a meaningful attempt to solve the entire problem at once. Instead, one must try to isolate those observables which depend predominantly on a single physical parameter, build up a hierarchy of such dependences, and from them construct a self-consistent picture of the physical structure and of the chemistry of at least the observed species. Figure 1 illustrates this procedure as we have applied it to the translucent clouds. The process must begin with knowledge of the chemistry of at least one species, in this case CO and its isotopomers as developed by van Dishoeck and Black (1988; hereafter vDB). With simple generalizations of the vDB model, the fractional abundances of ^{12}CO, ^{13}CO and $C^{18}O$ can be derived for a given density, temperature, radiation field, and extinction, as applies to a given depth within a cloud. The next step is to specify a physical structure, n(r) and T(r). For any such structure, we find by solving the equations of statistical equilibrium and radiative transfer for $C^{18}O$ that the ratio of line brightnesses $T_B(2\text{–}1)/T_B(1\text{–}0)$ depends strongly only on the central density n_0. Because CO is nonpolar, n_0 is uniquely determined independent of the ionization, since CO is not excited by electrons. For a given structure n(r) and n_0, the extinction A_{v0} from edge to center may be found. Since the temperature depends only on A_{v0} (e.g. Langer 1977), T_0 is then determined. If the chosen structures are hydrostatic equilibrium (h.e.) polytropes, n(r) and T(r) are completely determined at this point.

So far, a "standard" UV field ($I_{UV} = 1$) has been assumed. Because ^{12}CO is shielded less from the UV field than is H_2, the fractional abundance $X(^{12}CO)$ will depend on the value of I_{UV}, and hence so will the derived value of A_{v0}, which is derived by using the value of $X(^{12}CO)$ given by the vDB model of CO chemistry, in which $I_{UV} = 1$ is assumed. A_{v0} will be larger or smaller depending on whether I_{UV} is > 1 or < 1. In addition, because $C^{18}O$, ^{13}CO, and ^{12}CO are all shielded by differing amounts from the radiation field, their relative abundances will depend on the value of I_{UV}, hence so will the relative line brightnesses of ^{13}CO and $C^{18}O$, for a given value of $^{12}C/^{13}C$. The observed line intensity ratio of ^{13}CO to $C^{18}O$ is used to derive I_{UV}, and to refine the value of A_{v0}, by iteration. The value $^{12}C/^{13}C = 70$ is found suitable for all cirrus cloud cores but one.

The vDB CO chemistry models may be parameterized quite precisely by the following expressions:

$$X_r(^{12}CO) = (cn/I_{UV})/[10^{-3.6A_v} + (dn/I_{UV})]$$

This form represents the effects of photo- vs. gas-phase processes. Constants c and d are chosen so that $X_r(^{12}CO) \rightarrow 2.4 \times 10^{-4}$ at $A_v \rightarrow \infty$ (corresponding to a depletion factor 0.4 for carbon). n is the density at each zone in the model, at which the extinction is calculated via

$$A_v(R) = 5.3 \times 10^{-23} f(H/H2) \int_0^R n(r)dr$$

Other CO isotopomer abundances have the form

$$X_r(C^{18}O) = a(18)s_r(18)X_r(^{12}CO)$$

$$X_r(^{13}CO) = a(13)s_r(13)f_r(13)X_r(^{12}CO)$$

Here, a(18) and a(13) are constants that represent isotopic ratios (1/500 and $^{13}C/^{12}C$ respectively), the s_r functions represent differential shielding between the various isotopomers, and $f_r(13)$ represents fractionation of ^{12}CO into ^{13}CO. These functions are all illustrated in Turner (1993), including the effect of $^{12}C/^{13}C$ upon the s_r functions. All functions are evaluated for each of the 40 zones employed in the model clouds.

For a given chosen n(r), a model which explains the four observed lines of $C^{18}O$ and ^{13}CO for a given source is found by iteration until suitable values of n_0, T_0, A_{v0}, and I_{UV} are found. We can now introduce observations of polar molecules (H_2CO, NH_3, CS, C_3H_2, HC_3N) which respond to electron as well as neutral particle excitation. The electron abundance (set equal to that of C^+) is given by

$$X_r(e) = (1/500)(cI_{UV})(500/n)/[10^{2.30A_v} + (dI_{UV})(500/n)]$$

which is the appropriate generalization of the particular vDB model for which $n = 500$ cm^{-3}. Two observed transitions of each polar species then allow a determination of the central abundance X_0, and some indication of the radial distribution X_r of the species. Additional observed transitions would define X_r

more accurately. In Section 5 we describe results for X_r for H_2CO and NH_3. Calculations for the other species are in progress.

The goals of this project are i) to determine the dynamical status (i.e. equilibrium or not) of the cirrus cores and CB objects; ii) to derive reliable chemical abundances and distributions; iii) to compare the high-latitude cirrus cores with the low-latitude CB objects.

3. HYDROSTATIC EQUILIBRIUM POLYTROPIC MODELS

Equilibrium occurs when the gravitational force is balanced by the internal pressure. The equilibrium condition is described by solving simultaneously:

–(polytropic) equation of state: $p = K\rho^{1+1/N}$

–equation of hydrostatic equilibrium: $\nabla p = -\rho \nabla \varphi$

–Poisson's equation: $\nabla^2\varphi = -4\pi G\rho$ These three equations may be combined to yield

$$K(N+1)\nabla^2 \rho^{1/N} = -4\pi G\rho$$

A transformation of variables $r = \alpha\xi$, $\rho = \rho_0\theta^N$, $T = T_0\theta$, $p = K\rho_0^{1+1/N}\theta^{N+1}$, where $\alpha = [-K(N+1)/4\pi G]^{1/2}\rho_0^{[(1-N)/2N]}$, leads to the Emden equation

$$(\xi^m\theta')' = \xi^m\theta^N$$

All physical quantities are expressible in terms of ξ, θ, and the boundary values ξ_e, p_e, n_0, and T_0. The primes denote differentiation with respect to ξ. m = 0, 1, 2 for slab, cylindrical, and spherical geometries respectively. The polytropic index N is negative for an object whose temperature increases outward, as for a molecular cloud heated externally by the UV field. A value N = -3 ± 1 is appropriate for interstellar molecular gas (de Yong et al. 1980). We adopt N = -3. To include the effects of turbulent pressure (about equal to thermal pressure) a value N = -4 would be more appropriate, but such polytropes differ little from N = -3 ones, so the formal omission of turbulent pressure does not affect any conclusion.

Solutions of the Emden equation for N = -3 and for spherical geometry have n(r) decreasing with increasing r at a rate typically less than r^{-1} for $\xi \leq 15$, and T increasing with r. Radius, mass, and external pressure may be derived as

$$R = (kT_0/\mu)(2\pi G n_0 k T_0)^{1/2}\xi_e = 5.4 \times 10^{18}(T_0/n_0)^{1/2}\xi_e$$

$$M = 3.35 T_0^{3/2} n_0^{-1/2}\theta_e'\xi_e^2\ M_\odot$$

$$p_e = n_e k T_e = p_0\theta^{N+1}(\xi_e) = n_0 k T_0 \theta_e^{-2}$$

An important property of h.e. polytropes is their fundamentally limited column density, for a given n_0. For an optically thin line this translates into a limitation on the line intensity, given by

$$T_B \propto 5.4 \times 10^{18}(n_0 T_0)^{1/2}[\langle n\rangle / n_0)]\xi_e$$

where $\langle n\rangle$ is the average density through the polytrope, and T_B is the brightness temperature. The quantity $[\langle n\rangle / n_0)]\xi_e$ increases with increasing ξ_e until an

Table 1. h.e. MODELS

Source	n_0 (cm^{-3})	T_0 (K)	A_{v0} (magn)	I_{UV}	ξ_e	R (10^{18} cm)	p_e (K cm^{-3})	M ($M_\odot$)	θ (')
Cld 6	1900	15.5	0.97	0.35	1.2	0.58	18550	1.91	5.3
Cld 7	2600	15.0	0.98	0.35	1.0	0.41	27050	1.01	3.8
HMV	1400	17.9	1.21	1.0	2.0	1.22	10950	8.7	14.0
PℓIIIC	1500	17.5	0.92	0.5	1.2	0.70	16550	2.6	8.0
PℓIIIC1	1600	14.7	1.48	0.65	3.6	1.86	6200	17.9	13.3
PℓIIIC2	1200	14.2	1.54	0.45	10.0	5.87	1500	92.2	42.0
PℓIIID	1150	15.0	1.62	0.77	15.0	9.25	1000	183.	66.3
40 IRAS	1000	16.7	1.60	1.4	15.0	10.5	1000	231.	85.9
Cld 55B	1800	17.5	0.88	0.5	1.0	0.53	23950	1.5	6.1
Cld 55D	1800	18.5	0.68	0.3	0.7	0.38	28750	0.6	4.4

asymptotic value is reached, at $\xi_e \simeq 15$ for the hot-edged sphere; for larger ξ_e, $\langle n \rangle$ basically decreases at the same rate as the radius R is increased. This means that, for a given ratio $T_B(2\text{–}1)/T_B(1\text{–}0)$ of $C^{18}O$, which fixes n_0 and hence the column density, the model has a maximum possible intensity for both $C^{18}O$ lines, which increases as the line ratio increases.

In addition to polytropic h.e. models, we have for comparison calculated power law models of the form $n(r) = n_0 r^{-\alpha}$, with $\alpha = 0, 1/2, 1, 3/2, -2$. The values $\alpha = 3/2$ and 2 correspond to certain gravitational collapse conditions.

4. RESULTS FOR H.E. AND $R^{-\alpha}$ MODELS

Table 1 gives the results of polytropic h.e. models for 11 cirrus cloud cores. All objects are fit with a canonical value $^{12}C/^{13}C = 70$ except MCLD which requires a larger value. The total extinction is twice the value of A_{v0} given in Table 1. The predicted angular size (θ) is based on distances as known (or 100 pc otherwise). For values of $\xi_e \gtrsim 8$ the derived values of p_e and M are unreliable, because the column density (line intensities) depends only weakly on ξ_e while p_e and M are very strong functions of ξ_e. Sources so affected are Pl111C2, Pl111D, and Cld 40(IRAS). Values of A_{v0} and I_{UV} are well determined in all cases. With the exception of these three objects, values of p_e agree well with the modern value of ~25,000 K cm^{-3} derived by other methods (Cox 1990).

Table 2 compares solutions for r^0, h.e., and r^{-1} models with observations. h.e. models always have values intermediate between those of r^0 and r^{-1} models. The $r^{-\alpha}$ models with $\alpha > 1$ are ruled out because they produce excessive self-absorption of the ^{13}CO lines in the extensive, low-density outer regions, usually resulting in line-reversal, which is never observed. Values of A_{v0} and I_{UV} are nearly the same for all models. Within the observational uncertainties, both h.e. models and r^0 models reproduce the observed angular sizes. r^{-1}

Table 2. CIRRUS CORE PROPERTIES

Object	Masses ($M_\odot$)			Angular Diameters (')			
	r^0	h.e.	r^{-1}	r^0	h.e.	r^{-1}	Observed
Cld 6	1.8	1.9	68.4	4.9	5.3	16.4	5x7
Cld 7	1.3	1.0	79.5	4.4	3.8	15.2	6
HMV 13	6.8	8.7	71.6	10.7	14.0	25.2	>> 4
PℓIIIC	1.4	2.6	120.	5.7	8.0	27.2	>6
PℓIIIC1	9.0	17.9	128.	7.3	13.3	19.7	> 7
PℓIIIC2	21.8	(92.2)	203.	11.2	(42.0)	28.6	> 12x5
PℓIIID	92.6	(183.)	202.	20.1	(66.3)	40.5	9x7
40 IRAS	74.1	(231.)	117	21.3	(85.9)	35.2	> 4
Cld 55B	2.2	1.5	42.8	7.0	6.1	17.3	> 4
Cld 55D	1.0	0.6	27.4	5.4	4.4	13.9	> 4

models appear to overestimate source sizes, as well as producing considerably higher masses. This is a result of the strong self-absorption of the r^{-1} models, which weakens the emergent line emission. We consider r^{-1} models to be least satisfactory. We favor h.e. models because they appear to fit all objects despite the strong restriction on column density, as well as predicting values of p_e in accordance with those determined by other methods. It appears that the cirrus cores are in or close to hydrostatic equilibrium.

5. THE CHEMISTRY OF H_2CO AND NH_3 IN CIRRUS CORES

The 2 and 6 cm transitions of H_2CO and the (1,1) and (2,2) transitions of NH_3 have been observed in all 11 cirrus cores and 27 Clemens-Barvainis (CB) objects. Starting with the model parameters n_0, T_0, A_{v0}, and I_{UV} developed from the $C^{18}O$ and ^{13}CO observations, both H_2CO and NH_3 have been analyzed by solving the statistical equilibrium and radiative transfer equations in microturbulent approximation, including electron and neutral particle excitation. Polytropic h.e. models, and $r^{-\alpha}$ power law models are constructed.

a. H_2CO

Figure 2 shows typical results for H_2CO. The ratio $\Re \equiv T_B(6\ \text{cm})/T_B(2\ \text{cm})$ is plotted against $T_B(6\ \text{cm})$ for the various models of n(r), ξ referring to the h.e. models, and r^{-1} to the corresponding power law case. Three models of $X_r(H_2CO)$ are analysed: models 0, 1, and 2 refer respectively to $X_r(H_2CO) = \text{const} \times X_r(C^{18}O)$, $X_r(H_2CO) = X_0(H_2CO)$, and $X_r(H_2CO) = X_0(H_2CO) \times [T(r)/T_0]$. The value of $X_0(H_2CO)$ or of the constant increases along the curves from left to right. The general form of the curves arises from the influence of both electrons and opacity effects, which weaken the 6 cm absorption relative to the 2 cm absorption as the abundance increases. An error box for the observed

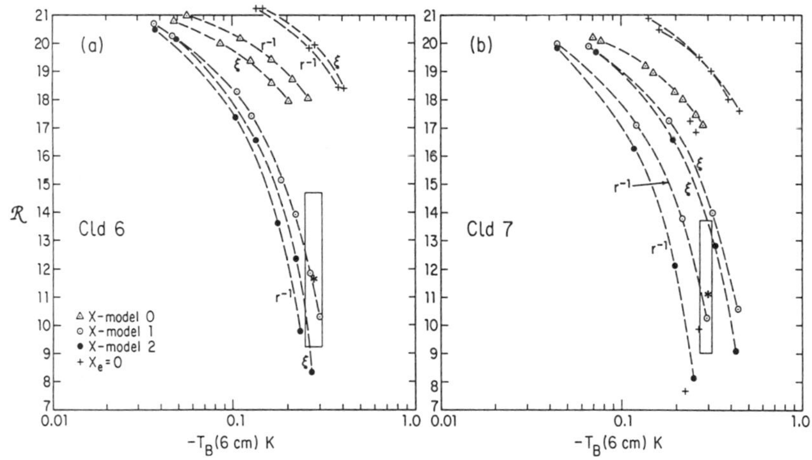

FIGURE 2

values of $\Re$ and of T_B(6 cm) is shown; the relatively large uncertainty in $\Re$ is due to the small intensities of the 2 cm transition.

Three conclusions are drawn from Fig. 2. First, the h.e. and r^{-1} models give very similar results, regardless of the X_r models. Thus, extremely precise observational results, beyond those currently feasible, are necessary to choose between n(r) models using H_2CO. Second, the value of $X_0(H_2CO) = (6 \pm 1.5) \times 10^{-9}$ for all 11 objects. This value is similar to that found in more massive cold dark clouds. Third, the cirrus cores show a definite preference for X_r model 1 or 2. Model 0 is ruled out for all the best determined objects.

The surprisingly large value $X_0(H_2CO)$ in the presence of strong UV radiation even at the centers of these objects ($A_{v0} \simeq 2$ mag) imposes severe restrictions on the formation mechanism for H_2CO. The abundance ratio $X_0(H_2CO)/X_0(C^{18}O) \geq 2 \times 10^{-2}$ is particularly well determined by the models. Gas phase chemistry has traditionally invoked the process $CH_3 + O \rightarrow H_2CO + H$ to form H_2CO, which yields

$$X_0(H_2CO)/X_0(C^{18}O) \leq 6 \times 10^{-11}[CH_3][O]/r_{ph}(H_2CO)[^{12}CO] \leq 2 \times 10^{-6}$$

at least four orders of magnitude less than the observed value. (The inequality arises from the choice of [CH_3] typical of much denser clouds where it is not subject to photodissociation.) Further, the strong preference for X-models 1 or 2 argues against gas-phase chemistry, for which the decrease in formation rate and increase of destruction rate (by photodissociation) with increasing r would indicate model 0. By contrast, models 1 or 2 would be expected for grain chemistry assuming that both formation (by photocatalysis) and destruction depend similarly on the UV strength.

That photocatalysis is a viable source of H_2CO in cirrus cores may be argued as follows. H_2CO is formed on grain surfaces by the process $CO + H + H \rightarrow H_2CO$ or CH_3OH. First consider small silicate grains. The accretion rate of CO on such grains is $\tau_{acc}(CO) = [n(CO)\sigma_g v S]^{-1} \simeq 9 \times 10^6$ sec, which is much less than the evaporation time of CO at relevant temperatures. Also, essentially all incident H atoms react with CO, because the timescales for incident H atoms to scan the grain surface and to tunnel through reaction barriers are much shorter than the evaporation time for H atoms when T $\lesssim$ 18 K. The formation rate for gas-phase H_2CO is the desorption rate from grains of H_2CO (or CH_3OH, which would likely first dissociate on the surface to H_2CO^*, then desorb as H_2CO). The desorption rate for a saturated species like H_2CO may be approximated (see Spitzer 1968) by $\tau_{desorb}^{-1} \simeq 5 \times 10^{-10} \exp(-2A_v)$ sec^{-1} mol^{-1}. Therefore the expected gas-phase H_2CO abundance is

$$X(H_2CO) \simeq 5 \times 10^{-10} n_{gr}(6300c)\exp(-2A_v)/[n(C^+)k + r_{ph}(H_2CO)]$$

where destruction is by reaction with C^+, and by photodissociation. n_{gr} is the number density of grains, whose typical size allows $\sim$ 6300 H_2CO or CH_3OH molecules to reside on the surface at full coverage. c is the fractional surface coverage, and has the value 0.001 to reproduce the observed value $X(H_2CO) \simeq 6 \times 10^{-9}$. Note that the desorption time scale $\tau_{desorp} = [5\times10^{-10}(6300c)\exp(2A_v)]^{-1}$ equals $\tau_{acc}(CO)$ when c = 0.26, which equals the H_2CO coverage on grains estimated by IR observations. The ratio of these two values of c further indicates that only $\sim$ 1/200 of the accreted CO need actually form CH_3OH or be desorbed by H_2CO to produce the observed H_2CO abundance.

At $A_{v0} \sim 2$ mag, it is possible that grains still retain an icy mantle. Icy grains are larger than silicate grains, and desorption rates are larger because of smaller binding energies, but the evaporation rate of H atoms is also larger. Assuming that these factors still result in a significant coverage by H_2CO or CH_3OH molecules, we may estimate the gas-phase abundance of H_2CO by

$$X(H_2CO) = 10^8 \exp(-2A_v) n_g \sigma_g Y p_i \eta_i / [n(C^+)k + r_{ph}(H_2CO)]$$

where $r_{ph}(H_2CO)$ is the photodesorption rate, $10^8 \exp(-2A_{v0})$ cm^{-2} sec^{-1} is the incident photon flux, $\sigma_g \simeq 10^{-9}$ cm^{-2}, Y is the probability that an incident photon is absorbed ($\sim$ 0.005 by experiment), p_i is the probability that an absorbed photon is absorbed by species i ($\sim$0.3, the coverage factor for H_2CO), and η_i is the probability that a photon absorbed by species i desorbs it. A value $\eta_i = 3 \times 10^{-4}$ is required to explain the observed value of X(H_2CO). Values of η_i are generally argued to be $\sim$0.1 for small molecules, smaller for larger molecules. We conclude that icy grains, if they can be maintained, can also produce the observed H_2CO.

b. NH_3

Figure 3 shows plots of $\Re \equiv T_B(1,1)/T_B(2,2)$ against $T_B(1,1)$ for two sources and for the same structural and abundance models as discussed for H_2CO. Electrons turn out to affect the NH_3 excitation only slightly, unlike H_2CO, but $\Re$ is very sensitive to temperature, hence to the particular X-models, since these weight T(r) differently throughout the object. Observational error boxes are shown, and again are large in the $\Re$ coordinate because of the weakness of the (2,2) lines.

Fig. 3 shows that (i) h.e. and r^{-1} models give very similar results regardless of the X-model. The n(r) models cannot be distinguished on the basis of these or

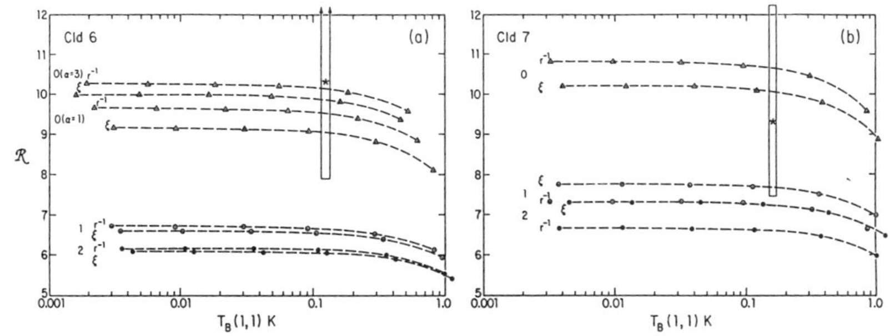

FIGURE 3

any feasible observations at the present time. (ii) The value of $X_0(NH_3)$ depends essentially only on $T_B(1,1)$ and is (6–8) $\times 10^{-9}$ for all cirrus cores, assuming an ortho/para ratio of 1 for NH_3. This value is only 2–3 times lower than that found in cold dense clouds where photodissociation is presumably unimportant. (iii) Contrary to H_2CO, the best-determined objects show a preference for X-model 0, and essentially rule out X-models 1 and 2. The fact that NH_3 is centrally peaked while H_2CO is not, is consistent with observed NH_3 linewidths which are about half the observed H_2CO linewidths.

As for H_2CO, the surprisingly large value of $X_0(NH_3)$ in the presence of strong UV radiation even at the centers of the cirrus cores imposes severe restrictions on the formation mechanism for NH_3. The preference for X-model 0 argues against grain processes, at least which involve UV processing, and in favor of gas-phase formation, whose rate would be highest in the dense cores where the destruction rate by photodissociation is smallest. This is consistent with laboratory work on photocatalysis, which suggests that NH_3 is not efficiently photocatalysed (Greenberg 1973) while H_2CO is efficiently photocatalysed (Breuer 1973). An explanation for these differences is that electronically excited surface species desorb efficiently. Thus CH_3OH, the end-product on the surface, is converted to H_2CO^* by a UV photon and desorbs as H_2CO, while the corresponding end-product, NH_3, converts to NH_2^* and dissociates as NH_2 rather than NH_3.

The two gas-phase reactions long considered as initiating the nitrogen-hydride species are (i) $N^+ + H_2 \rightarrow NH^+ + H$; and (ii) $N + H_3^+ \rightarrow NH_2^+ + H$. The first is endothermic by 186 K, and the second is calculated to have an activation barrier of ~ 1 eV.

Consider reaction (i). To overcome the endothermicity, models (e.g. Le Bourlot 1990) have invoked the rotational energy of ortho-H_2. N^+ is produced by the reaction $He^+ + N_2 \rightarrow N^+ + N + He$. Then for cold dark clouds, where NH_3 is destroyed principally by C^+ (cf. Le Bourlot 1990)

$$X(NH_3) = k'[He^+][N_2][f/(R+f)]/k[C^+] \simeq 10^{-8}$$

where the quantities f and R pertain to the efficiency with which N^+ is produced, and where f $\gg$ R typically holds. In cirrus cores, the same expression applies except that $r_{ph}(NH_3)$ replaces $k[C^+]$ as the dominant destruction agent of NH_3.

Now, with otherwise the same values, we find $X(NH_3) \leq 4 \times 10^{-11}$ for cirrus cores, where the inequality arises by adopting a value of $[N_2]$ typical of cold dark clouds; a much smaller value (likely by a factor of 1000) applies for the radiation fields found in the cirrus cores. By a very large factor, we can rule out reaction (i) in the case of cirrus cores.

Next consider reaction (ii). This reaction is certainly inapplicable if there is a 1 eV activation barrier, but if we assume there actually is no barrier and that a typical reaction rate of $\sim 10^{-9}$ cm^3 sec^{-1} applies, then we expect

$$X(NH_3) = k[N][H_3^+]/r_{ph}(NH_3) \simeq 1.6 \times 10^{-10}$$

where [N] and $[H_3{}^+]$ have been taken from the models of Langer and Graedel (1988) which assume no UV fields. Under UV conditions in the cirrus cores, all nitrogen is expected to be atomic, and [N] is $\sim$ 7 times larger than the Langer-Graedel value. $[H_3{}^+]$ is estimated to decrease about 2 times, if it has a small photodissociation rate, as is likely. In this case, $X(NH_3) \simeq 6.1\times10^{-10}$, a factor $\sim$ 14 times smaller than the observed value. Within the uncertainties, reaction (ii) thus appears capable of producing the observed NH_3 in cirrus cores. Note that reaction (ii) would produce an NH_3 abundance in cold dense clouds about 4.5 times the observed value, when C^+ is taken as the dominant destruction agent. Again this is within the uncertainties. We conclude that reaction (ii) must be taken seriously despite quantal calculations of an activation barrier. Experiments on this reaction are needed to resolve the situation.

Recently interstellar NH was discovered with an abundance 2×10^{-9} (Meyer and Roth 1991), a few times less than NH_3. If NH_3 were actually produced on grains, UV desorption as either NH_2 or NH would produce the observed NH abundance. The gas-phase reaction (i) has been shown by van Dishoeck (1992) to under-produce NH by a factor of $\sim$ 20. The gas-phase reaction (ii) can be shown to produce enough NH if it also produces the observed NH_3. Thus observations of NH serve mainly to confirm the conclusions we have drawn concerning NH_3 in cirrus cores.

6. THE PHYSICS AND CHEMISTRY OF SMALL GALACTIC PLANE CLOUDS

It is important to study a similar sample of small molecular clouds at low latitudes in order to answer several questions. First, is the steady-state condition of the cirrus cores a result of their small size ($\sim$10 $M_\odot$), or of their high latitude (where they may be less subject to perturbing forces such as shocks or shears)? Second, are there differences in the chemistry of high- and low-latitude clouds, again as a result of differences in perturbations, or perhaps of metallicity? These questions are emphasized further by the Lynds-type clouds, which cannot be fit by h.e. models, and whose chemistry (early-type) clearly differ from the cirrus core chemistry, which lacks ET-type species such as HC_3N.

We have studied a subset of 27 of the 240 small clouds selected optically by Clemens and Barvainis (1988) at low latitudes. The selection criteria are identical (except for latitude) to those of the cirrus cores: they are quiescent, of angular size 2′–6′, isolated, of small ellipticity, and optically selected as "opaque". As with the 11 cirrus cores, we have observed (and mapped) the

J=2–1 and J=1–0 transitions of $C^{18}O$ and ^{13}CO in all 27 CB objects, and find the line intensities and widths are similar to those of the cirrus cores. There is a somewhat wider range of ^{13}CO line brightnesses relative to $C^{18}O$ line intensities. We have analyzed the 27 objects with h.e., r^0, and r^{-1} structural models.

The results are as follows: (i) all objects (24) that can be fit by $r^{-\alpha}$ models can also be fit by h.e. models. (ii) the properties of the CB objects and cirrus cores are indistinguishable. For cirrus cores and CB objects respectively:

–mean distance is 100 pc, and 600 pc
–mean angular size is 4′–5′, and ~2′
–mean mass is 16.2, and 8.5 M_O
–mean external pressure is 12,200, and 15,900 K cm^{-3}
–mean extinction (edge to center) is 1.19, and 1.10 mag
–mean I_{UV} is 0.63, and 0.55.

None of these differences is significant. (iii) 17 of the objects can be fit with $^{12}C/^{13}C = 70$, the value adopted for all 11 cirrus cores, while 7 objects require $^{12}C/^{13}C > 70$, and 3 objects require $^{12}C/^{13}C > 115$, regardless of the structural model. There is no dependence on galactocentric distance, and the high–$^{12}C/^{13}C$ objects indicate regions where there is a dearth of ^{13}C (supernova-rich regions?) such as have been found recently in other galaxies. From a nucleosynthesis point of view, the ISM appears patchy.

Intensities of H_2CO and NH_3 lines (as well as of CS, not yet analyzed) are similar in both classes of objects. C_3H_2 lines are weaker in cirrus cores, and HC_3N (J=2–1) is not detectable in cirrus cores, and in only a few of the CB objects. The C_3H_2 and HC_3N results might imply that steady-state chemical conditions are more closely approximated in the cirrus cores, (C_3H_2 and HC_3N are classic "early-times" species), but the dominance of the radiation field in both classes of objects precludes any conclusions along these lines until analysis is done.

One conclusion seems apparent: all small clouds (~10 $M_\odot$, 2.5–4 mag total extinction) appear to be in or near hydrostatic equilibrium, independent of their galactic latitude. This contrasts with the denser, colder, more massive (~100 $M_\odot$) Lynds clouds, which have similar $C^{18}O$ line ratios but much larger line intensities, and lie well outside the intrinsic limits of the h.e. models.

REFERENCES

Breuer, H.D. 1973, in *Interstellar Dust and Related Topics*, eds. J.M. Greenberg and H.C. van de Hulst, D. Reidel, 399

Clemens, D.P., and Barvainis, R.E. 1988, ApJS, 68, 275 (CB)

Cox, D.P. 1990, in *The Interstellar Medium in Galaxies*, eds. H.A. Thronson and J.M. Shull (Dordrecht: Kluwer), 181

de Jong, T., Dalgarno, A., and Boland, W. 1980, A&AS, 91, 68

Greenberg, L.T. 1973, in *Interstellar Dust and Related Topics*, eds. J.M. Greenberg and H.C. van de Julst, D. Reidel, 413

Langer, W.D. 1977, ApJL, 212, L39

Langer, W.D., and Graedel, T.E. 1989, ApJS, 69, 241

Le Bourlot, J. 1991, A&A, 242, 235

Magnani, L., Blitz, L., and Mundy, L. 1985, ApJ, 295, 402

Meyer, D.M., and Roth, K. 1991, ApJL, 376, L49

Spitzer, L. 1968, *Physical Processes in the Interstellar Medium*, John Wiley & Sons, 210

Turner, B.E. 1993, ApJ, 405, 229

van Dishoeck, E.F. 1992, in *Astrochemistry of Cosmic Phenomena*, ed. P.D. Singh, Kluwer Academic, 148

van Dishoeck, E.F., and Black, J.H. 1988, ApJ, 334, 771 (vDB)

The First Symposium on the Infrared Cirrus and Diffuse Interstellar Clouds
ASP Conference Series, Vol. 58, 1994
Roc M. Cutri and William B. Latter (eds.)

PHOTOCHEMISTRY AND THE INTERSTELLAR RADIATION FIELD

EWINE F. VAN DISHOECK
Leiden Observatory, P.O. Box 9513, 2300 RA Leiden, The Netherlands

ABSTRACT The photodissociation processes controlling the chemistry of diffuse and translucent clouds are reviewed. Uncertainties in the intensity and shape of the interstellar ultraviolet radiation field are discussed, with special attention to the wavelength region 912–1100 Å where CO and H_2 are photodissociated and C is photoionized. The relation between diffuse, translucent and cirrus clouds is illustrated with models in which carbon is transformed from atomic to molecular form with increasing extinction.

INTRODUCTION

Diffuse, translucent and cirrus clouds are by definition clouds in which ultraviolet photons can penetrate. Both their chemical and physical structure is therefore controlled by photoprocesses. In terms of the photochemistry, it is convenient to distinguish two different regimes. In *diffuse clouds*, the transition of hydrogen from atomic to molecular form occurs but carbon is still mostly in atomic form as C^+. They typically have a total visual extinction $A_V^{tot} \leq 1$ mag and classical examples are the lines of sight studied with the *Copernicus* satellite, in particular those toward ζ Oph and ζ Per. In *translucent clouds*, hydrogen is mostly molecular and the transition of carbon from atomic (C^+ and C) to molecular (CO) form takes place. These clouds have A_V^{tot}=1–5 mag and examples are lines of sight toward highly–reddened stars such as HD 29647 (Crutcher 1985), HD 169454 (Jannuzi et al. 1988), HD 154368 (Gredel, van Dishoeck & Black 1991, 1993*a*, *b*), and high–latitude cirrus clouds such as that toward HD 210121 (Gredel et al. 1992; Welty & Fowler 1992). Such clouds are also thick enough for CO millimeter lines to be detectable and most of the high–latitude MBM cirrus clouds fall into this category (cf. Magnani, Blitz & Mundy 1985).

The depth to which ultraviolet photons can penetrate in the cloud and affect the chemistry depends on (i) the available amount of photons at the boundary of the cloud; (ii) the scattering properties of the grains as functions of wavelength; (iii) the competition between the atoms and molecules with the grains for the available photons, which, in turn, depends on the density of the cloud; and (iv) possible self–shielding of the molecules.

In this chapter, first the photodissociation processes of astrophysically important molecules are reviewed. Subsequently our knowledge of the intensity and shape of the interstellar radiation field is discussed. Finally, these ingredients are incorporated into representative models of diffuse and translucent clouds.

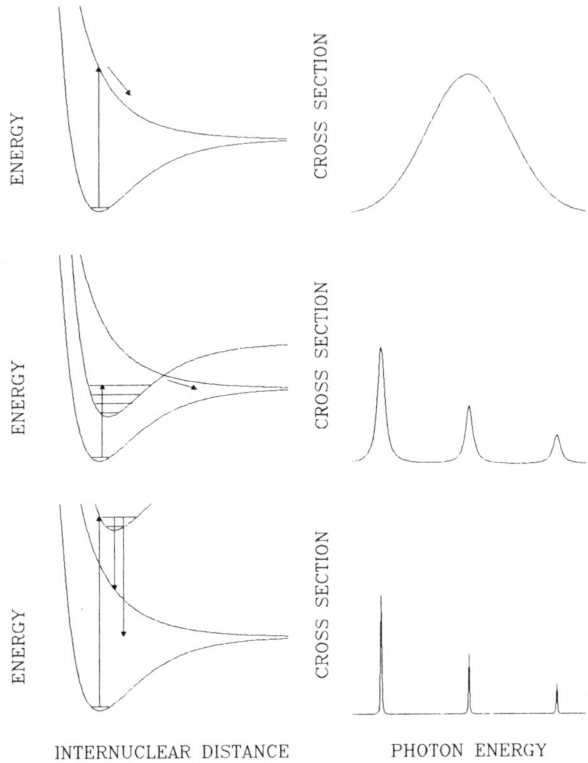

FIGURE 1 Photodissociation processes of diatomic molecules

PHOTODISSOCIATION PROCESSES

The processes by which photodissociation of simple astrophysical molecules can occur have been outlined in detail by van Dishoeck (1988) and Kirby & van Dishoeck (1988). A summary is presented in Figure 1. A general discussion of photodissociation has been given by Schinke (1993).

In the simplest process of *direct photodissociation*, a molecule absorbs a photon into an excited electronic state that is repulsive with respect to the nuclear coordinate(s). Since spontaneous emission back to the ground state is a comparatively slow process, virtually all of the absorptions lead to dissociation of the molecule. The resulting photodissociation cross section is continuous as a function of photon energy and peaks close to the vertical excitation energy. It is usually very small at the threshold, which corresponds to the dissociation energy of the molecule. The width is determined by the steepness of the repulsive potential: the steeper the curve, the broader the cross section.

In the process of *predissociation*, the initial absorption occurs into a bound excited electronic state, which subsequently interacts non–radiatively with a nearby repulsive electronic state. The strength of the interaction depends sensitively on the type of coupling and on the energy level involved, but the corresponding predissociation rates are typically comparable to or larger than the rates for spontaneous emission. The effective photodissociation cross section consists in this case of a series of discrete peaks. The strength of each peak reflects the product of the oscillator strength of the initial absorption and the dissociation efficiency of the level involved. The width is controlled by the sum of the radiative and predissociation rates and is generally large (15 km s^{-1} or more in velocity units) if the predissociation is rapid.

If the excited bound states are not predissociated, *spontaneous radiative dissociation* can still be effective through emission of photons into the continuum of a lower–lying repulsive state or the vibrational continuum of the ground electronic state. The efficiency of the process is in this case determined by the competition with spontaneous emission into lower–lying bound states. The photodissociation cross section again consists of a series of discrete peaks, but the peaks are not broadened and have widths determined by the total radiative lifetime (typically <0.1 km s^{-1}).

Because a molecule has many excited electronic states, in general all of these processes will occur. However, usually only one of them dominates the photodissociation of a molecule in interstellar clouds. For example, the photodissociation of simple hydrides like CH, CH^+, NH, OH, H_2O and CH_2 proceeds mostly through the direct process. On the other hand, the photodissociation of CO is controlled by predissociation processes, whereas that of H_2 occurs by spontaneous radiative dissociation. Whether the photodissociation is dominated by continuous or line processes has important consequences for the radiative transfer through the cloud, because in the latter case the lines can become optically thick if the abundance of the molecule is sufficiently high. It is no coincidence that the photodissociation of the two most abundant molecules in translucent clouds, H_2 and CO, occurs through line absorptions. Note, however, that the self–shielding is much more effective for H_2 than for CO, both because of the higher abundance of hydrogen compared with carbon, and because the H_2 ultraviolet absorption lines are intrinsically narrower than those of CO.

The photodissociation rate k_{pd} of a molecule by continuous absorption is given by

$$k_{pd}^{cont} = \int \sigma(\lambda) I(\lambda) d\lambda \;\; \mathrm{s}^{-1},$$

where σ is the cross section for photodissociation in cm^2 and I is the mean intensity of the radiation in photons cm^{-2} s^{-1} Å^{-1} as a function of wavelength λ in Å. For the indirect processes of predissociation and spontaneous radiative dissociation, the rate of dissociation by absorption into a specific level of a bound upper state u from lower level ℓ is

$$k_{pd}^{line} = \frac{\pi e^2}{mc^2} \lambda_{u\ell}^2 f_{u\ell} \eta_u x_\ell I(\lambda_{ul}) \;\; \mathrm{s}^{-1},$$

where $f_{u\ell}$ is the oscillator strength, η_u is the dissociation efficiency of the upper level which lies between 0 and 1, and x_ℓ is the fractional population in level ℓ.

Because the structure of diffuse clouds is controlled by the photodissociation of H_2 and CO, it is useful to recapitulate a few more details for these molecules. The photodissociation of H_2 occurs by absorptions into the Lyman and Werner bands at 912–1100 Å (Black & Dalgarno 1977; Black & van Dishoeck 1987). The lines have typical oscillator strengths $f \approx 0.0001$–0.03, and on average about 10% of the absorptions lead to dissociation. The strongest lines become optically thick at H_2 column densities of $10^{14} - 10^{15}$ cm^{-2} for Doppler broadening parameters b of 3 km s^{-1}. The photodissociation rate of H_2 is about 5×10^{-11} s^{-1} in the unattenuated interstellar radiation field, corresponding to a lifetime of about 700 yr. Inside the cloud, the lifetime becomes longer by several orders of magnitude because of the self–shielding process. The remaing 90% of the absorptions are followed by ultraviolet fluorescence back into the bound vibrational levels of the H_2 ground state. This fluorescence has recently been observed even in diffuse gas (Martin, Hurwitz & Bowyer 1990).

Significant progress has been made in the last 6 years in our understanding of the photodissociation processes of the important CO molecule. CO is an enormously stable molecule with a dissociation energy of 11.09 eV corresponding to a threshold of 1118 Å. The detailed laboratory spectra of Letzelter et al. (1987) established that no continuum absorption occurs longward of 912 Å, but that the CO photodissociation is dominated by line absorptions in about 30 bands, most of which are strongly predissociated. Subsequent experiments by Stark et al. (1991, 1993), Eidelsberg & Rostas (1990), Eidelsberg et al. (1991), Smith et al. (1991) and Levelt, Ubachs & Hogervorst (1992*a*, *b*) have extended this work to other isotopes and much higher spectral resolution. Although some uncertainties still persist in the oscillator strengths and predissociation rates of individual lines (Kirby & Cooper 1989; Stark et al. 1992; Chan, Cooper & Brion 1993), these are not expected to change much the overall photodissociation rate. The depth dependence of the CO photodissociation rate is affected not only by self–shielding, but also by shielding by H and H_2 because they absorb in the same wavelength region. ^{12}CO, in turn, can shield the less abundant ^{13}CO and C^{18}O species. Thus, a complete numerical simulation of the entire spectrum of ^{12}CO, ^{13}CO, C^{18}O, H_2 and H is required to compute correctly the attenuation at each depth into a cloud (van Dishoeck & Black 1988*b*; Viala et al. 1988*a*; Mamon, Glassgold & Huggins 1988; Abgrall et al. 1992). Simplified treatments have been given for CO by van Dishoeck & Black (1988*b*) and for H_2 by Federman, Glassgold & Kwan (1979) and van Dishoeck (1987).

INTERSTELLAR RADIATION FIELD

Estimates of the average intensity of the interstellar radiation field (ISRF) fall into two main categories (van Dishoeck & Black 1988*b*). The first method is to take the actual number of hot stars (O, B, A, ...) in a certain region and to calculate their summed radiation at a specific or average position using model atmospheres (e.g. Habing 1968; Witt & Johnson 1973; Jura 1974; Henry 1977; Hurwitz, Bowyer & Martin 1991). These models use observed ultraviolet fluxes of early–type stars for calibration, generally in the 1300–1500 Å region. For the critical wavelength region 912–1100 Å, the method relies almost entirely on model atmospheres. There is evidence that the often–used line–blanketed LTE atmosphere models (cf. Kurucz 1979) consistently overestimate the fluxes of O and B stars at $\lambda \leq 1100$ Å in comparison with calibrated, spectrophotometric observations (e.g. Cook, Cash & Snow 1989) (see Kudritzki & Hummer 1990 for

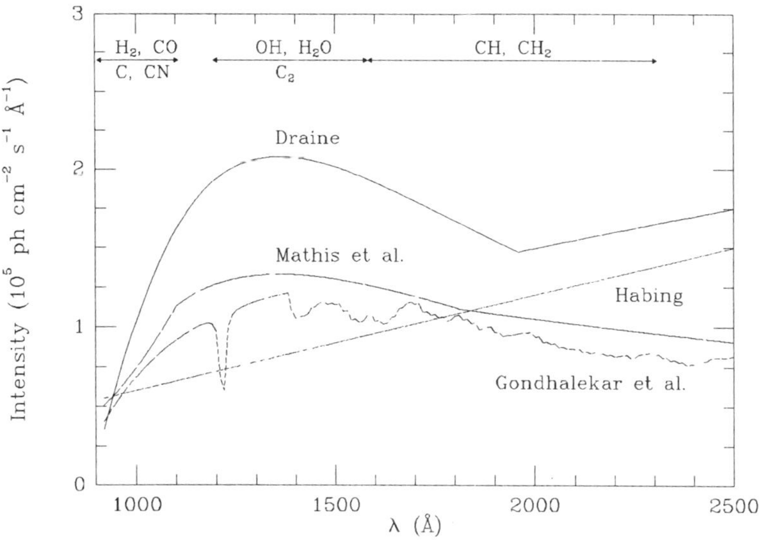

FIGURE 2 Various estimates of the average interstellar radiation field. The radiation field of Draine (1978) has been extended for $\lambda > 2000$ Å using the representation of van Dishoeck & Black (1982). The wavelength ranges where the photodissociation of some important interstellar molecules occurs are indicated.

a review of model atmospheres). In addition, H I absorption may be significant at 1025, 972 and <950 Å, especially for B stars. A new determination of the ISRF taking these effects into account appears timely and warranted.

The second method to infer the ISRF is to use direct measurements of the ultraviolet radiation from the sky at specific wavelengths. This method is not entirely independent from the first one because models of the spatial and spectral distribution of light from hot stars are often used to interpret very broadband measurements and to extrapolate the fluxes to wavelengths and regions of the sky that are unobserved. Also, the method suffers from uncertainties in the extinction corrections and the scattering properties of the grains (see below). The earlier observations were done at a variety wavelengths and positions with ultraviolet spectrometers aboard rockets, the *Apollo 17* and the *Voyager* spacecraft (see Paresce & Jacobsen 1980 for a review). This was followed by the TD–1 satellite, which performed an all–sky survey at 1565, 1965, 2365 and 2740 Å (Gondhalekar, Phillips & Wilson 1980), and by the STP 72–1 satellite, which obtained measurements in a broad band, 911–1050 Å, over about two–thirds of the sky (Opal & Weller 1984). Recent reviews have been given by Bowyer (1991) and Henry (1991). Direct starlight from early B stars dominates the measured flux at the shortest wavelengths. The contribution of diffuse radiation scattered by dust is of order 10–20% (Gondhalekar et al. 1980) and the true extragalactic component is negligibly small (Holberg 1986; Bowyer 1991). More recently, the Berkeley UVX experiment has measured the spectrum of the diffuse scattered emission in the 1415–1835 Å region at selected positions at both low and high

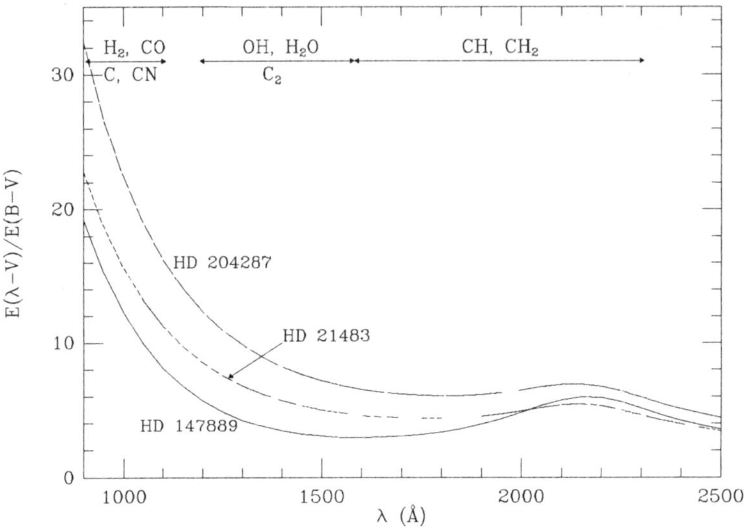

FIGURE 3 Observed extinction curves toward three reddened stars representing extreme cases. The wavelength ranges for photodissociation of various species are indicated.

latitudes (Hurwitz et al. 1991). Additional widely quoted references for the ISRF are those of Mathis, Mezger & Panagia (1983), who adopted the results of Gondhalekar et al. (1980) scaled by a factor 1.15, and of Draine (1978), who represented a variety of the early theoretical and observational work with a simple formula.

In Figure 2, the various determinations of the intensity of the average ISRF in the solar neighborhood are summarized. Although they generally agree within factors of two, their shapes are quite different. The wavelength regions where the photodissociation of some of the most important interstellar molecules occurs are indicated.

If a cloud is located close to a young, hot star, the intensity incident on the cloud boundary may be enhanced substantially compared with the average ISRF. Also, its wavelength dependence may be quite different, depending on its spectral type. At high latitudes where few early–type stars are available, observations indicate that the intensity of the ISRF at 1550 Å may be factors 5–10 lower. Moreover, the ratio of intensities at 975 and 1565 Å ranges from 0.03–1 depending on latitude (Gondhalekar et al. 1980; Opal & Weller 1984). Thus, the high–latitude clouds may be exposed to significantly less intense ultraviolet radiation, especially that capable of dissociating H_2 and CO and ionizing carbon.

The scattering parameters of the grains also play an important role in the penetration of photons. As discussed, for example, by Cardelli & Savage (1988) (see also Cardelli, this volume), the ultraviolet extinction curve shows substantial variations from place to place, especially at the shortest wavelengths. Figure 3 shows an example of three observed extinction curves which represent extreme cases. Again the largest variations are found in the 912–1100 Å region, and this can greatly affect the CO and CN photodissociation rates in the centers of the clouds (van Dishoeck & Black 1989). If such variations are due to different populations of grains, then similar effects are expected to occur for the albedo ϖ and the asymmetry parameter g as functions of wavelength. Very little is still known about these parameters at the shortest wavelengths (Roberge et al. 1991; Bowyer 1991) and the variations on the (photo)chemistry have yet to be explored in detail.

Photodissociation rates as functions of depth into a cloud using the Draine (1978) or Mathis et al. (1983) radiation field have been presented for a variety of molecules e.g. by van Dishoeck (1988) and Roberge et al. (1991). The depth dependence due to continuum extinction can be represented by a simple exponential function, with exponents that vary with grain scattering properties and that depend parametrically on cloud thickness.

CLOUD MODELS

Sophisticated depth–dependent models of diffuse and translucent clouds which include both line and continuum photoprocesses have been developed e.g. by Black & Dalgarno (1977), van Dishoeck & Black (1986, 1988*b*, 1989), Viala et al. (1988*a*, *b*) and Abgrall et al. (1992). They all adopt a homogeneous plane–parallel structure and have as input parameters: the total density $n_H = n(\mathrm{H}) + 2n(\mathrm{H_2})$ and kinetic temperature T at each depth; the total extent of the cloud, specified either by A_V^{tot}, N_H or $N(\mathrm{H_2})$; the overall scaling factor I_{UV} of the ISRF; the scattering properties of the grains $A(\lambda)$, $g(\lambda)$ and $\varpi(\lambda)$; the depletion factors δ_X of the various elements (carbon, oxygen, nitrogen etc.) from the gas phase; and the primary cosmic ray ionization rate ζ. At each depth z into the cloud, the photodissociation and photoionization rates are computed and the network of chemical reactions is solved at steady–state resulting in the local densities $n_i(z)$ of each species. The column densities are then obtained by integration over the whole cloud: $N_i = \int n_i(z)dz$. Note that only column densities can be compared directly with observations. The physical input parameters are usually obtained from diagnostic observations of e.g. the $\mathrm{H_2}$ and $\mathrm{C_2}$ rotational excitation and the C and $\mathrm{C^+}$ fine–structure excitation, whenever available. Recent reviews have been given by van Dishoeck & Black (1988*a*), van Dishoeck (1990, 1992) and Wagenblast & Williams (1993). More simplified approaches for specific molecules have been presented by Federman and co–workers (e.g. Federman & Lambert 1988).

An obvious extension of the above models is to calculate the temperature self–consistently within the models using the known heating and cooling processes (Black, this volume). Such models have been made by Roberge (1981), Le Bourlot et al. (1993) and Spaans (1993), but they often lead to too low temperatures at the edge of the cloud unless efficient heating by small grains or PAH's is taken into account. Time–dependent models have been presented e.g. by Wagenblast & Hartquist (1989) (see also Flower, this volume).

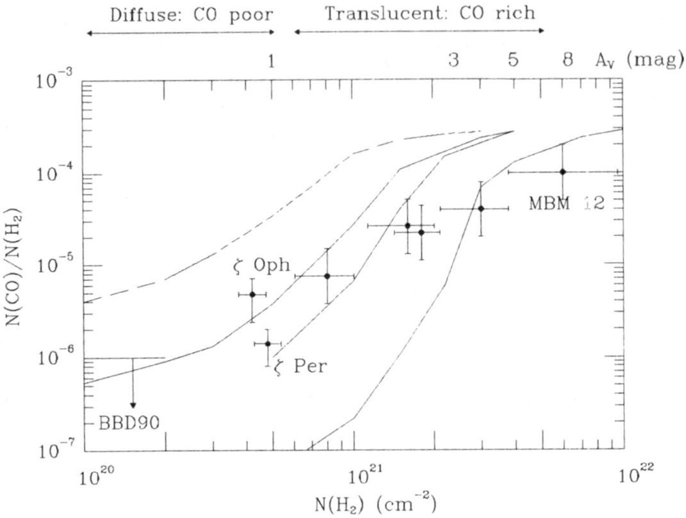

FIGURE 4 Computed and observed $^{12}CO/H_2$ column density ratios as functions of total H_2 column density. The full lines from left to right are model results for I_{UV}=0.5, 1.0 and 10, respectively, and $n_H \approx$500–2000 cm^{-3}. The dashed line is for I_{UV}=1 and $n_H = 10^4$ cm^{-3}. Observed values for a number of diffuse, translucent and high–latitude clouds are included.

CO abundance

Because the concentrations of H, H_2, C, C^+ and CO vary by several orders of magnitude with position into a translucent cloud, it is essential to use depth–dependent models for comparison with observations. The results of a series of models are shown in Figure 4, where the computed column density of CO w.r.t. that of H_2 is plotted as a function of the total H_2 column density, or, equivalently, the total extent of the cloud. It is seen that for typical conditions in translucent clouds, i.e. $n_H \approx 10^3$ cm^{-3} and I_{UV}=1, the CO/H_2 column density ratio increases very strongly around $A_V^{tot} \approx 1-2$ mag. The exact location of this transition and its sharpness depend sensitively on the physical parameters of the cloud, in particular the strength of the ISRF, the density and the gas–phase abundance of carbon. In particular, the transition shifts to much lower A_V^{tot} for higher densities and lower values of I_{UV}, parameters that are probably appropriate for high–latitude clouds. On the basis of such diagrams, the observations of CO in diffuse, translucent and high–latitude cirrus clouds can readily be understood, and some of them are included in Figure 4. There is no need to invoke a separate class of clouds with a different terminology for each set of data. Note that the high–latitude cirrus clouds appear to cover the full range of parameter space. Some of them are truly "diffuse", such as the clouds selected through IRAS 100 μm in which no CO millimeter emission has been detected (e.g. Blitz, Bazell & Désert 1990, indicated as BBD in Figure 4). Other clouds like MBM12 show CO antenna temperatures comparable to those found in dense clouds such as TMC–1 (Pound, Bania & Wilson 1990). The majority of the MBM clouds, as well as the Ursa Major cloud, probably fall in the middle regime, where carbon is being transformed into CO.

The sensitivity of the CO column density to the various parameters raises intriguing questions regarding the interpretation of observed CO maps: is the structure seen in such maps the result of small density variations, column density variations or variations in the radiation field or scattering properties of the grains (Gredel et al. 1992)? Small fluctuations can have large effects on the strengths of the CO millimeter lines, as long as these lines are not too optically thick. It is interesting to note that the CO column density of $\sim 10^{15}$ cm^{-2} for which the ultraviolet lines become self–shielding is comparable to that at which the millimeter lines become optically thick.

The ^{12}CO/^{13}CO column density ratio in diffuse and translucent clouds can also reveal much about their physical structure. It can deviate from the overall [^{12}C]/[^{13}C] ratio of about 60 (Stahl & Wilson 1992; Langer & Penzias 1993) by the action of two competing effects: isotope selective photodissociation, which will make the ratio larger than 60; and carbon–isotope exchange reactions, which will drive it to values smaller than 60. Isotope selective photodissociation occurs because the more abundant ^{12}CO molecule can shield some ultraviolet absorption bands of ^{13}CO but not all, resulting in a faster ^{13}CO rate inside the cloud (van Dishoeck & Black 1988*b*). On the other hand, the carbon–isotope exchange reaction $^{13}C^+ + {}^{12}CO \rightarrow {}^{13}CO + {}^{12}C^+$ is exoergic by about 36 K and proceeds very rapidly toward ^{13}CO at low temperatures. Thus, ^{13}CO is enhanced in regions of low temperature shielded from ultraviolet radiation in clouds where not all carbon has been converted into CO. This enhancement can become as large as an order of magnitude at temperatures as low as 10 K.

How do models of the ^{13}CO fractionation compare with observations? A recent series of *HST* ultraviolet data toward ζ Oph suggest ^{12}CO/^{13}CO ratios of at least 60 (Lyu, Smith & Bruhweiler 1993) and probably of order 150 (Sheffer et al. 1992; Lambert et al. 1993). These high values are indicative of isotope selective photodissociation in a cloud of relatively high temperature, although a model which reproduces ratios as high as 150 has yet to be devised. Unfortunately, no such high quality ultraviolet data yet exist for other clouds. The interpretation of millimeter observations suffers from the fact that the ^{12}CO lines are often optically thick whereas those of ^{13}CO are optically thin. A few high–latitude clouds have been found where even the ^{12}CO line is still optically thin and which show ^{12}CO/^{13}CO $\approx$ 60 or larger (Stark 1993). For a series of translucent clouds, Gredel et al. (1993*b*) have derived column density ratios ranging from 10 to more than 30. Comparison of the ^{12}CO and ^{13}CO antenna temperature maps with large scale optical emission shows a tendency for the ratio to be large if the optical emission is smooth, but small if it is "clumpy". This suggests the presence of local radiation "twilight" zones in the denser, clumpy regions where the ^{13}CO may be enhanced (Stark 1993).

Many (but not necessarily all) diffuse and translucent clouds show inhomogeneous or "fractal" structures (e.g. Falgarone, Phillips & Walker 1991; Gredel et al. 1993*b*). The main effect of an inhomogeneous distribution is that the ultraviolet radiation can penetrate deeper into the cloud through the lower density, less opaque parts. Thus, molecules are rapidly dissociated in the "interclump" gas, whereas they can still survive in the clumps themselves. The first models to treat the continuum radiative transfer in a clumpy medium were those of Boissé (1990), who showed that the effects of clumpiness can be explored by using different effective grain scattering parameters. A more detailed treatment of both line and continuum radiative transfer has been developed by Spaans (1993, see also this volume) and has been incorporated with the chemistry into two-dimensional clumpy clouds. As expected, a $C^+ \rightarrow CO$ transition zone occurs around each clump, and the formation rates of species such as CO, C_2 and CN are greatly enhanced in the high-density regions (see also van Dishoeck et al. 1991). Future observations of C I and C II absorption or emission with *HST*, *ISO* and/or ground-based telescopes may reveal more about the structure of translucent clouds.

The effects of non-Gaussian velocity distributions on the chemistry are more subtle. In general, they are not expected to affect the $C^+ \rightarrow CO$ transition because the CO and (saturated) H_2 ultraviolet lines are already intrinsically very broad (Jenkins et al. 1989). However, such turbulent motions can account for (part of) the energy needed to drive the formation of the CH^+ ion (Gredel, van Dishoeck & Black 1993*a*; see also Flower, this volume).

Other species

The steady-state models presented above can also reproduce the observed column densities of molecules such as CH, C_2, CN and OH if reasonable values for uncertain reaction rates are adopted. All results are very sensitive to the scaling factor I_{UV} for the ISRF, as well as its shape and the adopted extinction curve. Agreement with observations of some diffuse clouds is improved if I_{UV} is lower than the value inferred from the high-J population of H_2 under the assumption that ultraviolet pumping is the main contributor (Wagenblast 1992).

The steady-state models fail to reproduce the measured CH^+ column densities (see Flower, this volume), and also fall short of the observed NH abundance in diffuse clouds (Meyer & Roth 1991) and that of more complex species such as H_2CO, NH_3 and C_3H_2 in the thicker high-latitude cirrus cores (see Turner, this volume). On the other hand, the observed abundance of C_2H is surprisingly low (Gredel et al. 1993*b*). Possible alternative formation schemes include shock chemistry, grain surface chemistry and reactions in the turbulent interfaces of the clouds, but further theoretical and laboratory work on crucial reaction rates is needed to make more quantitative statements. On the observational side, more systematic studies of molecular abundances in a variety of translucent clouds are warranted. Recent observations of translucent clouds through millimeter absorption lines against extragalactic continuum sources such as 3C123 (Cox, Güsten & Henkel 1988) and BL Lac (Lucas & Liszt 1993) are of particular interest in this respect.

ACKNOWLEDGMENTS

Discussions with J.H. Black about many different aspects of diffuse and translucent clouds are appreciated. Support by the Netherlands Organization of Scientific Research (NWO) is gratefully acknowledged.

REFERENCES

Abgrall, H., Le Bourlot, J., Pineau des Forêts, G., Roueff, E., Flower, D.R., Heck, L. 1992, *A&A*, **253**, 525

Black, J.H., and Dalgarno, A. 1977, *ApJS*, **34**, 405

Black, J.H., and van Dishoeck, E.F. 1987, *ApJ*, **322**, 412

Blitz, L., Bazell, D., & Désert, F.X. 1990, *ApJ*, **352**, L13

Boissé, P. 1990, *A&A*, **228**, 483

Bowyer, S. 1991, *ARA&A*, **29**, 59

Cardelli, J.A., and Savage, B.D. 1988, *ApJ*, **325**, 864

Chan, W.F., Cooper, G., and Brion, C.E. 1993, *Chem. Phys.*, **170**, 123

Cook, T.A., Cash, W., and Snow, T.P. 1989, *ApJ*, **347**, L81

Cox, P., Güsten, R., and Henkel, C. 1988, *A&A*, **206**, 108

Crutcher, R.M. 1985, *ApJ*, **288**, 604

Draine, B.T. 1978, *ApJS*, **36**, 595

Eidelsberg, M., and Rostas, F. 1990, *A&A*, **235**, 472

Eidelsberg, M., Benayoun, J.J., Viala, Y., and Rostas, F. 1991, *A&AS*, **90**, 231

Falgarone, E., Phillips, T.G., and Walker, C. 1991, *ApJ*, **378**, 186

Federman, S.R., Glassgold, A.E., and Kwan, J. 1979, *ApJ*, **227**, 466

Federman, S.R., and Lambert, D.L. 1988, *ApJ*, 328, 777

Gondhalekar, P.M., Phillips, A.P., and Wilson, R. 1980, *A&A*, **85**, 272

Gredel, R., van Dishoeck, E.F., and Black, J.H. 1991, *A&A*, **251**, 625

Gredel, R., van Dishoeck, E.F., and Black, J.H. 1993*a*, *A&A*, **269**, 477

Gredel, R., van Dishoeck, E.F., and Black, J.H. 1993*b*, *A&A*, in press

Gredel, R., van Dishoeck, E.F., de Vries, C.P., and Black, J.H. 1992, *A&A*, **257**, 245

Habing, H.J. 1968, *Bull. Astr. Inst. Neth.*, **19**, 421

Henry, R.C. 1977, *ApJS*, **33**, 451

Henry, R.C. 1991, *ARA&A*, **29**, 89

Holberg, J.B. 1986, *ApJ*, **311**, 969

Hurwitz, M., Bowyer, S., and Martin, C. 1991, *ApJ*, **372**, 167

Jannuzi, B.T., Black, J.H., Lada, C.J., and van Dishoeck, E.F. 1988, *ApJ*, **332**, 995

Jenkins, E.B., Lees, J.F., van Dishoeck, E.F., and Wilcots, E.M. 1989, *ApJ*, **343**, 785

Jura, M. 1974, *ApJ*, **191**, 375

Kirby, K., and Cooper, D.L. 1989, *J. Chem. Phys.*, **90**, 4895

Kirby, K., and van Dishoeck, E.F. 1988, *Adv. At. Mol. Phys.*, **25**, 437

Kudritzki, R.P., and Hummer, D.B. 1990, *ARA&A*, **28**, 303

Kurucz, R.L. 1979, *ApJS*, **40**, 1

Lambert, D.L., Sheffer, Y., Gilliland, R.L., and Federman, S.R. 1993, *ApJ*, in press

Langer, W.D., and Penzias, A. 1993, *ApJ*, **408**, 539

Le Bourlot, J., Pineau des Forêts, G., Roueff, E., & Flower, D.R. 1992, *A&A*, **267**, 233

Letzelter, C., Eidelsberg, M., Rostas, F., Breton, J., and Thieblemont, B. 1987, *Chem. Phys.*, **114**, 274

Levelt, P.F., Ubachs, W., and Hogervorst, W. 1992*a*, *J. Chem. Phys.*, **97**, 7160

Levelt, P.F., Ubachs, W., and Hogervorst, W. 1992*b*, *J. Phys. II France*, **2**, 801

Lucas, R., and Liszt, H. 1993, *A&A*, in press.

Lyu, C–H., Smith, A.M., and Bruhweiler, F.C. 1993, *ApJ*, submitted

Magnani, L., Blitz, L., and Mundy, L. 1985, *ApJ*, **295**, 402

Mamon, G.A., Glassgold, A.E., and Huggins, P.J. 1988, *ApJ*, **328**, 797

Martin, C., Hurwitz, M., and Bowyer, S. 1990, *ApJ*, **354**, 220

Mathis, J.S., Mezger, P.G., and Panagia, N. 1983, *A&A*, **128**, 212

Meyer, D.M., and Roth, K. 1991, *ApJ*, **376**, L49

Opal, C.B., and Weller, C.S. 1984, *ApJ*, **282**, 445

Paresce, F., and Jakobsen, P. 1980, *Nature*, **288**, 119

Pound, M.W., Bania, T.M., and Wilson, R.W. 1990, *ApJ*, **351**, 165

Roberge, W.G. 1981, Ph. D. Thesis, Harvard University

Roberge, W.G., Jones, D., Lepp, S., & Dalgarno, A. 1991, *ApJS*, **77**, 287

Schinke, R. 1993, *Photodissociation Dynamics* Cambridge: Cambridge University

Sheffer, Y., Federman, S.R., Lambert, D.L., and Cardelli, J.A. 1992, *ApJ*, **397**, 482

Smith, P.L., Stark, G., Yoshino, K., Ito, K., and Stevens, M.H. 1991, *A&A*, **252**, L13

Spaans, M.C. 1993, *A&A*, submitted

Stahl, O., and Wilson, T.L. 1992, *A&A*, **254**, 327

Stark, G., Yoshino, K., Smith, P.L., Ito, K., and Parkinson, W.H. 1991, *ApJ*, **369**, 574

Stark, G., Smith, P.L., Ito, K., and Yoshino, K. 1992, *ApJ*, **395**, 705

Stark, G., Yoshino, K., Smith, P.L., Esmond, J.R., Ito, K., and Stevens, M.H. 1993, *ApJ*, **410**, 837

Stark, R. 1993, *A&A*, submitted

van Dishoeck, E.F. 1987, in *Astrochemistry*, M.S. Vardya and S.P. Tarafdar, Dordrecht: Reidel, 51.

van Dishoeck, E.F. 1988, in *Rate Coefficients in Astrochemistry*, T.J. Millar and D.A. Williams, Dordrecht: Kluwer, 49

van Dishoeck, E.F. 1990, in *The Evolution of the Interstellar Medium*, L. Blitz, San Francisco: A.S.P., 207

van Dishoeck, E.F. 1992, in *Astrochemistry of Cosmic Phenomena*, P.D. Singh, Dordrecht: Kluwer, 143

van Dishoeck, E.F., and Black, J.H. 1982, *ApJ*, 258, 533.

van Dishoeck, E.F., and Black, J.H. 1986, *ApJS*, **62**, 109.

van Dishoeck, E.F., and Black, J.H. 1988*a*, in *Rate Coefficients in Astrochemistry*, T.J. Millar and D.A. Williams, Dordrecht: Kluwer, 209

van Dishoeck, E.F., and Black, J.H. 1988*b*, *ApJ*, **334**, 771

van Dishoeck, E.F., and Black, J.H. 1989, *ApJ*, **340**, 273

van Dishoeck, E.F., Black, J.H., Phillips, T.G., and Gredel, R. 1991, *ApJ*, **366**, 141

Viala, Y.P., Letzelter, C., Eidelsberg, M., and Rostas, F. 1988*a*, *A&A*, **193**, 265

Viala, Y.P., Roueff, E., and Abgrall, H. 1988*b*, *A&A*, **190**, 215

Wagenblast, R. 1992, *MNRAS*, **259**, 155

Wagenblast, R., and Hartquist, T.W. 1989, *MNRAS*, **237**, 1019

Wagenblast, R., and Williams, D.A. 1993, in *Dust and Chemistry in Astronomy*, T.J. Millar and D.A. Williams, Bristol: Institute of Physics, 171

Welty, D.E., and Fowler, J.R. 1992, *ApJ*, **393**, 193

Witt, A.N., and Johnson, M.W. 1973, *ApJ*, **181**, 363

The First Symposium on the Infrared Cirrus and Diffuse Interstellar Clouds
ASP Conference Series, Vol. 58, 1994
Roc M. Cutri and William B. Latter (eds.)

CHEMICAL PROCESSES IN INTERSTELLAR SHOCKS

DAVID FLOWER
Physics Department, University of Durham, Durham DH1 3LE, UK

ABSTRACT I address the question "Is interstellar CH^+ produced in shocks?". A sophisticated model of the *o* Per interstellar cloud, incorporating a C-type shock, suggests that the answer is a qualified "Yes".

INTRODUCTION

The shock waves which are generated in the interstellar medium affect the chemistry in a number of ways. In particular, endothermic molecular reactions, or reactions with barriers, may be driven by the thermal energy of the postshock gas. Furthermore, in C-type shocks the kinetic energy associated with the drift of the ions through the neutrals (ambipolar diffusion) is often sufficient to produce the same effects. However, the unambiguous identification of such chemical effects in diffuse clouds has proved to be elusive.

One of the outstanding problems in interstellar chemistry has been to explain the column densities of CH^+ which are observed in diffuse clouds. Recognized as early as the 1950's (Bates and Spitzer 1951), the problem is to identify a formation mechanism which is sufficiently rapid to maintain the abundance of CH^+ in the face of its rapid removal by H_2,

$$CH^+ + H_2 \rightarrow CH_2^+ + H \tag{1}$$

and by electrons,

$$CH^+ + e^- \rightarrow C + H \tag{2}$$

Elitzur and Watson(1978) suggested that the endothermicity (4640 K) of the reaction

$$C^+ + H_2 \rightarrow CH^+ + H \tag{3}$$

might be overcome in the hot gas produced by the passage of a J-type shock. This suggestion was subsequently investigated by Draine and Katz (1986), Draine (1986), Pineau des Forets et al. (1986) and Flower et al. (1988), in the context of 3-fluid, C-type shocks. The outcome of these studies was that, whilst CH^+ column densities as high as those which are sometimes observed [$N(CH^+) > 10^{13}$ cm^{-2}] can be reproduced by the shock models, difficulties are encountered in simultaneously satisfying the constraints imposed by the observed column densities of OH, CH, and rotationally excited H_2 ($J > 2$). In addition, velocity differences that are expected between the ionized and neutral species, CH^+ and CH, have apparently not been observed (Crawford 1990; Lambert, Sheffer and Crane 1990; Hawkins and Craig 1991). As a consequence, the 'shock' hypothesis has been seriously questioned (Snow 1993; Gredel, van Dishoeck and Black

1993), and other explanations which invoke turbulence have been put forward (Duley et al. 1992).

The perceived failure of C-type shock models to satisfy the observational constraints stems from their inability to satisfactorily reproduce the column densities along lines of sight, such as that to ζ Oph, where N(CH^+) $> 10^{13}$ cm^{-2}. However, such regions may be atypical, and it is worth asking whether there are other diffuse clouds where the column densities which are observed can be satisfactorily reproduced by a model incorporating a C-type shock. With this question in mind, we shall consider the line of sight to *o* Per.

A SELF-CONSISTENT MODEL OF A DIFFUSE INTERSTELLAR CLOUD

The diffuse cloud along the line of sight to *o* Per was one of 4 clouds which were studied in detail by van Dishoeck and Black (1986). With the notable exceptions of CH^+ and H_2 (J = 3 and 4), the column densities of the observed species could be satisfactorily reproduced by a steady-state model of a cloud exposed to the ultraviolet background radiation field.

Recently, Heck et al. (1993) have attempted to improve significantly on the work of van Dishoeck and Black. They have taken advantage of refinements in the data pertaining to the dissociation and fluorescence of H_2 (Abgrall et al. 1987; Schmoranzer et al. 1990; Abgrall et al. 1992), induced by the ultraviolet radiation field. They have also made full use of the line atlas for CO (Eidelsberg et al. 1991), which enables the rate of dissociation of CO (and its isotopes) to be accurately calculated. Furthermore, rate coefficients for the collisional excitation of the fine structure transitions of C and O by ortho- and para-H_2 have become available (Schroeder et al. 1991; Jaquet et al. 1992) and were included in the model. Last, but not least, a C-type shock was incorporated in a completely self-consistent way.

From the computational standpoint, the most demanding aspect of the problem is the solution of the radiative transfer equation. In the spectral range $91.2 < \lambda < 126$ nm, there are several thousand lines of H_2, CO and its isotopes which absorb ultraviolet photons. Transitions of ortho- and para-H_2 that originate in their respective ground rotational states, J = 1 and J = 0, become saturated and broad, resulting in overlap with adjacent spectral lines. Heck et al. (1993) solved the transfer equation on a fixed grid of approximately 300 000 points in the wavelength range from the H Lyman limit to 126 nm, each line being described by its Voigt profile. The variation of the Doppler width with the kinetic temperature of the gas, which is significant through the shock, was taken fully into account, as were the Doppler shifts in the wavelengths of the absorption lines, owing to the acceleration of the gas by the shock wave. Rotationally excited states of H_2, particularly J = 2 and J = 3, become significantly populated in the shock-heated gas and contribute to the enhancement of the photodissociation rate within the C-type shock. The radiative transfer equation and the chemical conservation equations were solved simultaneously, proceeding from the exposed surface of the cloud ($\tau_v = 0$) towards its centre ($\tau_v = 0.5$); the cloud was assumed to be plane-parallel and isothermal (T = 80 K) prior to the arrival of the shock wave. The chemical network comprised about 500 reactions involving 60 atomic and molecular species of H, He, C, N, and O. The interstellar radiation field was taken from Mathis, Mezger and Panagia (1983).

At $\tau_v = 0.1$, the physical and chemical conditions in steady-state were adopted as the initial conditions of a C-type shock, which heats and compresses the gas. The magnetohydrodynamical and chemical conservation equations, together with the equation of radiative transfer, were numerically integrated through and behind the shock front (to $\tau_v = 0.5$). Owing to the compression, the gas density rises from $n_H = n(H) + 2n(H_2) = 20$ cm^{-3} in front to $n_H \approx 150$ cm^{-3} behind the shock. The shock speed was $v_s = 11$ km s^{-1}, and the initial value of the magnetic induction, transverse to the direction of flow, was 5 μG. In the first instance, solar system values of the elemental abundances were adopted: $n_{He}/n_H = 0.10$, $n_C/n_H = 3.6\ 10^{-4}$, $n_N/n_H = 1.1\ 10^{-4}$, and $n_O/n_H = 8.5\ 10^{-4}$ (Anders and Grevesse 1989). These values, and the cosmic ray ionization rate ($\zeta = 10^{-17}$ s^{-1}), were subsequently varied. Indeed, all the initial values, as well as the overall intensity of the incident ultraviolet radiation field, were varied to assess their influence on the computed column densities; the results presented below have been distilled from this investigation of the parameter space.

RESULTS

In columns 2 and 3 of Table 1 are given the column densities of the species observed along the line of sight to *o* Per and the values which derive from a representative model ('E' in their notation) of van Dishoeck and Black (1986). Whilst this model has many virtues, it fails to reproduce the observed values of N(CH^+) and N[H_2(J = 3 and 4)]. In column 4 are listed the values computed by Heck et al. (1993) when a C-type shock is incorporated in the model with the initial values specified above. The elemental abundances were taken equal to their solar system values, and $\zeta = 10^{-17}$ s^{-1}.

The introduction of a C-type shock immediately makes good the deficiencies of the steady-state model. The populations of the J = 3, 4 and 5 rotational levels of H_2 are greatly enhanced within the shock front and their column densities rise accordingly. The population of the J = 2 level almost thermalizes within the shock front, and the J = 4 → 2 rotational transition becomes the dominant cooling process in the hot gas. The intensity of this transition is then determined by the flux of kinetic energy associated with the shock, $\rho v_s^3 / 2$, where ρ denotes the mass density in front of the shock. The other major contribution of the C-shock is to the CH^+ column density, through reaction (3) above.

It may be seen from Table 1 that solar system abundances yield column densities of C^+, N and O (the main repositories of elemental carbon, nitrogen and oxygen) which are higher than observed. This conclusion is hardly surprising, as modest depletions of these elements, onto grains, are to be expected in diffuse clouds (and were introduced by van Dishoeck and Black). Depletion of these elements by a factor of approximately 0.4, to $n_C/n_H = 1.3\ 10^{-4}$, $n_N/n_H = 0.50\ 10^{-4}$, and $n_O/n_H = 3.5\ 10^{-4}$, yields the results in column 5 of Table 1.

Whilst the column densities of C^+, N and O are brought into agreement with the observations when the corresponding elements are depleted, the computed column density of OH falls significantly below the observed value. This problem is solved by increasing the cosmic ray ionization rate to the value $\zeta = 5\ 10^{-17}$ s^{-1}

TABLE 1 Observed and computed column densities (cm^{-2}) of atomic and molecular species along the line of sight to *o* Per. 'E' is a representative model of van Dishoeck & Black (1986). The remaining results are from Heck et al (1993), with varying heavy element depletion, δ, and cosmic ray ionization rate, ζ. Numbers in parentheses are powers of 10.

	Observed	E	$\delta = 1$	$\delta \approx 0.4$	$\zeta = 5 \times 10^{-17} s^{-1}$
H	$(7.4 \pm 1.1)(20)$	6.9(20)	6.4(20)	5.0(20)	5.5(20)
H_2	$(4.1 \pm 0.2)(20)$	4.1(20)	5.7(20)	5.2(20)	5.0(20)
H_2(J=0)	$(3.2 \pm 0.2)(20)$	3.2(20)	3.5(20)	3.5(20)	3.9(20)
H_2(J=1)	$(8.5 \pm 0.4)(19)$	8.8(19)	2.1(20)	1.7(20)	1.1(20)
H_2(J=2)	$(1.0 \pm 0.2)(18)$	5.9(17)	6.2(17)	1.6(18)	1.4(18)
H_2(J=3)	$(3.9 \pm 0.4)(17)$	2.7(16)	3.7(17)	4.7(17)	4.6(17)
H_2(J=4)	$(3.7 \pm 2.5)(16)$	2.5(15)	1.6(16)	1.0(16)	1.0(16)
H_2(J=5)	$(4.1 \pm 1.5)(14)$	4.0(14)	2.0(15)	8.9(14)	1.1(15)
H_2(J=6)	$<1(14)$	5.3(13)	2.2(13)	5.9(12)	7.0(12)
C	(3.3 - 6.4)(15)	3.3(15)	2.9(16)	3.3(15)	3.5(15)
C^+	(0.1 - 2.0)(17)	6.7(17)	6.1(17)	2.0(17)	2.0(17)
CH^+	$(6.0 \pm 0.5)(12)$	3.8(10)	1.1(13)	1.0(13)	7.6(12)
CH	$(1.9 \pm 0.1)(13)$	9.5(12)	1.0(14)	8.8(13)	7.2(13)
C_2	$(1.3 \pm 0.2)(13)$	3.2(12)	2.2(13)	6.3(12)	4.3(12)
CO	$(4.9 \pm 0.5)(14)$	1.4(14)	1.8(14)	6.4(13)	1.4(14)
CO^+	$< 8.0(13)$	1.6(10)	8.0(10)	6.6(10)	3.4(10)
N	$(7.9 \pm 3.3)(16)$	8.3(16)	2.0(17)	7.7(16)	7.7(16)
NH	$< 7.5(11)$	8.5(11)	2.8(10)	1.6(10)	7.6(10)
CN	$(1.6 \pm 0.1)(12)$	1.2(12)	2.0(12)	6.9(11)	5.4(11)
O	$(5.5 \pm 1.6)(17)$	6.9(17)	1.5(18)	5.4(17)	5.4(17)
OH	$(7.8 \pm 0.5)(13)$	7.2(13)	7.1(13)	3.2(13)	9.3(13)
H_2O	$< 2.6(13)$	5.4(11)	9.0(12)	4.3(12)	1.0(13)
H_2O^+	$< 3.6(13)$	3.0(11)	2.5(11)	3.5(11)	1.1(12)

which was favoured by van Dishoeck and Black, as may be seen from the last column of Table 1.

It is intriguing that both adjustments, to the cosmic ray ionization rate, ζ, and to the depletion factor, δ, affect the ratio of ortho- to para-H_2 [essentially, N(J =1) : N(J = 0)]. The reasons are to be found in the nuclear spin-changing reactions that interconnect ortho- and para-H_2,

$$H^+ + \text{para}-H_2 \leftrightarrow H^+ + \text{ortho}-H_2 \tag{4}$$

and

$$H + \text{para}-H_2 \leftrightarrow H + \text{ortho}-H_2 \tag{5}$$

Reaction (4) is endothermic, by 170.5 K, from left to right and, in cold gas, tends to convert ortho- to para-H_2; its rate coefficient is $k_4 = 2\ 10^{-10}$ cm^3 s^{-1}. On the other hand, reaction (5) has a barrier,

$$k_5 = 8\ 10^{-11} \exp(-3900/T)\ \text{cm}^3\ \text{s}^{-1},$$

TABLE 2 Percentage contributions to the column densities of $H_2(J)$ for the model in the last column of Table 1, from the preshock gas, the shock, and the postshock gas.

	Preshock	Shock	Postshock
$H_2(J=0)$	6	1	93
$H_2(J=1)$	55	14	31
$H_2(J=2)$	48	52	0
$H_2(J=3)$	2	98	0
$H_2(J=4)$	1	99	0
$H_2(J=5)$	2	98	0
$H_2(J=6)$	18	75	7

and proceeds at an appreciable rate only in the hot gas produced by the passage of the shock wave. Depletion of elemental carbon reduces the abundance of C^+, the major ion, and hence the strength of the collisional coupling between the ionized and the neutral fluids. As a consequence, the kinetic temperature attained within the shock and the rate of reaction (5) both fall. Depletion of elemental oxygen, on the other hand, results in an increase in the abundance of H^+, because the rate of its removal in the reaction

$$H^+ + O \rightarrow H + O^+ \tag{6}$$

decreases (k_6 = 1.0 10^{-9}exp(-227/T) cm^3 s^{-1}). Furthermore, the abundance of H^+ increases with the rate of cosmic ray ionization, ζ. The net effect of increasing ζ and depleting C and O is to reduce the ortho : para-H_2 ratio and bring it into better accord with the observed value.

Molecular hydrogen may be rotationally excited by ultraviolet pumping and fluorescence, by inelastic collisions, or following its formation on the surfaces of grains. All these mechanisms were incorporated in the model of Heck et al. A C-type shock enhances the rate not only of collisional excitation but also of ultraviolet pumping, as may be seen as follows. Transitions out of the J = 1 and J = 0 ground states of ortho- and para-H_2 become rapidly saturated in the preshock gas. Within the shock, collisional excitation to the J = 2 and 3 levels occurs. Because transitions out of these states are as yet unsaturated, the rate of ultraviolet pumping and hence of fluorescence and dissociation all increase. The percentage contributions to the column densities of the rotational levels $0 \leq J \leq 6$ of H_2 from the preshock gas, the shock and the postshock gas are given in Table 2.

Further constraints on the model are provided by observations of the column densities of individual fine structure levels of C, C^+ and O. In Table 3, the relative column densities are compared with those predicted by the model. The comparison is most meaningful for $C(^3P_J)$, where all three fine structure states have been measured. The rate coefficients for excitation of these fine structure states by ortho- and para-H_2 have been recently calculated (Schroeder et al. 1991). Table 3 shows that the major contribution to these column densities

TABLE 3 Contributions to the column densities of the fine structure levels of C, C^+ and O, expressed as percentages. The observed values are taken from van Dishoeck & Black (1986). Numbers in parentheses are powers of 10.

X(J)	Preshock	Shock	Postshock	X(J)/X	Observed
C(0)	1	24	75	74	84 ± 5
C(1)	0.5	24.5	75	22	11 ± 5
C(2)	0.5	31.5	68	4	5 ± 3
$C^+(1/2)$	28	4	68	99.5	96:
$C^+(3/2)$	35	10	55	0.5	4:
O(2)	27	4	69	100	100
O(1)	54	43	3	1.1(-4)	<1.6(-3)
O(0)	39.5	60	0.5	9.3(-5)	

arises in the compressed, postshock gas, where $n(H_2) = n_H/2 \approx 75$ cm^{-3}. The level of agreement with the observations lends further credibility to the model.

CONCLUDING REMARKS

Bearing in mind the complexity of the region being studied, a model of *o* Per which incorporates a C-type shock yields tolerable agreement with the column densities which are observed. The inclusion of a shock resolves a number of discrepancies which are apparent in the results from steady-state models. With the possible exception of CH, whose column density exceeds that measured, no new problems arise which are attributable to the presence of the shock. I conclude that, at least for this line of sight, the hypothesis that CH^+ is formed in a shock may well be correct.

REFERENCES

Abgrall, H., Launay, F., Roueff, E., and Roncin, J.-Y. 1987, *J. Chem. Phys.*, **87**, 2036

Abgrall. H., Le Bourlot, J., Pineau des Forets, G., Roueff, E., Flower, D. R., and Heck, L. 1992, *A&A*, **253**, 525

Bates, D. R., and Spitzer, L. 1951, *ApJ*, **113**, 441

Crawford, I. A. 1990, *MNRAS*, **241**, 575

Draine, B. T. 1986, *ApJ*, **310**, 408

Draine, B. T., and Katz, N. S. 1986, *ApJ*, **310**, 392

Duley, W. W., Hartquist, T. W., Sternberg, A., Wagenblast, R., and Williams, D. A. 1992, *MNRAS*, **255**, 463

Eidelsberg, M., Benayoun, J. J., Viala, Y., and Rostas, F. 1991, *A&AS*, **90**, 231

Elitzur, M., and Watson, W. D. 1978, *ApJ*, **222**, L141

Flower, D. R., Monteiro, T. S., Pineau des Forets, G., and Roueff, E. 1988, *Rate Coefficients in Astrochemistry*, eds. T. J. Millar and D. A. Williams (Dordrecht: Kluwer), 271

Gredel, R., van Dishoeck, E. F., and Black, J. H. 1993, *A&A*, **269**, 477

Hawkins, I., and Craig, N. 1991, *ApJ*, **375**, 642

Heck, E. L., Flower, D. R., Le Bourlot, J., Pineau des Forets, G., and Roueff, E. 1993, *MNRAS*, **262**, 795

Jaquet, R., Staemmler, V., Smith, M. D., and Flower, D. R. 1992, *J. Phys. B*, **25**, 285

Lambert, D. L., Sheffer, Y., and Crane, P. 1990, *ApJ*, **359**, L19

Mathis, J. S., Mezger, P. G., and Panagia, N. 1983, *A&A*, **128**, 212

Pineau des Forets, G., Flower, D. R., Hartquist, T. W., and Dalgarno, A. 1986, *MNRAS*, **220**, 801

Schmoranzer, H., Noll, T., Roueff, E., Abgrall, H., and Bieneck, R. 1990, *Phys. Rev. A*, **42**, 1835

Schroeder, K., Staemmler, V., Smith, M. D., Flower, D. R., and Jaquet, R. 1991, *J. Phys. B*, **24**, 2487

Snow, T. P. 1993, *ApJ*, **402**, L73

van Dishoeck, E. F., and Black, J. H. 1986, *ApJS*, **62**, 109

The First Symposium on the Infrared Cirrus and Diffuse Interstellar Clouds
ASP Conference Series, Vol. 58, 1994
Roc M. Cutri and William B. Latter (eds.)

A MULTI-TRANSITION CO AND NH_3 STUDY OF DENSE CORES IN THE INFRARED CIRRUS

ANDREAS HEITHAUSEN
I. Physikalisches Institut der Universität zu Köln, Zülpicher Str. 77, 50937 Köln

ABSTRACT Results from a multi–transition CO and NH_3 study of dense cores in MCLD 123.5+24.9, MCLD 126.6+24.5, and MBM 32 are presented. The data sets, consisting of maps in the ^{12}CO and ^{13}CO (1→0), (2→1), and (3→2) transitions obtained with the KOSMA 3 m telescope and of maps in the NH_3 (1,1) and (2,2) transitions obtained with the MPIfR 100 m telescope, were modelled using a spherically symmetric radiative transfer code, which allows variation of density, kinetic temperature, and abundance. The CO line radiation emerging from the cores can only be explained by an high ^{13}CO abundance, a power law index for the volume density of less than 1.5, and core masses close to 1 $M_\odot$.

INTRODUCTION

Despite the fact that infrared cirrus clouds are generally found to be low extinction objects which are gravitationally unbound, they harbour quite a number of dense condensations or cores (e.g. Mebold et al. 1987). Whether or not these cores are gravitationally bound objects is still a matter of controversy (cf. Turner et al. 1989, Großmann & Heithausen 1992). To solve this question it is necessary to study how dense cores in cirrus clouds differ from that in dark clouds.

OBSERVATIONS

To study the structure of dense cores in infrared cirrus clouds a multi–transition CO and NH_3 observations of the dense cores in MBM 32, MCLD 123.5+24.9, and MCLD 126.6+24.5 has been started. MBM 32 has been described in more detail by Schreiber et al. (1993), MCLD 123.5+24.9 by Großmann et al. (1990) and Großmann & Heithausen (1992), and MCLD 126.6+24.5 by Boden & Heithausen (1993). The data sets consist of maps in the ^{12}CO and ^{13}CO (1→0), (2→1), and (3→2) transitions obtained with the KOSMA 3 m telescope and of maps in the NH_3 (1,1) and (2,2) transitions obtained with the MPIfR 100 m telescope. Beamsizes for the CO(1→0) transitions are $3\overset{'}{.}8$, for (2→1) $2\overset{'}{.}2$, for (3→2) $1\overset{'}{.}2$, and for the NH_3 transitions $40''$.

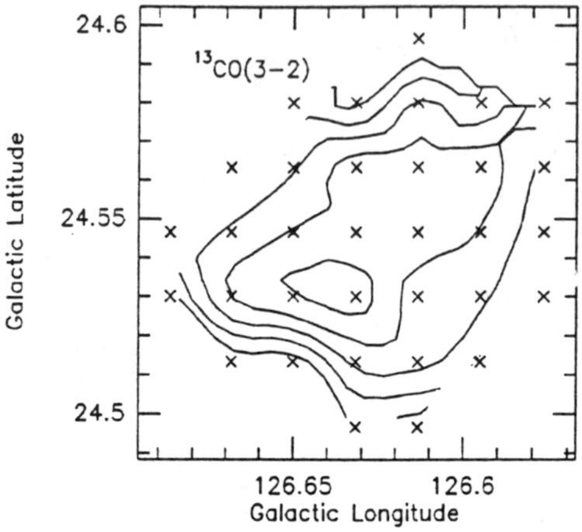

FIGURE I Integrated ^{13}CO (3→2) line for MCLD 126.6+24.5 obtained with the KOSMA 3 m telescope. Contours are every 0.05 K km s^{-1}, starting at 0.05 K km s^{-1}.

DISCUSSION

The cores are extended in all transitions and reveal structure even at the highest angular resolution, 40″ as obtained for the ammonia maps. The kinetic temperatures as determined from the NH_3 (1,1) and (2,2) data are 24 K for MBM 32 and less than 15 K for MCLD 123 and MCLD 126.

The multi–transition CO data could not be fitted applying the usually adopted LTE assumptions or LVG models which assume constant density and excitation temperatures over the whole cloud. A more realistic model has to be applied, which allows variation of the kinetic temperature, T_k, and H_2 density, $n(H_2)$, along the line of sight. This model divides the cloud into a number of concentric onion-type shells. Inside each shell, the physical parameters are kept constant. The temperature, $T_k(r)$, varies with radius according to

$$T_k(r) = T_{core}\, exp\left[\frac{N_{uv}^T - N|_r^R}{10^{21}\gamma}\right]$$

with $N|_r^R$ being the column density calculated from the outer cloud radius, R, to position r inside the cloud. N_{uv}^T is the column density necessary to shield the core of the cloud from the heating UV radiation; for $N|_r^R \geq N_{uv}^T$ the kinetic temperature is kept constant at T_{core}. The density, $n(r)$, varies according to a power law

$$n(r) = n(R)(r/R)^{-\alpha}$$

The abundances, $X(^{12}CO)$ and $X(^{13}CO)$, were kept constant over the cloud. For a more detailed description of the parameters see Gierens et al. (1992).

To compare the different transitions the data have been convolved with a 3.′8 beam, the lowest angular resolution for the CO observations. Some 30000 cloud models have been calculated systematically varying the different parameters listed in the above given equations. The values for the models representing best the observed line emissions are listed in table 1. For MBM 32 and

MCLD 126.6+24.5 the line radiation calculated for all CO transitions lies well within the range of uncertainty for the observed values, $\chi^2 \leq 0.5$. The model for MCLD 123.5+24.9 predicts weaker ^{12}CO (1→0) line emission and stronger ^{12}CO (3→2) line emission than observed, the other transitions are within the range of uncertainty of the observed values, $\chi^2 = 1.7$. Clearly, the data for this core need still more modelling. In the following I therefore only discuss the cores in MBM 32 and MCLD 126.6+24.5.

Adopting a distance of 100 pc for the cores I find beam averaged column densities of between 2 10^{20} cm^{-2} and 3 10^{20} cm^{-2} for the different cores. The densities in the clouds vary according to a power law from about 100 cm^{-3} at the edge of the cloud to a few 1000 cm^{-3} in the densest parts of the cores. The exponent of the power law is different for the cores and less than the value 1.5 found for giant molecular clouds (cf. Gierens et al. 1992) or dark clouds (Myers & Fuller 1992). The temperature inside a core was fixed to the value found from ammonia observations, it rises towards the outer side to 20 K for MCLD 126.6+24.5 and to about 60 K for MBM 32.

TABLE I Cloud parameter

	MBM 32	MCLD 123.5+24.9	MCLD 126.6+24.5
model #	102	108	43
$<n>$[cm^{-3}]	178	316	316
$<N>$[cm^{-2}]	1.8 10^{20}	1.0 10^{21}	3.2 10^{20}
α	0.25	1.25	1.0
T_{core}[K]	10	10	10
N^T_{max}[cm^{-2}]	1.0 10^{21}	0	0.5 10^{21}
γ	0.6	0.6	0.6
$X(^{12}CO)$	8 10^{-5}	8 10^{-5}	8 10^{-5}
$X(^{13}CO)$	8 10^{-6}	8 10^{-6}	8 10^{-6}
$X(C^{18}O)$	6.4 10^{-7}	6.4 10^{-7}	6.4 10^{-7}
r[pc]	0.24	0.77	0.24
M[M$_\odot$]	0.24	19.2	0.6
T_{min} [K]	39	10	10
$T_{core}(NH_3)$ [K]	24^{+10}_{-5}	≤ 15	≤ 15

Remarks: the model for MCLD 123 does not explain all observed transitions properly, whereas the models for MCLD 126 and MBM 32 predict line intensities for the different transitions which are well within the uncertainty of the observed values.

Table 1 lists the masses calculated from the models. The masses for MBM 32 and MCLD 126.6+24.5 are below 1 M$_\odot$. The comparison with the virial masses shows that the cores are gravitationally unbound and probably will disperse after a few 100000 years, confirming previous findings (Boden & Heithausen 1993, Großmann & Heithausen 1992).

ACKNOWLEDGMENTS

The KOSMA 3 m radio telescope at Gornergrat Süd Observatory is operated by the Universität zu Köln, and supported by the Deutsche Forschungsgemeinschaft through grant SFB 301, as well as by special funding from the Land Nordrhein-Westfalen. The observatory is administered by the Internationale Stiftung Hochalpine Forschungsstationen Jungfraujoch und Gornergrat, Bern, Switzerland.

REFERENCES

Boden, K.-P., Heithausen, A., 1993, *A&A* **268**, 255

Großmann, V., Heithausen, A., Meyerdierks, H., Mebold, U., 1990, *A&A* **240**, 400

Großmann, V., Heithausen, A., 1992, *A&A* **264**, 195

Gierens, K.M., Stutzki, J., Winnewisser, G., 1992, *A&A* **259**, 271

Mebold, U., Heithausen, A., Reif, K., 1987, *A&A* **180**, 213

Myers, P.C., Fuller, G.A., 1992, *ApJ* **396**, 631

Schreiber, W., Wouterloot, J.G.A., Heithausen, A., Winnewisser, G., 1993, *A&A* (submitted)

Turner, B.E., Rickard, L.J., Xu, L., 1989, *ApJ* **344**, 292

The First Symposium on the Infrared Cirrus and Diffuse Interstellar Clouds
ASP Conference Series, Vol. 58, 1994
Roc M. Cutri and William B. Latter (eds.)

INTERSTELLAR CYANOGEN MEASUREMENTS OF THE COSMIC MICROWAVE BACKGROUND TEMPERATURE IN DIFFUSE CLOUDS

KATHERINE C. ROTH[1,2,3] AND DAVID M. MEYER[2,3,4]
Dept. of Physics and Astronomy, Northwestern U., Evanston, IL 60208

ABSTRACT We report an indirect measurement of the cosmic microwave background temperature of $T_{\rm CMBR} = 2.729\,(+0.023, -0.031)$ K. This result is based on optical spectral observations of the CN rotational excitation in 5 diffuse clouds and a detection of 2.64 mm CN line emission toward the star HD 21483.

INTRODUCTION

The rotational excitation of cyanogen (CN) in diffuse clouds is due primarily to direct absorption of millimeter photons, the dominant source of which is the relic blackbody radiation predicted by big bang cosmologies. The CN excitation provides an indirect upper limit to the temperature of the cosmic microwave background radiation (CMBR) since local effects can increase the excitation temperature. Collisions with electrons and perhaps H_2 molecules are most likely the only significant local contributors to the rotational excitation of CN in diffuse clouds (Thaddeus 1972; Black and van Dishoeck 1991). Such collisional processes will produce CN line emission, the intensity of which reflects the amount of local excitation within the cloud (Penzias, Jefferts, and Wilson 1972; Crane et al. 1989). Therefore, it should be possible to directly measure these local effects in diffuse clouds and derive an accurate CMBR temperature.

Palazzi, Mandolesi, and Crane (1992) recently completed a study of CN excitation where the local collisional corrections were derived from particle density estimates within the absorbing gas. They found a weighted mean CMBR temperature of $T_{\rm CMBR} = 2.807 \pm 0.025$ K which is 2 σ above the recent results of the *COBE* team who measure $T_{\rm CMBR} = 2.726 \pm 0.010$ K (Mather et al. 1993). We have obtained absorption spectra of the CN (0,0) and (1,0) bands toward the stars ζ Ophiuchi, ζ Persei, HD 21483, HD 27778, and HD 154368. In addition,

---

[1]Now at Space Telescope Science Institute, 3700 San Martin Drive, Baltimore, MD 21218

[2]Visiting Astronomer, Cerro Tololo Inter-American Observatory. CTIO is operated by AURA, Inc. under contract to the National Science Foundation.

[3]Visiting Astronomer, Kitt Peak National Observatory. KPNO is operated by AURA, Inc. under contract to the National Science Foundation.

[4]Visiting Astronomer, National Radio Astronomy Observatory. NRAO is operated by Associated Universities, Inc. under contract to the National Science Foundation.

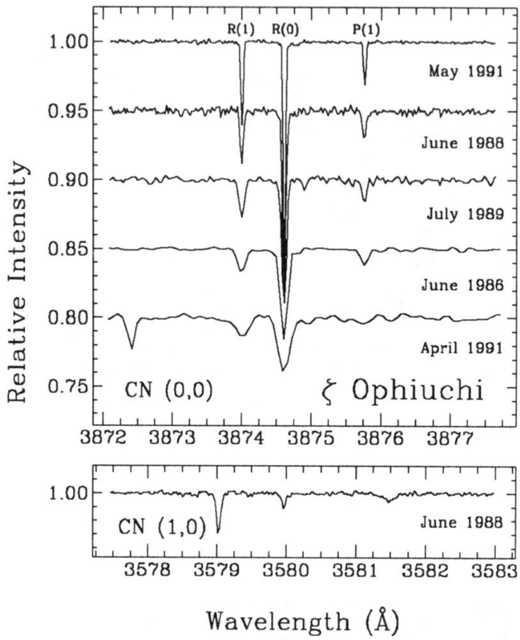

FIGURE 1 CN (0,0) and (1,0) bands for ζ Ophiuchi.

we have detected CN 2.64 mm line emission from the highest column density absorber toward HD 21483. We have derived excitation temperatures for each line of sight and corrected for local effects based on new and published millimeter emission measurements. We compare the resulting mean CMBR temperature to the Palazzi et al. and *COBE* results.

OBSERVATIONS

The CN optical data were collected over a 5 year period during ten separate observing runs at KPNO and CTIO. The majority of the observations utilized the 4 m telescope and echelle spectrograph combination, with 3 spectra having been obtained at the KPNO 0.9 m coudé feed. All the data were taken with CCD detectors, although several different chips including both RCA and blue-sensitive TI CCDs were used. The final spectra cover a wide range in both resolution and signal-to-noise ratio with typical values of $R \simeq (30 - 50) \times 10^3$ and S/N $\simeq$ 1000. To illustrate the range of data quality obtained, the various ζ Ophiuchi spectra are shown in Figure 1.

The CN millimeter data toward HD 21483 were obtained with the NRAO 12 m during two separate observing runs in 1989 November and 1990 March. The telescope was operated in total power mode with the 22 hr equally divided between on and off-source integration. We have made a 4 σ detection of CN 2.64 mm line emission at the same velocity as the observed absorption lines with a measured antenna temperature of $T_A = 33 \pm 8$ mK.

TABLE 1 Column Densities and Temperatures

STAR	b (km s^{-1})	N (10^{12} cm^{-2}) $J=0$	N (10^{12} cm^{-2}) $J=1$	T (K) T_{ex}	T (K) T_{CMBR}[a]
ζ Ophiuchi	0.884 ± 0.023[b]	$2.110 \pm ^{0.017}_{0.016}$	0.866 ± 0.014	2.735 ± 0.025	$2.735 \pm ^{0.025}_{0.040}$
ζ Persei	1.126 ± 0.047	$2.286 \pm ^{0.052}_{0.051}$	0.949 ± 0.054	2.751 ± 0.085	$2.751 \pm ^{0.085}_{0.090}$
HD 27778	$1.30 \pm ^{0.21}_{0.14}$	$10.87 \pm ^{0.74}_{0.69}$	$4.57 \pm ^{0.083}_{0.086}$	$2.767 \pm ^{0.093}_{0.099}$	$2.747 \pm ^{0.095}_{0.101}$
HD 21483	$1.310 \pm ^{0.038}_{0.034}$	$25.9 \pm ^{1.1}_{1.0}$	10.90 ± 0.10	$2.770 \pm ^{0.057}_{0.060}$	$2.695 \pm ^{0.060}_{0.063}$
HD 154368	$0.661 \pm ^{0.045}_{0.042}$[c]	$20.0 \pm ^{5.0}_{3.3}$	$7.90 \pm ^{0.26}_{0.24}$	$2.68 \pm ^{0.22}_{0.33}$	$2.65 \pm ^{0.22}_{0.33}$
		WEIGHTED MEAN TEMPERATURE			$< \mathbf{2.729} \pm ^{\mathbf{0.023}}_{\mathbf{0.031}} >$

[a]Local excitation correction adopted from this work (HD 21483), Crane et al. 1989 (ζ Ophiuchi), Palazzi et al. 1990 (HD 154368), and $N(0)$ interpolation (ζ Persei and HD 27778)

[b]b-value taken from Crane et al. 1986

[c]b-value found in part from Palazzi et al. 1990

DATA ANALYSIS

The optical data were reduced using the NOAO IRAF data reduction packages in a consistent manner similar to that described by Meyer and Roth (1990). The final database contains several (2–5) independent spectra of the CN (0,0) band toward each of the five stars and 1–2 spectra of the (1,0) band toward all but HD 154368. Equivalent widths were measured from single component Gaussian fits to each line with the associated errors including both statistical errors (Jenkins et al. 1973) and estimated continuum placement uncertainties. The measured equivalent widths for each star were consistent with each other in spite of the widely varying conditions of resolution and instrument configurations under which the spectra were obtained.

Column densities were derived from the weighted mean equivalent widths toward each star using a single-component Gaussian curve of growth. The Doppler b-values were derived from a combination of doublet-ratio analyses and direct measurement of the absorption profile FWHMs. For ζ Ophiuchi, we have elected to adopt the precise value of $b = 0.884 \pm 0.023$ km s^{-1} found by Crane et al. (1986). For the HD 154368 sight-line, our Doppler b-value is derived from both our saturation doublet-ratio data and the published value of Palazzi et al. (1990). The b-values we have used and the resulting ground and excited-state column densities toward our 5 stars are given in Table 1.

RESULTS

The CN rotational excitation temperature at 2.64 mm is given by the Boltzmann equation as

$$T_{ex} = 5.4416 \left\{ \ln \left[\frac{3N(0)}{N(1)} \right] \right\}^{-1} \tag{1}$$

where $N(0)$ and $N(1)$ are the column densities of the ground and first excited states, respectively. We have adopted the formalism of Palazzi et al. (1990) for converting 2.64 mm emission into a local temperature contribution $T_{\rm loc}$ defined such that $T_{\rm ex} = T_{\rm CMBR} + T_{\rm loc}$. Assuming that the CN emission uniformly fills the millimeter beam and side lobes, the 2.64 mm line we detect toward HD 21483 implies a local excitation correction of $T_{\rm loc} = 75 \pm 18$ mK. Similar studies have determined values for $T_{\rm loc}$ of $0.0\,(+31, -0)$ and 35 ± 9 mK toward ζ Ophiuchi and HD 154368, respectively (Crane et al. 1989; Palazzi et al. 1990). Assuming the amount of local excitation approximately scales with CN column density, we have adopted interpolated values for $T_{\rm loc}$ toward ζ Persei and HD 27778 of $0.0\,(+31, -0)$ and 20 ± 20 mK. These values have been used to produce the $T_{\rm CMBR}$ values given in the last column of Table 1.

Our observations provide 5 independent measurements of the CMBR temperature at 2.64 mm. These values are in excellent agreement with one another, even though these stars probe different Galactic sight-lines and the CN column densities in the absorbing gas clouds vary by more than an order of magnitude. Our weighted mean value yields a final CMBR temperature of $T_{\rm CMBR} = 2.729\,(+0.023, -0.031)$ K. This result is completely consistent with the *COBE* FIRAS measurement of $T_{\rm CMBR} = 2.726 \pm 0.010$ K reported by Mather et al. (1993) with no evidence of a statistical excess in the corrected CN excitation temperature. This result is a striking verification of the *COBE* findings since these two methods are fundamentally very different and they sample the CMBR both in the near-Earth environment and several light-years distant.

ACKNOWLEDGMENTS

It is a pleasure to acknowledge Isabel Hawkins and Ned Wright for helpful discussions and input. We also wish to thank the staff at KPNO, CTIO, and NRAO for their expert assistance with the observations. This work was supported in part through grants AST 90-21641 and AST 90-23372 to Northwestern University.

REFERENCES

Black, J. H., and van Dishoeck, E. F. 1991, *ApJ*, **369**, L9

Crane, P., Hegyi, D. J., Kutner, M. L., and Mandolesi, N. 1989, *ApJ*, **346**, 136

Crane, P., Hegyi, D. J., Mandolesi, N., and Danks, A. C. 1986, *ApJ*, **309**, 822

Jenkins, E. B., Drake, J. F., Morton, D. C., Rogerson, J. B., Spitzer, L., and York, D. G. 1973, *ApJ*, **181**, L122

Mather, J. C., et al. 1993, *ApJ*, submitted

Meyer, D. M. and Roth, K. C. 1990, *ApJ*, **349**, 91

Palazzi, E., Mandolesi, N., and Crane, P. 1992, *ApJ*, **398**, 53

Palazzi, E., Mandolesi, N., Crane, P., Kutner, M. L., Blades, J. C., and Hegyi, D. J., 1990, *ApJ*, **357**, 14

Penzias, A. A., Jefferts, K. B., and Wilson, R. W. 1972, *Phys.Rev.Lett*, **28**, 772

Thaddeus, P. 1972, *ARA&A*, **10**, 305

The First Symposium on the Infrared Cirrus and Diffuse Interstellar Clouds
ASP Conference Series, Vol. 58, 1994
Roc M. Cutri and William B. Latter (eds.)

HEAVY ELEMENTS IN THE INTERSTELLAR MEDIUM: COPPER, GALLIUM, GERMANIUM, KRYPTON, TIN

DANIEL E. WELTY
Astron. & Astrophys. Ctr., Univ. of Chicago, Chicago, IL 60637

LEWIS M. HOBBS
Yerkes Observatory, Univ. of Chicago, Williams Bay, WI 53191

DONALD C. MORTON
Herzberg Inst. of Astrophysics, Ottawa, ON, K1A 0R6 Canada

LYMAN SPITZER
Princeton Univ. Observatory, Princeton Univ., Princeton, NJ 08544

DONALD G. YORK
Astron. & Astrophys. Ctr., Univ. of Chicago, Chicago, IL 60637

INTRODUCTION

Over the past twenty years, observations of interstellar absorption lines obtained with the *Copernicus* and *IUE* satellites have yielded estimates of the gas phase abundances for many of the elements with atomic numbers $Z \leq 30$ (*i.e.*, up to Zn). While some elements (*e.g.*, C, N, O, S, Zn) seem to be present in the diffuse interstellar gas at nearly Solar abundances, relative to hydrogen, others (*e.g.*, Si, Fe, Cr, Ni, Al) seem to be deficient, typically by factors of 10 - 1000. The "missing" fractions of the latter elements are presumed to have been depleted into dust grains generally coextensive with the gas. Observations of apparent grain destruction in supernova remnants and high velocity clouds, signs of apparent grain growth in dense regions, and theoretical considerations of the physical and chemical processes likely to occur on grain surfaces all suggest that the size and composition of the dust grains should be functions of history and environment; the details of the evolution of the dust are poorly known, however (Jenkins 1989; Draine 1990).

The moderately high resolution (3.5 km s^{-1}) and high S/N (up to 1000) achievable with the *Hubble Space Telescope* Goddard High Resolution Spectrograph (GHRS) have already yielded improved information on the abundances and depletions of a number of elements in the interstellar clouds toward ζ Oph (Savage, Cardelli, & Sofia 1993) and HD 93521 (Spitzer & Fitzpatrick 1993). In addition, observations of a small number of bright, early-type stars with the GHRS have revealed weak interstellar absorption lines due to the heavy elements copper, gallium, germanium, krypton, and tin. Detections of Ga, Ge, and Kr toward the stars ζ Oph and ξ Per were reported by Cardelli, Savage, & Ebbets (1991). We report here the discovery of interstellar Sn (at Z=50, the heaviest element yet identified in the gas phase of the interstellar medium) and summa-

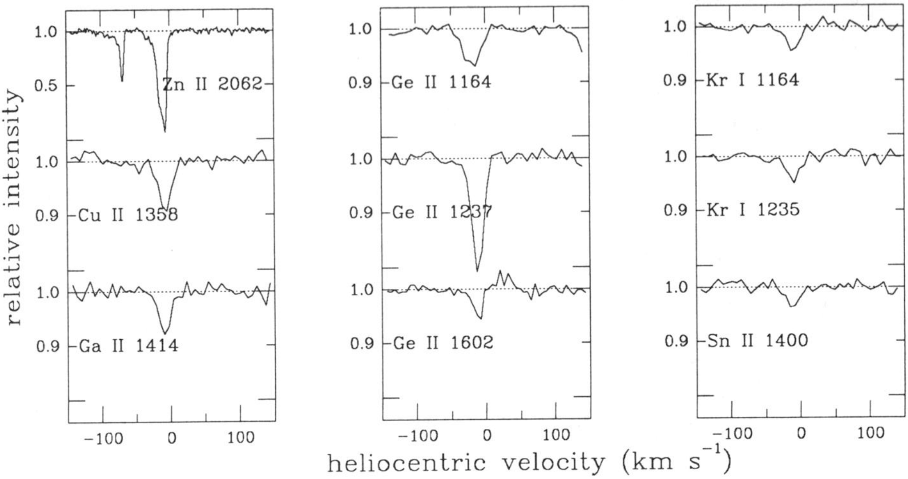

FIGURE I GHRS spectra of 1 Sco near lines due to Zn II (ECH-B; 3.5 kms^{-1}) and to Cu II, Ga II, Ge II, Kr I, and Sn II (G160M; 16-19 km s^{-1}). The G160M data have S/N $\sim$ 100-120, yielding 2σ equivalent width errors/limits of 1-2 mÅ for weak, unresolved lines. The Sn II λ1400 line has $W_\lambda = 4.0 \pm 1.8$ mÅ.

rize the most extensive and accurate interstellar abundance data available for these five heavy elements.

OBSERVATIONS

The observations were carried out using GHRS grating G160M and the small science aperture; a discussion of the observing strategy and data reduction procedures may be found in Hobbs *et al.* (1993). Normalized spectra for 1 Sco are displayed in Figure I; GHRS echelle-B data for Zn II (and Cr II) are included to give an indication of the component structures seen at higher resolution.

Although the 16-19 km s^{-1} resolution obtained with G160M does not permit the multiple interstellar components present along these lines of sight to be resolved, the weakness of the absorption lines considered here (generally < 10 mÅ) and the high quality of the data (S/N $\sim$ 100-120, yielding 2σ errors of 1-2 mÅ) have enabled us to obtain total line of sight column densities accurate to $\pm$0.1-0.2 dex in many cases. Since the lines observed are due to the ionization states dominant in neutral regions, the column densities can be compared with the total neutral hydrogen column densities and with Solar/meteoritic reference abundances to yield estimates for the depletions of the five elements. We find that Cu, Ga, Ge, and Kr are generally depleted; even for this relatively small sample, Cu, Ga, and Ge seem to show increasing depletion with increasing mean line of sight density $< n_H >$, as observed for many other elements. The derived abundances of Sn, however, are consistent with Sn being undepleted. The inferred logarithmic "dense cloud" depletions (*i.e.*, at log $< n_H >= 0.5$, representative of moderately dense clouds such as toward ζ Oph; Jenkins 1987) for the five elements are: Cu (-1.34), Ga (-1.13), Ge (-0.70), Kr (-0.25), and Sn (-0.08).

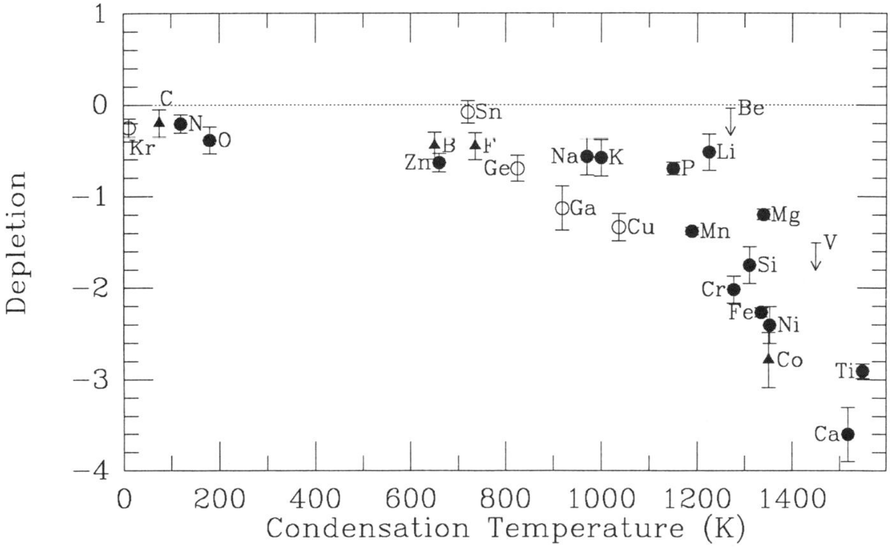

FIGURE II The "dense cloud" depletions versus the condensation temperature (T_C). Solid circles are adapted from Jenkins (1987); triangles are based on 1-3 lines of sight (B and Co from Federman *et al.* 1993); open circles are from GHRS data discussed here. The depletions seem more tightly correlated with T_C than with χ_{ion}, when allowance is made for various grain surface reactions thought to reduce the depletions of Na, K, Li, and P.

DISCUSSION

One major goal of interstellar abundance studies has been to identify the sites and processes in which the observed depletions of the various elements arise and are maintained. The depletions observed in various environments have thus been compared with such quantities as the condensation temperature in a gaseous, nearly Solar mixture of the elements (T_C; Field 1974), the first ionization potential (χ_{ion}; Snow 1975), and the adsorption energy on both graphite and silicate grain materials (Barlow 1978; Phillips, Pettini, & Gondhalekar 1984). In Figure II, we plot the "dense cloud" depletions for 26 elements versus T_C.

At first glance, the depletions seem to exhibit substantial scatter when plotted versus either T_C or χ_{ion}. It has been suggested, however, that the depletions of the alkali elements (Li, Na, K) might be significantly reduced via formation of saturated hydrides on the grain surfaces, followed by ejection from the grain and subsequent photodissociation (Watson & Salpeter 1972). Joseph (1988) has conjectured that the condensation of P in stellar outflows might be inhibited by the formation of PN, which would then photodissociate in the more diffuse interstellar gas. If such processes are operative to reduce the depletions of those four elements, then the remaining elemental depletions seem significantly better correlated with T_C than with χ_{ion}; several of the heavy elements (Ga, Ge, and Sn) are particularly useful discriminants.

Observation of these elements (and others which may be detectable) in additional lines of sight, particularly those with significantly larger and smaller $< n_H >$, should allow the influences of grain surface chemical processes and of shocks to be explored in more detail. For example, Barlow (1978) noted that Si, Ge, and Sn should form saturated tetrahydrides on the grain surfaces, releasing enough energy in each case to desorb the resulting molecule (but more likely for Si than for Sn). We observe Sn to be less depleted than Si, however, perhaps indicating that the Si is locked in a fairly stable, protected grain core. If some of these (and other) heavy elements beyond the iron peak are found to be generally not depleted, even in dense regions, then it might be interesting as well to examine their relative abundances near different supernova remnants, in order to investigate the respective formation sites of those elements thought to be created via the r and/or s-processes.

ACKNOWLEDGEMENTS

The observations reported here were obtained with the NASA/ESA *Hubble Space Telescope* at the Space Telescope Science Institute, which is operated by the Association of Universities for Research in Astronomy, Inc., under NASA contract NAS5-26555. Support for this work was provided by NASA through grant number GO-2251.01-87A from STScI.

REFERENCES

Barlow, M. J. 1978, *MNRAS*, **183**, 417

Cardelli, J. A., Savage, B. D., & Ebbets, D. C. 1991, *ApJ*, **383**, L23

Draine, B. T. 1990, in *The Evolution of the Interstellar Medium*, L. Blitz, PASP Conf. No. 12, 193

Federman, S. R., Sheffer, Y., Lambert, D. L., & Gilliland, R. L. 1993, *ApJ*, in press

Field, G. B. 1974, *ApJ*, **187**, 453

Hobbs, L. M., Welty, D. E., Morton, D. C., Spitzer, L., & York, D. G. 1993, *ApJ*, in press

Jenkins, E. B. 1987, in *Interstellar Processes*, D. J. Hollenbach & H. A. Thronson, Dordrecht: Reidel, 533

Jenkins, E. B. 1989, in *Interstellar Dust*, L. J. Allamandola & A. G. G. M. Tielens, Dordrecht: Reidel, 23

Joseph, C. L. 1988, *ApJ*, **335**, 157

Phillips, A. P., Pettini, M., & Gondhalekar, P. M. 1984, *MNRAS*, **206**, 337

Savage, B. D., Cardelli, J. A., & Sofia, U. J. 1992, *ApJ*, **401**, 706

Snow, T. P. 1975, *ApJ*, **202**, L87

Spitzer, L., & Fitzpatrick, E. L. 1993, *ApJ*, in press

Watson, W. D., & Salpeter, E. E. 1972, *ApJ*, **174**, 321

The First Symposium on the Infrared Cirrus and Diffuse Interstellar Clouds
ASP Conference Series, Vol. 58, 1994
Roc M. Cutri and William B. Latter (eds.)

INTERSTELLAR C_2 IN DENSE OPTICAL CLOUDS

BARRY L. LUTZ
Northern Arizona University

KARTIK SHETH
Grinnell College

RICHARD M. CRUTCHER
University of Illnois

Using the echelle spectrograph on the 4-m Mayall telescope at Kitt Peak National Observatory, we have observed the 2–0 vibration of the Phillips system of C_2 in nine moderately dense interstellar clouds in the directions of VI Cygni No. 12, ρ Ophiuchus, NGC 2024, HD 29647, BD +66 1675, HD 21483, HD 38563n, X Perseus, and HD 26571. B–V color excesses for these clouds range from 0.3 to 3.2 magnitudes. Equivalent measurements for the rotational lines identified in each cloud spectrum were reduced to column densities using a curve-of-growth analysis.

Rotational temperatures were derived by fitting Boltzman distributions to the observations, along with total column densities for C_2 in the clouds. In six of the clouds at least two temperatures are required to characterize the rotational population distribution of the C_2 molecule, while in the remaining three, a single temperature adequately describes the observations.

We have interpreted the observations of the six clouds that require at least two temperatures to characterize the rotational population distribution in terms of the radiative pumping models developed by van Dischoeck (Ph.D. thesis, University of Leiden, 1984). The lower temperatures derived from the two-temperature models range from 8 to 46 K and approximate the kinetic temperature of the gas; total gas densities estimated from van Dischoeck's models ranged from about 200 to 500 cm^{-3}.

A search for $^{12}C^{13}C$ in the direction of VI Cygni No. 12 was unsuccessful, but an upper limit derived from this high signal-to-noise (S/N = 750), yield a lower limit for the isotopic ratio of $^{12}C/^{13}C > 75$.

CHAPTER VII

CLOUD ENERGETICS AND DYNAMICS

The First Symposium on the Infrared Cirrus and Diffuse Interstellar Clouds
ASP Conference Series, Vol. 58, 1994
Roc M. Cutri and William B. Latter (eds.)

ENERGY BUDGETS OF DIFFUSE CLOUDS

JOHN H. BLACK
Steward Observatory, University of Arizona, Tucson, AZ 85721, USA

ABSTRACT Diffuse molecular clouds can be used as remote probes of the energetics of the interstellar medium. Important heating and cooling processes are reviewed. A summary is presented of recent observational results that pertain to the abundances and excitation of coolants in diffuse clouds. An empirical accounting is made of the heating and cooling in the well studied diffuse cloud toward ζ Oph.

INTRODUCTION

Diffuse interstellar clouds provide useful remote laboratories for the investigation of the energetics of the interstellar medium. The energy sources of these clouds include background starlight, cosmic rays, and diffuse X–rays. Dissipation of interstellar turbulence may be an important source of mechanical energy. In addition, there are shock waves that mediate the imbalances of pressure from place to place. In response to the heating by such sources, the clouds radiate through atomic and molecular emission lines that have been excited by collisions. The intensity of this radiation is regulated by internal densities, ionization, and chemical abundances, and it may be redistributed by effects of radiative transfer. The measured intensity of such line radiation is related to the rate of cooling of the gas. The balance between the processes of heating and cooling dictates the temperature of the material, which in turn affects its chemistry and ionization. The densities in diffuse and translucent clouds are usually low enough that the dust particles and gas are not strongly coupled to each other. In fact, the dust may independently absorb and re–process an order of magnitude more power than the gas. This can help account for the prominence of "cirrus" emission in the far–infrared spectra of otherwise inconspicuous regions.

Recent advances in astronomical techniques make it possible to do an improved accounting of the energy budget in diffuse clouds. Ultraviolet absorption–line spectroscopy of interstellar gas with *Hubble Space Telescope* (*HST*) can provide improved information on the abundances and excitation of coolants. Ground–based observations at mm and sub–mm wavelengths are now sensitive enough to probe the line emission of diffuse clouds. The *Cosmic Background Explorer* (*COBE*) has yielded important results on the global properties of the far–infrared and sub–mm radiation of gas and dust in the Galaxy.

This article begins with a brief summary of the heating sources and cooling mechanisms in diffuse clouds. Recent observational results on the abundances, excitation, and radiation of gaseous coolants are reviewed. The energy budget of one particularly well studied cloud, toward ζ Oph, is presented in detail.

HEATING, COOLING, AND TEMPERATURES

Basic concepts

The basic physics of heating and cooling processes can be found described in textbooks (e.g. Spitzer 1978) and in previous review articles (Dalgarno & McCray 1972; Black 1987). At the microscopic level, heating inside an interstellar cloud occurs when some energy source increases the energy of a particle above the mean thermal energy, causing the excess to be shared through elastic collisions, which drive the microscopic speed distribution toward higher kinetic temperature. For example, the photoionization of a neutral carbon atom by the average background starlight produces a photoelectron with almost 1 eV of kinetic energy, which is significantly larger than the mean energy of thermal electrons at a kinetic temperature $T_k = 100$ K. The rate of heating by this process is thus given approximately by the product of the density of carbon atoms, the rate of photoionization per carbon atom, and the mean excess kinetic energy per photoelectron. Cooling occurs when an inelastic collision process removes kinetic energy from the thermal energy distribution. Examples of such inelastic processes are rotational excitation of a molecule or collisional ionization of an atom. In the former case, the excited molecule may have a high probability of radiating a line photon before undergoing another collision: the net rate of cooling locally is proportional to the difference between the rate of upward collisions and the rate of corresponding downward collisions. In the latter case, the removal of kinetic energy from the gas is accompanied by a change of state. In practice, cooling by collisional ionization or by endoergic chemical reactions is unlikely to be important in typical diffuse clouds at low temperature ($T_k \lesssim 100$ K). The description of cooling can be complicated by effects of radiative transfer. A collisionally excited line photon may escape its point of origin only to be re-absorbed elsewhere before it can escape the cloud. If that absorption event produces an excited molecule that is collisionally de-excited before it can radiate, then the macroscopic energy budget of the cloud has not been changed. In other words, a proper evaluation of the local cooling rate must include a reduction for the excitation caused by line radiation created elsewhere in the cloud, and an accounting of the global cooling rate must incorporate the non-local effects of radiative transfer. When the heating source is ultraviolet radiation, for example, the flux will also be attenuated with depth into a cloud. The interplay among internal structure and the effects of radiative transfer on heating and cooling radiation cause gradients in kinetic temperature to exist.

Temperatures

The notion of "kinetic temperature" has already been invoked as though it has a well defined meaning in interstellar clouds. By kinetic temperature, we mean the parameter that characterizes the Maxwellian speed distributions of atoms, molecules, and electrons. Remarkably, even in low-density (10–1000 atoms cm^{-3}) diffuse clouds, it appears that elastic collisions are frequent enough to maintain Maxwellian distributions at nearly the same temperature for all components of the gas—at least to a first approximation (cf. Spitzer 1978). There are interesting, unresolved questions about the behavior of interstellar matter on the length scale where the microscopic and macroscopic realms overlap. For example, at $T_k \approx 25$ K, a hydrogen molecule lives almost 3 months between elastic collisions when the density is $n(H_2) = 100$ cm^{-3}, which means that the thermalization length (i.e. the microscopic scale length) is no smaller than 2 AU$/n(H_2)$. In principle, there may be some minor species that do not thermalize, and it may be that interstellar turbulence superimposes a superthermal tail onto the average speed distribution of atoms and molecules. For example, if interstellar CH^+ is produced by energetic processes (i.e. formed hot), it may be destroyed by rapid reactions with H and H_2 just as quickly as elastic collisions with those species would otherwise tend to thermalize it (Gredel, van Dishoeck & Black 1993). We begin with the expectation that diffuse clouds may be close to *thermal balance*; that is, that the rates of heating and cooling equilibrate on time scales short compared with the ages of diffuse clouds and that there is a well defined microscopic kinetic temperature at every point in a cloud. This is not the same thing as *local thermodynamic equilibrium* (LTE), which would require that everything about the physical and chemical state be specified by the same value of temperature (i.e. the radiation at all wavelengths, the ionization of the gas, the abundances of molecules, and their distribution among internal energy states). In general, the radiation and ionization of the interstellar medium will be far out of equilibrium with the microscopic kinetic motions.

The departures from equilibrium are responsible for the variety of other kinds of "temperatures" that are often used to characterize the interstellar medium. The dust particles may have a temperature $T_{\rm D}$ that is determined by the radiative balance. Visible and ultraviolet starlight is absorbed and thermalized internally so as to balance the infrared radiation of the particle. Although the classical dust grains (sizes $\gtrsim 0.1$ μm) will probably have a well defined average temperature, smaller particles are thought to suffer such large excursions of internal energy upon the absorption of a single photon that their state must be described statistically through a temperature distribution function (e.g. Purcell 1976, Andriesse 1978, Sellgren 1984). The interstellar radiation field can also be described in terms of a brightness temperature $T_{\rm rad}$. For each frequency, this is the temperature at which the Planck function equals the local specific intensity, $I_\nu = B_\nu(T_{\rm rad})$. For a typical location in interstellar space in the solar neighborhood, the average radiation field has stellar sources in the ultraviolet, visible, and near–infrared, and is dominated by thermal emission from interstellar dust in the far–infrared, until the cosmic background radiation (CBR) takes over at mm and submm wavelengths. For example, the mean intensity due to the combined effects of O and B stars corresponds to a radiation temperature $T_{\rm rad} \approx 3000$ K at $\lambda = 100$ nm. At the other extreme, $T_{\rm rad} = T_{\rm CBR} = 2.73$ K at $\lambda \approx 1$ mm. The internal excitation of an atom or molecule can be described by excitation temperatures $T_{\rm ex} = T_{ji}$: for any pair of energy states $i \leq j$ separated by an energy ΔE_{ji}, the ratio of number densities is given by

$$\frac{n_j}{n_i} = \frac{g_j}{g_i} \exp\Big(-\frac{\Delta E_{ji}}{kT_{ji}}\Big)$$

where g_i is the statistical weight of state i. Finally, it is important to keep in mind that the microscopic kinetic temperature is often smaller than the "Doppler" temperature that would characterize the observable line widths of well resolved spectral lines in the diffuse interstellar medium. These superthermal motions have long been called "turbulent" and there is now a growing body of work that describes the character of the interstellar turbulence (cf. Falgarone 1993). Thermal pressure in diffuse interstellar clouds is often a tiny fraction of the total pressure, which is dominated by turbulent motions.

Clouds and their structure

There are several ways of describing what is meant by an interstellar cloud; in particular, the term *diffuse cloud* is used in different ways by different investigators. Empirically, an interstellar cloud means a relatively isolated concentration of interstellar gas and dust. When it is observed spectroscopically with high resolution in Doppler velocity (better than 10 km s^{-1}), the cloud can usually be identified as an isolated Doppler–velocity component. Although the densities of diffuse clouds are usually found to be low ($\lesssim 10^3$ cm^{-3}), the empirical defining characteristic is small line–of–sight column density or extinction. In this sense, "diffuse cloud" often refers to one in which the total visual extinction $A_V \lesssim 1$ mag. Similar, but somewhat thicker, clouds with $A_V \approx 1 - 5$ mag have been arbitrarily labelled "translucent clouds". Interstellar clouds with larger extinctions ($A_V > 5$ mag) are usually rather different chemically and physically from the thin ones. Some investigators make a physical distinction between *diffuse* clouds that are not self–gravitating and gravitationally bound clouds (cf. Elmegreen 1989, 1993). Diffuse clouds of low extinction are vulnerable to ultraviolet starlight, which enforces chemical distinctions upon them: in typical diffuse clouds, the photon–dominated chemistry maintains C^+ in greater abundance than either C or CO and allows H_2 to outnumber H only slightly in a column–averaged sense. The translucent clouds appear to represent the transition between the largely atomic diffuse clouds and the fully molecular giant molecular clouds or dark clouds. The ambiguous terminology derives in part from effects of observational selection and the varieties of technique: for example, the infrared cirrus cloud Number 80 in the catalogue of Désert, Bazell, & Boulanger (1988) is also the high–latitude molecular cloud G56.5–44.1 and furthermore corresponds to the absorbing cloud labelled according to the background star HD 210121 (Gredel, van Dishoeck, de Vries, & Black 1992).

Even small, diffuse clouds show evidence of complex internal structure. Absorption line observations provide information on a one–dimensional distribution of atoms and molecules along the line of sight. The analyses of extensive absorption–line data suggest indirectly that there may be significant gradients in temperature and molecular abundances even through the thin, diffuse clouds (cf. van Dishoeck & Black 1986). Recent observations at very high resolution (better than 1 km s^{-1}) demonstrate that the common diatomic species CN, CH, and CH^+ in the ζ Oph diffuse "clouds" have different distributions in radial velocity (Lambert, Sheffer, & Crane 1990). Another view of related small–scale structure comes from the recognition that the intensity of the CO $J = 1 \rightarrow 0$ rotational line emission in this region varies significantly on a scale smaller than 0.5 pc (Liszt 1992). Small–scale structure has important implications for our understanding of the heating and cooling of interstellar clouds. First, there is simply the evidence that the kinetic temperature of the gas may vary from values around $T_k \approx 100$ K where the heating rates are highest to $T_k \approx 10 - 30$ K where the densities (and associated cooling rates) are high and the intensity of starlight has been reduced by extinction. Second, if we are trying to measure the column densities and excitation of coolants from saturated absorption lines, then knowledge of the line–of–sight velocity distribution or dispersion is essential for accurate determinations. Third, the degree of penetration of ultraviolet starlight (and the size of associated heating rates) may depend severely on how non–uniform the medium is (see e.g. the contributed paper of Spaans 1993).

Heating processes

The underlying energy sources for interstellar matter include starlight, cosmic rays, magnetic fields, and turbulent motions. It appears that the most effective heating mechanisms for most of the diffuse and translucent clouds are the photoelectric effect and the dissipation of turbulence. An evaluation of the photoelectric heating rate requires information both on the spectrum of ultraviolet starlight and the optical properties of the dust particles. The mean Galactic background spectrum is illustrated in Figure I. The upper panel displays the intensity averaged over all directions appropriate for a location near the Sun. This combines the mean Galactic spectrum in the submm and far–infrared regions derived from *COBE* spectrometer measurements by Wright et al. (1991) and the mean background in the infrared, visible, and ultraviolet computed by Mathis et al. (1983). The lower panel shows the same spectrum presented as a radiation temperature. Note that the 2.73 K CBR dominates this spectrum between 2 GHz and 0.6 THz. The integrated intensity in this spectrum is $4\pi \int I_\nu d\nu = 0.053$ ergs s^{-1} cm^{-2} from $\nu = 10^8$ Hz to the Lyman limit of atomic hydrogen. The corresponding energy density of radiation is 1.1 eV cm^{-3}.

An important recent development concerns the contribution of a population of very small grains or very large molecules to the photoelectric heating. The large heating efficiency of small grains has been recognized for some time (e.g. Watson 1972, Jura 1976). There is now compelling evidence that large molecules (or small grains) are a significant component of the interstellar medium. At least one of the hypotheses about their identity (polycyclic aromatic hydrocarbons or PAHs) permits quantitative estimates of their contribution to photoelectric heating (d'Hendecourt & Léger 1987; Lepp & Dalgarno 1988; Verstraete et al. 1989; Bakes & Tielens 1993). The amount of energy absorbed by PAHs can also be related to the infrared radiation that they are claimed to emit through a fluorescence process: the average power of this PAH emission is estimated to be 1.4×10^{-24} ergs s^{-1} per H in the interstellar medium (Désert et al. 1990), which is 25 times larger than the energy budget of the gas in the ζ Oph diffuse clouds (see below).

The diffuse interstellar bands have not yet been definitively identified with PAHs or anything else. Jenniskens (1992) has presented an extensive catalogue of all known diffuse bands with an average absorption strength ($W_\lambda/E(B-V)$, an equivalent width per unit color excess) for each. Based upon this band strength, an average gas/extinction ratio, and the interstellar radiation field shown in Fig. 1, the total power absorbed by all diffuse bands in Jenniskens' catalogue is 6.67×10^{-27} ergs s^{-1} per hydrogen. Whether any of this goes into heating the gas is not known, but the absorbed energy may have a profound effect upon the carrier of the diffuse bands.

The interstellar turbulence is becoming better characterized as a result of observations of the small–scale structure in molecular clouds (Falgarone 1993;

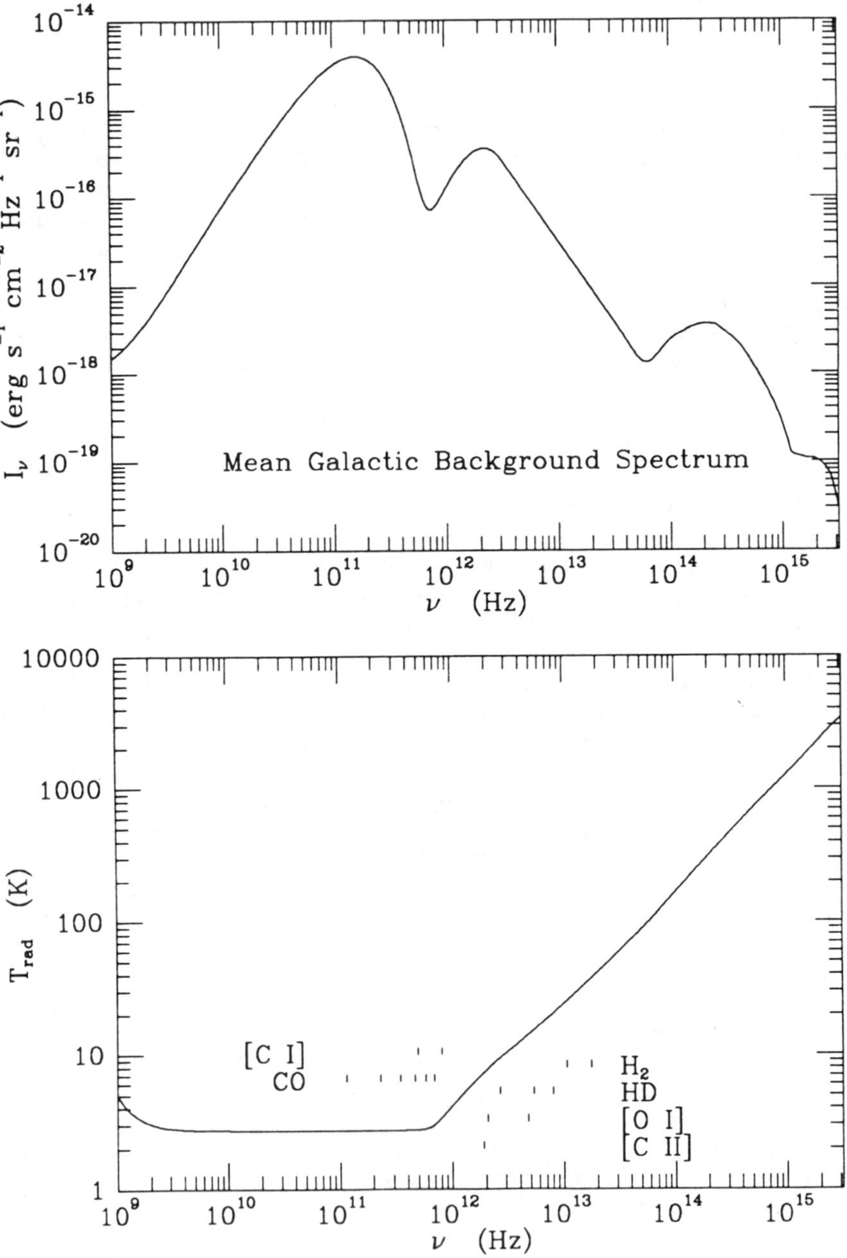

FIGURE I Mean Galactic Background Spectrum

Falgarone and Phillips 1990, 1991; Falgarone, Phillips, & Walker 1991; Falgarone, Puget, & Pérault 1992). As mentioned above, the small–scale structure appears to be an important aspect of interstellar clouds; it manifests itself through observable fluctuations in line intensity and in Doppler velocity. Given assumptions about the dissipation scale–length and the velocity field of the turbulence, it is possible to estimate a rate of heating by dissipation of turbulence in diffuse clouds. It appears that it may be possible to relate the turbulent heating rate to observable chemical consequences of the turbulence (Spaans et al. 1993). Aside from these empirically based estimates, there has also been some progress in numerical simulations of interstellar turbulence (Falgarone 1993). There have also been discussions of the global dissipation rate in the Galaxy based upon a quasi–linear theory of turbulence driven by gravitational coupling between the stars and the interstellar gas (Deiss, Just, & Kegel 1990).

Cooling Processes

Collisional excitation of low–lying energy levels in abundant atoms and molecules followed by emission of spectral lines will usually dominate the cooling of the gas in diffuse clouds. Table I lists a selection of such levels and their energies in temperature units. Calculations of the rates of cooling depend upon theoret-

TABLE I Low–lying energy levels of abundant coolants

Species	Term	Level J	Energy E_J/k(K)	Species	Term	Level J	Energy E_J/k(K)
C	$2s^2 2p^2\ ^3P$	0	0.00	^{12}CO	$X^1\Sigma^+$	0	0.00
		1	23.62			1	5.53
		2	62.46			2	16.60
C^+	$2s^2 2p\ ^2P$	1/2	0.00			3	33.19
		3/2	91.21			4	55.32
O	$2s^2 2p^4\ ^3P$	2	0.00			5	82.97
		1	227.71			6	116.16
		0	326.58	^{13}CO	$X^1\Sigma^+$	0	0.00
H_2	$X^1\Sigma_g^+$	0	0.00			1	5.29
		1	170.49			2	15.87
		2	509.88			3	31.73
		3	1015.12			4	52.89
HD	$X^1\Sigma^+$	0	0.00			5	79.33
		1	128.38			6	111.05
		2	384.26				

ical studies of the collisional excitation processes. The cross sections and rates for collisional excitation of rotational levels in H_2, HD, and CO are discussed in some detail by Flower (1990), along with some of the older results on fine–structure excitation in C, C^+, and O. Within the last 5 years, there have been some significant improvements in the atomic data for C, C^+, and O. The recommended references are listed in Table II. In order to calculate in detail the local cooling rate by any of these processes, it is necessary (1) to specify the number density of the coolant, (2) to compute the steady–state populations in each of

TABLE II References for collisional cross sections and rates

Species	Reference	Species	Reference
e + C	Johnson et al. (1987)	$H_2 + C^+$	Flower (1990)
He + C	Staemmler & Flower (1991)	e + O	Berrington (1988)
H_2 + C	Schröder et al. (1991)	H + O	Launay & Roueff (1977)
$e + C^+$	Keenan et al. (1986)	H_2 + O	Jaquet et al. (1992)
$H + C^+$	Launay & Roueff (1977)		

its energy states, and (3) to determine the difference between upward collisional energy transfer and downward collisional energy transfer taking into account the non–local effects of radiative transfer on the escape of the cooling radiation. Note that the collisional processes must compete with absorption and stimulated emission in the background radiation field as well as with the "trapped" line radiation itself. As indicated in the lower panel of Figure I, the lower cooling lines of CO and [C I] will feel the 2.7 K CBR, while the transitions of [C II], [O I], H_2, and HD must be excited above a hotter background ($T_{rad} \approx 7 - 35$ K) in order to effect much cooling. H_2 will not couple strongly to the infrared background, however, because the absorption probabilities are negligibly small in its electric quadrupole transitions. On the other hand, H_2 is effectively pumped by electronic transitions in the ultraviolet, which will produce some fluorescent infrared emission that merely redistributes starlight energy and produces no direct cooling of the interstellar gas (Black & Dalgarno 1976; Black & van Dishoeck 1987).

AN EMPIRICAL ENERGY BUDGET OF THE ζ OPH CLOUD

In order to illustrate an analysis of heating and cooling in a diffuse cloud, the energy budget of the well studied diffuse molecular cloud toward ζ Oph is described in some detail. This cloud has been observed through a variety of visible and ultraviolet absorption lines and through a few emission lines at cm, mm, and submm wavelengths. The total extinction is $A_V \approx 1.0$ mag, while the total column densities of hydrogen are $N(\mathrm{H}) = 5.25 \times 10^{20}$ cm^{-2}, $N(\mathrm{H_2}) = 4.2 \times 10^{20}$ cm^{-2}, and $N_H = N(\mathrm{H}) + 2N(\mathrm{H_2}) = 1.36 \times 10^{21}$ cm^{-2} (Morton 1975; Spitzer et al. 1974; Spitzer & Morton 1976). We can use various observations of column densities and excitation of coolants to estimate the mean rate of cooling of the gas in this cloud. Generally, the observable properties refer to column–averaged abundances. The effects of temperature and density upon excitation are also weighted in complicated ways according to the line–of–sight variations of T_k and n_H. Thus we cannot determine local cooling rates at any point in such a cloud, except indirectly through comparison with theoretical models that allow for variations with depth in a self–consistent manner. Even so, the empirical estimates of column–averaged cooling rates provide a useful approximate picture of the global energy budget of a diffuse cloud. In what follows, the rates of cooling and heating will all be normalized to the total column density of hydrogen nuclei, N_H, and will thus have dimensions of power per H nucleus. Indirect treatments of thermal balance based upon absorption–line observations of coolants have previously been applied to diffuse clouds (e.g. Pottasch et al. 1979). With recent results from the high-resolution spectrograph on *Hubble Space Telescope* (*HST*), it is now possible to do this more accurately.

Cooling rates

The column densities of the ground-term fine-structure levels of C^+ toward ζ Oph are $N(C^+ \ J = 1/2) = 1.6 \times 10^{17}$ cm^{-2} and $N(C^+ \ J = 3/2) = 2.35 \times 10^{15}$ cm^{-2}. These are the values that give the best agreement among the recent measurement of the C II] 2325 Å line (Cardelli et al. 1993), the velocity distribution function of Savage et al. (1992), and the older observations of the saturated 1334 and 1335 Å lines (Morton 1975). This re-analysis is necessary because no *HST* results have been published for the upper fine-structure level. The corresponding excitation temperature for the two fine-structure levels is $T_{ex} = 18.8$ K, and the implied optical depth of the [C II] 158 μm fine-structure transition is $\tau \approx 1.1/\Delta V$, where ΔV is the appropriate line width (full-width at half-maximum, FWHM) in km s^{-1}. The cooling rate is $4\pi \int I_\nu d\nu / N_H \approx 4.5 \times 10^{-26}$ ergs s^{-1} per hydrogen, based on the assumptions that the radiation is isotropic and that all of the excess excitation represents cooling. The observed excitation and optical depth are in harmony with various combinations of density and temperature: $(n_H, T_k) = (1.5 \times 10^4$ cm^{-3}, 20 K), (820, 30), (345, 40), (210,50), (150,60), and (77, 100), based upon a detailed calculation of the excitation and radiative transfer with an adopted electron fraction $n(e)/(n(\mathrm{H}) + n(\mathrm{H_2})) \approx 2 \times 10^{-4}$ (Black & van Dishoeck 1991). These solutions are local properties, while the observations provide column-averaged properties. Because various other indicators of density suggest that most of the gas in this cloud has $n_H < 10^3$ K (e.g. van Dishoeck et al. 1991), most of the excited C^+ must arise in parts of the cloud where $T_k > 30$ K. Crutcher (1977) has observed the C 166α radio recombination line toward ζ Oph. His measured intensity is consistent with the preceding analysis of C^+ if we allow for the probably large variations of column density and density within the 21$'$ beam diameter of his observations (cf. variations of CO line intensity by factors of 2–5 over $\pm 10'$ according to Liszt 1992).

The excitation of neutral carbon has not yet been re-done with *HST* data. Using the column densities in $J = 0$, 1, and 2 of de Boer & Morton (1979), we find cooling rates of 1.9×10^{-28} and 2.1×10^{-28} ergs s^{-1} per hydrogen in the [C I] $J = 1 \rightarrow 0$ and $2 \rightarrow 1$ fine-structure transitions, respectively. This appears to be consistent with the measured intensity in the $1 \rightarrow 0$ transition of Keene, Blake, & Phillips (1987), if allowance is made for the large baseline uncertainty and the complicated cloud structure within the 3$'$ beam of that observation. With the assumption of uniform conditions (density, temperature, abundances, etc.), the existing data on the column-averaged excitation of C suggest a temperature in the range $T_k \approx 42 - 58$ K and a density $n_H \approx 160 - 325$ cm^{-3}.

From the *HST* observation of O in its $J = 2$ ground level (Savage et al. 1992) and the older upper limit on column density in $J = 1$ (Morton 1975), we find an upper limit on the cooling in [O I] $J = 1 \rightarrow 2$ of $< 1 \times 10^{-26}$ ergs s^{-1} per hydrogen, which is not very helpful. Put another way, for the estimated mean density, and any temperature $T_k \leq 100$ K, we can predict that the column density in $J = 1$ is $N(\mathrm{O}\ J = 1) < 2.4 \times 10^{12}$ cm^{-2}. The corresponding limit for $J = 0$ is another factor of 2 times smaller.

Although part of the rotational excitation of H_2 in diffuse clouds is radiative and does not affect the kinetic energy budget, the observed column densities in the $J = 2$ and 3 levels (Spitzer et al. 1974) provide an upper limit on the cooling through collisionally excited $J = 2 \rightarrow 0$ and $3 \rightarrow 1$ rotational lines. The inferred cooling rates are 5.5×10^{-27} and 4.8×10^{-27} ergs s^{-1} per hydrogen, respectively.

The analysis of ultraviolet absorption lines of CO by Sheffer et al. (1992) yields a total column density $N(^{12}CO) = (2.2 \pm 0.4) \times 10^{15}$ cm^{-2} and a rotational excitation temperature of $T_{ex} = 4.2 \pm 0.3$ K in at least two velocity components. They find the abundance of ^{13}CO in this cloud to be 150 times smaller. For these conditions, the total energy loss in all rotational lines of ^{12}CO and ^{13}CO is 6.8×10^{-29} ergs s^{-1} per hydrogen. Alternatively, the observed emission in the $J = 1 \rightarrow 0$ and $2 \rightarrow 1$ lines and the upper limit on the intensity of $3 \rightarrow 2$ (Le Bourlot et al. 1989, van Dishoeck et al. 1991) indicate a cooling rate of $(1.2-1.9) \times 10^{-28}$ ergs s^{-1} per hydrogen. The factor–of–two difference between these two estimates is within the range of the uncertainties: the radio observations sample a larger area of the sky ($\approx 10 - 20''$) than the ultraviolet absorption measurements, while the adoption of an uncertain, but uniform, excitation temperature for such subthermal excitation translates into a large uncertainty in line intensity. It is important to keep in mind that CO accounts for only 1.3% of the gas–phase carbon in this cloud; however, CO can become the predominant form of gaseous carbon in translucent clouds only a few times thicker, where the line emission of CO will become a much more important contributor to the total cooling.

The amount of energy radiated in the $J = 1 \rightarrow 0$ rotational transition of HD is 1.8×10^{-29} ergs s^{-1} per hydrogen. The abundances of other potential coolants appear to be too low to affect the cooling. From the above discussion, the cooling of the ζ Oph diffuse cloud seems to be dominated by the collisionally excited line radiation of [C II] and H_2. The total cooling rate is estimated to be 6×10^{-26} ergs s^{-1} per hydrogen.

Heating rates

Ultraviolet starlight probably provides most of the heating, through several mechanisms. The photoelectric effect in dust grains is sensitive to the temperature and density of the gas, which regulate the electrical potential of the particles (Draine 1978). The same dust particles also attenuate the ultraviolet light, so that even in a cloud as thin as $A_V = 1$ mag, the heating rate decreases by a factor of 3 or more from edge to center, for dust with the optical properties of the Draine & Lee (1984) model. The heating by conventional grains may lie in the range $(2 - 5) \times 10^{-26} I_{UV}$ ergs s^{-1} per hydrogen, where I_{UV} is a scaling factor for the radiation field relative to that of Draine (1978) at a wavelength of 1000 Å. Bakes & Tielens (1993) have investigated the heating from very small carbon grains and PAH molecules, based upon an extension of the interstellar grain size spectrum to the molecular scale. For the conditions appropriate for the boundary of the ζ Oph cloud, their analysis suggests $5 \times 10^{-26} I_{UV}$ ergs s^{-1} per hydrogen. This rate will be reduced deeper inside the cloud with increasing attenuation. The balance between atomic and molecular hydrogen in a diffuse cloud is maintained by the processes of formation of grain surfaces, which may release as much as 1.5 eV per nascent molecule, and of spontaneous radiative dissociation by starlight at wavelengths $\lambda = 912 - 1110$ Å, which contributes approximately 0.4 eV per dissociated atom pair. The dissociation rate varies severely with depth through a diffuse cloud: the rates of the two processes are in balance in steady state, so the combined heating through formation and dissociation of H_2 is estimated to range from $6 \times 10^{-27} I_{UV}$ near the edge to $2 \times 10^{-28} I_{UV}$ ergs s^{-1} per hydrogen at the center of the ζ Oph cloud. Photoionization of neutral atoms is a minor heating source: in the case of C, the yield is approximately 1 eV per photoelectron, for an estimated rate of $1 \times 10^{-28} I_{UV}$ ergs s^{-1} per hydrogen. Heating by cosmic rays and the secondary electrons that they spawn will heat the ζ Oph cloud at a rate $\approx 1 \times 10^{-28}(\zeta_0/10^{-17})$ ergs s^{-1} per hydrogen, where ζ_0 is the hydrogen ionization rate in s^{-1}. The heating by dissipation of turbulence in the ζ Oph cloud may be at least 10^{-27} ergs s^{-1} per hydrogen. In total, the heating that depends on UV starlight appears able to provide quite enough energy, $\approx 7 \times 10^{-26} I_{UV}$ ergs s^{-1} per hydrogen, to balance the estimated losses through cooling. Heating by turbulent dissipation and penetrating cosmic rays will add a smaller amount, but with different depth–dependence and a larger relative contribution deep inside the cloud.

Summary

The foregoing comparison of heating and cooling is deceptive in the sense that it refers only to estimates of column–averaged rates. The real test of our understanding of the thermal balance will be to evaluate the rates of all these processes as functions of depth through a realistic model, to determine the resulting kinetic temperature at which the local heating and cooling balance, and then to verify that the computed temperature profile is consistent with all observations. The growing body of relevant observations will surely stimulate improved efforts to do such modeling.

ACKNOWLEDGMENTS

Research in astrochemistry at the University of Arizona is supported by the NASA Origins of Solar Systems program through grant NAGW–2255. Further support for the interpretation of ultraviolet absorption line observations of interstellar clouds is provided by STScI for program GO–2415.

REFERENCES

Andriesse, C.D. 1978, *A&A*, **66**, 169

Bakes, E.L.O., & Tielens, A.G.G.M 1993, *ApJ*, submitted

Berrington, K.A. 1988, *J.Phys.B*, **21**, 1083

Black, J.H. 1987, in *Interstellar Processes*, D.J. Hollenbach & H.A. Thronson, Jr., editors, (Dordrecht: Reidel), p. 731

Black, J.H., & Dalgarno, A. 1976, *ApJ*, **203**, 132

Black, J.H., & van Dishoeck, E.F. 1987, *ApJ*, **322**, 412

Black, J.H., & van Dishoeck, E.F. 1991, *ApJ*, **369**, L9

Cardelli, J.A., Mathis, J.S., Ebbets, D.C., & Savage, B.D. 1993, *ApJ*, **402**, L17

Crutcher, R.M. 1977, *ApJ*, **217**, L109

Dalgarno, A., & McCray, R.A. 1972, *ARA&A*, **10**, 375

de Boer, K.S., & Morton, D.C. 1979, *A&A*, **71**, 141

Deiss, B.M., Just, A., & Kegel, W.H. 1990, *A&A*, **240**, 123

Désert, F.X., Bazell, D., & Boulanger, F. 1988, *ApJ*, **334**, 815

Désert, F.X., Boulanger, F., & Puget, J.-L. 1990, *A&A*, **237**, 215

Draine, B.T. 1978, *ApJS*, **36**, 595

Draine, B.T., & Lee, H.M. 1984, *ApJ*, **285**, 89

Elmegreen, B.G. 1989, *ApJ*, **338**, 178

Elmegreen, B.G. 1993, *ApJ*, in press

Falgarone, E. 1993, invited paper, this volume

Falgarone, E., & Phillips, T.G. 1990, *ApJ*, **359**, 344

Falgarone, E., & Phillips, T.G. 1991, in *Fragmentation of Molecular Clouds and Star Formation*, IAU Symposium 147, E. Falgarone, F. Boulanger, & G. Duvert, editors, (Dordrecht: Kluwer), p. 119

Falgarone, E., Phillips, T.G., & Walker, C.K. 1991, *ApJ*, **378**, 186

Falgarone, E., Puget, J.-L., & Pérault, M. 1992, *A&A*, **257**, 715

Flower, D. 1990, *Molecular Collisions in the Interstellar Medium*, (Cambridge: The University Press)

Gredel, R., van Dishoeck, E.F., & Black, J.H. 1993, *A&A*, **269**, 477

Gredel, R., van Dishoeck, E.F., de Vries, C.P., & Black, J.H. 1992, *A&A*, **257**, 245

d'Hendecourt, L., & Léger, A. 1987, *A&A*, **180**, L9

Jaquet, R., Staemmler, V., Smith, M.D., & Flower, D.R. 1992, *J.Phys.B*, **25**, 285

Jenniskens, P. 1992, PhD Thesis, Leiden

Johnson, C.T., Burke, P.G., & Kingston, A.E. 1987, *J.Phys.B*, **20**, 2553

Jura, M. 1976, *ApJ*, **204**, 12

Keenan, F.P., Lennon, D.J., Johnson, C.T., & Kingston, A.E. 1986, *MNRAS*, **220**, 571

Keene, J., Blake, G.A., & Phillips, T.G. 1987, *ApJ*, **313**, 396

Lambert, D.L., Sheffer, Y., & Crane, P. 1990, *ApJ*, **359**, L19

Launay, J.M., & Roueff, E. 1977, *A&A*, **56**, 289

Le Bourlot, J., Gérin, M., & Pérault, M. 1989, *A&A*, **219**, 279

Lepp, S., & Dalgarno, A. 1988, *ApJ*, **335**, 769

Liszt, H.S. 1992, *ApJ*, **390**, 226

Mathis, J.S., Mezger, P.G., & Panagia, N. 1983, *A&A*, **128**, 212

Morton, D.C. 1975, *ApJ*, **197**, 85

Purcell, E.M. 1976, *ApJ*, **206**, 685

Pottasch, S.R., Wesselius, P.R., & van Duinen, R.J. 1979, *A&A*, **74**, L15

Savage, B.D., Cardelli, J.A., & Sofia, U.J. 1992, *ApJ*, **401**, 706

Schröder, K., Staemmler, V., Smith, M.D., Flower, D.R., & Jaquet, R. 1991, *J.Phys.B*, **24**, 2487

Sellgren, K. 1984, *ApJ*, **277**, 623

Sheffer, Y., Federman, S.R., Lambert, D.L., & Cardelli, J.A. 1992, *ApJ*, **397**, 482

Spaans, M. 1993, contributed paper, this volume

Spaans, M., Black, J.H., & van Dishoeck, E.F. 1993, in preparation

Spitzer, L., Jr., 1978, *Physical Processes in the Interstellar Medium*, (New York: Wiley–Interscience)

Spitzer, L., & Morton, W.A. 1976, *ApJ*, **204**, 731

Spitzer, L., Cochran, W.D., & Hirschfeld, A. 1974, *ApJS*, **28**, 373

Staemmler, V., & Flower, D.R. 1991, *J.Phys.B*, **24**, 2343

van Dishoeck, E.F., & Black, J.H. 1986, *ApJS*, **62**, 109

van Dishoeck, E.F., Black, J.H., Phillips, T.G., & Gredel, R. 1991, *ApJ*, **366**, 141

Verstraete, L., Léger, A., d'Hendecourt, L., Dutuit, O., & Deforneau, D. 1990, *A&A*, **237**, 436

Watson, W.D. 1972, *ApJ*, **176**, 103

Wright, E.L. et al. 1991, *ApJ*, **381**, 200

The First Symposium on the Infrared Cirrus and Diffuse Interstellar Clouds
ASP Conference Series, Vol. 58, 1994
Roc M. Cutri and William B. Latter (eds.)

DYNAMICS OF INTERSTELLAR CLOUDS: TURBULENCE, SHOCKS

E. FALGARONE
Radioastronomie Millimétrique, Ecole Normale Supérieure, 24 rue Lhomond, 75005 Paris, France

ABSTRACT All interstellar clouds are turbulent with supersonic velocities, except perhaps dense cores on the verge of forming stars. Their structure should thus conceal the signatures of two different dissipative processes: one due to shock waves which locally compress and heat the cloud, the other due to the property of intermittency of turbulence, *i.e.* the uneven distribution in space and time of the regions of strong shear or vorticity which are at the origin of sporadic and local high dissipation rates of kinetic energy.

In interstellar clouds, including cirrus clouds, the signatures of the former are rare (except in regions highly perturbed by hot stars, supernovae remnants...) while those of the latter phenomenon exist. This may be understood in the broad frame of supersonic turbulence, and bear some consequences on the physics and chemistry of interstellar clouds.

INTRODUCTION

A persisting problem in the dynamics of molecular clouds is the onset of star formation. Since it is now recognized that support against gravitational instability is supplied by turbulence, either purely hydrodynamical or (most likely) magneto-hydrodynamical, clues to the onset of collapse may be brought by the understanding of the dissipation mechanisms of the turbulent kinetic energy of interstellar clouds.

To this respect, cirrus clouds (at least, their molecular counterpart when it exists) are not peculiar clouds. They have low average column densities but they contain dense cores (Turner 1994) and might even form stars, although negative searches of T Tauri stars in high latitude clouds (hereafter HLCs) have been reported (Magnani, Caillault and Armus 1990). We show below that most of what is known of molecular clouds apply to them. They are part of the self–similar structure of the interstellar matter, and are highly turbulent. However, their location in the Galaxy (at high galactic latitude) make them be the closest clouds to the Sun which allows observations with unprecedented spatial resolution. Such clouds probably exist throughout the Galaxy but are difficult to detect when blended with higher column density (therefore brighter) structures (Waller and Boulanger 1994).

Cirrus clouds or HLCs are therefore ideal sites to understand the dynamics of interstellar matter, down to the smallest scales, which is a point of special relevance in the study of the processes which dissipate the kinetic energy.

CIRRUS CLOUDS AS ELEMENTS OF THE HIERARCHY OF INTERSTELLAR CLOUDS

Structure

Cirrus clouds, when molecular and bright enough in CO molecular lines to be compared to other clouds, have often been included in samples of molecular clouds not associated with star–formation activity. In the presentation of their large–scale CO survey of the second quadrant, Heithausen et al. (1993) display histograms of cloud sizes and CO integrated brightnesses for cirrus clouds and the subset of Lynds clouds included in the surveyed area. They show no significant difference. The difference between the cloud masses inferred from the surveys of Magnani et al. (1985) and Heithausen et al. (1993) can be explained by the better sensivity of the latter.

The internal structure of cirrus clouds have also been studied in terms of the properties (internal velocity dispersion, mass and size) of their substructures as seen in CO and isotopes, *i.e.* the regions of bright emission above a background level, identified in space and velocity space. Several studies are exclusively devoted to high latitude clouds like those of MBM 12 alias L1457 (Pound, Bania and Wilson, 1990; Zimmermann, 1993) or the HLC around the star HD 210121 (Gredel et al. 1992), others include parts of HLCs (Falgarone and Pérault 1988; Falgarone, Puget and Pérault 1992; Falgarone et al. 1994). In scatter diagrams like that of the internal velocity dispersion versus size, the points associated with HLCs or cirrus clouds lie at the border of the range spanned by all other clouds. This is illustrated in Figure I adapted from Falgarone, Puget and Pérault (1992). In their paper, these authors compare data obtained by different groups from different data sets, with their own small scale results, all the analysis being conducted in about the same way. Data sets are from Herbertz et al. (1991) for NGC1499 (small stars), from Carr (1987) for a part of Cepheus barely affected by star formation (asterisks), and the others are described in Falgarone and Pérault (1987). The points corresponding to HLCs (MBM 12 and Ursa Major) are the large heavy symbols, and the two large stars correspond to small scale structures of a cloud edge (Falgarone and Phillips 1994).

It is remarkable that the structures of HLCs at the smallest scales in the diagram span almost the same range of velocity dispersions (0.2 to 2 km s^{-1}) as the rest of the clouds, including the largest.

CO lineshapes of HLCs are also similar to those of many non–starforming molecular clouds. The lines are usually very broad and often reveal non–Gaussian wings. Such profiles can be seen in Magnani et al. (1990), in Gredel et al. (1992), in Falgarone et al. (1994). Although many interpretations have been advocated for the origin of such wings (see Elmegreen 1994), Falgarone and Phillips (1990) have proposed that it may be a signature of the intermittent nature of turbulence.

The density of the gaseous component which excites the rotational transitions of CO is also comparable to that found in clouds of larger column density. This property is inferred from the comparison of the intensities of the two lowest transitions of CO and ^{13}CO. An example is shown in Falgarone, Puget and

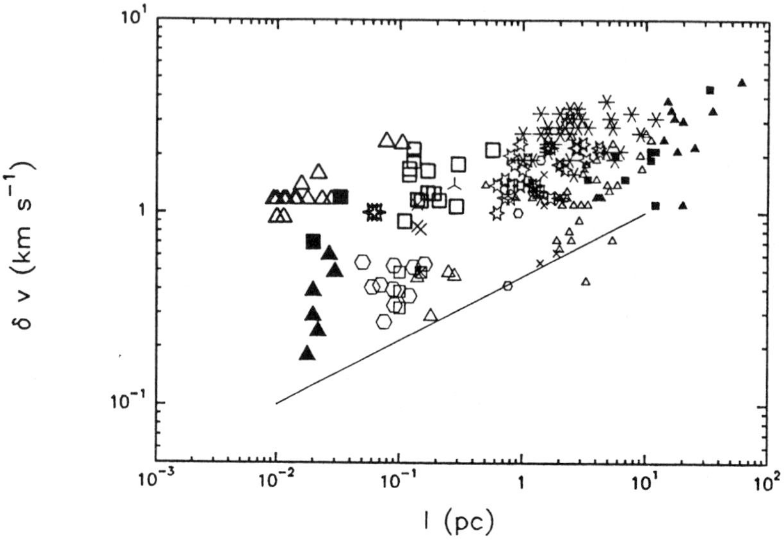

FIGURE I Non–thermal velocity dispersion versus size of an ensemble of structures drawn for a variety of data sets. The references and corresponding symbols are given in Falgarone, Puget and Pérault (1992). The data points corresponding to HLCs are the heavy large squares and triangles: Ursa Major (solid symbols), MBM 12 (empty symbols) from references given in the text. The slope (1/3) of the Kolmogorov law for incompressible turbulence is shown.

Pérault (1992). Spectra of the ^{13}CO integrated emission of two fields in Ursa Major and in a non–starforming area of the Taurus–Auriga complex are compared. The peak line temperature in the former is weak, only $T_A^* = 0.8$K and is $T_A^* = 3.2$K in the latter. However, the line ratios of the two lowest rotational transitions are the same within the error bars, $r = 0.5$, in both fields and stay the same all across the lineprofiles.

In summary, the molecular part of cirrus clouds and HLCs, as traced by the lowest rotational transitions of CO and ^{13}CO, appear to have the same dynamical properties as most other molecular clouds not associated with active starforming regions. In spite of their low average column density, the density of the gas component is as large as in other clouds as shown by the line ratios.

Gravitational binding and stability

As far as their dynamics is concerned and in particular their gravitational stability, the molecular counterpart of HLCs and cirrus clouds have always been observed far out of virial balance between their internal kinetic energy T and their gravitational energy W. This departure can be expressed in terms of the dimensionless parameter $\alpha = 2T/ \mid W \mid = 5\sigma^2 R/GM$ used by Bertoldi and McKee (1992) in which σ is the internal velocity dispersion (predominanttly non–thermal) and R and M are the radius and total gas mass of a given structure. From the results of the CO(J=1-0) survey of Magnani, Blitz and Mundy (1985) on 22 high latitude complexes, an average value of $\alpha \sim 100$ is derived

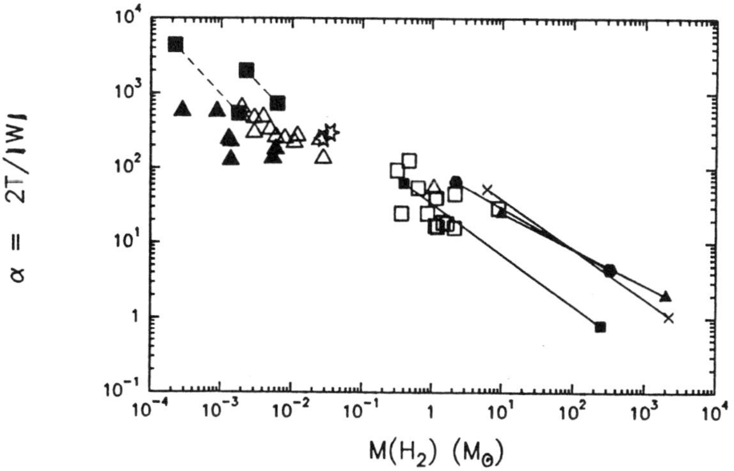

FIGURE II The virial parameter α as a function of the mass of substructures in HLCs. The solid lines are least square fits obtained by Bertoldi and McKee (1992) in the substructures of four molecular clouds (see text for the symbols). The hexagons correspond to structures in a cloud edge.

for a median cloud mass $M = 18M_\odot$ estimated with the CO/H_2 conversion factor $N(H_2) = 2.5 \times 10^{20}\, W_{CO}$ of Bloemen et al. (1986). Pound et al. (1990) have studied the internal structure of one of these HLCs (MBM12 or L1457/58) in the ^{13}CO(J=1-0) transition. Figure II displays the values of α (large empty squares) for the 14 substructures they have isolated in their space–velocity maps. The same HLC has been studied at higher angular resolution by Zimmermann (1993). The substructures isolated in his ^{13}CO(J=2-1) map which have sizes larger than 0.01pc are the large empty triangles. The large solid squares correspond to the two substructures of Ursa Major previously mentioned, which lie at a sharp CO transition region. The connected symbols stand for two mass determinations, one with $M(H_2)$ only, estimated by using the conversion factor $N(H_2) = 5 \times 10^{19}\, W_{CO}$ found by Heithausen et al. (1993) to be valid for HLCs, the other takes $M(HI)$ into account, the atomic gas mass being determined from the measurements of de Vries et al. (1987). The solid triangles represent substructure in a bright region of the same HLC seen in ^{13}CO(J=1-0) (Falgarone and Pérault 1988). On this figure, we also show the least square fits obtained by Bertoldi and Mc Kee (1992) for all the structures (but the most massive) found in the ρ Oph filaments by Loren (1989a) (solid squares), in the Rosette by Blitz and Stark (1986) (solid triangles), in Cep OB3 by Carr (1987) (solid hexagons) and in Orion B by Bally et al. (1987) (crosses).

This figure shows how far the small scale structures of HLCs are from being bound by self-gravity. Furthermore, it seems to support the finding of Loren (1989b) of a power-law relation between α and the structure mass. The samples of substructures selected in each HLC are quite small and it is therefore difficult to determine a slope for MBM12 or Ursa Major, taken individually. On the other hand, if such a relation does exist, it means that there is some lower limit to masses of cloud substructures in virial balance between self-gravity and kinetic energy. Given the law found by Loren (1989b), $\alpha \propto M^{-2/3}$, and the data plotted on Figure II, this limit may be of the order of a few Solar masses. This dependence has been recently justified theoretically by Bertoldi and McKee (1992). There is indeed a simple explanation to that power law behaviour. Introducing the internal average pressure $P_i = 1/3\,\rho\sigma^2$ considered as the internal kinetic energy density in any form (thermal or turbulent), the virial parameter can be written $\alpha = 15\,RP_i/\rho GM$. Bertoldi and McKee also introduce the largest gravitationally stable mass M_{cr} for isothermal sheres immersed in an external pressure P_e

$$M_{cr} = a\frac{\sigma^4}{P_e^{3/2}G^{1/2}} = \frac{a}{\rho^2}\frac{P_i^2}{P_e^{1/2}G^{3/2}}$$

where a is a coefficient close to 1. This expression shows that it is only when $P_i \sim P_e$ (*i.e.* when self–gravity is not dominant) that $M_{cr} \propto (P/G)^{3/2}$ and $\alpha \propto (M_{cr}/M)^{2/3}$.

The distribution of the data points in Figure II along a $M^{-2/3}$ slope means that for all these structures self–gravity is not dominant over the other confining processes and furthermore that M_{cr} does not vary by more than one order of magnitude from one complex to another within the Solar Neighborhood (although the differences noticed by Bertoldi and McKee between the ρ Oph filaments, the Rosette, Orion B and Cep OB3 may be real).

In all the above mass determinations, based on the intensity of tracers like CO or ^{13}CO and CO/H_2 conversion factors, biases may be large. HLCs and cirrus clouds have a low average column density and structures poorly shielded from the ambient UV field contain a smaller fraction of H_2 relative to HI than other clouds. However it seems unlikely that masses there could be underestimated by several orders of magnitudes (see the example of the Ursa Major structures discussed above). These structures also lie in the transition region $A_v \sim 1$, where the CO/H_2 abundance varies by orders of magnitude. This has been extensively discussed in van Dishoeck and Black (1988) and subsequent reviews (van Dishoeck 1993; 1994). The fact that conversion factors seem to apply (see the discussion in Heithausen et al. 1993) may be related to the high gas density in the CO emitting regions, however small they may be. If as large as $\sim 10^4$ cm^{-3}, pathlengths as small as 0.015 pc are sufficient to self–shield CO above the typical 0.5 mag already provided by H_2 and HI. Low brightness structures are beam diluted emission of still smaller (unresolved structures) dense enough to possess the standard (or close to standard) CO/H_2 ratio.

Molecular HLCs and cirrus clouds, as most of the substructures of other clouds, are not bound by self–gravity but confined by an external pressure too large to be thermal, $P/k \sim$ a few 10^4 K cm^{-3} (Falgarone et al. 1992). They are among the most turbulent clouds (out of star forming regions) with supersonic but Alfvénic velocities. Turbulence there have several possible origins. HLCs are on the average poorly shielded from the ambient UV field. UV heating may drive convection at velocities large enough to induce turbulence (Bréart de Boisanger et al. 1992). Turbulence may be driven also by MHD waves generated by supernovae explosions or by the passage of a pressure wave from a supernova (Pound et al. 1990). They have comparable magnetic and kinetic energy densities (Goodman et al. 1994), much larger than their gravitational energy density. Signatures of supersonic turbulence are therefore expected to be present in these clouds (as in all others) and we discuss below recent results of numerical simulations of compressible turbulence which might be a useful framework to understand the various processes at work in compressible turbulent media.

WHAT DO WE KNOW ABOUT SUPERSONIC TURBULENCE?

Recent three-dimensional numerical simulations of compressible turbulence with an unprecedented resolution (512^3 grid points) by Porter, Pouquet and Woodward (1994) shed a new light on the evolution of shocks and energy transfers between the various modes. For a detailed description of the method used (PPM) see Porter, Pouquet and Woodward (1992; 1993). The Mach number of the simulation is moderate, only 1.1 (rms) at t=0. The fluid is a perfect gas with $\gamma = 4/3$. The evolution of the flow is not forced. Energy is fed into the fluid at t=0 and the system subsequently evolves freely under the constraint of these initial conditions. The time evolution is described by the authors as comprising three main phases. Their duration is conveniently expressed in terms of the acoustic time $\tau_{ac} = L/c_S$ which is the sound crossing time of the energy containing scale, L.

A first short phase, during which compressional waves steepen, ends at $t \sim 0.3\tau_{ac}$ with shock formation. The ratio of the maximal to minimal densities reaches $\rho_{max}/\rho_{min} \sim 60$. Shock interactions are at the origin of the second phase in which the vorticity generated by these interactions allows most of the kinetic energy to be transferred to non compressible modes. The energy dissipation rate is subsequently considerably reduced which explains the long duration of the second phase, up to $t \sim 3\tau_{ac}$. In this phase the flow has most of the properties of incompressible turbulence: an inertial range of scales has developed with a power spectrum for the solenoidal (incompressible) part of the flow (which contains more than 90% of the whole energy) close to that predicted by Kolmogorov $E(k) \propto k^{-5/3}$ although the ratio of the maximal to minimal densities is still $\rho_{max}/\rho_{min} \sim 5$. The probability distribution function (PDF) of vorticity in this phase shows a conspicuous non-Gaussian tail, illustrative of the intermittency of the flow (Figure III). At the end of this phase only a few shocks have survived. The last phase is that of the decay of turbulence.

Clearly, such simulations cannot approach the complexity of the dynamics of the interstellar clouds, which are strongly coupled to the magnetic field, far from being polytropic structures (the radiative cooling depends on the gas density, temperature and velocity field), without mentioning self–gravity and chemical fluctuations.

However, as will be seen below, the non-linear coupling of the velocity modes which gives rise to turbulence may be such a universal process that it generates velocity patterns which seem to be found even in the interstellar medium.

To test the relevance of the results of such simulations to the interstellar medium, Falgarone et al. (1994) have computed synthetic line profiles from the velocity fields computed by Porter et al. (1994) at three different epochs, the first, at $t = 0.5\tau_{ac}$ just after the shock formation,the second at $t = 1.2\tau_{ac}$ in a phase of the evolution when non-compressible modes have already gained a large fraction of the energy and the third one, at $t = 2.4\tau_{ac}$ when the inertial

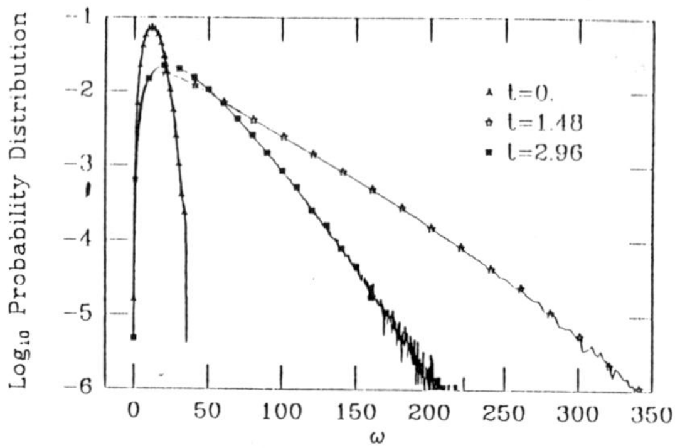

FIGURE III Probability distribution function of the vorticity at three different epochs expressed in units of τ_{ac} (from Porter, Pouquet, Woodward, 1992).

range is the largest just before decay starts, and in which the energy in the incompressible modes largely encompasses that in the compressible ones.

The synthetic spectra have been computed in the optically thin approximation and are simply histograms of the line of sight projection velocity. An example of a map of 256 synthetic spectra at $t = 1.2\tau_{ac}$ is shown in Figure IV (lower panel) and is displayed with a sample of real spectra taken in a HLC, Ursa Major.

The similarity between the observed line profiles with the synthetic profiles is striking. All the diversity of the observed lineshapes is found in the simulated profiles: the double–peaks, the asymmetries with respect to the centroid velocity and the broad linewings. Other examples of spectra maps can be found in the literature, although few of such maps are published (see for example Gredel et al. 1992). The spectra map shown in Figure IV is indeed quite representative of those of non–starforming interstellar clouds.

It is encouraging to see that these calculations which include no radiative transfer yet produce lineshapes which resemble the observed spectra more closely than many radiative transfer codes do. As said before, this comparison is far from being a final answer to the question of the origin of rotational lineshapes in molecular clouds but we consider it as a powerful indication that turbulence, taken here as a most general concept, is a viable mechanism for generating the type of velocity field, and most importantly its correlations at all scales, required to explain the shapes of lineprofiles in the interstellar clouds.

It supports the idea that the dynamics of interstellar clouds and HLCs in particular is dominated by turbulence, likely MHD turbulence, given the intensity of the magnetic field and the amounts of kinetic energy and magnetic energy in clouds (Goodman et al. 1994). According to the results of Porter, Pouquet and Woodward discussed above, small shocks as well as zones of large vorticity are expected to be present in interstellar clouds. MHD shock signatures such as those discussed by Draine (1980), Draine, Roberge and Dalgarno (1983), Draine and Katz (1986) and Pineau des Forêts et al. (1986) should be present, but to a lesser extent than predicted, because of the efficient transfer of energy

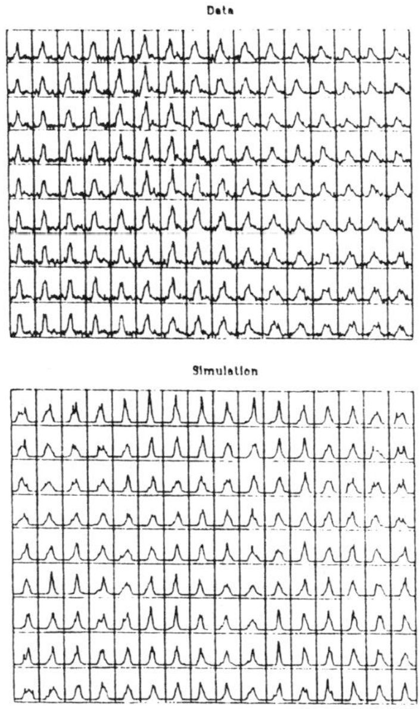

FIGURE IV Upper panel: $^{12}CO(J=2\text{-}1)$ map of a field in Ursa Major obtained at the CSO telescope. The beam sampling is 15" and the velocity resolution 0.06 km s^{-1}. Lower panel: Synthetic profiles computed with the turbulent velocity field of Porter et al. (1993) ignoring any radiative trapping

from compressible to incompressible modes. Specific signatures of the zones of high vorticity should be present too and we discuss below what these signatures might be.

SPECIFIC OBSERVATIONAL SIGNATURE OF THE INTERMITTENCY OF TURBULENCE

Probability distribution of the gas kinetic temperature

The zones of high vorticity in turbulent flows are also closely associated with those of large dissipation rate of kinetic energy and the two subsets of space where these quantities are concentrated are highly structured and intermittent (Brachet, 1990; Vincent and Meneguzzi 1991). The dissipation of the turbulent kinetic energy via viscous processes eventually turns this energy into heat. It is therefore possible to compute the rate at which the gas in the dissipation regions is heated by this process, as well as its equilibrium temperature (Falgarone and Puget 1994).

The local dissipation rate of kinetic energy is

$$\epsilon = \eta/2 \sum_{i,j} (\partial_j v_i + \partial_i v_j)^2$$

where $\eta = \rho\nu$ is the dynamical viscosity and ρ the density. The energy dissipation rate averaged over a volume of size δl is approximated by $\epsilon_{\delta l} = \eta(\mid \delta v \mid /\delta l)^2$

where δl is the lag over which the velocity increments δv are measured, in flow experiments for example. The detailed justification for this approximation is discussed in Falgarone and Puget (1994). The term $\epsilon_{\delta l}$ is therefore a local heating source which is found to exceed by orders of magnitude all the other heating terms due to UV photons, coupling to dust grains, cosmic rays, estimated following Black (1987) for instance.

The experiment of Gagne (1987) in flows with large Reynolds numbers provide measurements of the velocity increments δv over a large range of scales δl, down to only a few times the dissipation scale. The probability distribution functions (PDFs) of the dissipation rates averaged over various sizes δl are derived from his PDFs of velocity increments. His results have been scaled to the interstellar medium turbulence since the statistical flow properties depend only on the kinetic energy transfer rate $(\delta v)^3/\delta l$ and the dissipation scale, l_D. In atomic clouds, the dissipation scale is estimated to be

$$l_D \sim 10\mathrm{AU}(\frac{n_H}{10^2\mathrm{cm}^{-3}})^{-3/4}(\frac{T_k}{30\mathrm{K}})^{3/8}$$

Thermal balance between heating and radiative cooling processes is reached because the zones of intermittency have a lifetime almost an order of magnitude larger than their own anticipated turnover timescale $\sim \delta v/\delta l$ (see for example the experiment of Douady, Couder and Brachet 1991). The lifetime of an intermittent structures of high vorticity is estimated to be comparable to the turnover time of the largest scales, which is about 2×10^6 yr in an HI cloud. This lifetime is much larger than the cooling time obtained for radiative processes, $\tau_{cool} = E/\Lambda \sim 10^4$ yr for HI at $T = 50K$, where Λ is the radiative cooling rate taken from Dalgarno and McCray (1972). Balance is thus reached between the heating rate $\epsilon_{\delta l}$ provided locally by the intermittent large dissipation of kinetic energy and the radiative cooling rate of the HI gas.

The resulting PDF of the gas kinetic temperature is shown on Figure V (in logarithmic intervals) for two different densities which bracket those of HI clouds and two values δl of the size of the dissipative structures, expressed in units of the dissipation length, l_D.

The remarkable result is that a full range of temperatures is found with different probabilities which may be interpreted as volume filling factors within the entire cloud at any time. Temperatures as large as $T_k = 10^4$K are found to occur with non negligible probabilities (resp. filling factors) of 10^{-2} to 10^{-3}.

Such large temperatures are unexpected given the low average kinetic energy of cold HI clouds. This result can be understood as follows: kinetic energy is fed into the small intermittent structures over timescales which largely exceed their own turnover timescale and the amount of energy accumulated during that time in these structures, dissipated locally and turned into heat largely exceeds the average kinetic energy.

Possible consequences on chemistry

Many chemical reactions with activation barriers of several thousands Kelvins can therefore take place in regions of intermittent large dissipation rate. Preliminary calculations by E. Roueff (private communication) show that, as expected, the abundances of molecules like OH, O_2, H_2O, CH^+ in these regions are or-

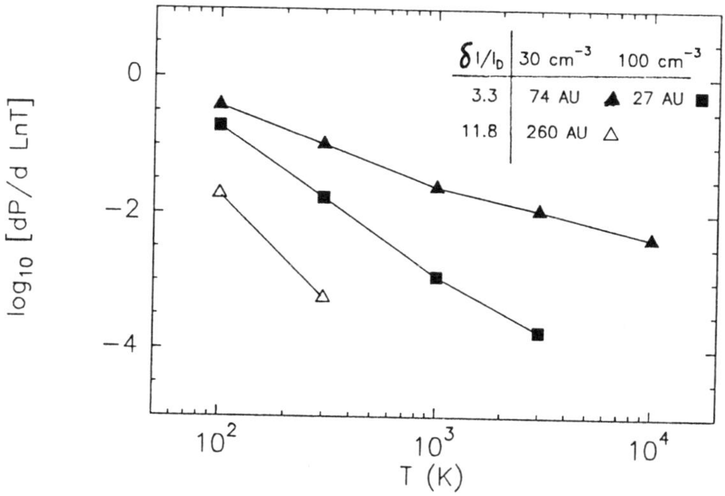

FIGURE V Probability distribution function of the gas kinetic temperature in the intermittent regions of dissipation of turbulence in atomic HI gas (from Falgarone and Puget 1994)

ders of magnitude larger than in the surrounding gas. A fraction as small as $N_{warm}/N_{tot} \sim 0.03$ of warm gas at $T_k = 10^3$K on a line of sight is sufficient to reproduce the observed value $N(CH^+)/E_{B-V} \sim 4 \times 10^{13}$cm^{-2}mag^{-1} found by Gredel, van Dishoeck and Black (1993) in translucent clouds. The corresponding volume filling factor of warm gas depends on the geometry of the dissipation regions.

SUMMARY

As long as their molecular counterpart (as traced by CO) is considered HLCs and cirrus clouds are alike all other molecular clouds which are not in an active phase of star formation. They are likely to be shaped by MHD turbulence given the large predominance of kinetic energy above self–gravitating energy. To this respect, they are among the most turbulent clouds. In order to determine the relative importance of shocks fronts compared to other dissipative structures like regions of high vorticity and shear, results of three–dimensional numerical simulations of compressible turbulence may be relevant. Such recent simulations reveal that shock fronts form first and very rapidly transfer most of the kinetic energy to solenoidal (incompressible) modes which prevents a rapid dissipation of kinetic energy. Once most of the energy has been transferred to incompressile modes, dissipation occurs in an intermittent way, in regions of high vorticity and shear which can thus reach locally very high kinetic temperature. A PDF of kinetic temperatures of the gas is derived from known PDFs of velocity gradients in terrestrial flows. This assumes that the statistical properties of turbulence are universal.

An interesting consequence of the existence of tiny localized regions of large kinetic temperature ($T_k > 10^3$K) is the possibility of overcoming there the activation barriers of many chemical reactions, in particular making possible a rapid formation of CH^+. Another interesting issue is to determine which would be the contribution of these warm regions within cold atomic clouds to the warm neutral medium (WNM).

REFERENCES

Bally, J., Langer, W.D., Stark, A.A. and Wilson, R.W.1987, *ApJ*, **312**, L45

Bertoldi F. and McKee C.F. 1992 *ApJ*, **395**, 140

Black, J.H. 1987 in *Interstellar processes*, D.J. Hollenbach and H.A. Thronson, Dordrecht:Reidel, 731

Blitz, L. and Stark, A.A. 1986, *ApJ*, **300**, L89

Bloemen, J.B.G.M. et al. 1986 *A&A*, **154**, 25

Brachet, M.E., 1990 *C. R. Acad. Sci. Paris* **311**, 775

Bréart de Boisanger,C., Chièze, J.P. and Meltz, B. 1992, *ApJ*, **401**, 182

Carr, J.S. 1987 *ApJ*, **323**, 170

Dalgarno, A. and McCray, R.A., 1972 *ARA&A*, **10**, 375

de Vries, H.W., Heithausen, A., and Thaddeus, P. 1987 *ApJ*, **319**, 723

Douady, S., Couder, Y. and Brachet, M.E. 1991, *Phys.Rev.Lett*, **67**, 983

Draine, B.T. 1980, *ApJ*, **241**, 1021

Draine, B.T., Roberge, W.G., and Dalgarno, A. 1983, *ApJ*, **264**, 485

Draine, B.T. and Katz, N. 1986, *ApJ*, **306**, 655

Draine, B.T., and McKee, C.F. 1993 *ARA&A*, **31**, in press

Elmegreen, B.G. 1994, this volume

Falgarone, E. and Pérault M. 1987 in *Physical processes in Interstellar clouds*, G.E. Morfill and M. Scholer, Dordrecht: Kluwer, 59

Falgarone, E. and Pérault 1988 *A&A*, **205**, L1

Falgarone, E. and Phillips T.G. 1990, *ApJ*, **359**, 344

Falgarone, E. and Puget, J.-L. 1994, *A&A*, submitted

Falgarone, E., Puget, J.-L., and Pérault, M. 1992, *A&A*, **257** 715

Falgarone, E. and Phillips, T.G. 1994, *A&A*, submitted

Falgarone E., Lis, D.C., Phillips, T.G., Pouquet A., Porter, D.H. and Woodward, P.R. 1994 in preparation

Gagne, Y. 1987, Thèse d'Etat, Université de Grenoble.

Goodman, A.A., Myers, P.C., Güsten, R., Heiles, C. 1994, this volume

Gredel R., van Dishoeck, E.F., de Vries, C.P. and Black, J.H. 1992 *A&A*, **257**, 245

Gredel R., van Dishoeck, E.F., and Black, J.H. 1993 *A&A*, **269**, 477

Heithausen, A., Stacy, J.G., de Vries, H.W., Mebold, U. and Thaddeus, P. 1993 *A&A*, **268** 265

Herbertz, R., Ungerechts, H. and Winnewisser, G. 1991 *A&A*, **249**, 483

Hollenbach, D. and McKee, C.F. 1989 *ApJ*, **342**, 306

Loren, R. 1989a *ApJ*, **338**, 902

Loren, R. 1989b *ApJ*, **338**, 925

Magnani, L., Blitz, L. and Mundy L. 1985, *ApJ*, **295**, 402

Magnani, L., Caillault, J.-P., and Armus, L., 1990, *ApJ*, **357**, 602

Magnani, L., Carpenter, J., Blitz, L. Kassin, N.E. and Nath, B.B. 1990, *ApJS*, **73**, 747

Myers, P.C. and Goodman A.A 1988 *ApJ*, **326**, L27

Pineau des Forêts, G., Flower, D.R., Hartquist, T.W., Dalgarno, A. 1986, *MNRAS*, **220**, 801

Porter, D.H., Pouquet, A., and Woodward, P.R. 1992, *Phys. Rev. Lett.*, **68**, 3156

Porter, D.H., Pouquet, A., and Woodward, P.R. 1993, *Theor. Comp. Fluid Dynamics* in press

Porter, D.H., Pouquet, A., and Woodward, P.R. 1994, *Phys. Fluids A* in press

Pound, M.W., Bania, T.M. and Wilson, R.W. 1990 *ApJ*, **351** 165

Turner, B.E. 1994 this volume

van Dishoeck, E.F. and Black J.H. 1988 *ApJ*, **334** 771

van Dishoeck E.F. in *Astrochemistry of Cosmic Phenomena* P.D. Singh, Dordrecht: Kluwer, 1992, 143

van Dishoeck, E.F. 1994, this volume

Vincent, A. and Meneguzzi, M. 1991, *J. Fluid Mech.*, **225**, 1

Waller, W.H. and Boulanger, F.B. 1994, this volume

Zimmermann, T. 1993, PhD Thesis, University of Köln, Germany.

The First Symposium on the Infrared Cirrus and Diffuse Interstellar Clouds
ASP Conference Series, Vol. 58, 1994
Roc M. Cutri and William B. Latter (eds.)

MAGNETIC EFFECTS IN DIFFUSE CLOUDS

B. G. ELMEGREEN
IBM Research Division, T.J. Watson Research Center, Yorktown Heights, NY 10598

ABSTRACT The structure and internal motions of interstellar cirrus and other diffuse clouds have characteristics that suggest a relatively strong magnetic field is present. Possible formation scenarios also involve magnetism. This paper reviews some of these characteristics and suggests ways in which magnetic field measurements can be used as diagnostics for cloud physical processes. Topics include: (a) environmental influences on diffuse cloud structure; (b) the ubiquity of filaments; (c) structure and turbulence from magnetic Rayleigh-Taylor and Kelvin-Helmholtz instabilities; (d) filamentary structure from rotation and twisting magnetic fields; (e) field line orientations in clouds; (f) structure and velocity distribution functions from strong magnetic waves and non-linear processes, and (g) the minimum size for magnetic waves.

1. INTRODUCTION

1.1 Defining Cloud Types

This discussion will be limited to *diffuse* clouds, as opposed to self-gravitating clouds, using a distinction based on the dimensionless ratio $P/(G\sigma^2)$, which is proportional to the ratio of the pressure inside a cloud to the self-gravitational energy density. The mass column density in the cloud is denoted by σ. Diffuse clouds have $P/(G\sigma^2) >> 1$ and self-gravitating clouds have this ratio much less than 1. A criterion based on $P/(G\sigma^2)$ is introduced because the name *diffuse* refers to a lack of regular structure, i.e., to a nebulous or haphazard structure, and once gravity becomes important, clouds get pulled together into roundish, structured objects. The criterion is also introduced to avoid confusion with another distinction between cloud types that is based on chemistry or molecule formation (i.e., molecular versus atomic clouds). Extensive discussions of these cloud types are in Elmegreen (1985; 1993a,b).

1.2 What $P/(G\sigma^2) >> 1$ implies for diffuse clouds

The importance of pressure compared to self-gravity for a diffuse cloud implies that *most diffuse clouds form by pressure-related events*, as during the expansion of a shell, the destruction of a self-gravitating cloud following star formation, thermal instabilities, compression in steepened non-linear magneto-acoustic waves (supersonic turbulence) and so on. It also implies that *diffuse clouds should be sensitive to their environments*, i.e., to pressure variations and gradients, which make them easily put out of dynamical equilibrium. Thus pressure related instabilities such as Rayleigh-Taylor or Kelvin-Helmholtz instabilities, and thermal instabilities (for some temperature ranges) should be important for cloud structures.

This sensitivity to environment means that much of the internal kinematic energy in a diffuse cloud comes from outside (Joncas et al. 1992). For example, there is a correlation in the Magnani, Blitz and Mundy (1985) clouds between the internal clump-to-clump velocity dispersion and the cloud velocity (Elmegreen 1988), which implies that the large scale turbulence inside the clouds is related to the clouds' bulk motion relative to the surrounding gas. This correlation is analogous to the situation in a shock front, where the thermal speed of the atoms behind the front is proportional to the shock speed. In the case of the MBM clouds, the concept of atoms as the building blocks of the material should be replaced by a concept of gas clumps as the building blocks. Then the collection of a generally clumpy ambient medium into a dense shell or ridge by some mild pressure disturbance leads to a velocity dispersion of the clumps (rather than individual atoms) that is proportional to the relative speed of the ridge. The highly agitated diffuse or cirrus clouds that are observed at high latitudes may be regarded as active pressure fronts which are still accumulating ambient clumpy gas and continuously agitated by the individual impacts of these ambient clumps. The fronts do not have to be pushed from behind by some currently high pressure to show these characteristics, they need only have been pushed in the past. A theoretical discussion of such fronts is in Elmegreen (1988).

The MBM clouds could also be dense patches compressed by supersonic turbulence in the surrounding medium, or they could be post-shock material of some other nature, and the clumps inside of them could be formed there by turbulent processes. They are likely to have a correlation between velocity dispersion and bulk velocity in these cases too.

If the clouds and clumps in the interstellar medium are relatively long-lived (static), bounded for example by a high thermal pressure in the intercloud medium, and they are not transient density spikes bounded by the variable ram pressure from interstellar turbulence, then the accumulation of clumps into a dense clumpy shell seems to require a magnetic field. If a clumpy interstellar medium had no magnetic field and a mild intercloud breeze or pressure gradient pushed on it, the clumps would accelerate separately in inverse proportion to their column densities (acceleration = pressure divided by column density) and they would scatter. The collection of ambient clumpy gas into a clumpy shell or ridge requires a similar acceleration for each clump, and this is most easily accomplished if the clumps are connected together by intertangled magnetic field lines (see Figure 1). Then the clumps with low column densities will not be able to get far ahead of the clumps with high column densities because magnetic tension in the field lines that connect these two types of clumps will hold them near to each other. This same magnetic tension also provides the force for transmitting the momentum from the accreting clumps to the rest of the clumps in the layer, thereby maintaining a uniform clump-to-clump velocity dispersion in the layer. Without such field forces, the accretion of a new clump could tear a hole in the moving ridge, affecting primarily the area that the new clump actually touches. The new clump might even go entirely through the ridge if it meets no other clump that can share its momentum or if it has a large column density. This would be particularly true for non-magnetic pressure fronts that

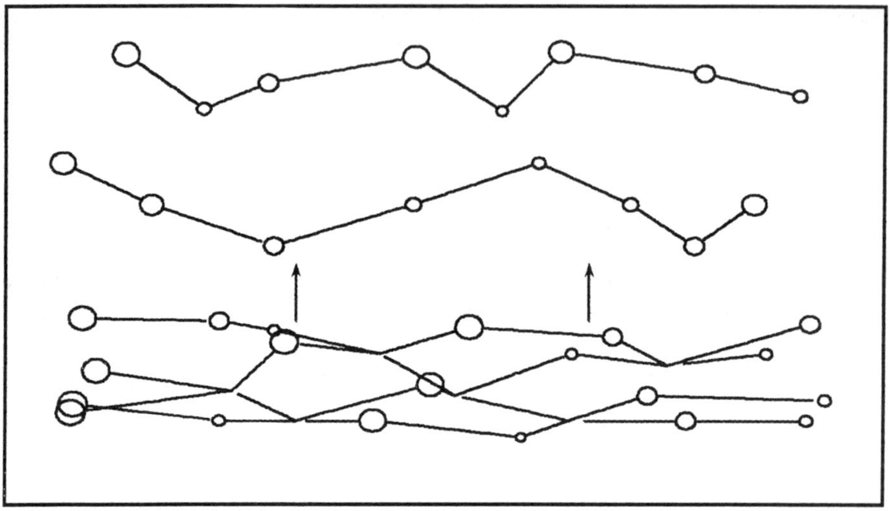

FIGURE 1 A diagram of magnetically tangled clumps ahead of (top) and behind a moving pressure front.

have moved less than one clump mean free path. Magnetic pressure fronts can smoothly accumulate the ambient clumpy medium even over a short distance, much less than the mean free path, because the *effective* mean free path for *magnetic* clump interactions is much smaller than the geometric mean free path (Clifford and Elmegreen 1983).

These considerations imply that super-virial clouds need not have short lives; they can be confined by the same pressure that makes them move and gives them their shape. This is basically a ram pressure from converging motions of some type (shock-like or turbulent) in accreting gas.

An alternative point of view to this picture is that the clumps are relatively short-lived turbulent splash regions that are destroyed or smoothed together when they enter a pressure front but then reform into new clumps inside the front. It could be that many clumps and even whole diffuse clouds are such splash regions (Section 5). In this case, young, high-pressure shells or filaments will have greater clump-to-clump velocity dispersions than low pressure or ambient regions because there is more turbulent energy in the shell than outside as a result of the compressive energy put into it at the pressure front. This is the same energy source for clump motions as in the previous picture of long-lived clumps. In either case, the whole cloud can last much longer than the internal crossing time.

2. FILAMENTS

Except for self-gravitating regions in the interstellar medium, much of the large-scale, morphologically distinct structure seen in HI or at 100 μm seems to be filamentary. This is apparently true also for shell-like structures, which are in fact either entirely filamentary in bulk or are shell-like but composed of filaments (Heiles 1989). Even many of the HI and cirrus knots or clumps could be parts of filaments, viewed along the filament axes (Verschuur 1991a,b). The relative fraction of the diffuse gas in the form of filaments as opposed to sphere-like clumps or hollow shells is unknown, but it could be larger than a casual glance at cirrus maps or velocity-channel maps would suggest. A discussion of several filament formation mechanisms is in Lazarian (1993).

If a curved filament looks like a cloud when the filamentary axis is on the line of sight, then the column density contrast between the apparent cloud and the rest of the filament to the side is determined from geometry to be

$$\frac{N_{\rm cloud}}{N_{\rm filament}} = \left(\frac{2R_{\rm curvature}}{R_{\rm cloud}} - 1\right)^{1/2}. \tag{1}$$

Verschuur (1991a) measured $N_{\rm cloud}/N_{\rm filament} \sim 5$ to 10, so $R_{\rm curvature}/R_{\rm cloud} \sim 25$. This value of 25 is also approximately the ratio of the curvature to the radius of the filaments mapped by Verschuur (1991a) in his figure 5, so the identification of the clouds with kinked filaments seems reasonable. To be sure that a diffuse cloud is really an isolated globular feature and not part of a filament, the density contrast between the cloud and the surrounding gas that connects to it has to be very high, perhaps 100 or more. Most cloudy objects are not truly isolated from other gas with density contrasts this large unless they are self-gravitating clouds.

Some filaments viewed from the side show sinusoidal velocity structure (Shuter et al. 1987; Heiles 1988; Verschuur 1991a,b). If the field lines are parallel to the filament, then this velocity structure can result from magnetic waves where the restoring force is magnetic tension, as in an Alfvén wave. In this case the wave travels along the filament and the point of maximum velocity occurs at the same point on the filament where the geometric distortion has its greatest extent, but this velocity maximum is out of phase with the positional displacement, i.e., the velocity is largest (and the tension smallest) when the filament is straight and the velocity is smallest or zero (and the tension largest) at the time when the filament has its greatest displacement.

If the field lines are perpendicular to the filament, then a sinusoidal velocity structure can arise from compressional magnetic waves travelling along the filament, provided the filament is viewed at an angle so the line of sight picks up a component of this parallel velocity. In this case, the filament can remain nearly straight during the whole oscillation (although there could be perpendicular oscillations at the same time), and there will be a correlation between density and velocity along the filament. The same field line orientation could also allow a non-magnetic motion perpendicular to the filament that is driven by an external pressure gradient. Such motions are observed in the L204 filament studied by McCutcheon et al. (1986). In this case, the velocity is greatest where the filament displacement is greatest as if all parts of the filament began to move sideways at the same time and the lowest density regions moved the fastest. This is reminiscent of a Rayleigh-Taylor instability, which would distort a pressurized filament in the same way, although it could also result simply from a variable *initial* column density along the filament, rather than an instability (see below).

2.1 Structure from Rayleigh-Taylor and Kelvin-Helmholtz Instabilities

The Rayleigh-Taylor instability (RTI) operates when external pressures accelerate a cloud. Such accelerations should be common in the ISM. Consider first a case with no magnetic fields. When an initially flat and uniform cloud or filament is accelerated as a whole by an external pressure gradient, the acceleration will have a value that is equal to the pressure difference across the cloud divided by the mass column density. If there is a small perturbation in column density or pressure, then the acceleration will vary with position and the cloud will deform. This deformation can do either of two things depending on the type of pressure. If the accelerating pressure is locally isotropic and perpendicular to the surface everywhere (as would be the case for a thermal pressure), then the deformation of the cloud causes the external pressure to pinch in at the convex parts and push out at the concave parts, leading to a transverse flow of gas inside the layer that adds mass to the convex, or lagging, parts. Another way to say this is that the cloud gas falls into the valley of the deformation (the convex part) because of the effective gravity from the acceleration. The result of this gas motion is that the lagging convex parts of the layer acquire a higher column density and so accelerate even less than they did with the initial perturbation alone, and the leading concave parts decrease in column density and accelerate more than they did initially. Thus the deformation gets larger until the external pressure eventually pokes a hole or drives a bubble through the cloud.

A second possibility arises when the pressure is mostly a ram pressure from accreting material, and the accreting side is a shock front. Then the flow inside the layer is away from the convex part and into the concave part (e.g., Kimura and Tosa 1988) as dictated by the flow in an oblique shock, although strong self-gravity in the layer can lead to a net inflow at some dense region even if there is an outflow close to the shock (Yoshida and Habe 1992). The flow away from the convex part tends to remove the initial perturbation in column density, but it does not necessarily stabilize the cloud. The increasing column density in the concave part will cause it to accelerate more slowly, and eventually it becomes the convex part; similarly the convex part from the first growth of the perturbation loses matter, decelerates more quickly and becomes the concave part. In this way, the layer can oscillate, and if there is enough time, the oscillation amplitude can grow (Vishniac 1983; Ryu and Vishniac 1987; Elmegreen 1989). This is an overstability, but it may be damped in the non-linear regime and not lead to a puncture or other major structural change in the layer (Mac Low and Norman 1993).

The McCutcheon et al. (1986) observation discussed at the end of Section 2.0 raises an interesting point about RTIs in supersonic flows. If the pressure gradient that drives an RTI is so large that the motions parallel to the gradient quickly become supersonic, and if the transverse motions in the pressurized layer remain slow because of jump conditions or cooling at the shock, then these transverse motions may not be fast enough to sustain the instability or give it much growth. The familiar RTI for subsonic flows grows so easily because the transverse flow that changes the column density is essentially infinitely mobile, having a maximum flow speed (nearly sonic) that is much larger than the deformation speed. For the supersonic RTI, the initial density irregularities and their amplitudes may determine the extent and shape of the deformation more than any exponentially growing amplitudes from infinitesimal perturbations. The result is then simply an initial value problem rather than an instability.

When a magnetic field is present, the RTI will still operate when there is a pressure gradient across a cloud, and this can produce major deformations or punctures in the cloud both parallel and perpendicular to the field (depending on field orientation). The magnetic RTI should also grow in the presence of a purely gravitational acceleration from the rest of the cloud or from the galactic plane. The growth consists of an increasing collection of material in the lagging or lower potential regions, a decrease of material in the leading or higher potential regions, a deformation of the field lines to follow this material, and a differential acceleration in the lagging and leading parts, or the lower and upper parts in the case of real gravity.

The magnetic RTI differs from the non-magnetic RTI in the direction parallel to the field when this direction is perpendicular to the acceleration. Recall that in the non-magnetic case the transverse gas motions that lead to the change in local column density and acceleration are caused by the pinch and expansion pressures at the convex and concave surfaces, respectively. In the magnetic case the pinch and expansion occur throughout the whole volume of the cloud and not just at the surface because of gradients in the magnetic pressure and tension from curved field lines. Sometimes the driving force for this volume effect is called magnetic buoyancy because the field lines can exert a pressure without any associated mass and so they can be visualized as a lighter fluid rising up to poke holes in the heavier fluid, which is the gas.

There are several important points about RTIs that are relevant to the present discussion of interstellar diffuse clouds. First of all, gravitational forces are commonly resisted by magnetic forces in the interstellar medium, and gas in various forms is commonly accelerated by pressure gradients, so the RTI should be widespread, perhaps even pervasive to the extent that most gas in both self-gravitating clouds and diffuse clouds is constantly reacting to an RTI. On the scale of the galactic thickness, the RTI contributes to the formation of giant cloud complexes (Parker 1966; Mouschovias, Shu and Woodward 1974; Elmegreen 1982; Matsumoto et al. 1990; Elmegreen 1991a; Shibata and Matsumoto 1991; Nakamura, Hanawa and Nakano 1991; Hanawa, Nakamura and Nakano 1992; Hanawa, Matsumoto and Shibata 1992; Giz and Shu 1993) and to the onset of turbulence (Asseo et al. 1978). Inside individual clouds, self-gravity can lead to deformations of the field and gas and the formation of clumpy structure parallel to the field, as shown for magnetic, self-gravitating filaments by Nakamura et al. (1993). The RTI can presumably drive turbulence inside clouds also. In C-type shock fronts with a transverse magnetic field, the accelerating escape of field lines through the partially ionized gas away from the driving pressure can lead to an analogous instability ahead of the front even at a constant shock velocity (Wardle 1990). Diffuse clouds are particularly prone to the RTI because of pressure variations in their environment.

The second point about the RTI is that in many cases it should form thin clumpy filaments rather than randomly positioned clumps. The magnetic RTI forms structure on a scale parallel to the magnetic field that is given by $k^{-1} \sim v_{\rm A}^2/g$, where k is the wavenumber, $v_{\rm A} = B/(8\pi\rho)^{1/2}$ is the Alfvén speed, and g is the acceleration from gravity or a pressure gradient, in which case $g = \Delta P/\sigma$

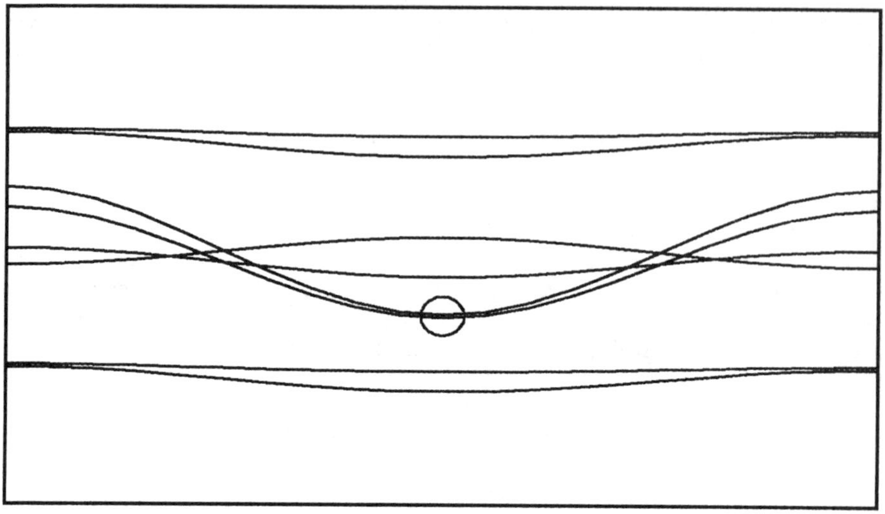

FIGURE 2 A single filament separating from other gas as a result of shear-like motions in the Rayleigh-Taylor instability. This image views, from an angle 20 degrees above the plane, hypothetical field lines with vertical and horizontal displacements (based on Parker 1967).

for external pressure difference across the cloud ΔP and mass column density σ. The length scale perpendicular to both the magnetic field and gravity is c^2/g for local sound or turbulent speed c because the field offers no resistance to shear-like motions in this perpendicular direction. This difference between the parallel and perpendicular length scales ($c << v_A$) implies that an intercloud wind should make thin filaments that are parallel to the field in the cloud. A schematic diagram of such a filament is in Figure 2.

After some substitutions, the characteristic length parallel to the field for pressure-driven accelerations is

$$k^{-1} \sim \frac{v_A^2}{g} \sim \frac{\sigma_{\rm ave} v_A^2}{\Delta P} \sim \frac{\sigma_{\rm ave} v_A^2}{P} \sim \frac{\sigma_{\rm ave} v_A^2}{\rho_{\rm ave} v_A^2} \sim \frac{\sigma_{\rm ave}}{\rho_{\rm ave}} \sim R \qquad (2)$$

Here the pressure differential across the cloud is denoted by ΔP; this will be approximately equal to the internal cloud pressure P when it is large, and P is approximately equal to the average density times the square of the Alfvén speed. Actually P should be set equal to $\rho_{\rm ave}$ times the sum of v_A^2 and the summed squares of the turbulent and sound speeds, but the magnetic energy is often comparable to the turbulent energy in pressurized regions, and greatly exceeding the thermal energy, so this use of v_A^2 alone is probably good to within a factor of 2. The whole cloud radius is R. This expression indicates that the magnetic RTI should make filaments that permeate the whole cloud (i.e., extend for a length comparable to the length of the field lines in the cloud to within a factor of 2).

Perpendicular to the field, the length for shear motions and field line separation in the RTI is

$$k^{-1} \sim \frac{c^2}{g} \sim \frac{\sigma_{\rm ave}c^2}{\Delta P} \sim \frac{\sigma_{\rm ave}c^2}{P} \sim \frac{\sigma_{\rm ave}c^2}{\rho_{\rm local}c^2} \sim \frac{\sigma_{\rm ave}}{\rho_{\rm local}}. \tag{3}$$

Here we have used the substitution $P \sim \rho_{\rm local}c^2$ because the average pressure on a large scale is likely to be about the same as the pressure on a small scale. The length perpendicular to the field is shorter than the length parallel because the characteristic velocity for pressure and momentum transfer is much less perpendicular (using c) than parallel (using $v_{\rm A}$) to the field for shear motions. For compressive motions, both velocities are about the same ($v_{\rm A}$), and the two length scales are about the same, but the RTI is essentially driven by shear-like motions.

Equation (3) suggests that the characteristic length for structure perpendicular to the field is inversely proportional to the local density, which can vary from point to point depending on the local thermal or turbulent speed c. This implies that the condensations that form in a RTI might have a correlation between size and density in the sense that *size times density = constant.* The constant of proportionality, $\sigma_{\rm ave}$, suggests that the (filamentary) clumps formed by the RTI have an *angular filling factor of about unity.* These properties of clumps in clouds are occasionally observed. For the RTI, they can be present without self-gravity. They would probably follow from other instabilities also, as long as they are driven by pressure and resisted by inertial forces or other forces that are proportional to the inertial force.

The characteristic lengths for gravity-driven RTIs are $k^{-1} \sim v_{\rm A}^2/g$ and c^2/g for parallel and perpendicular directions, but in these cases there are no local density dependences or size-density relations. For virialized c, this relation is simply $k^{-1} \sim R$. However, the virial theorem alone gives nk^{-1} = constant for a uniform pressure boundary because then $\sigma \sim (P/G)^{1/2}$ is about constant so $M \propto R^2$ and $\rho \propto 1/R$ for any scale R at which the virial theorem is satisfied.

A cloud that is structured by RTIs would not necessarily look like the usual model of an RTI at the onset of its growth, i.e., consisting of long tongues of material pointing in the direction of the pressure gradient or gravity. If the RTI has been operating for a long time, particularly in the presence of self-gravity and virialized motions, it should develop over a wide range of scales repeatedly and may simply drive turbulence, producing random structures that are typical of turbulence.

Kelvin-Helmholtz instabilities (KHI) should be important when the intercloud medium moves past the cloud in a wind or post-shock flow. This is the same situation that drives the RTI, so KHI should be ubiquitous for diffuse clouds too. KHIs occur at the edges of diffuse clouds where the intercloud wind passes parallel to the cloud surface. The RTI deforms the edge where the wind is perpendicular to the surface. Both the KHI and the RTI will tear a cloud to pieces, possibly generating a turbulent clumpy structure (e.g., Klein et al. 1990). Because they are both related to pressures and operate with shear-like motions rather than compressive motions, the scaling arguments given above should apply to both. The KHI has been applied to interstellar matter in a variety of ways (e.g., Fleck 1983, 1989; Hunter and Whitaker 1989; Mac Low et al. 1993; Murray et al. 1993).

3. STRUCTURE FROM ROTATION AND HELICAL FIELDS

Rotating clouds or pieces of clouds can twist the intervening field lines and pinch the material into a filament. Observations of twisting filaments were reported by Uchida et al. (1990,1991). The confinement pressure from a helical field with radius R and wavelength λ for the twist is

$$P = \frac{B \cdot \nabla B}{4\pi} \sim -\frac{\pi R B_z^2 \mathbf{e}_R}{\lambda^2} \tag{4}$$

for average field strength parallel to the cloud B_z. This confinement pressure exceeds the expansion pressure from $\nabla B^2/(8\pi)$ when $\lambda < 2\pi R$, which implies that only a tightly wrapped helix can be confined by the field.

Two clouds with radius R and separation L have to turn relative to each other $L/2\pi R$ times to pinch significantly the intervening field lines. If L is the cloud-to-cloud separation on a field line, which is probably comparable to the collision mean free path $1/(n\pi R^2)$, then

$$\frac{L}{2\pi R} \sim \frac{2}{3\pi f} \sim 4 \tag{5}$$

for volume filling factor $f \sim 5\%$. Thus typical interstellar clouds have to turn relative to each other 4 times to pinch the intervening field lines and make a filament.

Do such field line twists occur very often? The field line tension resists significant twisting so perhaps the formation of filaments by field line twisting will not happen very often. We can estimate the condition for filament formation from the equation of motion of two clouds on the same magnetic flux tube. The force per unit volume that is exerted on a thin disk of gas through which the field lines are directed is $B \cdot \nabla B/4\pi$, which is approximately a force per unit area $B_\theta B_z/(4\pi)$ for field components B_z along the average direction between the clouds and B_θ in the azimuthal direction around the cylinder that connects the clouds. The total torque on the cloud is this average force per unit area integrated over the area $2\pi r dr$ and lever arm r. The moment of inertial of the cloud is approximately the mass surface density integrated over the area $2\pi r dr$ times r^2. Thus the equation of motion, which sets the angular acceleration equal to the torque divided by the moment of inertia, is

$$\frac{d^2\theta}{dt^2} \approx \frac{B_z B_\theta}{3\pi\sigma R}. \tag{6}$$

This equation assumes that the field strengths are uniform over the face of the cloud and that the field lines that emerge from the cloud are not divergent.

There are two basic types of solutions for this torsional motion, exponential and sinusoidal, depending on how much of the angular momentum goes into the intercloud medium. If a cloud just begins to spin, then the surrounding gas will spin up with it and the cloud will slow down exponentially. In this case, $B_\theta \sim -B_z R(d\theta/dt)/v_{A,i}$ for intercloud Alfvén speed $v_{A,i}$. Thus the solution for the motion is an exponential decay from the initial velocity with a decay rate of approximately $B_z^2/(3\pi\sigma v_{A,i})$.

The sinusoidal solution occurs if all of the intercloud gas between two clouds on a magnetic flux tube spins with the clouds and already has all of the angular momentum it can get. Then $B_\theta \sim -B_z\theta R/L$ for cloud separation L, and the oscillation frequency is $\omega \sim B_z/(3\pi\sigma L)^{1/2}$. This oscillatory solution should be common when the mass in clouds greatly exceeds the mass in intercloud gas (actually, when the summed moments of inertia of the clouds exceeds the total moment of inertial of the intercloud medium that is connected to these clouds on magnetic flux tubes).

The maximum angle of the wrapping of the field around the helix is given by the initial angular velocity of the cloud divided by ω, and this maximum wrap angle gives the ratio between the θ and axial components of the field:

$$\begin{aligned}\frac{B_\theta}{B_z} &= \frac{\theta_{\rm wrap}R}{L} = \frac{(d\theta/dt)_0}{B_z}\left(\frac{3M_{\rm cloud}}{L}\right)^{1/2} \approx \left(\frac{E_{\rm rot}}{E_{\rm B}({\rm betweenclouds})}\right)^{1/2} \\ &= 1.4\left(\frac{(d\theta/dt)_0}{1{\rm kms}^{-1}{\rm pc}^{-1}}\right)\left(\frac{B_z}{10^{-6}{\rm Gauss}}\right)^{-1}\left(\frac{M_{\rm cloud}}{M_\odot}\right)^{1/2}\left(\frac{L}{\rm pc}\right)^{-1/2}\end{aligned} \tag{7}$$

where $E_{\rm rot}$ is the rotational energy of the cloud and $E_{\rm B}$ is the energy in the magnetic field between the clouds. (This identification of B_θ/B_z with the square root of the energy ratio depends on the geometric structure of the field lines between the clouds, because the field strengths are really energy densities. We assume here as an approximation that the field lines have only θ and z components between the clouds; i.e., that they do not bow out).

For the parameter values chosen in equation (7), the gas between two rotating clouds can be significantly pinched so that it makes a filament. The ratio of tension (inward) to compression (outward) depends on the ratio B_θ/B_z, and is

$$\frac{\rm tension}{\rm compression} = \frac{2(B_\theta/B_z)^2}{1+(B_\theta/B_z)^2}. \tag{8}$$

If the rotational energies of two clouds on a flux tube are comparable to the magnetic energy between the clouds, then the field lines can become twisted so much that they pinch down the material between the clouds and make a filament.

Such pinching would seem to be important in a turbulent medium where the kinetic energy density, which is divided and constantly transforming between rotational and compressional energy of the gas, is comparable to the magnetic energy. Then turbulent vortices should make magnetic filaments with helical fields, and the rotational energy of the turbulence should oscillate between the kinetic energy of spinning gas and twisted field line energy. Indeed, Verschuur pointed out to me at this conference that the IRAS cirrus picture used for the conference announcement contains a filament between two bright clouds, and perhaps this is an example of the magnetic twist. If it is, then observations should show a rotation of the two clouds relative to each other with a rotational axis in the direction of the filament, which should also be the direction of the average magnetic field in the filament.

4. PARALLEL AND PERPENDICULAR FIELD ORIENTATIONS

Magnetic fields are generally more cohesive for shear-like motions parallel than perpendicular to the field direction, so they tend to generate filaments, as discussed in Section 2a. These filaments should have nearly parallel field structures *if they are young.* Even helical fields should appear to have a parallel field structure if the front and back components of the helix cancel in the averaging process during the observations.

Pressure-swept layers with $P >> B_0^2/8\pi$ for ambient field strength B_0 will usually put the magnetic field nearly parallel to the layer. Then the layer can become filamentary parallel to **B** because of shear-type (e.g., non-compressive) instabilities in the pressure front. Gravitational instabilities in the front might also pull the field lines together and make filaments (which then collapse into globules). An example of a filament with a nearly parallel field orientation is in Cygnus (McDavid 1984). The Ophiuchus filament is another example (Vrba, Strom and Strom 1976); in this case the cloud looks like a comet with a filamentary tail formed by external pressures from the Sco-Cen association.

Sometimes the filamentary structure is perpendicular to the field direction, as in Taurus (Moneti et al. 1984) and L204 (McCutcheon et al. 1986). This orientation can arise when the gas is swept up parallel to **B** with a relatively low pressure, $P < B_0^2/8\pi$. This process really forms sheets which may appear to be filaments when seen edge-on or which may break up into filaments later. It is possible that P from a sweeping pressure disturbance is rarely less than $B_0^2/8\pi$ in the ambient medium, so another explanation for perpendicular field structure seems required.

The most important rule for field orientations inside clouds and filaments is that the field always wants to become straight as it joins with the background field. Magnetic tension, which comes from the $B \cdot \nabla B/4\pi$ term in the equation of motion for a magnetic fluid, tends to straighten the field. This implies that cloud orientations, which generally have some fixed relation to the direction of the pressure source that made the cloud (e.g., a supernova), and field orientations, which are governed primarily by the background field at late times (after relaxation to straightness) should be *unrelated.* Thus the field inside a relatively old or quiescent cloud or filament should have the same direction as the far distant field regardless of the orientation of the cloud. This preference for background orientation allows regions with many irregular filaments to have a nearly uniform field structure, as observed by Goodman et al. (1990).

A schematic diagram of initially parallel field orientations followed by a straightening with the tensional force, is shown in Figure 3. On the left, a cloud that is made by compression initially has a near-parallel field structure and an orientation determined mostly by the pressure source. After the field straightens out because of its tensional force (on the right), the field becomes nearly parallel to the background field but the cloud orientation need not change much, especially if kinked field lines emerge from the cloud at all positions along its length, as in a shock front, rather than at only the ends, as in a twisted filament.

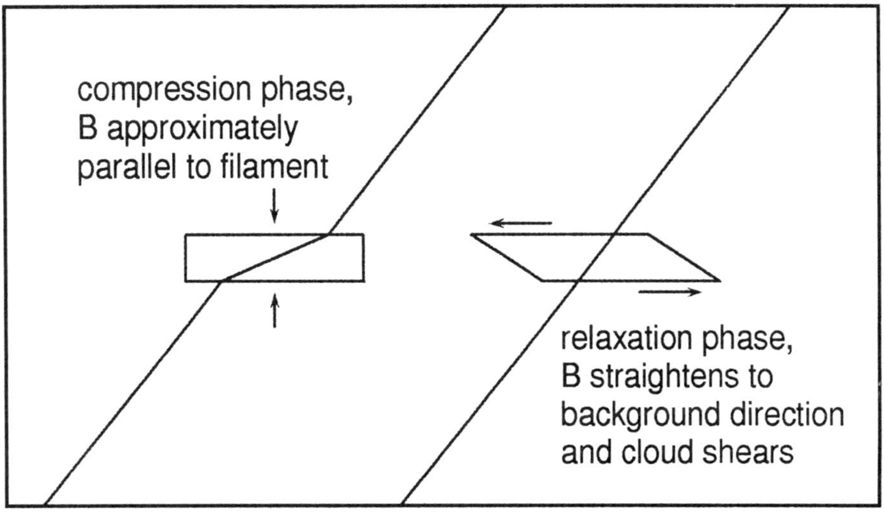

FIGURE 3 The tendency for magnetic field lines to straighten relative to the background field orientation is shown. A young region is on the left and an old region on the right.

5. STRUCTURE FROM STRONG MAGNETIC WAVES

5.1. Supersonic turbulence and density structure

Strong magnetic and sonic waves should steepen to make density spikes and they should interact to make dense splash regions in the converging flow (Elmegreen 1990; Carlberg and Pudritz 1990; Fatuzzo and Adams 1993). We refer to the region of increasing density between two approaching flows as a splash region in analogy with the splash of water when kinetic energy is converted into potential at the convergent point. For a compressible gas, the kinetic energy of the convergent flow is converted into compressional energy in the density spike.

The peak density is given approximately by pressure equilibrium for strong magnetic waves: $\rho_{\rm peak} \sim \rho_{\rm ave}(v_{\rm A}/c_{\rm local})^2$, where $c_{\rm local}$ is the local turbulent speed. Smaller densities should occur in regions with oblique convergence of turbulent eddies.

A region with supersonic turbulence should have time variable clumpy structure because the supersonic motions are continuously transformed into compressional energy at the convergent points in the flow, and then the pressure gradients in these clumps transform the compressive energy back into motion. Clump confinement is then from parallel wave pressure or ram pressure, and not from a warm static interclump medium. The interclump medium, which still must be present, can be cold, but the important point is that it is moving at a high speed, possibly supersonically.

Clumps that are formed by a converging flow persist for a time equal to the travel time *between* the clumps, and not the wave travel time inside the clumps. This was shown by a 1-D simulation in Elmegreen (1991b). This time scale implies that clouds whose large internal motions make them look unbound and short-lived can in fact be bound by the convergent flow around them (or by the motion of the clouds through the ambient medium – see Section 1b), and they can be relatively long-lived, for as long as the convergent flow or relative motion lasts. This long life could be several million years, instead of several hundred thousand years, in the case of the high-dispersion cirrus and high latitude clouds.

The inevitable density gradients around tiny clumps or self-gravitating cores in a supersonically turbulent region should correspond to regions with varying streaming velocities, scaling perhaps as $v \propto \rho^{-1/2}$ (e.g. Elmegreen 1990; Fatuzzo and Adams 1993). This correspondence between density and streaming velocity suggests that broad line wings in clouds with supersonic turbulence may result from emission in the lower (average) density regions (Elmegreen 1990). Note that the CO(2-1)/CO(1-0) ratio is about the same in the low average density regions as in the high average density regions (Falgarone and Phillips 1991), so the local density, where the gas is actually emitting, may be high everywhere, but the density of the smeared out clumps, which is the same as the density averaged over a wavelength, can vary with position by several orders of magnitude.

With this density-velocity correlation, the core-wing line profiles that are observed in molecular clouds (Falgarone and Phillips 1990) are the result of a combination of a Gaussian core from the vortical motions or incompressible part of the turbulence (as is typical for incompressible turbulence in the laboratory), plus whatever Gaussian structure is in the individual clumps, and a spectral wing emission from the higher velocities in the lower density regions. The observed profiles were fit to this composite structure in Elmegreen (1990), assuming that the systematic wave or streaming velocity scales inversely with some power of the density and that the density varies with position inside the telescope beam. These characteristic core-wing line profiles were also found in the recent 2D magneto-hydrodynamic simulations by Yue et al. (1993).

In general, there should be a superposition of rotational energy and compressional energy inside a supersonically turbulent region, and so a superposition of these two velocity distributions, Gaussian and Gaussian+wing, gives the composite line profile. Regions with more compressional energy and proportionally less rotational energy should have relatively strong line wings, and regions with more of the rotational than compressional energy should have relatively weak line wings. In all cases, the ratio of the wing width to the Gaussian core depth is about 3, as given by Elmegreen (1990).

We might expect that much of the stirring in an interstellar cloud is from explosive or compressive events, such as self-gravity, stellar pressures, intercloud pressures, and so on, which implies that the energy input is largely in the form of compressive energy. Rotational energy can also be applied to a cloud by the twist mechanism discussed in Section 3, or in the shear region of an oblique shock, or in some other way. If a region is left alone after this initial energy input, then the compressive energy presumably dissipates and flows into rotational energy, as shown in the Porter et al. (1992) simulations discussed at this meeting by E. Falgarone. Thus the relative amplitude of the line wings should reveal something about the time since the compressive turbulent energy was applied. Young regions with recent compressional stirring should have relatively high intensity line wings, and old regions without much compressional energy available should have weaker wings. There should also be a simultaneous variation of the core-intercore density ratio, which should decrease with time as the turbulence subsides (i.e., becomes sub-sonic). As this core-intercore density ratio decreases, the line wings become stronger but less broad until a Gaussian is reached. Thus the relative strength of the wings should be a measure of how much the cloud is being agitated by compressive or self-gravitational forces.

5.2. Minimum Alfvén wavelengths

The minimum size of a magnetically formed density feature is the ratio of the Alfvén speed divided by the the ion-neutral collision frequency (Elmegreen and Fiebig 1993):

$$k^{-1} = \frac{0.5v_{\rm A}}{< n_{\rm g}\sigma_{\rm g} > c + n_{\rm i} < \sigma_{\rm in}c >} \approx \frac{0.006{\rm pc}}{n_4(1+0.75n_4^{-1/2})} \tag{9}$$

where the numerical evaluation is for the Mathis, Rumpl, and Nordsieck (1977) grain size distribution, $n(a) \propto a^{-3.5}$ between 50 Ångstroms and 0.1 microns, and it assumes that all grains gyrate and that $B \propto n^{1/2}$ giving $v_{\rm A} \sim 1$ kms^{-1}. The length scale given here is very small, comparable to the smallest scales yet observed, and it is also decreasing with increasing density as $1/n$, which is similar to the way the size of a density feature decreases with increasing density. Thus the number of minimum wavelengths inside a density feature is about constant, given with these approximations by

$$Rk_{\rm max} \approx 16\left(1+0.75n_4^{-1/2}\right). \tag{10}$$

Actually, the minimum length ($k_{\rm max}^{-1}$) is larger than this because the largest grains cannot gyrate around the field (viscous forces exceed the magnetic forces). A better assessment of the minimum length was calculated by Elmegreen and Fiebig (1993). For the MRN grain distribution, which has very tiny grains, some grain coupling to the field is present for all densities up to 10^{10} cm^{-3}, and magnetic waves can be a factor of 10 smaller than the structure scale. Thus non-linear magnetic waves and wave-related compressions can create, or be important for, even the finest structures that are observed in diffuse and cirrus clouds, wherever tiny grains exist.

Note that the product $Rk_{\rm max}$ is about constant for a wide range of densities, which implies that all the associated scales have about the same range of spatial frequencies in Fourier (k) space. Thus they can have the same structural properties regardless of scale.

Magnetic coupling to neutral matter is even stronger at length scales smaller than the Jeans length, where clumps are unbound, because then it is likely that the magnetic field strength stops increasing as $n^{1/2}$ but becomes more constant. In that case $v_{\rm A}$ decreases as $n^{-1/2}$ and $Rk_{\rm max}$ increases as $n^{1/2}$. Large $Rk_{\rm max}$ implies strong coupling.

REFERENCES

Asseo, E., Cesarsky, C.J., Lachieze-Ray, M. & Pellat, R. 1978, *ApJ*, **225**, L21

Carlberg, R.G. & Pudritz, R.E. 1990, *MNRAS*, **247**, 353

Clifford, P. & Elmegreen, B.G. 1983, *MNRAS*, **202**, 629

Elmegreen, B.G. 1982, *ApJ*, **253**, 655

Elmegreen, B.G. 1985, in *Protostars and Planets II*, ed. D.C Black & M. S. Matthews, (Tucson: University of Arizona Press), p.33

Elmegreen, B.G. 1988, *ApJ*, **326**, 616

Elmegreen, B.G. 1989, *ApJ*, **340**, 786

Elmegreen, B.G. 1990, *ApJ*, **361**, L77

Elmegreen, B.G. 1991a, *ApJ*, **378**, 139

Elmegreen, B.G. 1991b, in *Physics of Star Formation and Early Stellar Evolution*, ed. C.J. Lada & N. Kylafis, Kluwer Academic Publishers, Dordrecht, p. 35

Elmegreen, B.G. 1993a, in *Protostars and Planets III*, ed. E.H. Levy & J.I. Lunine, University of Arizona Press, Tucson, 1993, p. 97

Elmegreen, B.G. 1993b, *ApJ*, **411**, 170

Elmegreen, B.G. & Fiebig, D. 1993, *A&A*, **270**, 397

Falgarone, E. & Phillips, T.G. 1990, *ApJ*, **359**, 344

Falgarone, E. & Phillips, T.G. 1991, in *Fragmentation of Molecular Clouds and Star Formation*, eds. E. Falgarone, F. Boulanger & G. Duvert (Dordrecht: Reidel), p. 119.

Fatuzzo, M. & Adams, F.C. 1993, *ApJ*, **412**, 146

Fleck, R.C., Jr. 1983, *ApJ*, **270**, 507

Fleck, R.C., Jr. 1989, *AJ*, **97**, 783

Giz, A.T. & Shu, F.H. 1993, *ApJ*, **404**, 185

Goodman, A.A., Bastien, P., Myers, P.C. & Menard, F. 1990, *ApJ*, **359** 363.

Hanawa, T., Matsumoto, R. & Shibata, K. 1992, *ApJ*, **393**, L71

Hanawa, T., Nakamura, F. & Nakano, T. 1992, *PASJ*, **44**, 509

Heiles, C. 1988, *ApJ*, **324**, 321

Heiles, C. 1989, *ApJ*, **336**, 808

Hunter, J.H., Jr. & Whitaker, R.W. 1989, *ApJS*, **71**, 777

Joncas, G., Boulanger, F. & Dewdney, P.E. 1992, *ApJ*, **397**, 165

Kimura, T. & Tosa, M. 1988, *MNRAS*, **234**, 51

Lazarian, A. 1993, *MNRAS*, submitted.

Klein, R.I., McKee, C.F. & Collela, P. 1990, in *The Evolution of the Interstellar Medium*, ed. L. Blitz, San Francisco, Astronomical Society of the Pacific, p. 117

Mac Low, M.-M., Klein, R. I., McKee, C. F., Norman, M. L. & Stone, J. M. 1993, in Evolution of Interstellar Matter and Dynamics of Galaxies, ed.

B.W. Burton, P.O. Lindblad & J. Palous (Cambridge: Cambridge U. Press), p. 93.

Mac Low, M.-M. & Norman, M. L. 1993, *ApJ*, **407**, 207

Magnani, L. A., Blitz, L. & Mundy, L. 1985, *ApJ*, **295**, 402

Mathis, J.S., Rumpl, W. & Nordsieck, K.H., 1977, *ApJ*, **217**, 425

Matsumoto, R., Horiuchi, T., Hanawa, T. & Shibata, K. 1990, *ApJ*, **356**, 259

McCutcheon, W.H., Vrba, F.J., Dickman, R.L. & Clemens, D.P. 1986, *ApJ*, **309**, 619

McDavid, D. 1984, *ApJ*, **284**, 141

Moneti, A., Pipher, J.L., Helfer, H.L., McMillan, R.S. & Perry, M.L. 1984, *ApJ*, **282**, 508

Mouschovias, T. Ch., Shu, F.H. & Woodward, P. 1974, *A&A*, **33**, 73

Murray, S.D., White, S.D.M., Blondin, J.M. & Lin, D.N.C. 1993, *ApJ*, **407**, 588

Nakamura, F., Hanawa, T. & Nakano, T. 1991, *PASJ*, **43**, 685

Nakamura, F., Hanawa, T. & Nakano, T. 1993, *PASJ*, **4**, 1.

Parker, E.N. 1966, *ApJ*, **145**, 811

Parker, E.N. 1967, *ApJ*, **149**, 517

Porter, D.H., Pouquet, A. & Woodward, P.R. 1992, *Theor.Comp.Fluid Dyn.*, **4**, 13

Ryu, D. & Vishniac, E.T. 1987, *ApJ*, **313**, 820

Shuter, W.L.H., Dickman, R.L. & Klatt, C. 1987, *ApJ*, **322**, L103

Shibata, K. & Matsumoto, R. 1991, *Nature*, **353**, 633

Uchida, Y., Mizuno, A., Nozawa, S. & Fukui, Y. 1990, *PASJ*, **42**, 69

Uchida, Y., Fukui, Y., Monishima, Y., Mizuno, A., Iwata, T. & Takaba, H. 1991, *Nature*, **349**, 140

Verschuur, G. 1991a, *Ap&SS*, **185**, 305

Verschuur, G. 1991b, *Ap&SS*, **185**, 137

Vishniac, E.T. 1983, *ApJ*, **274**, 152

Vrba, F.J., Strom, S.E. & Strom, K.M. 1976, *AJ*, **81**, 958

Wardle, M. 1990, *MNRAS*, **246**, 98

Yoshida, T. & Habe, A. 1992, *Prog.Theor.Phys.*, **88**, 251

Yue, Z.Y., Zhang, R., Winnewisser, G. & Stutzki, J. 1993, *Ann.Phys.*, **2**, 9

The First Symposium on the Infrared Cirrus and Diffuse Interstellar Clouds
ASP Conference Series, Vol. 58, 1994
Roc M. Cutri and William B. Latter (eds.)

THE RELATIONSHIP BETWEEN DIFFUSE AND DENSE GAS IN THE GALAXY

G. R. KNAPP
Department of Astrophysical Sciences, Princeton University, Princeton, NJ08544

ABSTRACT The spatial distribution of molecular and atomic gas, both globally in the Galaxy and on scales down to 0.1 pc in the local interstellar medium shows that three distinct states of the cold interstellar medium can be identified: dense, self-gravitating molecular clouds, diffuse molecular gas and diffuse atomic gas; these last two are not gravitationally bound but may be in pressure equilibrium. The denser the ISM the more likely it is to be molecular; the observations suggest that the maximum possible HI surface density is something like 2×10^{21} cm^{-2}.

1. INTRODUCTION

The gas in the interstellar medium exists in several distinct physical phases. The ionized gas (the hot coronal gas and the HII regions) contains a negligible fraction of the total mass of interstellar gas but has a thermal energy content which is comparable to that of the cold neutral atomic and molecular gas. This review discusses the spatial, chemical and evolutionary relationships of the cold neutral diffuse and dense gas on several scales in the Galaxy.

2. OBSERVATIONAL PROBES

Diffuse cold interstellar gas can be studied by several observational probes, including: (1) colors and spectroscopy of early-type stars; (2) the HI 21 cm line in emission and absorption (cf. Kulkarni and Heiles 1988). This line is probably the single most powerful probe of the diffuse interstellar medium, but its interpretation is not straightforward because the ubiquity of the Galactic HI makes disentangling its three dimensional structure very difficult, especially on small scales; (3) emission at 12 μm, 25 μm, 60 μm and 100 μm from interstellar dust observed by IRAS, which allows the possibility of mapping the distribution of interstellar dust and the gas to dust ratio. The interpretation of the IRAS data is difficult because emission at 60 μm and 100 μm is in the exponential régime since the dust temperature is about 18 K, and because much of the short wavelength emission arises from small grains which are intermittently heated to high temperatures by the absorption of single photons (Draine and Anderson 1985); (4) millimeter and submillimeter wavelength emission from several molecules, particularly CO. CO emission has been detected from interstellar clouds of surprisingly low mass and density (Magnani, Blitz and Mundy 1985; Knapp and Bowers 1988; Jannuzzi *et al.* 1988; Guhathakurta, this conference). While, in general, CO line emission is a reasonably good tracer of molecular gas in denser regions of the interstellar medium, the chemistry of the diffuse medium is dominated by photochemistry and molecular hydrogen can exist at lower total column densities than can CO because H_2, being more abundant, more easily self-shields than does CO (van Dishoeck and Black 1989). There is thus a range of column densities ($\sim 10^{20} - 10^{21}$ cm^{-2}) in the ISM where gas is potentially undetectable with either the CO or the HI line.

The volume filling factor of the diffuse atomic interstellar medium is generally thought to be about 0.5, although this conclusion may change radically when detailed maps of the structure of the HI become available (e.g. Joncas *et al.* 1992).

The dense neutral interstellar medium is probed by long-wavelength IR emission from dust and by emission in many molecular lines. The workhorse molecule is again CO. The observations show that the dense medium is in self-gravitating clouds ranging in mass up to $> 10^6$ $M_\odot$ (e.g. Sanders, Scoville and Solomon 1985) which contain about half of the total mass of the interstellar medium. The dense interstellar medium is overwhelmingly in molecular form.

The interpretation of CO observations is complicated by the high opacity of the CO rotational transitions and lack of detailed knowledge of the CO abundance, which make the calculation of the molecular hydrogen column density from CO line intensities uncertain. The interpretation is further complicated by the presence of at least two types of cloud, cold (T $\sim$ 5 K) and warm (T $\sim$ 20 K) (e.g. Chiar *et al.* 1993) where the higher temperatures are presumably due to the presence of embedded stars.

The observations summarized above suggest that we can identify four phases of the cold neutral medium: the HI clouds, the warm neutral medium, the diffuse molecular gas, and the dense molecular gas. Observations of the temperatures and densities of the diffuse atomic gas show that the HI clouds are not self-gravitating but are in pressure equilibrium with the warm atomic gas and with the hot coronal gas (McKee and Ostriker 1977; Kulkarni and Heiles 1988). The pressure in the diffuse gas is about $P/k \sim 500$ K cm^{-3} (e.g. Jenkins, Jura and Loewenstein 1983). The dense molecular gas (with $n(H_2) \geq 10^3$ cm^{-3} and M $>$ a few hundred $M_\odot$) is in self-gravitating clouds.

What about the diffuse molecular clouds? They appear not to be self-gravitating; the observed line widths are larger by factors of several than the values calculated from the observed cloud masses under the assumption that the clouds are in virial equilibrium (e.g. Magnani, Blitz and Mundy 1985). This is the case for all of the phases of the diffuse cold medium, whose internal velocity fields appear to be dominated by turbulent rather than thermal motion.

The diffuse atomic clouds, diffuse molecular clouds and the warm neutral medium all have values of $n(H)\sigma^2 \sim 15$ cm^{-3}(km s^{-1})^2, where n(H) is the density of hydrogen nuclei and σ is the one dimensional velocity dispersion of the observed line profile. This would correspond to a pressure $P/k \sim 2000$ K cm^{-3}, much higher than the thermal pressure in the diffuse gas but sufficient to support the gas layer at its observed scale height in the gravitational field of the Galactic disk (cf. Cox 1990; Lockman and Gehman 1991; Merrifield 1993). The relationships among gas pressure, gas motion, the vertical structure of the gas in the Galactic disk and the physical state of the gas are thus quite complex. This will be discussed further in Section 6.

3. THE LARGE SCALE GALACTIC DISTRIBUTION

The Galaxy-wide distribution of atomic and molecular gas can be calculated from detailed maps of the Galactic HI and CO (Weaver and Williams 1974a,b; Kerr *et al.* 1986; Gordon and Burton 1976; Cohen, Dame and Thaddeus 1986; Clemens *et al.* 1986; Sanders *et al.* 1986). These observations show that (outside the Galactic center region) the gas in the Galactic disk seems to be about 50% atomic and 50% molecular. The ratio of molecular to atomic gas is not constant but is higher where the total gas density is higher. Gordon and Burton (1976), for example, show that the total gas surface density $\Sigma(H_2) + \Sigma(HI)$ falls roughly exponentially with distance R from the Galactic center between $R/R_o = 0.4$ and 2 with about the same scale length as the mass of the Galactic disk. The ratio of the molecular to atomic gas falls rapidly with R, however, with most of the molecular gas lying between R/R_o = 0.4 to 0.7 (the 'molecular ring').

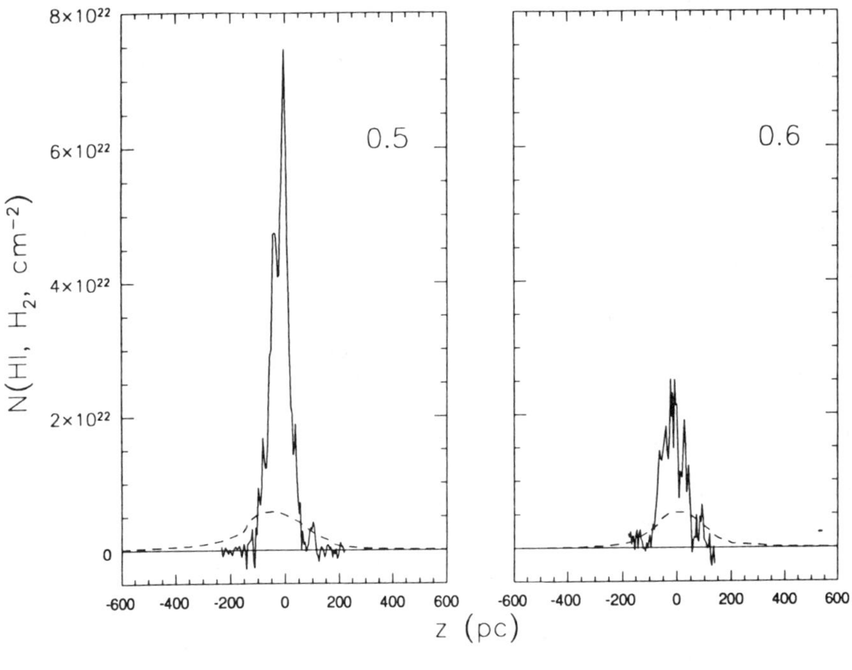

Fig. 1. HI (dashed line) and H_2 (solid line) column densities at R/R_o = 0.5 and 0.6 from emission at the tangent points at $l = 30°$ and $l = 36.9°$. The HI data are from the survey by Weaver and Williams (1974a,b). The H_2 column densities are calculated from the CO(1-0) data of Knapp, Stark and Wilson (1985).

Likewise, the z-distribution of the gas shows that the atomic gas layer is at least twice as thick as the molecular gas layer. This is illustrated in Figure 1, which shows the z-distribution of the atomic and molecular gas at two distances R/R_o from the Galactic center, found from observations of gas at the tangent points. The HI data are from Weaver and Williams (1974a,b) and the CO data from Knapp, Stark and Wilson (1985), where the H_2 column density is calculated from the CO line intensity I_{CO} using $N(H_2) = 2 \times 10^{21}$ I_{CO} cm^{-2}. (The observations were made with very different beam sizes so the numbers in Figure 1 should not be taken too literally).

These observations of the CO and HI distributions show a second important point, however; the HI density is approximately constant at 0.5 cm^{-3} throughout the Galaxy regardless of the total gas density. This suggests that there is a threshold density for the formation of molecular gas.

Sodroski *et al.* (1987) and Grabelsky *et al.* (1987) have examined the z-distribution of the CO, HI and 100 μm emission in the Galaxy and have shown that the 100 μm emission traces the gas and that the latitude distributions of the CO, HI and 100 μm emission show the same deviations from $b = 0°$ throughout the Galaxy. In particular, both the diffuse and dense gas layers trace the warping of the Galactic plane beyond R_o (cf. Kerr 1983).

Overall, then, the diffuse and dense gas in the disk of the Galaxy have the following properties; the HI and the CO have the same midplane deviations, the molecular gas dominates in the inner regions of the Galactic disk and at low latitudes, and the HI midplane density is constant at $\sim$ 0.5 cm^{-3} between 0.2 R_o and 1.1 R_o, beyond which it falls slowly and the gas is mostly atomic.

4. SPATIAL RELATIONSHIPS

The interstellar medium is exceedingly tangled. Recent observations show that the structure of the cold interstellar medium is better described by filaments and holes in a general distribution of gas than by clouds embedded in a lower density medium (Bregman *et al.* 1993). Both the atomic and molecular gas are tightly confined to the Galactic disk, but the atomic gas is more widely distributed in the z and R directions and the molecular gas is concentrated to the regions where the total gas density is highest.

Several authors (e.g. Wannier *et al.* 1983; Elmegreen and Elmegreen 1987; Chromey *et al.* 1989) have investigated the relationship between molecular and atomic gas on the scale of clouds and cloud complexes. These authors show that giant molecular clouds are often associated with large envelopes of atomic gas. Often, however, the velocity extent of the HI is larger than that of the molecular gas and the HI does not seem to be gravitationally bound.

Maps of the HI emission in the Galactic plane made with sufficiently high spatial and velocity resolution (a few arcminutes and better than 1 km s^{-1}) show large numbers of self-absorption features which closely coincide in both spatial and velocity extent with large molecular clouds (e.g. Bania and Lockman 1984). These maps show that the actual amount of HI in a molecular cloud is very small, a few percent at most. If the hydrogen in these clouds was originally in atomic form the clouds are several $\times 10^6$ years old (cf. the discussion of H_2 formation by Hollenbach and McKee 1979).

Detailed mapping of high latitude diffuse clouds has some chance of disentangling the detailed structure since the lines of sight are relatively simple, and this work has been done by several groups (e.g. de Vries *et al.* 1987; Désert, Bazell and Boulanger 1988; Boulanger and Pérault 1988; Heiles, Reach and Koo 1988; Guhathakurta and Tyson 1989; Herter, Shupe and Chernoff 1990; Joncas *et al.* 1992; Verschuur *et al.* 1992; Guhathakurta, this conference). These authors have compared maps of the high latitude gas made with IRAS, the 21 cm line, the CO line(s) and scattered light. These maps show that the gas and the IR emission correlate very well. For some clouds which are of too low density for the detection of CO emission, the HI and 100 μm emission correlate well. For clouds dense enough to contain detectable CO, on the other hand, the CO line flux correlates much better with the 100μm flux than does the HI. These results show that in detail, as well as globally, the molecular gas is preferentially found in the regions of highest total density.

5. THE MASS OF DIFFUSE AND DENSE GAS

As we saw in Section 3, the masses of molecular and atomic gas deduced from 21 cm and CO line measurements of the Galaxy are about 2.5×10^9 $M_\odot$ in each component. However, some of the CO emission must be contributed by the diffuse non-gravitating molecular medium which we have identified as a separate phase.

The total molecular gas content of a galaxy is usually taken to be proportional to the CO line intensity on the grounds that, since the CO line is optically thick, the total observed intensity is just proportional to the number of molecular clouds present. However there are at least two problems with this: (1) the ^{12}CO line is enormously optically thick, so that all other things being equal the same ^{12}CO line intensity will be produced by a large range of column densities and (2) since the ^{12}CO line is optically thick the line temperature depends on the excitation temperature of the cloud, so that clouds with embedded stars will be brighter.

The CO isotopic lines are considerably less optically thick than are the ^{12}CO lines. Indeed, observations of ^{12}CO and ^{13}CO rotational line emission from low mass diffuse clouds have shown that I(^{12}CO/I(^{13}CO) is larger for diffuse than for dense clouds and that I(^{12}CO)/N(H_2) is smaller for diffuse clouds than for dense clouds (de Vries, Heithausen and Thaddeus 1987; Knapp and Bowers 1988), which is consistent with the much lower optical depths for these clouds. Polk *et al.* (1988) used these properties of diffuse clouds to investigate the global ^{12}CO and ^{13}CO emission from a cloud ensemble and found that this global ratio is consistent with about 20% of the mass in the molecular ring being in the form of low optical depth gas. Within the solar circle, then, the mass of dense molecular gas is about 2×10^9 $M_\odot$, the mass of HI is about 10^9 $M_\odot$ and the mass of diffuse molecular gas about 0.5×10^9 $M_\odot$.

The question of the relative amounts of cold (5 K) and warm (20 K) molecular gas can in principle be investigated by looking at the statistics of the distribution of the peak antenna temperature T_A^*. However, one must be careful because low values of T_A^* could be due to low temperature, to low optical depth or to low density. Two recent papers discuss the relative amounts of cold and warm CO by comparing the emission in several transitions. Chiar *et al.* (1993)

compare large scale maps of the $^{12}CO(2-1)$ and $^{12}CO(1-0)$ emission in regions of the molecular ring and show that about 60% of the ^{12}CO emission arises from cool ($T < 10$ K) clouds which are presumably not forming stars. Wright *et al.* (1991) analyzed the large scale fluxes of the CO(2-1), CO(3-2), CO(4-3) and CO(5-4) lines as observed by the FIRAS instrument on the COBE satellite. They find that a two-temperature model best fits the data, with 88% of the CO being at 5 K and 12% at 19 K. These results suggest that more than half of the CO is in cold clouds, and give the following very rough distribution of mass in the inner galaxy ($R < R_o$): the total mass of gas is about 3.5×10^9 $M_\odot$. Of this, about 30% is in HI, 15% in diffuse unbound molecular clouds, 45% in cold gravitationally bound molecular clouds and 10% in warm, star-forming molecular clouds.

6. THE EVOLUTIONARY RELATIONSHIP BETWEEN THE DIFFUSE AND DENSE GAS

The relative constancy of the HI density with R may therefore be due to some threshold for the formation of H_2. The mean HI surface density within R_o is $\Sigma(HI) = 5.5 \times 10^{20}$ cm^{-2}. For a normal gas to dust ratio, (with $N_H/A_V = 2 \times 10^{21}$ cm^{-2} mag^{-1}) this corresponds to $A_v = 0.3^m$, which is just about the value of the visual extinction at which the hydrogen in an interstellar cloud undergoes the transition between atomic and molecular gas (van Dishoeck and Black 1989; Elmegreen 1993).

Lockman and Gehman (1991) and Merrifield (1993) show that the HI layer can be held up in the gravitational field of the Galactic disk with the value of the turbulent velocity given by the HI line width. The vertical acceleration in the Galactic disk close to the Galactic plane is

$$\ddot{z} = -4\pi G\rho_o z \ \ \text{cm s}^{-2} \tag{1}$$

where ρ_o is the total midplane mass density. The oscillation frequency is then

$$\omega^2 \simeq 4\pi G\rho_o$$

giving a period of 10^8 years for $\rho_o = 0.1$ $M_\odot$ pc^{-3} (cf. Kuijken and Gilmore 1989). If the gas is atomic at high z and is converted to molecular gas at low latitudes, the neutral gas will undergo the transition between the atomic and molecular phases on a free-fall timescale, about 3×10^7 years.

Suppose f is the fraction of the gas which is in the form of HI and that the mean value of f at height z is $\bar{f}(z)$. The rate of conversion between the molecular and atomic phases at height z is then

$$\frac{df}{dt} = \frac{1-f}{\tau_d} - f\frac{\bar{f}(z)\rho_g(z)}{\rho_g(0)\tau_f} \tag{2}$$

where τ_d and τ_f are the molecule destruction and formation timescales and $\rho_g(z)$ is the total gas density at z. In a steady state

$$\frac{1-\bar{f}}{\tau_d} = \frac{\bar{f}^2}{\tau_f}$$

or, if $\bar{f}$ is small,

$$\bar{f}^2 \sim \frac{\tau_f}{\tau_d}$$

giving

$$\bar{f} \sim \frac{1}{\sqrt{\rho_g}}$$

since $\tau_f \sim \rho_g^2$ while $\tau_d \sim \rho_g$.

The structure of the gas atmosphere in the gravitational field given by equation (1) is

$$\rho_g(z) = \rho_g(0)e^{-z^2/2h^2} \tag{3}$$

where

$$h^2 = \frac{\sigma_z^2}{4\pi G\rho_o}$$

and σ_z is the vertical velocity dispersion, assumed to be constant with z. Since the total gas surface density

$$\Sigma_g = \rho_g h \sim \frac{\rho_g}{\sqrt{\rho_o}}$$

the fraction of the gas in the form of HI is

$$f \sim \rho_o^{-1/4}\Sigma_g^{-1/2}$$

so that

$$\Sigma(HI) = f\,\Sigma_g \sim \rho_o^{-1/4}\Sigma_g^{1/2}$$

Since the gas and mass surface density distributions in the Galactic disk have roughly the same scale length, this gives

$$\Sigma(HI) \sim \Sigma_g^{1/4}$$

i.e. $\Sigma(HI)$ depends only very weakly on Σ_g, in agreement with the observations. These simple considerations suggest that the conversion between atomic and molecular gas is rapid and depends most strongly on the local column density; the disk of a galaxy can maintain an HI layer whose optical depth to the ultraviolet radiation which destroys molecules is $\sim 1^m$. If the opacity is higher than this the gas is mostly molecular. The maximum HI surface density in a disk is then about 2×10^{21} cm^{-2}.

ACKNOWLEDGEMENTS

Many thanks to Bill Latter and Roc Cutri for the invitation to this delightful conference, and to Jim Gunn for many valuable conversations. This research is supported by NSF grant AST89-21700 to Princeton University.

REFERENCES

Bania, T.M., and Lockman, F.J. 1984, ApJS, 54, 513.

Boulanger, F., and Pérault, M. 1988, ApJ, 330, 964.
Bregman, J.N., Kelson, D.D., and Ashe, G.A. 1993, ApJ, 409, 682.
Chiar, J., Kutner, M.L., Verter, F., and Leous, J. 1993, ApJ, (in press).
Chromey, F.R., Elmegreen, B.G., and Elmegreen, D.M. 1989, AJ, 98, 2203.
Clemens, D.P., Sanders, D.B., Scoville, N.Z. and Solomon, P.M. 1986, ApJS, 60, 297.
Cohen R.S., Dame, T.M., and Thaddeus, P. 1986, ApJS, 60, 695.
Cox, D.P. 1990, in 'The Interstellar Medium in Galaxies', ed. H.A. Thronson and J.M. Shull (D. Reidel: Kluwer), p181.
Désert, F.X., Bazell, D., and Boulanger, F. 1988, ApJ, 334, 815.
van Dishoeck, E.F., and Black, J.H. 1988, ApJ, 334, 771.
Draine, B.T., and Anderson, N.E. 1985, ApJ, 292, 494.
Elmegreen, B.G. 1993, ApJ, (in press).
Elmegreen, B.G., and Elmegreen, D.M. 1987, ApJ, 320, 182.
Gordon, M.A., and Burton, W.B. 1976, ApJ, 208, 346.
Grabelsky, D.A., Cohen, R.S., Bronfman, L., Thaddeus, P., and May, J. 1987, ApJ, 315, 122.
Guhathakurta, P., and Tyson, J.A. 1989, ApJ, 346, 773.
Heiles, C.E., Reach, W.T., and Koo, B.-C. 1988, ApJ, 332, 313.
Herter, T., Shupe, D.L., and Chernoff, D.F. 1990, ApJ, 352, 149.
Hollenbach, D.L., and McKee, C.F. 1979, ApJS, 41, 555.
Jannuzzi, B.T., Black, J.H., Lada, C.J., and van Dishoeck, E.F. 1988, ApJ, 332, 995.
Jenkins, E.B., Jura, M., and Loewenstein, M. 1983, ApJ, 270, 88.
Joncas, G., Boulanger, F., and Dewdney, P.E. 1992, ApJ, 397, 165.
Kerr, F.J., Bowers, P.F., Jackson, P.D., and Kerr, M. 1986, A&AS, 66, 373.
Kerr, F.J. 1983, in 'Serendipitous Discoveries in Radio Astronomy', ed. K. Kellerman and B. Sheets, NRAO Charlottesville, p294.
Knapp, G.R., and Bowers, P.F. 1988, ApJ, 331, 974.
Knapp, G.R., Stark, A.A. and Wilson, R.W. 1985, AJ, 90, 254.
Kulkarni, S.R., and Heiles, C.E. 1988, in 'Galactic and Extragalactic Radio Astronomy' ed. K.I. Kellerman and G.L. Verschuur, Springer-Verlag, p 95.
Kuijken, K., and Gilmore, G. 1989, MNRAS, 239, 605.
Lockman, F.J., and Gehman, C.S. 1991, ApJ, 382, 182.
Magnani, L., Blitz, L., and Mundy, L. 1985, ApJ, 295, 402.
McKee, C.F., and Ostriker, J.P. 1977, ApJ, 218, 148.
Merrifield, M. 1993, MNRAS, 261, 233.
Polk, K.S., Knapp, G.R., Stark, A.A. and Wilson, R.W. 1988, AJ, 332, 432.
Sanders, D.B., Clemens, D.P., Scoville, N.Z., and Solomon, P.M. 1986, ApJS, 60, 1.
Sanders, D.B., Scoville, N.Z., and Solomon, P.R. 1985, ApJ, 289, 373.
Sodroski, T.J., Dwek, E., Hauser, M., and Kerr, F.J. 1987, ApJ, 322, 101.
Verschuur, G.L., Rickard, L.J., Verter, F., Pound, M.W. and Leisawitz, D. 1992, ApJ, 390, 514.
de Vries, H.W., Heithausen, A., and Thaddeus, P. 1987, ApJ, 319, 723.
Wannier, P.G., Lichten, S.L., and Morris, M. 1983, ApJ, 268, 727.
Weaver, H., and Williams, D.R.W. 1974a, A&AS, 8, 1.
Weaver, H., and Williams, D.R.W. 1974b, A&AS, 17, 1.
Wright, E.L. *et al.* 1991, ApJ, 381, 200.

The First Symposium on the Infrared Cirrus and Diffuse Interstellar Clouds
ASP Conference Series, Vol. 58, 1994
Roc M. Cutri and William B. Latter (eds.)

OBSERVED EFFECTS OF THE UV FIELD ON CIRRUS

R.J. LAUREIJS
Space Science Dept., ESTEC, POB 299, 2200 AG Noordwijk, The Netherlands

G. HELOU
IPAC, Caltech

Y. FUKUI, A. MIZUNO, K. IMAOKA
Dept. of Radioastronomy, Nagoya University

F.O. CLARK
Geophysics Directorate, Phillips Laboratory

ABSTRACT IRAS images, CO observations, and A_B from starcounts have been used for a large scale study of the L134 high lattitude cloud complex. These tracers reveal a cloud morphology which is affected by the UV field from nearby sources. Both dense regions and a higher density contrast are observed on the shadow side of the clouds. We infer that the anisotropic radiation field affects the cloud's density distribution: a weaker field causes a smaller halo on the shadow side.

THE UV FIELD IN THE L134 COMPLEX

As part of a large scale study of the L134 complex of clouds (Laureijs et al., 1993) we have analyzed possible effects of the UV field acting upon the clouds. The dominant UV sources nearest to the L134 complex ($d_\odot$ =110 pc) are the Sco OB2 association ($d_\odot$ =170 pc) and the star ζ Oph ($d_\odot$ =190 pc). Using the stellar parameters of Habing (1968) and assuming no extinction from intervening dust we derive a radiation density of 1.3 10^{-16} erg cm^{-3}Å^{-1} at 1000 Åin the vicinity of L134. This is 3 times the canonical value of 4.9 10^{-17} erg cm^{-3}Å^{-1} (Mathis et al., 1983). Assuming that dust extinction is negligible between the stars and the L134 complex, we infer that

i. the energy density is at most 4 times the canonical ISRF, if we add the presence of a diffuse galactic UV component, and

ii. the UV field illuminating the L134 complex is highly anisotropic and comes from Sco OB2 and ζ Oph.

OBSERVED UV ANISOTROPY IN L134

We have plotted different tracers along a slice through L134 in Fig. 1. The slice points approximately in the direction of Sco OB2. The radiation field is expected to decrease monotonically towards positive offsets with the steepest gradient. The center of the slice (at 0') is about 1' south of the position of the ammonia core in L134.

The infrared tracers show that both I_{12} and I_{25} are brighter and more extended on the illuminated side (panel a). The maxima of I_{60} and I_{100} are also on the illuminated side (panel b). On the shadow side of the cloud there is little far infrared emission and no evidence for secondary emission peaks as seen at 12 and 25 μm.

The morphologies of CO, ^{13}CO and C^{18}O are markedly different from each other (panel c). CO and ^{13}CO peak close to the 100 μm emission maximum. The ammonia core lies on the declining flanks of the CO and ^{13}CO profiles on the shadow side of the cloud. In contrast, the C^{18}O profile peaks sharply near the ammonia position.

We have plotted $T_R^*(CO)$ to analyze the gas temperature (panel d). $T_R^*(CO)$ gradually decreases from about 16 to 10 K. Assuming optically thick and thermalized CO, the inferred kinetic temperature drops 7 K, from 20 K on the illuminated side down to 13 K on the shadow side. We note that the drop in $T_R^*(CO)$ is less steep than the decline in CO. This indicates that the lines are both weaker and narrower on the shadow side.

The ratio of integrated ^{13}CO and C^{18}O line emission, W_{13}/W_{18}, is high on the illuminated side and remains low on the shadow side (panel h). The peak in the ratio is due to ^{13}CO fractionation, causing the relative abundance of ^{13}CO to rise with respect to the C^{18}O abundance. The absence of a clear fractionation peak on the shadow side is a strong indicator that the UV radiation field must be weak on the shadow side (van Dishoeck and Black, 1988).

In summary, the most noticeable observations are:

1. the asymmetric profiles of $T_R^*(CO)$ and W_{13}/W_{18} in L134 most clearly indicate anisotropic illumination by the external UV field. The lines are narrower on the shadow side.
2. the infrared emission is stronger and more extended on the illuminated side of the cloud. The infrared brightness profiles are consistent with the notion that the clouds have a larger diffuse layer on the illuminated side.

Similar trends are observed in other clouds in the complex

1.3 ANISOTROPIC UV FIELD: EFFECTS ON MORPHOLOGY

The molecular, infrared, and the extinction data clearly exhibit a regular pattern in L183, L134, L169, and L1780. Diffuse material is widespread on the eastern and southern sides of the clouds, whereas the dense cores are found on the opposite sides. The gas temperature shows a gradient along the same direction. The variation in W_{13}/W_{18} is indicative an anisotropic radiation field. These measurements are consistent with the position of the main UV sources. The alignment of the density gradient and the major symmetry axes of the clouds along the direction of the field readily suggest a physical connection.

Analysis of the hydrostatic equation of state of a cloud excluding magnetic fields was carried out by Falgarone and Puget (1985). In the model, there are three regions: 1. an isothermal core, 2. an intermediate layer with lower density and reasonable UV shielding, and 3. a halo where the gas is indirectly heated

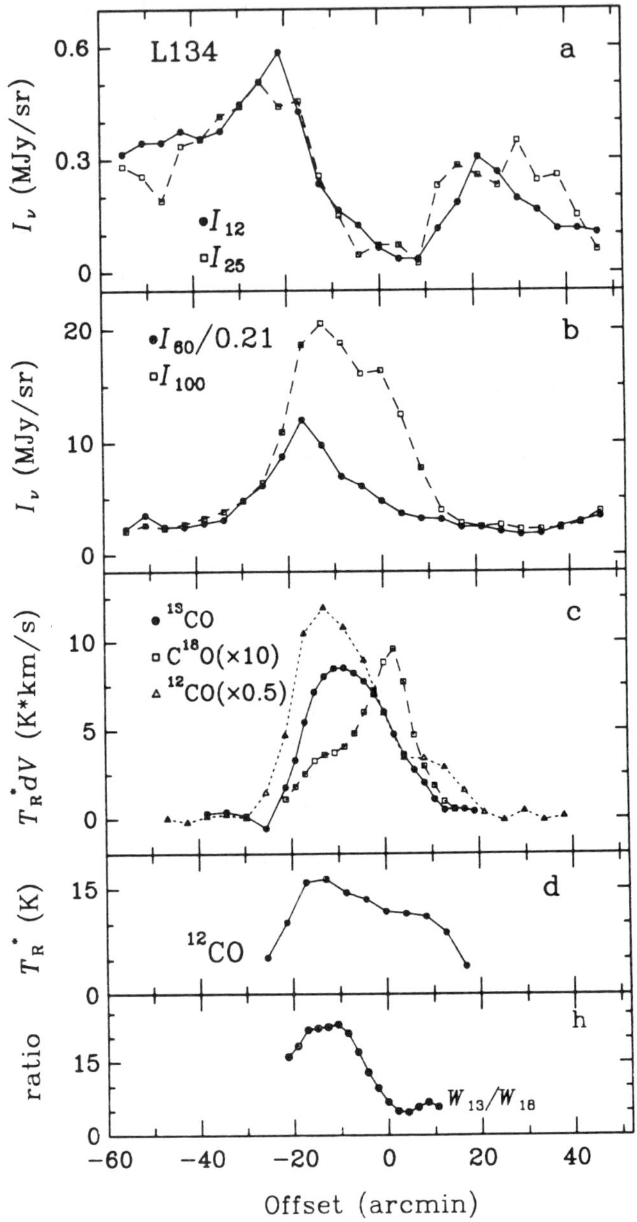

FIGURE I The variation of different tracers along a slice through L134. Position 0 corresponds to the position of the ammonia core in the cloud. The position angle of the slice is 335 degrees. The tracers are IRAS 12 and 25 μm (a), IRAS 60 and 100 μm (b), integrated line emission for different CO isotopes (c), CO peak temperature (d), and the ratio of integrated line emission of ^{13}CO and C^{18}O (h).

by the UV field. The halo acts as an absorbing layer protecting the core from the radiation field. Wild variations in the UV field affects only the outer cloud regions and the temperature in the central regions is nearly constant for a wide range of parameters.

The size of the halo is predicted to be a sensitive function of the UV field strength. The weaker the field, the smaller the halo becomes. For a single cloud in an anisotropic field, we expect a larger halo in direction of the UV field. Such a morphology is observed for the clouds in the L134 complex. The slices clearly show a decrease in density and an increase in velocity dispersion in the direction of the UV source. This mechanism offers a natural explanation for the dense regions in the NW corners of L183, L134, and L1780.

We conclude that the anisotropy of the radiation field can have a significant influence on the morphologies of apparently quiescent high latitude clouds. The morphologies are affetced in such a way that CO and ^{13}CO (or 60 and 100 μm) images can be used to infer the direction of the radiation field.

ACKNOWLEDGMENTS

RJL acknowledges the Phillips Laboratory for the hospitality during his stay.

REFERENCES

Falgarone, E, Puget, J. L. 1985, *A&A*, 157

Habing, H. 1968, *B.A.N.*, **19**, 421

Laureijs, R. L., Fukui, Y., Mizuno, A., Imaoka, K. Helou, G., Clark, F. O. 1993, *in preparation*

van Dishoeck, E. F., Black, J. H. 1988 *ApJ*, **334**, 771

The First Symposium on the Infrared Cirrus and Diffuse Interstellar Clouds
ASP Conference Series, Vol. 58, 1994
Roc M. Cutri and William B. Latter (eds.)

MODELING TRANSIENT HEATING IN THE CHAMAELEON DIFFUSE CLOUD: VSGS, PAHS, HACS?

S. D. DOTY AND C. M. LEUNG
Physics Department, Rensselaer Polytechnic Institute, Troy, NY 12180

D. C. LIS
G. W. Downs Laboratory of Physics, Caltech, Pasadena, CA 91125

ABSTRACT We model IRAS surface brightness observations of a highly regular diffuse cloud in the Chamaelon region. The excess mid-infrared emission ($\lambda < 30\mu m$) suggests that transient heating of very small grains (VSGs), large polycyclic aromatic hydrocarbons (PAHs), or hydrogenated amorphous carbons (HACs) may be present. Using detailed, self-consistent radiative transfer models (for both the classical and transiently heated particles) we interpret the IRAS observations and study the effects of PAHs and HACs on the energy spectrum.

INTRODUCTION

Transient heating of small ($a < 30$Å) particles, or isolated regions of larger particles (DW, Duley & Williams, 1988), has long been the attributed source of the excess mid- and near-infrared emission seen in reflection nebulae, and diffuse clouds. The correlation of the diffuse emission with the UIR bands has suggested that the particles responsible for the emission are hydrocarbons. However, the difficulty of current grain models to accurately account for the position, strength, and width of the UIR bands, does not resolve the question of whether these particles are large, independant molecules (PAHs), hydrocarbons embedded in a larger particle (HACs), or some other as yet unidentified constituent. To explore the effects of different grain properties on the emergent spectrum, and hence provide constraints on the source of the NIR emission, we have constructed self-consistent radiative transfer models of a source. We compare the results with IRAS observations, and make suggestions for future observations.

SOURCE & MODEL PROPERTIES

Laureijs et al (1989) studied the diffuse cloud DC 300.2 - 16.9 in the Chamaeleon region. Their analysis suggested that at least three different grain components and transient heating of small particles were necessary to explain the large and extended 12 and 25μm emission. Due to its highly regular geometry and available surface brightness data, this source is an ideal candidate for studying small particle properties through modeling.

In our model, we adopt a grain model similar to that of Desert et al (DBP, 1990): we allow for 3 different grain compositions, consisting of 13 different grain sizes. We consider classical silicate grains (100Å$< a <$ 0.1μm), very small graphite grains (17Å$< a <$ 100Å), and smaller PAH/HAC-like particles (5Å$< a <$ 12Å). The size and density distributions are taken to be power laws. In order to fit the observed data, we must deplete the density of the smallest VSGS from the power law in the outer portions of the cloud. The radiation field is taken from Mathis et al (1983), with an enhancement ($< 2\times$) in the UV - Visible region.

The opacities for the PAH particles are taken from the DBP model. For comparison, we also construct models using HAC-like particles (instead of PAHs). The HAC data are taken from measurements by Ogmen & Duley (1987), and Duley (1993), for $\lambda < 17\mu m$, with a λ^{-1} emissivity law for $\lambda > 17\mu m$. Due to a discrepancy in the data regarding the ratio $\tau_{3.4}/\tau_V$, we consider three models: PAH, HAC A (large $\tau_{3.4}/\tau_V$), and HAC B (small $\tau_{3.4}/\tau_V$) – each having similar composition for larger grains. Compared to the PAH data, the HAC data have much larger emissivities in the near- and mid-infrared.

Physically, two approaches may be taken regarding our "HAC-like" particles. They can be considered to be simply a set of truly small particles with optical properties that differ from the PAHs. Or they can be thought of as the individual, isolated sites that DW envision will undergo local transient heating, superposed on the emission from larger grains. In either case, our particle sizes ($a \sim$ 10Å) correspond to the size of the emitting particle/region.

RESULTS

Cloud Properties

We find that all three models can adequately fit the IRAS surface brightness of DC 300.2 - 16.9. However, as each model uses different grain properties, the physical parameters differ somewhat among the models. The results are shown in Table I below.

TABLE I Physical Parameters for PAH & HAC Models

Parameter	PAH	HAC A	HAC B
τ_V (total)	9%	10%	7%
τ_V (surface)	64%	66%	40%
τ_{UV} (surface)	69%	54%	91%
Mass	21%	15%	1%

Notice that although the total visible optical depth due to the smallest particles is consistently about 7 – 10%, the optical depth at the surface and mass percentages may vary widely in order to fit to the IRAS data. This is due to the difference of the absorption cross sections in the UV - NIR, where Q_{abs}(HAC B)$>$ Q_{abs}(HAC A)$\gg$ Q_{abs}(PAH). The enhanced absorption means that for a cloud to absorb the same amount of energy from the ISRF, M_d(HAC B)$<$ M_d(HAC A)$<$ M_d(PAH) is required.

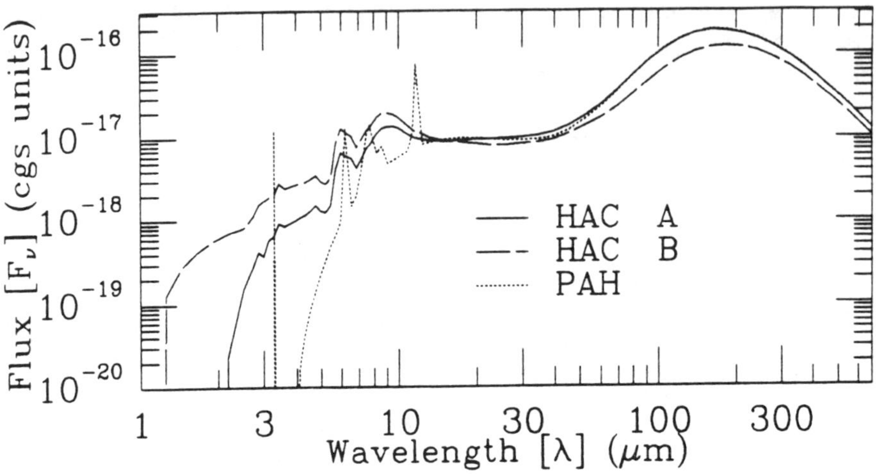

FIGURE I Flux Spectrum of model cloud for PAH, HAC A, & HAC B

Energy Spectrum

While all three models can fit the IRAS data, large ($> 100\times$) differences exist in the near infrared ($\lambda < 7\mu m$) as shown in Figure I due do the enhanced emissivity of the HAC-like particles at short wavelengths.

K & M Bands

To discriminate among the three models, we can exploit the differences in the NIR emission using the K, and M photometric bands. Using canonical detector responses and atmospheric transmission curves, we have produced synthetic observations of this source at these three bands. The results for all three models are shown in Figure II, where we plot the synthetic surface brightnesses for the K and M bands.

Implications of Model Parameters

Although further observations will be necessary to constrain possible grain models, our work has suggested several interesting trends. The 12 and $25\mu m$ emission require transient heating of small particles or small regions of larger particles. The grains reponsible for this emission are concentrated (with respect to particles of other sizes) at an intermediate cloud radius. This could possibly be due to an intermediate region where the cloud is too opaque for effective photodestruction, and not yet dense enough for much coagulation/collisional destruction of these transiently heated particles. Furthermore, we require an enhanced external radiation field. Our models cannot distinguish between a global (i.e. different radiation fields), or a local effect (nearby stars). These appear to be interesting questions, which merit further investigation.

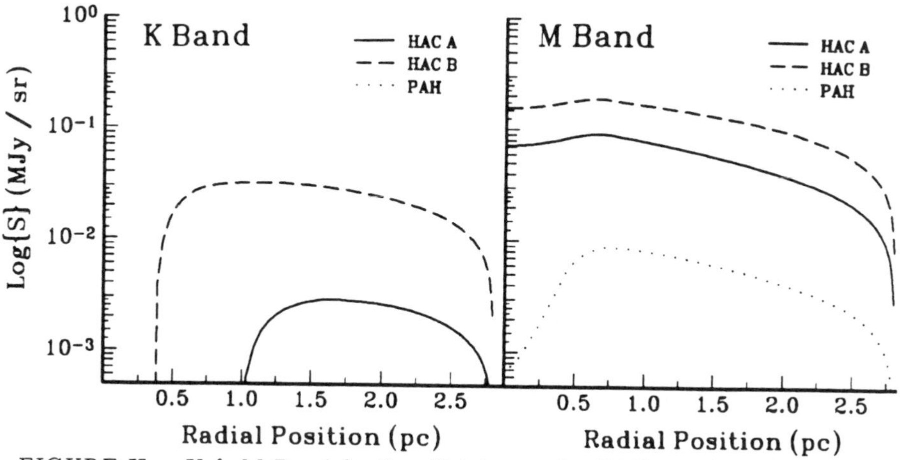

FIGURE II K & M-Band Surface Brightness for PAH, HAC A, & HAC B

CONCLUSIONS

We have modeled the IRAS surface brightness data of a diffuse cloud in Chamaeleon, using a detailed, self-consistent radiative transfer code. Although the contribution of small particles to cloud mass and optical depth must be changed to make different grain models fit the observed IRAS surface brightnesses, good fits can be obtained. The effects of the optical properties become evident shortward of $7\mu m$, where the differences in the energy spectrum of PAH and HAC-like particles can be greater than two orders of magnitude. We propose that these differences could be observed through photometry in the K, and M bands, thus providing important constraints on the optical properties of the transiently heated particles.

ACKNOWLEDGMENTS

This work has been supported by grants from NASA and the Air Force. S. D. Doty has been supported by a DoD Fellowship.

REFERENCES

Desert, F.-X., Boulanger, F., and Puget, J. L. 1990, *A&A*, **237**, 215

Duley, W. W., 1993, private communication

Duley, W. W., and Williams, D. A., 1988, *MNRAS*, **231**, 969

Laureijs, R. J., Chlewicki, G., Clark, F. O., and Wesselius, P. R., 1989, *A&A*, **220**, 226

Mathis, J. S., Mezger, P., and Panagia, N., 1983, *A&A*, **128**, 212

Ogmen, M., and Duley, W. W., 1988, *ApJ*, **334**, L117

The First Symposium on the Infrared Cirrus and Diffuse Interstellar Clouds
ASP Conference Series, Vol. 58, 1994
Roc M. Cutri and William B. Latter (eds.)

PHOTOELECTRIC HEATING OF INTERSTELLAR GAS BY PAHS AND SMALL GRAPHITIC GRAINS.

E.L.O BAKES
Peyton Hall, Princeton University, Princeton, NJ 08544

A.G.G.M. TIELENS
NASA Ames Research Center, Moffet Field, CA 94035

ABSTRACT We present an *ab initio* model of the photoelectric charging of polycyclic aromatic hydrocarbon molecules (PAHs) and very small graphitic grains in the diffuse interstellar medium. We show that this model agrees with the observations by Pottasch et al. (1979) for the cooling of the diffuse interstellar medium and that the charging effects have important consequences for interpreting observations of the diffuse interstellar bands (DIBs) using laboratory methods. We find that around half of the gas heating in the grain population comes from grains up to 15Å in radius. The other half originates from grains between 15–100Å.

INTRODUCTION

The dominant heating mechanism of the interstellar gas is thought to be the photoelectric effect on dust grains. Dust grains absorb FUV photons, which can lead to the ejection of an electron from the grain. The electron carries excess kinetic energy which will be transferred to the gas as thermal energy. This study addresses the problem of explaining the amount of heating in the diffuse ISM observed by Pottasch et al. (1979).

THE PHOTOELECTRIC HEATING RATE

The energy carried away by a photoejected electron depends on the work function and the Coulomb barrier. Thus, the photoelectric heating rate depends strongly on the grain charge. The charge itself is a balance between photoelectric ejection and grain–electron recombination. While previous studies have considered a typical radius of grain at an average charge for each grain size, we consider a grain size distribution plus a charge distribution. We have developed a model for the photoelectric ejection from grains which is valid for a large range of grain radius. The model parameters are chosen in such a way that it reproduces the measured photoelectric yield for small PAHs as well as that for bulk graphite (Feuerbacher et al. 1973, Verstraete et al. 1990). This model is combined with the model of Draine and Sutin (1987) for electron–grain recombination, which emphasizes the importance of short range Coulomb focussing for grains of radius < 100Å and gas temperatures < 100 K. The resulting photoelectric heating rate is illustrated in Figure I for typical conditions in diffuse interstellar clouds (G_o=1, n_H= 25 cm^{-3}, electron density $n_e=7.5 \times 10^{-3}$ cm^{-3} and gas temperature T=100K). Half of the total gas heating results from the photoelectric effect on species with radius up to 15Å. $\times 10^5$ carbon atoms). The other half comes from grains of radius 15–100Å (1500–4.75 $\times 10^5$ carbon atoms). Larger grains do not contribute appreciably to the gas heating. The increase in the heating rate with decreasing grain size reflects the decreasing average grain charge. Essentially, the ionization rate decreases proportional to the volume (i.e. N_c) while the recombination rate decreases only linearly with size (i.e. Coulomb focussing). The increased contribution by small grains is also partly due to the increased photoelectric yield of small species compared to bulk materials (Watson 1972).

For our purposes, FUV extinction is dominated by small grains (a<300Å) and the total absorbed energy is about 10^{-24} erg s^{-1} (H atom)$^{-1}$. The total heating rate for very small graphitic grains and PAHs is thus given by

$$\Gamma_{pe} = 10^{-24}\, \varepsilon\, G_o\, n_H \text{ erg s}^{-1} \text{ (H atom)}^{-1}$$

where G_o=FUV flux normalized to the Habing field for the solar neighborhood, ε is the photoelectric heating efficiency and n_H is the hydrogen nucleus density. We have numerically calculated the photoelectric heating efficiency for a wide range of astrophysical conditions. A good fit to the results is given by

$$\varepsilon = \frac{3 \times 10^{-2}}{1 + 2 \times 10^{-4} G_o T^{1/2}/n_e}$$

This equation is applicable to ISM at temperatures less than 4000 K. At higher temperatures, electron–grain recombination cooling becomes more important. Further details of this may be found in Bakes and Tielens (1993). The analytic fit to the numerical results is shown in Figure II. As the efficiency curve illustrates, when ionization increases and grain–electron recombination decreases (i.e. $G_o T^{1/2}/n_e$ increases), the neutral fraction decreases, resulting in a drop in the gas heating rate. The calculated heating rate for the diffuse ISM of 2.4 $\times$ 10^{-26} n_H ergs s^{-1} (H atom)$^{-1}$ is comparable to the heating rate derived from UV absorption studies of the level populations of the dominant cooling ion, [CII] (Pottasch et al. 1979, Grey, Lequeux and Boulanger 1993).

PAHS AND THE DIFFUSE INTERSTELLAR BANDS

In Figure III, we show the grain charge distribution for coronene, a 100 and 1000 carbon atom grain where the electron sticking coefficient for coronene is 10^{-3} and 1 for the grains in the diffuse ISM. Whilst coronene cations dominate, this is not true for other small PAHs if their electron affinity is greater than 1eV. Thus, negatively charged PAHs could be important contributors to the DIBs. The increase in hydrogen nucleus density increases the grain–ion recombination rate and allows greater charge neutralization on the grains. Thus, the most probable charge can shift from 1 to 0 in coronene and from 0 to –1 in larger grains. The grain charge distribution is sensitive to the physical conditions in the ISM and this should be considered when interpreting correlations amongst the DIBs.

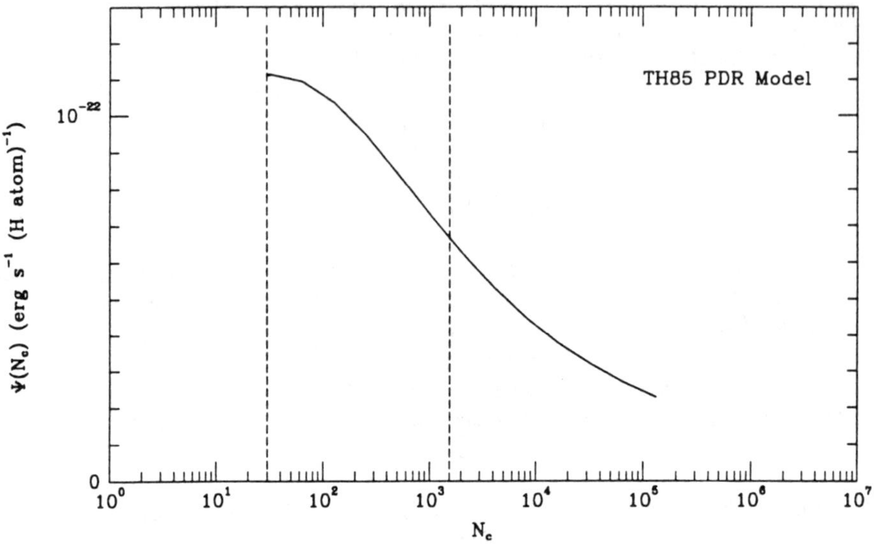

FIGURE I The contribution to the photoelectric heating rate by grains with N_C carbon atoms. The contribution is plotted in such a way that in this linear–logarithmic plot equal areas under the curve represent equal contributions to the total heating rate. For orientation purposes, note that the IR–emission features are due to species with 20–300 carbon atoms. The 25μm cirrus results from species with 300–1500 carbon atoms and the 60μm cirrus originates in grains with $> 10^5$ carbon atoms.

REFERENCES

Bakes, E.L.O., Tielens, A.G.G.M. (1993) *submitted to ApJ*

Draine, B.T., Sutin, B. 1987, *ApJ* **320**, 8033

Feuerbacher, B., Willis, R. F. & Fitton, B. 1973, ApJ **181**, 101

Grey, C., Lequeux, J., Boulanger, F. 1993, *A&A* **266**, 457

Pottasch, S.R., Wesselius, P.R., van Duinen, R.J. 1979, *A&A* **74**, L15

Verstraete, L., Léger, A., d'Hendecourt, L., Dutuit, O., Defourneau, D. 1990, *A&A* **237**, 436

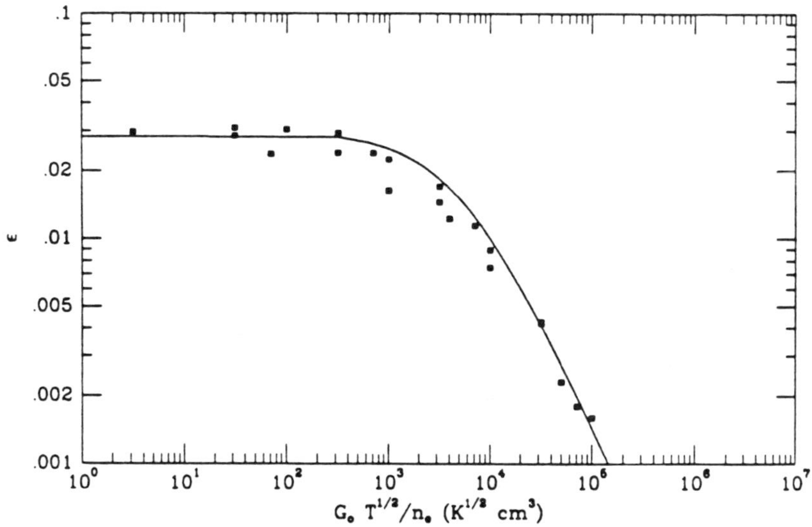

FIGURE II The photoelectric heating efficiency curve is shown here, where the line is an analytic fit to the numerical results, represented by points.

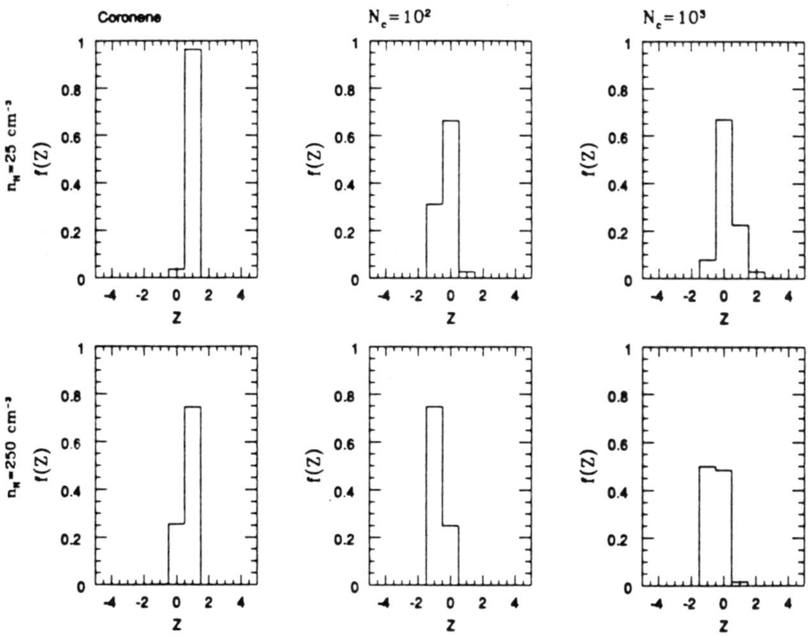

FIGURE III A comparison of the most probable charges for the coronene molecule and a planar grain of 100 carbon atoms and a spherical grain of size 1000 carbon atoms in the diffuse ISM. The top row of figures shows the charge distribution for typical DIM conditions while the second row illustrates the effect of increasing the hydrogen nucleus density.

The First Symposium on the Infrared Cirrus and Diffuse Interstellar Clouds
ASP Conference Series, Vol. 58, 1994
Roc M. Cutri and William B. Latter (eds.)

FAR–INFRARED COLORS OF SHOCKED INTERSTELLAR MEDIUM

ZHONG WANG
100–22 IPAC, California Institute of Technology, Pasadena, CA 91125

ABSTRACT Shock heating plays an important role in the infrared emission of diffuse interstellar clouds. By studying a sample of shocked regions surrounding a supernova remnant, it is possible to evaluate the radiative energy distribution in the infrared as a function of the preshock density of the ambient medium. Infrared colors derived from the beam-corrected IRAS maps of IC 443 do indeed show such variations in a systematic way. In particular, we find evidence that fast ($v > 50$ km s^{-1}) shocks developing in a low density gas correspond to higher S(60)/S(100) and lower S(12)/S(25) ratios. The former clearly indicates a warmer dust temperature, while the latter is probably due to selective destructions of small grains in the shock.

INTRODUCTION

When an interstellar cloud is shocked by strong kinematic disturbance, the gas and dust are heated and partially ionized (Draine & McKee 1993). The infrared emission from such interactions can be seen in many regions of our own Milky Way, and it could be an important contributor to the very high bolometric luminosities of infrared bright galaxies. Yet the energy spectrum of such shock emission, which presumably depends on the physical conditions of the pre-shock gas, remains relatively unexplored. Observationally, things can be quite complicated as it is not always easy to distinguish between emission from interstellar shocks and those from other mechanisms such as irradiation of high energy photons (e.g., Draine & Woods 1991).

It is therefore worthwhile to investigate shocked regions with simple structures. One ideal candidate is the bright emission regions around nearby supernova remnants (SNRs). In this contribution, we describe a study of the far-infrared (FIR) emission in the shocked regions around IC 443, making use of the IRAS maps combined with a new image reconstruction technique. Our focus is to analyze the shock's far-infrared color dependence on the ambient (preshock) gas.

THE SHOCKED REGION

IC443 is one of the brightest Galactic SNRs fully mapped by IRAS in all four bands. This SNR is sufficiently extended that its emission are mostly resolved. It is located slightly off the Galactic Plane ($b = 3.0°$), in a region of highly inhomogeneous ambient gas known for its extreme density variations. A substantial fraction of the shell of IC 443 overlaps with a large molecular cloud (cf. Cornett, Chin & Knapp 1977; Burton *et. al.* 1988) Molecular line surveys (Dickman *et. al.* 1992) and near-infrared spectral line imaging (Burton, Spyromilio, & Wang 1993) provide convincing evidence that the shell is interacting with the surrounding clouds, generating shocks that are quite unique in their color among Galactic FIR emitting sources (Arendt 1989).

In the four IRAS bands (12, 25, 60, and 100 μm), the shell-like structures of this SNR are clearly seen. The emission peaks which were initially identified as possible point sources (Huang *et. al.* 1986) are actually extended regions of molecular shockfront and have clear counterparts of intense molecular line emission (Braun & Strom 1986; Dickman *et. al.* 1992; Burton *et. al.* 1993 — see Fig. 1). Of particular interest to this study is that these regions clearly have different FIR colors than other parts of the SNR where the bright X-ray, radio continuum, and optical filaments are found.

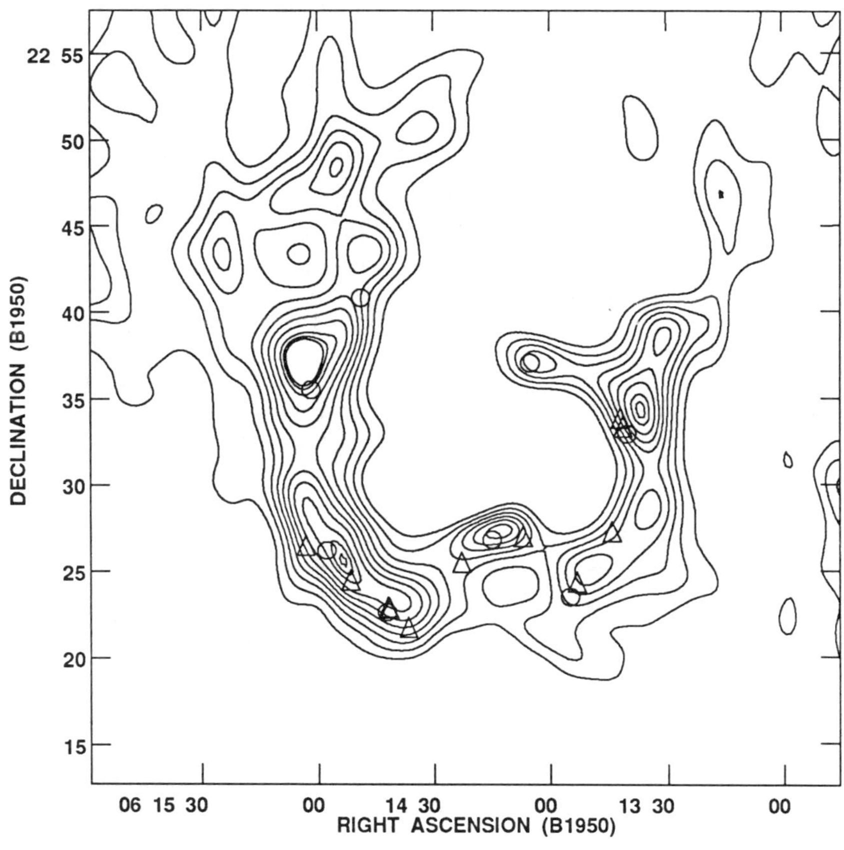

Figure 1. Positional correspondence between the peak FIR emission and molecular shocks. The contours are the 100 μm dust emission of IC 443, based on the deconvolved IRAS map (see text). Regions where molecular line emission from shocked gas is detected are marked by circles (Dickman *et. al.* 1992) and triangles (Wang & Burton 1993).

To further investigate such color variations, the IRAS data were carefully flat-fielded and corrected for zodiacal light. A newly developed Maximum Correlation Method (MCM) algorithm, known as HiRes (Aumann, Fowler & Melnyk 1990), was applied to the images. We used different number of iterations in each band, such that the effective beams for all four bands are similar (Wang & Burton 1993). The final images have a resolution of about $1.1' \times 1.9'$, (Fig. 1 & 2), a significant increase over that of earlier maps (e.g., Braun & Strom 1986; Arendt 1989).

RESULTS: THE IRAS COLOR VARIATION

In order to better understand the color variations around the SNR shell, we need to refer to the known physical conditions in the surrounding regions from observations in other wavelengths. These comparisons tell us the following:

1) Dust emission is found well correlated with the SNR shockfront as identified by a combination of multi-wavelength data, though it appears to be more wide-spread, appearing both in regions of fast, ionizing atomic shocks (the NE side) and of slower, nonionizing molecular shocks (the H_2 S(0) emitting ridge, see Burton *et. al.* 1988). This suggests that the total FIR energy generated, or the rate at which kinetic energy is converted to thermal is relatively insensitive to the density variations in the shells;

2) More important, however, is that the FIR *color* changes drastically from one part of the shell to the other, following the density variations in the ambient medium. As can be seen from Figure 2, in the NE where atomic shocks are prominent, there is clearly a low S(12)/S(25) flux ratio, and a high S(60)/S(100) ratio comparing to the rest of the FIR bright regions (SE and SW) where molecular shocks dominant;

3) Comparing with the FIR colors of other SNRs (cf. Arendt 1989), the integrated color of IC 443 is about the average, consistent with it being a mid-aged SNR. On the other hand, we find that the FIR color at the fast shocks strongly resembles that of younger SNRs, while the color of molecular shock regions are similar to the old SNRs;

4) Based on the 60 and $100\mu m$ flux densities alone, we find that the regions of lowerest thermal temperature form an almost-complete ring. This large-scale structure overlaps with where the mm/submm and NIR lines of shocked molecular gas are found, indicating a close correlation between molecular clouds and shocked dust emitting at low temperatures.

DISCUSSION

These findings illustrate clearly that the energy distribution of dust emission in interstellar shocks depends critically on the preshock density, n_0. Where n_0 is low, the shock speed is high and so is the heating rate. The FIR (60 and 100 μm) color will then reflect a high dust temperature. Conversely, lower dust temperatures are found where the ambient gas density is high. In IC 443, the latter case is typically regions dominated by molecular gas. As pointed out by Draine & McKee (1993), non-ionizing (C-type) shocks below ~ 45 km s^{-1} are

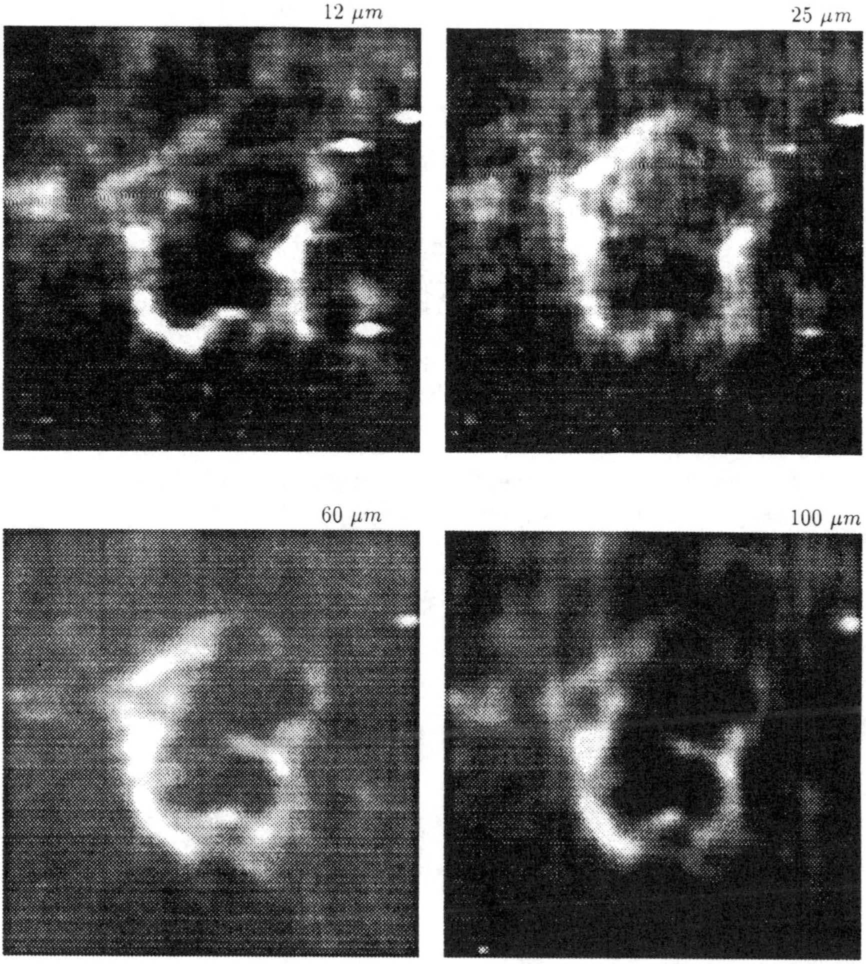

Figure 2. Greyscale maps of the deconvolved IRAS images of IC 443. This is an one-degree field centered on the supernova remnant, north is at top, and east is to the left. Grey levels are linear but the scale of each map is set differently. These images have been processed such that the effective resolution is very similar in all four bands. Clear differences between the relative intensities of the four bands can be seen, for example, by comparing the NE side with the SE and SW sides. In other wavelengths, the NE side is very bright in optical emission lines as well as in X-ray and radio continuum emission; while the SE and SW sides are most bright in the H_2 S(0) line and millimeter/submillimeter molecular line emissions. Some of the faint vertical stripes visible in the background of the maps are due to artifacts in the HiRes processing.

naturally found in molecular gas and are expected to be bright in the infrared (see also Wang & Scoville 1992).

The other thing that is clear from the maps in Figure 2 is that an enhanced ratio of S(60)/S(100) is usually accompanied by a depressed S(12)/S(25). Actually, there is a similar trend in the IRAS flux densities of galaxies (Helou 1986), which has been accounted for in terms of changing conditions in the ionizing radiation (Désert 1986). In the case of SNR shocks, the explanation may well be related but also somewhat different. It is likely that the color ratio change among different shocked regions in the SNR is directly related to the amount of very small grains (VSG) that survive at the shock front. Faster shocks in the diffuse gas tend to destroy most VSG which is mainly responsible for the emission at the shortest ($12\mu m$) wavelength.

We should also mention that, for shocks in dense gas, line emissions could alter the colors measured using IRAS flux density ratios. Burton *et. al.* (1990) have detected a strong [OI]$\lambda 63\mu m$ line at one of the molecular emission peaks in IC 443, and estimate that it may contribute substantially to the local broad-band $60\mu m$ flux density. If this is generally the case, then the IRAS data, even with the beam-corrected maps, can not be used to get effective dust temperatures directly. More extended FIR spectroscopic observations are therefore needed to help us better address the color variations of dust emission in the shocked ISM.

Acknowledgments I would like to thank Dr. M. G. Burton of the Anglo-Australian Observatory for his contribution to this study.

REFERENCES

Arendt 1989, Ap.J. Suppl. Ser., 70, 182

Aumann, H. H., Fowler, J. W., & Melnyk, M 1990, A.J., 99, 1674

Braun, R. & Strom, R. 1986, A&A, 164, 193

Burton, M. G., Geballe, T. R., Brand, P. W. J. T., & Webster, A. S. 1988, MNRAS, 231, 617

Burton, M. G., Hollenbach, D. J., Hass, M. R., & Erickson, E. F. 1990, Ap.J., 355, 197

Burton, M. G., Spyromilio, J. & Wang, Z. 1993, *in preparation*

Cornett, R. H., Chin, G., & Knapp, G. R., 1977, A&A, 54, 889

Désert, 1986, F.X., in *Light on Dark Matter*, ed. F. P. Israel, 213

Dickman, R. L., & Snell, R. L., Ziurys, L. M., & Huang, Y. L., 1992, Ap.J., 400, 203

Draine, B. T. & McKee, C. F. 1993, Ann. Rev. of A. & A. (in press)

Draine, B. T. & Woods, D. T. 1991, Ap.J., 383, 621

Helou, G. 1986, Ap.J. Lett., 311, L33

Huang, Y. L., Dickman, R. L., & Snell, R. L. 1986, Ap.J. 302, L63

Wang, Z., & Burton, M. G. 1993, *in preparation*

Wang, Z. & Scoville, N. Z. 1992, Ap.J. 386, 158

The First Symposium on the Infrared Cirrus and Diffuse Interstellar Clouds
ASP Conference Series, Vol. 58, 1994
Roc M. Cutri and William B. Latter (eds.)

2-D MONTE CARLO MODELS OF CLUMPY INTERSTELLAR CLOUDS

MARCO SPAANS
Sterrewacht Leiden, P.O. Box 9513, 2300 RA Leiden, the Netherlands

INTRODUCTION

Interstellar clouds are illuminated by the diffuse galactic background. The dust which is ubiquitously present in the clouds attenuates a part of this UV background leading to an increase in, for example, radiative lifetimes of molecular species or a decrease in the heating rate as a function of depth into the cloud. The inclusion of a chemical network then yields observational quantities like column densities and emission line intensities. So far most investigations have been based on the assumption of homogeneity. However, nowadays a large (and growing) amount of data has been gathered in different molecular lines which indicate a highly clump-like structure of interstellar clouds on different scales (Falgarone and Pérault 1987). A comparison of homogeneous and clump-like models with the observations thus seems an inescapable necessity (Boissé 1990).

CLUMPY STRUCTURE

The hierarchy of the clump size distribution extends all the way down to the smallest (presently) accessible scales of 0.01 pc (Falgarone and Pérault 1988). We model the clumpy structure which is observed in a large number of clouds by adopting a two-phase model which consists of a high and low density phase. The ratio between the dense and the diffuse phase may be freely chosen, but observations (Mezger et al. 1992) seem to indicate a value of the clump-interclump density ratio of about 100. The precise choice for this ratio does not influence any of our results described below, as long as it is substantial, i.e. a low and high density phase may be distinguished. The construction of the clumpy medium uses the (observable) extinction curve as a boundary condition, in that each observer will measure the same curve for a particular realization along his line of sight. The clumpy structure is characterized by a volume filling factor, F, the volume of the cloud which is in the high density phase. We have also included a gradient between the high and low density phases in order to achieve a more realistic density structure.

RADIATIVE TRANSFER

We use a Monte Carlo method to solve the transfer equation appropriate for the penetration of diffuse UV radiation into clumpy interstellar clouds, subject to attenuation by coherent, nonconservative, anisotropic scattering from grains over the full 912-5000 Å wavelength range. We are thus not limited to any geometry or to any specific angular distribution of the incident radiation field. We use the grain properties of Roberge et al. (1991). For the present investigation we have taken a slab geometry, with the radiation field incident on both sides. Table 1 shows a comparison between clumpy and homogeneous models for $\lambda = 1400$ Å with $<T>=<I/I_0>$, the ratio between the intensity at a certain depth A_λ and the incident intensity.

TABLE I Homogeneous versus Clumpy Radiative solutions

	F=1	F=0.5	F=0.2
A_λ	$<T>_H$	$<T>_1$	$<T>_2$
2.5	7.10×10^{-2}	1.03×10^{-1}	1.79×10^{-1}
5.0	8.96×10^{-3}	1.90×10^{-2}	5.99×10^{-2}
7.5	1.23×10^{-3}	3.79×10^{-3}	2.32×10^{-2}
10.0	1.73×10^{-4}	7.75×10^{-4}	9.41×10^{-3}
15.0	3.57×10^{-6}	3.39×10^{-5}	1.63×10^{-3}
20.0	7.42×10^{-8}	1.49×10^{-6}	2.89×10^{-4}

The line transfer problem for H_2 and CO is treated analogously in the one-line approximation according to table 2 where the wavelength sampling is now appropriate for a Voigt profile. The effective albedo, ω and opacity, τ are given by

$$d\tau_\lambda = \eta_\lambda(\tau)^{-1} d\tau, \tag{1}$$

$$\omega_{eff,\lambda} = \eta_\lambda(\tau)\,\omega_\lambda, \tag{2}$$

$$\eta_\lambda(\tau) = \frac{\alpha_d}{\alpha_d + \alpha_{H2}(\lambda)}. \tag{3}$$

Notice that the effective albedo will be approximately zero so that geometric shielding of the interclump medium by the clumps may be important in the 912-1100 Å wavelength range. The H/H_2 transition is first iterated to convergence after which the same is done for CO with inclusion of the shielding of CO by H_2. This is accomplished by shifting the lines 2 Å and giving them an effective coupling of 4 (= the number of H_2 lines for every CO line). These numbers were derived according to a weighting scheme using the oscillator strengths, efficiencies and relative positions of all the H_2 and CO lines (Spaans 1993) and were found to be consistent with more detailed calculations (Black and Van Dishoeck 1987) in the homogeneous limit.

TABLE II Parameters for the one-line Approximation

	H_2	CO
N_{eff}	135	33
λ_{eff}(Å)	1000.0	1002.0
f_{eff}	3.5×10^{-3}	2.0×10^{-2}
$A_{eff}(s^{-1})$	5.7×10^{8}	3.0×10^{11}
η_{eff}	0.127	1.0

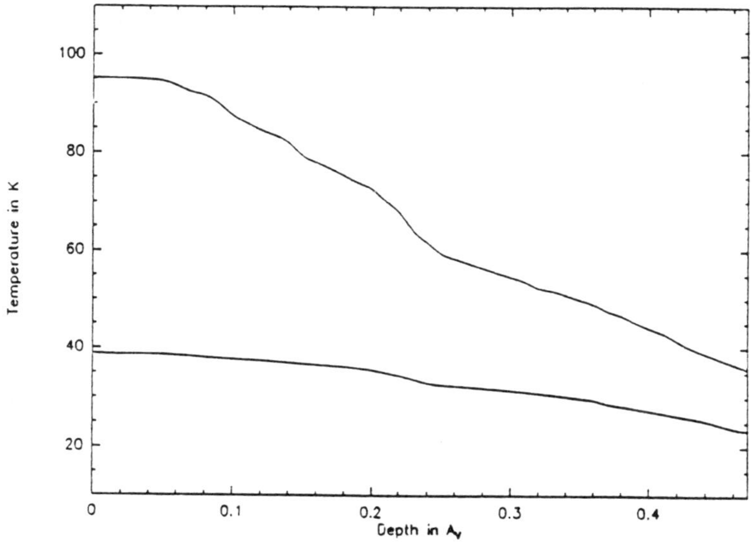

FIGURE I The temperature structure for a homogeneous diffuse cloud with central density $n_0 = 300$, $I_{UV}=7$ and $\zeta = 4 \times 10^{-17}$. It is assumed that 15 % of the total carbon abundance is in the form of PAHs. The bottom curve is without the contribution of PAHs.

RESULTS AND CONCLUSIONS

We show calculations of a generic model with an average density of 10^3 cm^{-3}, a volume filling factor F=25%, a total extent of A_V=4 (2 pc), and illuminated by the average interstellar UV field, with I_{UV}=1. The dominant sources of heating like photoelectric emission of grains and cooling by fine-structure emission by C^+ and OI are included. We have also included the contribution of PAHs, following the description of Bakes (1993). Table 3 shows a comparison between clumpy and homogeneous models.

The presence of clumpiness may increase the internal UV field with more than an order of magnitude at $A_V \sim 5$ with respect to the homogeneous case. The angular distribution of the incident radiation field is also important, especially in those parts of the spectrum where the radiation is scattered mainly in the forward direction. The contribution to the heating by PAHs is significant at the edge of the cloud, but is not found to solve the heating problem for diffuse and translucent clouds (figure 1). The fluctuations in measured column densities as a function of line of sight depends on the clumpiness as well as on the particular molecule. For realistic values of the filling factor this effect largely averages out. The CO abundance is very sensitive to the intrinsically higher density inside the clumps for the diffuse and translucent part of parameter space. Because CO is just becoming optically thick a small shift in the transition layer with extinction will result in a large increase in column density (Van Dishoeck and Black 1988). The C^+ emission extends much further into the cloud and has a smooth underlying component, since the UV radiation may leak through the holes in the density distribution, in agreement with recent observations (Howe et al. 1991; Stacey et al. 1993).

TABLE III Column Densities for Clumpy versus Homogeneous Models

Parameters and Species	Translucent Cloud			ζ Oph		
	Hom	Clumpy	$\sigma(N)$[%]	Hom	Clumpy	Observed
n_H	1000	1000		500	500	
T_0	50	50		20	30	
F	1	0.25		1	0.6	
A_V	5.1	5.0		0.86	0.90	1.0
b	3.0	3.0		1.2	3.0	
δ_C	0.4	0.4		0.50	0.50	0.30
δ_O	0.5	0.5		0.62	0.70	
I_{UV}	1.0	1.0		4.0	4.0	
H_2	4.0(21)	4.1(21)	5.17	4.2(20)	4.5(20)	(4.2±0.3)(20)
H	1.5(20)	2.6(19)	57.3	5.3(20)	5.2(20)	(5.2±0.2)(20)
C	2.6(17)	9.6(16)	10.4	3.1(15)	2.8(15)	(3.2±0.6)(20)
C^+	7.9(17)	2.7(17)	5.82	4.5(17)	4.4(17)	(9.3±4.5)(16)
CO	3.0(17)	1.1(18)	16.5	2.0(14)	1.2(15)	(2.0±0.3)(15)
CH	5.2(14)	5.5(14)	48.0	2.1(13)	2.5(13)	(2.5±0.1)(13)
O	3.1(18)	2.2(18)	14.2	7.1(17)	7.0(17)	(7.1±0.5)(17)
OH	1.8(15)	1.0(15)	22.6	4.5(13)	4.0(13)	(4.8±0.5)(13)
H_2O	2.5(15)	7.1(14)	14.8	5.5(11)	2.9(11)	≤2.2(13)

REFERENCES

Bakes, E., 1993, *in preparation*

Black, J.H., Van Dishoeck, E.F., 1987, *ApJ*, **322** 412

Boissé, P., 1990, *A&A*, **228** 483

Falgarone, E., Pérault, M., 1987 in Physical Processes in Interstellar Clouds. ed. G.E. Morfill & M. Scholer (Dordrecht:Reidel)

Falgarone, E., Pérault, M., 1988, *A&A lett.*, **205** L1

Howe, J.E., Jaffe, D.T., Genzel, R., Stacey, G.J., 1991, *ApJ*, **373** 158

Mezger, P.G., Sievers, A., Zylka, R., Haslam, C.G.T., Kreysa, E., Lemke, R., 1992, *A&A*, **265** 743

Roberge, W.G., Jones, D., Lepp, S., Dalgarno, A., 1991, *ApJ*, **77** 287

Spaans, M., 1993, *in preparation*

Stacey, G.J., Jaffe, D.T., Geis, N., Genzel, R., Harris, A.I., Poglitsch, A., Stutzki, J., Townes, C.H., 1993, *ApJ*, **404** 219

Tielens, A.G.G.M., Hollenbach, D.J., 1985a, *ApJ*, **291** 722

Van Dishoeck, E.F., Black, J.H., 1988, *ApJ*, **334** 771

The First Symposium on the Infrared Cirrus and Diffuse Interstellar Clouds
ASP Conference Series, Vol. 58, 1994
Roc M. Cutri and William B. Latter (eds.)

THE MAGNETIC FIELD IN URSA MAJOR

A.A. GOODMAN & P.C. MYERS
Harvard-Smithsonian Center for Astrophysics

R. GÜSTEN
Max-Planck-Institut für Radioastronomie

C. HEILES
UC Berkeley

INTRODUCTION

This article summarizes, in primarily graphical form, our study of the magnetic field in the Ursa Major high-latitude cloud complex. Myers et al. (1993) presents these results along with Zeeman observations of additional high-latitude clouds in greater detail. Earlier Zeeman observations in high-latitude cloud positions are reported in Heiles (1989) and Verschuur (1989).

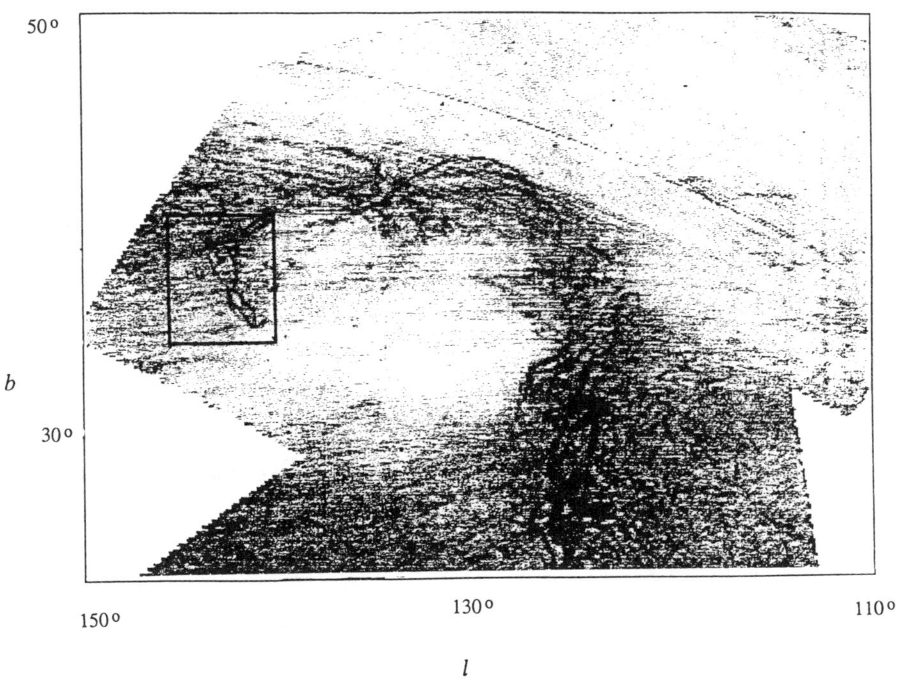

Figure 1 100-μm IRAS image of the NCP Loop region.

OBSERVATIONS AND RESULTS

We have mapped the line-of-sight component of the magnetic field, $B_{||}$, in the Ursa Major complex of diffuse clouds, via measurements of the Zeeman effect in the 21- cm line of H I at the 85-foot telescope at Hat Creek. This set of observations represents the first beam-sampled map of $B_{||}$ in *any* extended cloud.

The Ursa Major cloud complex is a piece of the large-scale H I structure known as the North Celestial Pole (NCP) Loop, and it includes an elongated structure, estimated to be about 10 pc long, oriented perpendicular to the loop and pointing "down" toward the galactic plane, as well as gas which appears to be part of the loop itself. **Figure 1** shows a 100 μm IRAS image of the NCP Loop, with a box around the "Ursa Major complex."

Maps of the complex in ^{12}CO and 100 μm emission (de Vries, Heithausen, & Thaddeus 1987) show that the brightest H I is approximately coincident with the brightest molecular and dust emission, implying that the H I we are observing is likely to be spatially co-located with dense gas in these clouds. **Figure 3** shows the map of the line- of-sight field strength in Ursa Major derived from our Zeeman observations, accompanied by maps of H I brightness and CO integrated intensity. $B_{||} > 4$ μG throughout most of the region observed, and the field is seen to vary on scales finer than 0.4 pc. Observations at the Effelsberg 100-m telescope imply that the field may also vary on still finer scales (Myers et al. 1993.) The greatest field strength measured in the region is 19 ± 4 μG, greater than typical "strong" fields measured in diffuse clouds (Heiles 1989) by a factor ~ 2. **Figure 2**, shows the H I Stokes *I* and *V* profiles for the position in Figure 3 where $B_{||} = 19 \pm 4$ μG: the Zeeman effect is clearly visible as an S-shaped curve in the *V*-spectrum.

Figure 3 shows that there is structure in the field which bears a weak resemblance to the structure in the H I emission itself. **Figure 4** shows the variation of $B_{||}$ and of IRAS 100 μm flux–which should be a good tracer of total column density, along the two "cuts" labeled in Figure 3. The field strength appears to be very well-correlated with column density along the filamentary cloud (Cut 2), but poorly correlated in the direction tangent to the NCP loop (Cut 1).

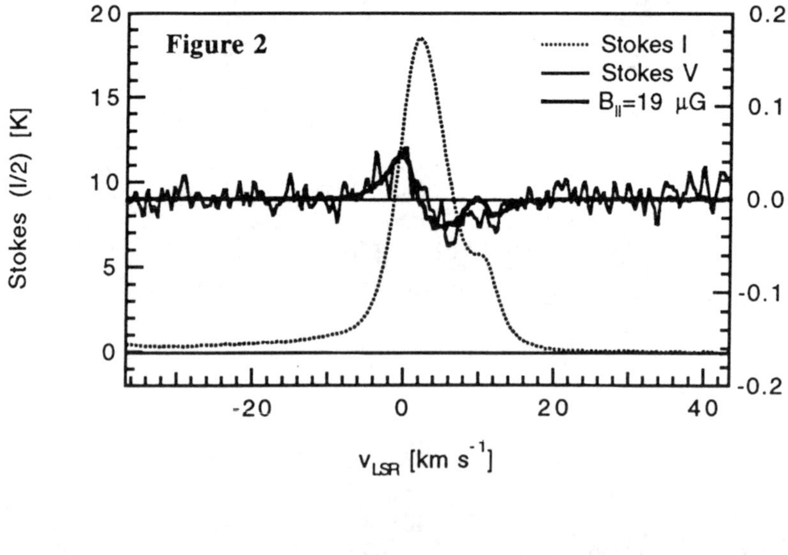

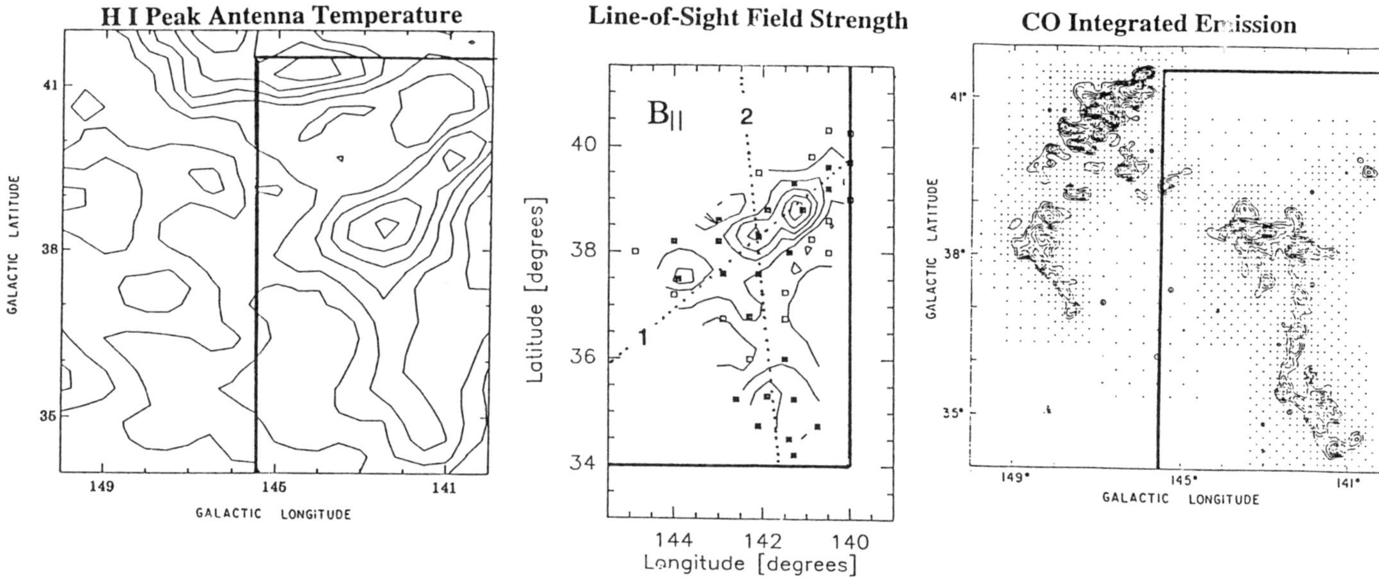

Figure 3: *H I Antenna Temperature, line-of-sight field strength, and CO integrated emission in the Ursa Major Complex.*

All H I data are from the Hat Creek 85-foot telescope (FWHP beam~ 36'), and the CO map is from deVries et al. 1987. In the center panel, the contours of $B_{||}$ are at 3, 6, 9, 12, 15, and 18 μG. Positions marked with filled squares are detections, open squares are upper limits contoured at 2σ. Note that 1 degree corresponds to about 2 pc for a distance of 110 pc.

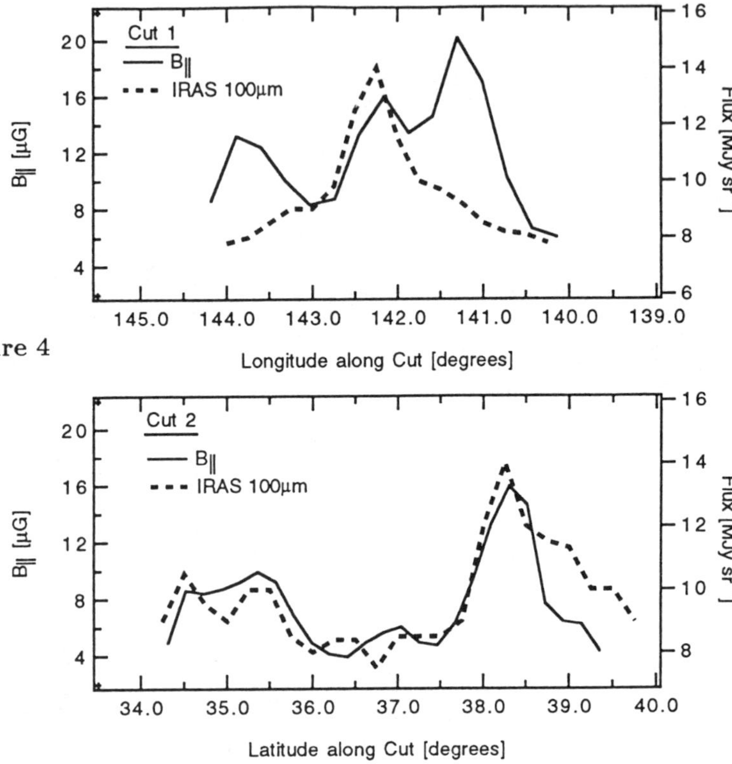

Figure 4

DISCUSSION

The magnetic and kinetic energy densities, **M** and **K**, in the Ursa Major Complex are comparable ($\sim 2 \times 10^{-11}$ erg cm^{-3}), and they are greater than the gravitational energy density by two orders of magnitude. The similarity of **M** and **K** in the diffuse clouds discussed here, as well as in denser, self-gravitating clouds, suggests a strong relationship between **M** and **K** in interstellar clouds, which can be described as MHD turbulence or as a spectrum of Alfvén waves.

ACKNOWLEDGMENTS

We thank Bill Reach for Figure 1, Nancy Levenson for the H I map in Figure 3, and Marc Pound and Ralph Pudritz for helpful consultations.

REFERENCES

de Vries, H.W., Heithausen, A. & Thaddeus, P. 1987, *ApJ*, 319, 723

Heiles, C. 1989, *ApJ*, 336, 808

Myers, P.C., Goodman, A.A., Heiles, C., & Güsten, R. 1993, *ApJ*, in prep.

Verschuur, G.L. 1989, *ApJ*, 339, 163

The First Symposium on the Infrared Cirrus and Diffuse Interstellar Clouds
ASP Conference Series, Vol. 58, 1994
Roc M. Cutri and William B. Latter (eds.)

THE OBSERVATION AND INTERPRETATION OF CORRELATED VELOCITY STRUCTURES IN A HIGH-LATITUDE TRANSLUSCENT MOLECULAR CLOUD

T.N. LAROSA
NAS/NRC Resident Research Assoc. NASA/MSFC

LORIS MAGNANI
Dept. of Physics, Univ. of Georgia

S.N. SHORE
GHRS Science Team, NASA/GSFC

The high-latitude transluscent molecular cloud MBM16 has been mapped in the optically thin 6 cm transition of H_2CO. The H_2CO adsorption was found to be distributed primarily in spatially distinct, large-scale structures, which exhibit velocity coherence on scales of 0.5 pc. These structures are not gravitationally bound and are most likely transient. Their observation suggests that the cloud velocity field may be organized by a shear flow of scale length the size of the cloud. The implications of this result for the dynamics and formation of the cloud will be discussed.

INDEX

Object Index

General Index